T0226124

High-Temperature Ordered
Intermetallic Alloys IX

MATERIALS RESEARCH SOCIETY
SYMPOSIUM PROCEEDINGS VOLUME 646

High-Temperature Ordered Intermetallic Alloys IX

Symposium held November 27–29, 2000, Boston, Massachusetts, U.S.A.

EDITORS:

Joachim H. Schneibel
Oak Ridge National Laboratory
Oak Ridge, Tennessee, U.S.A.

Kevin J. Hemker
Johns Hopkins University
Baltimore, Maryland, U.S.A.

Ronald D. Noebe
NASA Glenn Research Center
Cleveland, Ohio, U.S.A.

Shuji Hanada
Tohoku University
Sendai, Japan

Gerhard Sauthoff
Max-Planck-Institut für Eisenforschung
Düsseldorf, Germany

Materials Research Society
Warrendale, Pennsylvania

CAMBRIDGE UNIVERSITY PRESS
Cambridge, New York, Melbourne, Madrid, Cape Town,
Singapore, São Paulo, Delhi, Mexico City

Cambridge University Press
32 Avenue of the Americas, New York NY 10013-2473, USA

Published in the United States of America by Cambridge University Press, New York

www.cambridge.org
Information on this title: www.cambridge.org/9781107412903

Materials Research Society
506 Keystone Drive, Warrendale, PA 15086
http://www.mrs.org

First published 2001
First paperback edition 2013

Single article reprints from this publication are available through
University Microfilms Inc., 300 North Zeeb Road, Ann Arbor, MI 48106

CODEN: MRSPDH

ISBN 978-1-107-41290-3 Paperback

CONTENTS

*Invited Paper

TITANIUM ALUMINIDES II
AND METAL SILICIDES I

IRON ALUMINIDE, IRIDIUM AND
OTHER ORDERED INTERMETALLIC ALLOYS

METAL SILICIDES II

*Invited Paper

HIGH-TEMPERATURE ORDERED
INTERMETALLIC ALLOYS

NICKEL ALUMINIDES

*Invited Paper

TITANIUM ALUMINIDES III

Author Index

Subject Index

PREFACE

High-temperature structural intermetallics continue to be an active field of research. These proceedings document the research presented at Symposium N, "High-Temperature Ordered Intermetallic Alloys IX," held November 27–29 at the 2000 MRS Fall Meeting in Boston, Massachusetts. The invited and contributed papers in these proceedings provide a representative cross-section of the world-wide research currently carried out on this topic.

The most "popular" material continues to be γ-TiAl which is beginning to find applications as a light-weight, high-strength, oxidation resistant structural material. Aluminides with $L1_2$ and B2 structures, which, in a way, started the field of intermetallics some 20 years ago, attract less interest than they used to. On the other hand, there is renewed interest in structural silicides with melting points on the order of 2000°C. Intermetallics with "functional" properties (i.e., properties other than mechanical properties) also started to receive some interest in this symposium. Examples include intermetallics with thermoelectric properties as well as intermetallics with shape memory effects at unusually high temperatures above 1000°C. The research described in this symposium covers a large spectrum of research ranging from very basic studies to actual applications and is therefore expected to be of interest for a wide range of readers.

Joachim H. Schneibel
Kevin J. Hemker
Ronald D. Noebe
Shuji Hanada
Gerhard Sauthoff

May 2001

ACKNOWLEDGMENTS

It is a pleasure to acknowledge financial support of this symposium from the following organizations:

> Chrysalis Technologies Incorporated
> NASA Glenn Research Center (HOTPC Program Office)
> Oak Ridge National Laboratory

The organizers of this symposium would like to thank all of the participants, all of the authors who invested many hours in putting together the papers for the proceedings, and all of those who made the review process happen.

MATERIALS RESEARCH SOCIETY SYMPOSIUM PROCEEDINGS

MATERIALS RESEARCH SOCIETY SYMPOSIUM PROCEEDINGS

Prior Materials Research Society Symposium Proceedings available by contacting Materials Research Society

Titanium Aluminides I

Mat. Res. Soc. Symp. Proc. Vol. 646 © 2001 Materials Research Society

Recent Advances in Development and Processing of Titanium Aluminide Alloys

Fritz Appel, Helmut Clemens and Michael Oehring
Institute for Materials Research, GKSS Research Centre,
Max-Planck-Str., D-21502 Geesthacht, GERMANY

ABSTRACT

Intermetallic titanium aluminides are one of the few classes of emerging materials that have the potential to be used in demanding high-temperature structural applications whenever specific strength and stiffness are of major concern. However, in order to effectively replace the heavier nickel-base superalloys currently in use, titanium aluminides must combine a wide range of mechanical property capabilities. Advanced alloy designs are tailored for strength, toughness, creep resistance, and environmental stability. Some of these concerns are addressed in the present paper through specific comments on the physical metallurgy and technology of gamma TiAl-base alloys. Particular emphasis is placed on recent developments of TiAl alloys with enhanced high-temperature capability.

1. INTRODUCTION

Titanium aluminide alloys exhibit unique mechanical properties combined with low density and good oxidation and ignition resistance. Thus, they are one of the few classes of emerging materials that have the potential to be used in demanding structural applications at elevated temperatures and in hostile environments. A vast amount of efforts has been expended over the past ten years in attempts to optimize the composition and microstructure of the alloys [1-5]. Currently the alloy design is focused on γ(TiAl)-base alloys, which are slightly lean in Al and microalloyed with several third elements. These have led to complex alloys with the general composition (in at.%)

$$\text{Ti-(46-49)Al-(0-4)Cr,Mn,V-(0.5-2)Nb-(0-1)Si,B,C.} \qquad (1)$$

The major phases in alloys of this type are γ(TiAl) and α_2(Ti$_3$Al). In general, a reduction of Al content tends to increase the strength level, but is harmful for ductility and oxidation resistance. Additions of Cr, Mn and V up to levels of 2 % for each element have been shown to enhance ductility. The role of various other third elements is to improve other desired properties such as oxidation resistance (Nb, Ta) and creep strength (W, Mo, Si, C) [1]. Boron additions greater than 0.5 at.% are effective in refining the grain size and stabilizing the microstructure [6]. The addition of third elements not only changes the relative stability and transformation pathways of the phases, but also brings new phases into existence. It must be mentioned that the full details of the high-temperature phase equilibria and sequence of phase evolution with temperature and ternary or higher additions are not yet fully understood. By thermomechanical treatments a broad variety of microstructures can be generated in the alloys described by (1), which are often characterized in terms of the volume fraction of lamellar colonies and equiaxed γ grains; these are fully-lamellar, nearly-lamellar, duplex and near gamma structures. The general features of the correlations between alloy chemistry, microstructure and properties have been outlined in various review papers which should be consulted for additional background and information [1-6]. The established database indicates that the alloys are viable materials for engineering applications [7]. However, in spite of the potential technological impact of γ(TiAl)-base alloys, the useful range of application is limited by their susceptibility to brittle fracture, which persists up to 700 °C and the rapid loss of strength at higher temperature. Thus, in most mechanical properties the titanium aluminides are inferior to the nickel-base superalloys currently in use, even if the comparison is made on a strength to weight basis. This impedes practical use of the titanium aluminides and is the driving force for the current research and development.

As with many other materials, the alloy attributes that are desirable for high temperature service are counter to those that are considered to be desirable for low temperature toughness and damage tolerance. Thus, the balance of the mechanical properties has to be carefully chosen by the alloy design. This needs a detailed understanding of the relevant failure processes and their correlation with alloy chemistry and microstructure. At the same time, the capability for the alloy design is strongly related to the constraints of acceptable processing routes. These concerns are addressed in the present paper through global commentary on the physical metallurgy of γ (TiAl)-base alloys and the associated processing technologies. Particular emphasis is paid on recent developments of TiAl alloys with enhanced high temperature capability. In order to limit the scope of this review, consideration will be given in separate sections to the following topics:
- deformation behaviour, dislocation multiplication and mobilities,
- creep properties,
- implementation of strengthening mechanisms,
- processing.

2. DEFORMATION BEHAVIOUR

Titanium aluminides are relatively brittle materials, exhibiting little plasticity at ambient temperatures. Typical of such deformation behaviour is that the gliding dislocations are either too low in density or too immobile to allow the specimen to match the superimposed strain rate. At elevated temperatures titanium aluminides suffer from insufficient creep resistance and structural changes. Such behaviour is often associated with dislocation climb and the operation of diffusion assisted dislocation sources [3]. Thus, the factors governing the multiplication and mobility of the dislocations might be important in several different ways and will now be considered.

2.1 Generation of perfect and twinning partial dislocations

In two-phase alloys deformation of the γ(TiAl) phase is mainly provided by ordinary dislocations with Burgers vector b = 1/2<110] and order twinning 1/6<11$\bar{2}$]{111} [3, 8]. Ordinary 1/2<110] dislocations have a compact core structure [9], which suggests that cross slip and climb are relatively easy. Multiplication of these dislocations can therefore take place through the operation of dislocation sources incorporating stress driven cross slip or climb as has been observed in disordered metals [10]. At room temperature multiplication has been found to be closely related to jogs in screw dislocations, which were probably generated by cross slip (figure 1a) [3, 11].

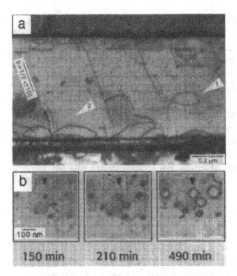

Figure 1. *Dislocation generation in a two-phase titanium aluminide alloy of composition Ti-48Al-2Cr. (a) Initial stage of multiplication of an ordinary 1/2<110] dislocation by cross glide. The dipole arms trailed at a high jog (1) in a screw dislocation are widely separated so that they could pass each other and may act as single ended dislocation sources. Note the emission of the dislocation loops (2) from the interface, which have been discussed in the text. Compression at T = 300 K to strain ε = 3 %. (b) Nucleation and growth of prismatic dislocation loops during in situ heating inside the TEM, following pre-deformation of the sample at room temperature to strain ε = 3%.*

At elevated temperatures multiplication of ordinary dislocations occurs through the operation of Bardeen-Herring climb sources [3] as demonstrated in figure 1b by a sequence of micrographs, part of a TEM in situ study. Apparently, the critical vacancy concentration required to operate a Bardeen-Herring source is relatively low. Processing routes of titanium aluminides often involve thermal treatments followed by rapid cooling, which certainly leads to large vacancy supersaturations. Under such conditions Bardeen-Herring sources can probably operate throughout the entire period where annealing out of excess vacancies takes place.

A mechanism common to both low and elevated temperatures is the emission of ordinary dislocations from the mismatch structures of lamellar interfaces, which is certainly supported by coherency stresses [3, 11-14]. As with many other materials, twins in γ(TiAl) preferentially nucleate at lattice defects with a favorable atomic configuration, which can be rearranged into an embryonic twin [15]. In two-phase titanium aluminides the lamellar interfaces seem to be the prevalent sites for twin formation. The twins are nucleated at misfit dislocations with a Burgers vector out of the interfacial plane which provide a Burgers vector component parallel to the twinning shear direction [15]. There are one-plane ledges in the twin/matrix interface, suggesting that the twins have grown through the propagation of Shockley partial dislocations.

The relative contributions of dislocation glide and twinning depends on the aluminium concentration, the content of ternary and quaternary elements, and the deformation conditions. There is growing evidence that the activation of superdislocations in the γ(TiAl) phase of two-phase alloys requires significantly higher shear stresses so that the different dislocation glide systems often cannot simultaneously operate. Thus, there are many more restrictions upon possible deformation modes in γ(TiAl) than for disordered metals with face centered cubic structures. In grains or lamellae which are unfavorably oriented for 1/2<110] glide or twinning, significant constraint stresses can be developed due to the shape change of deformed adjacent grains. It is now well established that γ(TiAl) is prone to cleavage fracture on {111} planes [11, 16, 17], which at the same time are the glide planes and twin habit planes. Thus, blocked dislocation glide and twinning may easily lead to the nucleation of cracks. Once nucleated on {111} planes, the cracks can rapidly grow to a critical length. For an alloy design towards improved ductility, thus, the activation of glide involving c-components of the tetragonal unit cell of γ(TiAl) is of major concern.

2.2 Dislocation mobilities

Information about the factors governing the dislocation mobility in α_2(Ti$_3$Al) + γ(TiAl) alloys have been obtained by TEM observations and analyzing the deformation behaviour in terms of thermodynamic glide parameters. When the effects of temperature T and strain rate s = dε/dt are coupled by an Arrhenius type equation, the total flow stress σ can be described as [18-20]

$$\sigma = \sigma_\mu + \sigma^* = \sigma_\mu + (f/V)(\Delta F^* + kT \ln s/s_0), \qquad (2)$$

with

$$\Delta F^* = \Delta G + V\sigma^*/f. \qquad (3)$$

σ_μ is the long-range internal stress, σ^* is the thermal stress part and V is the activation volume. ΔF^* is the free energy and ΔG the Gibbs free energy of activation. s_0 contains the density of mobile dislocations, the attempt frequency of the dislocations and the slip path of the dislocations after successful activation and is considered to be constant. k is the Boltzmann constant and f = 3.06 is a Taylor factor to convert σ and s to average shear quantities. The activation parameters V, ΔG and ΔF^* involved in eqs. 2 and 3 were determined by strain rate and temperature cycling tests according to the method proposed by Schöck [18]. V can be described as the number of atoms that have to be coherently thermally activated for overcoming the glide obstacles by the dislocations. It is therefore expected that V will undergo a significant change when there is a change of the mechanism that controls the glide resistance of the dislocations. The characteristic temperature dependence of σ and V estimated at the beginning of deformation is demonstrated in figure 2 [20, 21]. The flow stress is almost independent of temperature up to about 1000 K and then decreases. In contrast 1/V passes through a broad minimum between 800 and 1000 K indicating that significant changes in the micromechanisms controlling the dislocation velocity occur. The relevant processes have been investigated in several studies. Accordingly, at room temperature the dislocation velocity is controlled by a combined operation of localized pinning, jog dragging and lattice friction [3, 20], while locking of dislocations due to

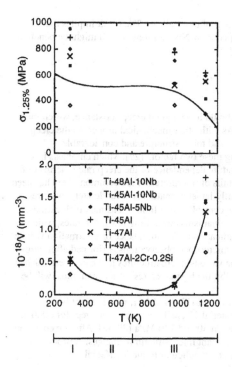

Figure 2. *Dependence of the flow stress σ and the reciprocal activation volume 1/V of binary and niobium containing alloys on the deformation temperature. The drawn lines refer to the values of a Ti-47Al-2Cr-0.2Si alloy with a near gamma microstructure. Values estimated at strain ε = 1.25 % and strain rate s = 4.16 x 10⁻⁴ s⁻¹.*

the formation of defect atmospheres occurs in the temperature range 420 to 770 K [22, 23]. There is increasing evidence that the defect atmospheres in two-phase alloys are formed by antisite defects, i. e. Ti atoms situated on Al sites [24]. The strong increase of the reciprocal activation volume above 900 K indicates that a new thermally activated process becomes important. For the materials investigated here this temperature is just the brittle ductile transition, where the flow stress degrades. In most cases, the activation enthalpy is close to 3 eV [3, 20, 21], which is in reasonable agreement with the self-diffusion energy of γ(TiAl) [25]. This is indicative of a diffusion assisted dislocation mechanism. Climb of ordinary 1/2<110] dislocations has been recognized on different γ-base titanium aluminide alloys by post mortem [26] and in situ TEM studies [3, 20]. Dislocation climb is known to be a stress driven thermally activated process. Thus, the strength properties of the material become strongly rate dependent and degrade at low strain rates. These features are also characteristic of the high temperature deformation of the titanium aluminides and particularly harmful for their creep resistance. An important point is the observation that the activation enthalpy ΔH of the alloys containing a large addition of Nb [21] is significantly higher than that of other alloys. The result implies that diffusion assisted deformation processes are impeded by Nb additions, which might be beneficial for the design of creep resistant TiAl alloys [21].

3. CREEP PROPERTIES

For the intended applications γ(TiAl)-base should have a good creep resistance, which must be present in the earlier stages of deformation. As with other mechanical properties the creep characteristics are sensitive to alloy composition and microstructure and considerable improvements have been achieved by optimizing these two factors [27]. Much emphasis has been placed on describing the stress and temperature dependence of the creep rate in terms of the Dorn equation [28]. While differences in the details of interpretation are common, there has been general agreement about the fact that creep of fully-lamellar materials at moderately low stresses ($\sigma \leq 150$ MPa) and temperatures (T < 750 °C) is controlled by climb of ordinary dislocations. However, rather less is known about the micromechanisms controlling low creep rates, which apparently are more relevant for engineering applications. This imbalance of information will be addressed in the present section in that electron microscope observations performed after long-term creep will be discussed [29-32]. Among the various microstructures that can be established in two-phase alloys fully-lamellar structures exhibit the best creep resistance, and, thus, will be mainly considered here.

The alloy investigated had the composition (in at.%) Ti-48Al-2Cr, which might be considered as a model alloy for fully-lamellar materials. Tensile samples were crept for 6000 to 13400 hours at 700 °C and stresses ranging between 80 and 140 MPa [29-32]. After creep testing the materials were found to have undergone significant microstructural changes. A prominent feature is the formation of multiple-height ledges perpendicular to the interfacial

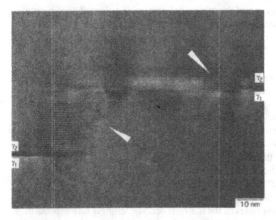

Figure 3. *Structural changes observed after long-term creep of a Ti-48Al-2Cr alloy at T = 700 °C, stress $\sigma_a = 140$ MPa, t = 5988 hours to strain $\varepsilon = 0.69$ %. Note the formation of extended ledges at a γ/γ interface.*

boundaries (figure 3). The ledges had often grown into zones which extended over about 10 nm. The formation of these ledges can formally be rationalized by twining and antitwinning operations parallel to the (111) plane of the interfaces [32]. When the slabs grow further it is apparently energetically favorable to nucleate a γ grain. Two-phase alloys with the base line composition Ti-(45-48)Al-2Cr suffer from dissolution of the α_2 phase and coarsening of the γ lamellae. This is because the kinetics of the decomposition of the α phase into α_2 and γ lamellae during cooling after processing is sluggish and the volume fraction of gamma is less than at equilibrium. Thus, during long term creep, dissolution of the α_2 phase and formation of γ grains occur and finally lead to a complex conversion of the lamellar morphology to a fine spheroidized microstructure. There is ample evidence that misfit dislocations and interface ledges play an important role for achieving the phase equilibrium in that they provide the required change in the stacking sequence and serve as paths for easy diffusion. More on this subject is provided in separate studies [30, 31]. In view of these observations an alloy design towards improved high temperature strength and creep resistance should rely on systems with stable phases and microstructures. Equally important are a reduced diffusivity of the material and the implementation of additional glide resistance, in order to impede dislocation climb processes.

4. ALLOY DESIGN TOWARDS IMPROVED STRENGTH

In view of the anticipated applications several studies have been performed in order to strengthen γ(TiAl) base alloys by solid solution and precipitation hardening. From the engineering viewpoint the challenge is to establish these mechanisms without compromising desirable low-temperature properties, such as ductility and toughness.

Recently, it has been shown that a significant strengthening effect can be achieved when Nb is added with an amount of 5-10 at.% to polycrystalline two-phase alloys [21, 33, 34]. Nb is also a commonly added element because of its ability to improve the oxidation resistance [35, 36]. Despite the extensive body of investigations broadly confirming these results, there is some controversy about the nature of the strengthening effect of Nb additions, i. e. whether or not it arises from solid solution hardening. Thus, the origin of the hardening mechanism has been studied by a systematic variation of the Al and Nb contents. The investigations involve tests on binary alloys with the equivalent Al contents, which are thought to ascertain the effects of off-stoichiometric deviations on microstructure and yield strength [21]. As demonstrated in figure 2 the strengthening effect is almost independent of the Nb content and the values of the reciprocal activation volume estimated for the binary alloys are very similar to those of the Nb containing alloys. Thus, the strengthening effect of large Nb additions has been attributed to the related changes of the microstructure. This view is supported by recent ALCHEMI studies (atom location by channeling enhanced microanalysis) of site occupation, which revealed that Nb solely occupies the Ti sublattice [37, 38]. However, the Nb containing alloys exhibit at room temperature an appreciable ductility, whereas the binary alloys of the same Al content do not. This finding indicates that significant changes of the deformation mechanism occurred due to the Nb additions. In Nb containing alloys an abundant activation of twinning has been recognized and the superdislocations were found to be widely dissociated. As planar faults, twins and dissociated superdislocations often coexist in the same grain or lamella, it is speculated that the stacking fault energies of γ(TiAl) are lowered by the Nb additions and that twin nucleation originates from the superposition of extended stacking faults on alternate {111} planes [15, 39].

Appreciable improvements in strength and creep resistance have also been achieved by precipitation hardening [40, 41]. The strengthening effect critically depends on the size and the dispersion of the particles. In this respect carbides, nitrides and silicides appear to be beneficial as the optimum dispersion can be achieved by homogenization and ageing procedures. Utilizing this method, in a Ti-48.5Al-0.37C alloy a fine dispersion of Ti_3AlC precipitates were generated which is characterized by a length l_p=22 nm and width (along <100]) d_p=3.3 nm of the precipitates [41]. The precipitates are elongated along the c axis of the γ matrix and exhibit strong coherency stresses due the differences in crystal structure and lattice constants. These structural features give rise to a strong glide resistance for all types and characters of dislocations, which persists up to temperatures of about 750 °C and leads to significantly improved high temperature strength and creep resistance. Figure 4 demonstrates, e.g., the interaction of deformation twins with the precipitates in some detail [41]. The high glide resistance is manifested by the strong bowing-out of the twinning partial dislocations. In the local region of the precipitates the shape of the twin/matrix interface is much less regular and the twins are often deflected. These processes often lead to fragmentation of the twins, i.e. islands of untwinned regions occur. However, as with the TiAl-Nb alloys, twin nucleation seems to be relatively easy. Twin nuclei were often found together with widely separated superdislocations giving rise to the speculation that the twins originate form overlapping stacking faults [15]. In this way a fine dispersion of deformation twins is formed, which provides shear components in c direction of the tetragonal cell of γ(TiAl) and, thus, is beneficial for room temperature ductility and toughness.

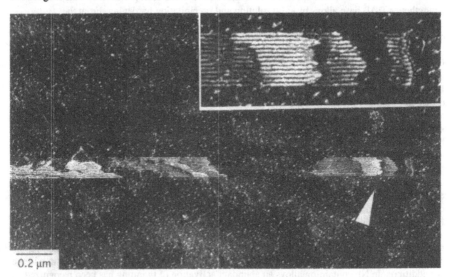

0.2 µm

Figure 4. *Immobilization of a deformation twin due to the interaction with Ti_3AlC perovskite precipitates. Note the high density of the precipitates, which become evident by strain contrast, and the pinning of the twinning partial dislocations by the precipitates. Ti-48.5Al-0.37C, compression at T = 300 K to strain ε = 3 %.*

However, carbides and nitrides seem to be susceptible to coarsening by Ostwald ripening, thereby reducing their effectiveness in impeding dislocation motion. In the case of carbides, implementation of H phase particles Ti_2AlC could be a possible alternative [40], because this phase is thermodynamically more stable and apparently provides also a good creep resistance. However, coarse platelets of H-phase tend to be located at lamellar colony boundaries, which is certainly harmful for low temperature ductility. Borides are virtually stable up to the melting point of γ(TiAl), however, metallurgical techniques to refine their relatively coarse dispersion appear to be less efficient.

Nevertheless, in conclusion of this section it may be summarized that an alloy design based on relatively large Nb additions together with a fine dispersion of perovskite or H-phase precipitates seems to be a suitable technique for expanding the service range for titanium aluminides towards higher temperatures and stresses.

5. PROCESSING

γ(TiAl) alloys are available in all conventional product forms: ingot forgings, extrusions and sheet. The structural applications outlined in the introductory section may require the material to be formed into complex geometries with large dimensions. Processing routes generally follow those of conventional titanium alloys with some special alterations. Yet titanium aluminides are generally difficult to process due to their solidification behaviour, susceptibility to degradation from contamination and intrinsic brittleness. This holds particularly for the novel high strength material described in the previous section, for which the hardening mechanisms have to be implemented within the constraints of technically acceptable processing routes. In view of the available coverage of this article attention is concentrated on ingot metallurgy and wrought processing.

5.1 Ingot quality

Vacuum arc melting (VAR) is currently the most widely used practice for preparing ingots from elemental or master alloying additions. In order to ensure a reasonable chemical homogeneity throughout ingots of 200 -300 mm diameter the meltstocks are usually double-or triple-melted [42]. The peritectic solidification reactions occurring in the composition range (45-49) at.% Al give rise to an unavoidable micro-segregation, the extent of which depends on the nominal Al level and the content of refractory elements. Al is rejected to the interdendritic region while refractory elements, in particular those stabilizing the β phase, are concentrated in the dendritic cores. The differences in concentration are as large as a few at.% and vary on a length scale of about 1 mm [43]. Rapid solidification processing generally reduces segregation, refines microstructure and, thus, produces a more homogeneous material consolidation. As ingot size increases, cooling becomes slower and the as-cast grain size increases, thereby exacerbating the problems associated with segregation. Heat treatments in the ($\alpha+\gamma$) phase field are mostly ineffective to mitigate the chemical gradients [44, 45]. Annealing in the α or ($\alpha+\beta$) phase fields may lead to significantly faster homogenization kinetics, but mostly results in rapid grain growth.

5.2 Primary ingot break-down

Significant improvements in the chemical homogeneity and refinement of microstructure can be achieved by hot working and the associated dynamic recrystallization. There is an intimate correlation between alloy chemistry, hot working conditions and the evolution of the microstructure, which has to be considered for the alloy design and processing. The microstructural evolution has been systematically studied on a series of binary and technical alloys with aluminium contents ranging between 45 and 54 at. % [46]. Not surprisingly, the degree of dynamic recrystallization increases with strain, however, no substantial recrystallization occurs below strains of about 10%. There is also a marked effect of the aluminium concentration on the recrystallization behaviour, which is manifested in the observation that the recrystallized volume fraction is at maximum for aluminium contents of 48-50 at.% (figure 5). The recrystallization behaviour of two-phase alloys is also assisted by the presence of boride particles [46]. Boron is known to significantly refine the as-cast microstructure, which is generally a good precondition for homogeneous hot working and recrystallization [6]. In addition, particle stimulated dynamic recrystallization may occur, when dislocations are accumulated at the particles during hot working [46].

Primary ingot break-down has been accomplished on an industrial scale utilizing forging and extrusion [1, 5, 43-53]. Typical conditions for large-scale isothermal forging are T=1000-1200°C at strain rates $s = 10^{-3} - 10^{-2}$ s^{-1}. 50 kg-billets have been successfully forged within this processing window to height reductions of 5:1. The as-forged structure appears banded, consisting of stringers of α_2 particles in a fine-grained γ matrix. In two-phase alloys it is also common to observe lamellar colonies with lamellae lying in the plane of forging. These colonies are probably undeformed remnants from the cast structure.

Extrusion is usually carried out at temperatures around the α-transus temperature [1, 5, 49, 51-53]. Under these conditions (typically 1250 –1380 °C) severe oxidation and corrosion occurs, thus, the work piece has to be encapsulated. In most cases conventional Ti alloys or austenitic

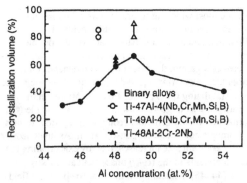

Figure 5. *Dependence of the volume fraction of recrystallized grains on the aluminium content of binary and technical alloys [46]. Deformation at T = 1000 °C and s = 5 x 10^{-4} s^{-1} to strain ε = 75 %.*

steels are used as can material. At the extrusion temperature the can materials have a significantly lower flow stress than the TiAl billet [54]. This flow stress mismatch is often as high as 300 MPa and leads to inhomogeneous extrusion and cracking. These problems can largely be overcome by a novel can design involving radiation shields as an effective thermal insulation [55]. This reduces the heat transfer from the work piece to the can and enables controlled dwell periods between preheating and extrusion. Taking advantage of this concept, extrusion processes have been widely utilized for TiAl ingot break down. For example, 80 kg ingots were uniformly extruded into a rectangular shape with a reduction of the cross section of 10:1 (figure 6) [43]. The alloy had the composition Ti-45Al-10Nb and represents a new family of high strength materials that have been described in the previous section. Extrusion above the α-transus temperature T_α resulted in a refined nearly-lamellar microstructure with a colony size of 30-50 mm as demonstrated in figure 7a. Extrusion below T_α led to duplex microstructures with coarse and fine-grained banded regions (figure 7b). These structural inhomogeneities are associated with a significant variation in the local chemical composition, which is manifested at a length scale comparable to or slightly smaller than that of the as-cast material [43]. This observation provides supporting evidence that the dynamic recrystallization during hot working is strongly affected by local composition. The coarse-grained bands probably originate from the prior Al-rich interdendritic regions where no α_2 phase was present. Thus, grain growth following recrystallization is not impeded by α_2 grains. On contrary, the fine-grained bands or lamellar colonies are formed in Al-depleted core regions of the dendrites.

The structural and chemical inhomogeneities of primarily processed TiAl products provide severe limitations for the reliability of components or secondary processing. Other quality issues concern cavities, internal wedge cracks and surface connected cracks. Thus, further improvement of hot working procedures is of major concern. In view of the results discussed above, this requires a tight optimization of the alloy composition and hot working parameters, respectively, and to find novel engineering solutions for increasing the amount of imparted strain energy, in order to achieve a more homogeneous recrystallization. Advances in the manufacturing of components have been described in recent conference proceedings and review articles [1, 2, 5, 43, 44, 52, 56], the reader is referred to these papers for further details.

Figure 6. *Ingot break-down of an engineering alloy with the base line composition Ti-45Al-(5-10)Nb+X by canned extrusion at T_α-ΔT. The ingot, originally of 192 mm diameter and 700 mm height, was canned using austenitic steel, sealed in vacuum, heated up and then extruded. A reduction ratio of 10:1 and a rectangular die were used. The final TiAl extrusion was rectangular with cross section dimensions of 100 x 30 mm².*

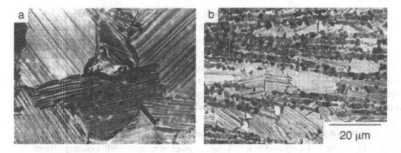

Figure 7. *Back-scattered electron images of a Ti-45Al-10Nb alloy extruded to 7:1 reduction. (a) Nearly lamellar microstructure observed after extrusion at $T_\alpha + \Delta T$. (b) Duplex structure with a banded morphology observed after extrusion at $T_\alpha - \Delta T$.*

5.3 Mechanical properties of wrought material

The refined microstructure established after hot working generally results in a significant strengthening, when compared with cast material. The increase in yield strength can be rationalized in terms of Hall-Petch relations although quantitative descriptions are often difficult due to the complexity of the microstructures [57]. Figure 8 shows the dependence of the density compensated yield stress on temperature for forged and extruded γ base alloys, which have been developed at GKSS. Extremely high yield stresses in excess of 1000 MPa were obtained on Ti-45Al-(5-10)Nb derivative alloys after extrusion to a reduction ratio of 7:1 [43]. For example on one alloy variant (figure 8) at room temperature in tension a fracture stress of 1100 MPa at a

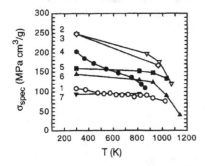

Figure 8. *Temperature dependence of density adjusted yield stresses for forged and extruded gamma-base titanium aluminide alloys. (1) Forged Ti-47Al-2Cr-0.2Si, near gamma microstructure, (2) extruded Ti-45Al-(5-10)Nb, duplex microstructure,(3) Ti-45Al-(5-10)Nb+X, duplex microstructure. For comparison the values of nickel base superalloys and conventional titanium alloys are given, with (4) IMI 834, (5) René 95, (6) Inconel 718, (7) IN 713 LC.*

plastic strain of $\varepsilon=2.5\%$ was determined. This combination of room temperature strength and ductility is the best ever reported on γ(TiAl) base alloys.

6. CONCLUSIONS

During the past five years significant progress has been achieved in the physical metallurgy of titanium aluminide alloys. Compositions of alloys have been identified that are capable of carrying stresses in the duplex microstructural condition in excess of 700 MPa at service temperatures of 700 °C and strain rates $s = 10^{-5}$ s^{-1}. Even higher stresses can be attained under these loading conditions if the alloys are transformed to fully-lamellar microstructures. Thus, wrought alloys of this type can be an attractive alternative to the heavier nickel base superalloys in certain ranges of stress and temperature. The future and promise of γ(TiAl) base alloys and manufacturing of components lies in innovative processing methods designed to achieve better performance. Specifically, substantial improvements are required in ingot quality and hot working procedures to obtain homogeneous and defect-free material with desired microstructures.

ACKNOWLEDGMENTS

The authors would like to thank Ms. E. Tretau, U. Brossmann, U. Christoph, St. Eggert, U. Lorenz, J. Müllauer, and J. Paul for helpful discussions and continuous support. The financial support of this research by the Helmholtz-Gemeinschaft (HGF-Strategiefonds) is gratefully acknowledged.

REFERENCES

1. Y-W. Kim and D.M Dimiduk, *Structural Intermetallics,* eds. M. V. Nathal, R. Darolia, C. T. Liu, P. L. Martin, D. B. Miracle, R. Wagner, M. Yamaguchi, (TMS, Warrendale, PA, 1997), p. 531.
2. *Gamma Titanium Aluminides 1999*, eds. Y-W. Kim, D. M. Dimiduk, M. H. Loretto, (TMS, Warrendale, PA, 1999).
3. F. Appel and R. Wagner, *Mater. Sci. Eng.* **R22**, 187-268, 1998.
4. M. Yamaguchi, H. Inui, K. Itoh, *Acta Mater.* **48**, 307 (2000).
5. H. Clemens, H. Kestler, *Advanced Eng. Mater.* **2**, 551 (2000).
6. J. A. Christodoulou, H. M. Flower, *Advanced Eng. Mater.* **2**, 631 (2000).
7. D. M. Dimiduk, *Mater. Sci. Eng.* **A263**, 281 (1999).
8. M. Yamaguchi and Y. Umakoshi, *Progress in Materials Science* **34**, 1-148 (1990).
9. K. J. Hemker, B. Viguier and M. J. Mills, *Mater. Sci. Eng.* **A164**, 391 (1993).
10. J. P. Hirth and J. Lothe, *Theory of Dislocations*, 2nd edn. (Krieger Publishing, Malabor), (1992).
11. F. Appel, U. Christoph and R. Wagner, *Phil. Mag. A* **72**, 341 (1995).
12. F. Appel, P. A. Beaven and R. Wagner, *Acta Metall. Mater.* **41**, 1721 (1993).
13. F. Appel, U. Christoph and R. Wagner, *Interface Control of Electrical, Chemical and Mechanical Properties*, eds. S. P. Murarka, K. Rose, T. Ohmi, and T. Seidel, Mater. Res. Soc. Symp. Proc. (MRS, Pittsburgh, PA, 1994), Vol. **318**, p. 691.
14. F. Appel and U. Christoph, *Intermetallics* **7**, 1273 (1999).

15. F. Appel, in *Advances in Twinning*, eds. S. Ankem and C. S. Pande (TMS, Warrendale, PA, 1999), pp. 171-186.
16. M. H. Yoo, C. L. Fu, J.K. Lee, *Twinning in Advanced Materials* (TMS, Warrendale, PA, 1994), p. 97.
17. M. H. Yoo, C. L. Fu, *Metall. Trans.* A, **29A,** 49 (1998).
18. G. Schöck, *Phys. Stat. Sol.* **8**, 499 (1965).
19. A. G. Evans and R. W. Rawlings, *Phys. Stat. Sol.* **34**, 9 (1969).
20. F. Appel, U. Lorenz, M. Oehring, U. Sparka and R. Wagner, *Mater. Sci. Eng.* A233, 1 (1997).
21. J. D. H. Paul, F. Appel and R. Wagner, *Acta Mater.* **46**, 1075 (1998).
22. M. A. Morris, T. Lipe and D. G. Morris, *Scripta Mater.* **34**, 1337 (1996).
23. U. Christoph, F. Appel and R. Wagner, *High-Temperature Ordered Intermetallics VII*, eds. C.C. Koch, C.T. Liu, N.S. Stoloff, A. Wanner, Mater. Res. Soc. Symp. Proc. (MRS, Pittsburgh, PA, 1997), Vol. 460, p. 77.
24. U. Christoph, F. Appel, this volume.
25. S. Kroll, H. Mehrer, N. Stolwijk, Ch. Herzig, R. Rosenkranz and G. Frommeyer, *Z. Metallkunde* **83**, 8 (1992).
26. B. K. Kad and H. L. Fraser, *Phil. Mag. A*, **69**, 689 (1994).
27. J. Beddoes, W. Wallace and L. Zhao, *International Materials Reviews* **40**, 197 (1995).
28. T.A. Parthasaraty, M.G. Mendiratta and D.M. Dimiduk, *Scripta Mater.* **37**, 315 (1997).
29. M. Oehring, P.J. Ennis, F. Appel and R. Wagner, *High-Temperature Ordered Intermetallics VII*, Mater. Res. Soc. Symp. Proc. (MRS, Pittsburgh, PA 1997), Vol. 460, p. 257.
30. M. Oehring, F. Appel, P. J. Ennis and R. Wagner, *Intermetallics* 7, 335 (1999).
31. F. Appel, M. Oehring and P. J. Ennis, *Gamma Titanium Aluminides 1999*, eds. Y.-W. Kim, D. M. Dimiduk, M. H. Loretto (TMS, Warrendale, PA, 1999), p.603.
32. F. Appel and R. Wagner, *Atomic Resolution Microscopy of Surfaces and Interfaces* ed. D.J.Smith, Mater. Res. Soc. Symp. Proc. (MRS, Pittsburgh, PA, 1997),Vol. 466, p. 145.
33. S.-C. Huang, *Structural Intermetallics* , eds. R. Darolia, J. J. Lewandowski, C. T. Liu, P. L. Martin, D. B. Miracle, M. V. Nathal (TMS, Warrendale, PA, 1993), p. 299.
34. G. Chen, W. Zhang, Y. Wang, J. Wang and Z. Sun, *Structural Intermetallics*, eds. R. Darolia, J. J. Lewandowski, C. T. Liu, P. L. Martin, D. B. Miracle, M. V. Nathal (TMS, Warrendale, PA, 1993), p. 319.
35. H. Nickel, N. Zheng, A. Elschner and W. Quadakkers, *Microchim. Acta* **119**, 846 (1995).
36. L. Singheiser, W.J. Quadakkers and V. Shemet, *Gamma Titanium Aluminides 1999* , eds. Y-W. Kim, D. M. Dimiduk, M. H. Loretto (TMS, Warrendale, PA, 1999), p. 743.
37. E. Mohandas and P. A. Beaven, *Scripta Metall. Mater.* **25**, 2023 (1991).
38. C. J. Rossouw, C. T. Forwood, M. A. Gibson and P. R. Miller, *Phil. Mag. A* **74**, 77 (1996).
39. F. Appel, U. Lorenz, J.D.H. Paul and M. Oehring, *Gamma Titanium Aluminides 1999*, eds. Y.-W. Kim, D. M. Dimiduk, M. H. Loretto (TMS, Warrendale, PA, 1999), p. 381.
40. W.H. Tian and M. Nemoto, *Gamma Titanium Aluminides*, eds. Y-W. Kim, R. Wagner and M. Yamaguchi (TMS, Warrendale, PA, 1995), p. 689.
41. U. Christoph, F. Appel and R. Wagner, *Mater. Sci. Eng.* **A239-240**, 39 (1997).
42. P. McQuay, V.K. Sikka, *Intermetallic Compounds*, Vol. 3, Progress, eds. J.H. Westbrook, R.L. Fleischer (J. Wiley, Chicester, 2001), in press.
43. F. Appel, U. Brossmann, U. Christoph, S. Eggert, P. Janschek, U. Lorenz, J. Müllauer, M. Oehring, J. D. H. Paul, *Adv. Eng. Mater.* **2**, 699 (2000).

44. S. L. Semiatin, J. C. Chesnutt, C. Austin, V. Seetharaman, *Structural Intermetallics 1997*, eds. M. V. Nathal, R. Darolia, C. T. Liu, P. L. Martin, D. B. Miracle, R. Wagner, M. Yamaguchi (TMS, Warrendale, PA, 1977), p. 263.
45. C. Koeppe, A. Bartels, J. Seeger and H. Mecking, *Metall. Trans.* **24A**, 1795 (1993).
46. R. M. Imayev, G. A. Salishchev, V. M. Imayev, M. R. Shagiev, A. V. Kuznetsov, F. Appel, M. Oehring, O. N. Senkov, F. H. Froes, *Gamma Titanium Aluminides 1999*, eds. Y-W. Kim, D. M. Dimiduk, M. H. Loretto (TMS, Warrendale, PA, 1999), p. 565.
47. Y-W. Kim, *Acta Metall. Mater.* **40**, 1121 (1992).
48. R.M. Imayev, V.M. Imayev and G.A. Salishchev, *J. Mater. Sci.* **27**, 4465 (1992).
49. P.L. Martin, C.G. Rhodes and P.A. McQuay, *Structural Intermetallics*, eds. R. Darolia, J.J. Lewandowski, C.T. Liu, P.L. Martin, D.B. Miracle, and M.V. Nathal (TMS, Warrendale, PA, 1993), p. 177.
50. H. Clemens, P. Schretter, K. Wurzwallner, A. Bartels and C. Koeppe, *Structural Intermetallics*, eds. R. Darolia, J.J. Lewandowski, C.T. Liu, P.L. Martin, D.B. Miracle, and M.V. Nathal (TMS, Warrendale, PA, 1993), p. 205.
51. C.T. Liu, P.J. Maziasz, D.R. Clemens, J.H. Schneibel, V.K. Sikka, T.G. Nieh, J. Wright and L.R. Walker, *Gamma Titanium Aluminides* , eds: Y-W. Kim R. Wagner, M. Yamaguchi (TMS, Warrendale, PA, 1995), p. 679.
52. H. Clemens, H. Kestler, N. Eberhardt, W. Knabl, *Gamma Titanium Aluminides 1999*, eds. Y-W. Kim, D. M. Dimiduk, M. H. Loretto (TMS, Warrendale, PA, 1999), p. 209.
53. M. Oehring, U. Lorenz, R. Niefanger, U. Christoph, F. Appel, R. Wagner, H. Clemens and N. Eberhardt, *Gamma Titanium Aluminides 1999*, eds. Y-W. Kim, D. M. Dimiduk, M. H. Loretto (TMS, Warrendale, PA, 1999), p. 439.
54. S.L. Semiatin and V. Seetharaman, *Scripta Metall. Mater.* **31**, 1203 (1994).
55. Of patent application by F. Appel, U. Lorenz, M. Oehring and R. Wagner, DE 1974257A1, FR Germany.
56. H. Clemens, Z. Metallkde. **90**, 569 (1999).
57. D. M. Dimiduk, P. M. Hazzledine, T. A. Parthasarathy, S. Seshagiri, M. G. Mendiratta, *Metall. Trans.* A, **29** (1998), 37.

Mat. Res. Soc. Symp. Proc. Vol. 646 © 2001 Materials Research Society

DEVELOPMENT OF A CREEP RESISTANT TiAl-BASE ALLOY

S.C. Deevi[*], W.J. Zhang, C.T. Liu[#] and B.V. Reddy
Research Center, Chrysalis Technologies Incorporated, Richmond, VA 23234, U.S.A.
[#]Metals and Ceramics Division, Oak Ridge National Laboratory, Oak Ridge, TN 37831, U.S.A.

ABSTRACT

In this paper, the mechanical and physical properties of a Ti-47Al-4(Nb, W, B) alloy (CTI-8) developed by Chrysalis Technologies are presented. The properties of CTI-8 alloy are compared with other TiAl-base alloys and currently used materials for high temperature applications. The CTI-8 alloy exhibits excellent creep resistance, good high temperature strength and better physical properties than the currently used materials. The comparisons suggest that CTI-8 has a great potential in aerospace and automotive industries.

INTRODUCTION

TiAl-base alloys have a low density of ~4 g/cm^3 and are promising candidates for a number of critical high-temperature applications. Despite their low density, the creep and oxidation resistances of conventional TiAl alloys are inferior to those of superalloys that they can replace. In order to improve the creep resistance, the traditional approach has been to produce fully-lamellar microstructure by extrinsic strengthening of lamellar interfaces [1]. The disadvantage of fully-lamellar microstructure is its poor room-temperature ductility due to the relatively large lamellar colony size (generally larger than 100 μm) [2]. The creep resistance of TiAl alloys can also be improved by intrinsic strengthening such as alloy design [1]. By following intrinsic strengthening approach, good creep resistance may be maintained even with fine-grained nearly-lamellar microstructure. Fine-grained lamellar microstructure gives rise to good room temperature ductility. As a result, a good balance of creep resistance and ductility can be achieved.

Following the alloying approach, we developed several TiAl-base alloys containing Nb, W and B additions. In this paper, we discuss the CTI-8 alloy based on Ti-47Al-4(Nb, W, B). We compare both the mechanical and physical properties of CTI-8 alloy with other TiAl-base alloys and with currently used materials for high temperature applications, for example, turbine blades, automotive valves and turbocharger rotors. The details on composition, microstructure and data resources of the TiAl-base alloys used for comparison are described in Table 1.

Table 2 provides a list of currently used superalloys for turbine and automotive engine components. Inconel 718 is the most widely used high-temperature superalloy. Ti-6Al-4V is the most widely used titanium alloy. Inconel 617 is the alloy of choice for thermal protection components in space vehicles. Heat resistant steel 21-4N is commonly used as a valve material in automotive engines. Inconel 751 is used in high performance valves. Inconel 713C is mostly used for turbocharger wheels when the design requires higher heat resistance than the valves. The physical properties of these alloys were taken from refs. 11-13 and are compared with CTI-8 alloy.

---

[*] Corresponding author, E-mail address: Seetharama.C. Deevi@pmusa.com

Table 1. Composition and microstructure of TiAl-base alloys for comparison

Alloy	Composition	Microstructure (processing)	Data resources
GE	Ti-47Al-2Nb-2Cr	duplex (cast)	3, 5
XD47	Ti-47Al-2Nb-2Mn	nearly-lamellar (cast)	3, 4, 6
TAB	Ti-47Al-4(Cr,Nb,Mn,Si,B)	nearly-lamellar (extruded)	7
K5	Ti-46.5Al-2Cr-3Nb-0.2W	fully-lamellar (forged)	8, 9
ABB	Ti-47Al-2W-0.5Si	nearly-lamellar (cast)	10

Table 2. Alloys currently used for turbine and automotive engine components

Alloy	Composition	Applications
Inconel 718	Ni-19Cr-18Fe-5.1Cb+Ta-3Mo-0.9Ti-0.5Al	aerospace engine
Ti-6Al-4V	Ti-6Al-4V	wide usage
Inconel 617	Ni-22Cr-12.5Co-9Mo-1.2Al-1.5Fe-0.5Mn-0.5Si	space vehicle
Inconel 713C	Ni-13Cr-4Mo-2Cb-6Al-0.7Ti	turbocharger wheel
Inconel 751	Ni-15.5Cr-7Fe-2.3Ti-1.2Al-0.9Nb	exhaust valve
21-4N	Fe-21Cr-4Ni-9Mn-0.5C-0.4N	exhaust valve

EXPERIMENTAL DETAILS

The material used for this study is a Ti-47Al-4(Nb, W, B) alloy (CTI-8). The CTI-8 alloy was vacuum cast, HIP'ed and then aged at 1000°C for 2 hrs. Constant load tensile creep tests were conducted in air at 760°C at initial stresses ranging from 100 to 500 MPa. All test temperatures were controlled to within ±1°C of the intended test temperature. The extensometer has a strain resolution of 10^{-5}. Tensile tests were conducted in air at a strain rate of about 1×10^{-4}/s. Static Young's modulus was measured by tensile tests. A dual push-rod dilatometer of Theta Instruments was used to measure the linear thermal expansion from room temperature to 1100°C. The thermal conductivity was measured using the Kohlrausch apparatus. The specific heat was measured using a standard Perkin-Elmer Model DSC-2 (Differential Scanning Calorimeter) using Sapphire as a reference material.

RESULTS AND DISCUSSION

The optical microstructure of the cast CTI-8 is shown in Fig. 1. It has a nearly lamellar microstructure with the equiaxed γ grain size of about 40 μm and the lamellar grain size of approximately 500 μm.

The minimum creep rates of CTI-8 and other TiAl alloys are presented in Fig. 2a as a function of stress. As can be seen, the creep rates of CTI-8 are lower than that of other TiAl alloys at stresses less than 200 MPa, which are the stresses that most of the envisaged applications are expected to impose on TiAl alloys. Especially to note here is that the minimum creep rate of CTI-8 is close to 10^{-10}/s at 760°C/100MPa. To our knowledge, this is the lowest creep rate obtained so far for any TiAl-base alloys. In terms of the time to 0.5% creep strain, CTI-8 is superior to other TiAl alloys, see Fig. 2b. The time to 0.5% creep strain of CTI-8 is nearly two times longer than that of the creep resistant ABB alloy, and three times longer than that of the fully-lamellar K5 alloys. The excellent creep resistance of CTI-8 is believed to be due to the absence of β phase, good microstructural stability and low diffusivity of refractory elements [14].

Figure 1. Optical microstructure of cast CTI-8 alloy

Figure 3a compares the specific yield strengths of CTI-8 with currently used materials in the temperature range 20-1000°C. It is clearly seen in this plot that CTI-8 becomes much stronger at higher temperatures relative to these alloys. The specific yield strength of CTI-8 at 850°C is higher than that of Inconel 718 by about 50 MPa-cm^3/g. The Young's modulus of CTI-8 is comparable to that of superalloys and heat resistant steel 21-4N at service temperatures, and much higher than that of Ti-6Al-4V alloys. In terms of specific Young's modulus, CTI-8 is much stronger than currently used alloys (Fig. 3b). The specific modulus of CTI-8 is 50% greater than conventionally used alloys. A high elastic modulus is usually considered to be beneficial in aerospace applications, since service loads lead to smaller elastic deflections. High specific stiffness is another advantage of CTI-8 alloy.

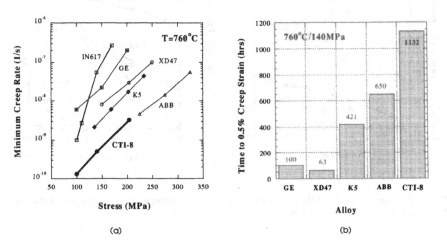

Figure 2. Comparison of (a) the minimum creep rates at 760 °C and (b) the time to 0.5% creep strains at 760 °C/140MPa between CTI-8 and other TiAl-base alloys.

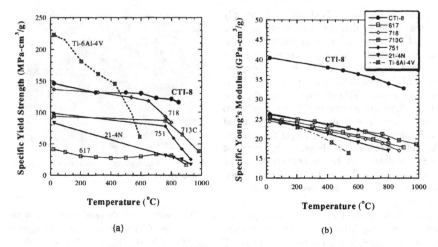

Figure 3. Comparison of (a) the specific yield strengths and (b) the specific Young's modulus of CTI-8 with currently used materials.

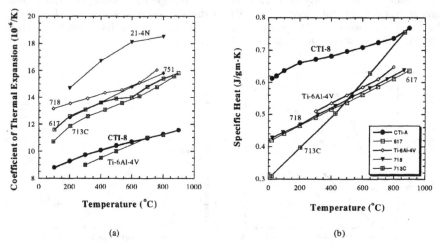

Figure 4. Comparison of (a) the coefficient of thermal expansion and (b) specific heat between CTI-8 and currently used materials.

The coefficient of thermal expansion of CTI-8 alloy is much lower than that of currently used superalloys and heat resistant steel 21-4N and is close to that of Ti-6Al-4V, see Fig. 4a. Meanwhile, the specific heat of CTI-8 is higher than that of currently used materials, see Fig. 4b. A lower thermal expansion reduces thermal stress and a higher specific heat reduces the thermal gradient, in both cases, resulting in improved thermal shock and fatigue resistance. As shown in Fig. 5, the thermal conductivity of CTI-8 is comparable to that of currently used superalloys and heat resistant steel 21-4N in the service temperature. Therefore, similar thermal gradients are expected in the components when replacing the superalloys with CTI-8 alloy.

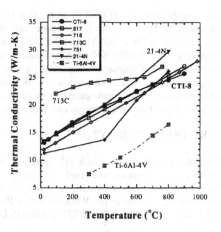

Figure 5. Comparison of thermal conductivity of CTI-8 with currently used materials.

Finally, the ductility of CTI-8 can be greatly improved after extrusion. The room temperature ductility of extruded CTI-8 with a nearly lamellar microstructure is as high as 3.3%, higher than those of other wrought and cast TiAl-base alloys, see Fig. 6a. The important point is that the extruded CTI-8 maintains the good high temperature strength of cast CTI-8 alloy (Fig. 6b), indicating the intrinsic strengthening effect of alloying addition. Creep tests of extruded CTI-8 are currently underway and will be communicated soon.

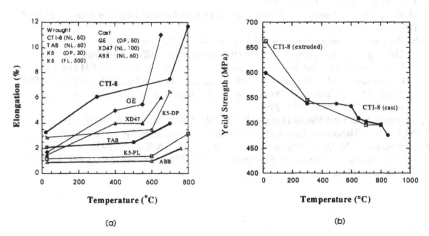

(a)   (b)

Figure 6 (a) Comparison of the ductility of extruded CTI-8 with other TiAl alloys and (b) the yield strengths of extruded and cast CTI-8 as a function of temperature.

CONCLUSION

The Ti-47Al-4(Nb,W, B) (CTI-8) alloy exhibits excellent creep resistance at the stresses of the envisaged service conditions in aerospace and automotive engines. The minimum creep rate of CTI-8 is 1.2×10^{-10}/s at 100 MPa/760°C. The time to 0.5% creep strain of CTI-8 is nearly two times longer than that of ABB alloy and three times longer than that of fully-lamellar K5 alloy at 140MPa/760°C. Compared to the currently used materials, the specific yield strength of CTI-8 is higher at the service temperatures. Physical properties of CTI-8 are similar to or better than currently used alloys for aerospace and automotive applications. CTI-8 has much higher specific modulus, lower thermal expansion and higher specific heat.

ACKNOWLEDGEMENTS

The authors would like to thank Felecia Logan, Travis Wood and Yenew Kassaye for technical assistance in the experimental work. The authors also thank Prof. D.H. Sastry for many helpful discussions. Research at Oak Ridge National Laboratory (ORNL) was sponsored by Chrysalis Technologies, Inc. under the Work for Other Program at ORNL managed by UT-Battelle Inc. under contract DE-AC05-00OR 22725 with US Department of Energy.

REFERENCES

1. J. Beddoes, W. Wallance and L. Zhao, *Inter. Mater. Rev.*, **40**, 197 (1995).
2. Y-W. Kim, *Intermetallics*, **6**, 623 (1998).
3. D.E. Larson, *Mater. Sci. Eng. A*, **213**, 128 (1996).
4. Y-W. Kim, *J. Metals*, **7**, 33 (1994).
5. M.M. Keller, P.E. Jones, W.J. Porter and D. Eylon, in *Gamma Titanium Aluminides 1995*, TMS, 441 (1995).
6. B. Skrotzki, in *Gamma Titanium Aluminides 1999*, TMS, 619 (1999).
7. M. Oehring et al., in *Gamma Titanium Aluminides 1999*, TMS, 439 (1999).
8. Y-W. Kim and D.M. Dimiduk, in *Structural Intermetallics 1997*, TMS, 531 (1997).
9. S.W. Schwenker and Y-W. Kim, in *Gamma Titanium Aluminides 1999*, TMS, 985 (1999).
10. V. Lupic, M. Marchionni, G. Onofrio, M. Nazmy and M. Staubli, in *Gamma Titanium Aluminides 1999*, TMS, 349 (1999).
11. Aerospace Structural Metals Handbook, CINDAS/USAF, Ho CY ed., 1997, USA.
12. INCO Alloys Handbook, Inco Alloys International, 1988, Huntington, WV, USA.
13. S. Isobe and T. Noda, in Structural Intermetallics 1997, TMS, 427 (1997).
14. W.J. Zhang and S.C. Deevi, to be presented at 2001 TMS Annual Meeting, Feb. 11-15, 2001, New Orleans, USA.

Mat. Res. Soc. Symp. Proc. Vol. 646 © 2001 Materials Research Society

Creep Behavior and Microstructural Stability of Lamellar γ-TiAl (Cr, Mo, Si, B) with Extremely Fine Lamellar Spacing

Wolfram Schillinger[1], Dezhi Zhang[2], Gerhard Dehm[2],
Arno Bartels[1] and Helmut Clemens[3]
[1]Materials Science and Technology, TUHH, Hamburg, GERMANY
[2] Max-Planck-Institut für Metallforschung, Stuttgart, GERMANY
[3]Institute for Materials Research, GKSS-Research Center, Geesthacht, GERMANY

ABSTRACT

γ-TiAl (Cr, Mo, Si, B) specimens with two different fine lamellar microstructures were produced by vacuum arc melting followed by a two-stage heat treatment. The average lamellar spacing was determined to be 200 nm and 25-50 nm, respectively. Creep tests at 700°C showed a very strong primary creep for both samples. After annealing for 24 hours at 1000 °C the primary creep for both materials is significantly decreased. The steady-state creep for the specimens with the wider lamellar spacing appears to be similar to the creep behavior prior to annealing while the creep rate of the material with the previously smaller lamellar spacing is significantly higher. Optical microscopy and TEM-studies show that the microstructure of the specimens with the wider lamellar spacing is nearly unchanged, whereas the previously finer material was completely recrystallized to a globular microstructure with a low creep resistance. The dissolution of the fine lamellar microstructure was also observed during creep tests at 800 °C as manifested in an acceleration of the creep rate. It is concluded that extremely fine lamellar microstructures come along with a very high dislocation density and internal stresses which causes the observed high primary creep. The microstructure has a composition far away from the thermodynamical equilibrium which leads to a dissolution of the structure even at relatively low temperatures close to the intended operating temperature of γ-TiAl structural parts. As a consequence this limits the benefit of fine lamellar microstructures on the creep behavior.

1 INTRODUCTION

The mechanical properties of γ-TiAl based alloys are strongly dependent on the microstructure. The creep resistance of fully lamellar microstructures is significantly influenced by the lamellar spacing. Recent studies report on the creep behavior of γ-TiAl based alloys with fully lamellar microstructures with an average lamellar spacing in the range of about 135 to 1200 nm [1, 2, 3]. All those studies conclude that the creep resistance increases with decreasing average lamellar spacing especially at high stresses. The present study describes the adjustment of extremely fine lamellar microstructures and their creep behavior as well as their thermal stability.

2 EXPERIMENTAL

A 63 mm diameter ingot was prepared by double vacuum arc melting using commercially pure charge materials and master alloys. The analyzed composition of the ingot was Ti-45.52Al-1.44Cr-1.72Mo-0.25Si-0.27B (in at. %); the oxygen, nitrogen, hydrogen and carbon impurity levels were 1260 wppm, 56 wppm, 23 wppm and 125 wppm, respectively. In order to investigate the evolution of the microstructure and its influence on the creep behavior a series of heat treatments were conducted:

$$
\begin{array}{rcl}
A & : & 1350\ ^\circ C/2\ h/air\ cooling \rightarrow 700\ ^\circ C/1\ h/air\ cooling \\
B & : & 1350\ ^\circ C/2\ h/oil\ quenching \rightarrow 700\ ^\circ C/1\ h/air\ cooling \\
A + HT & : & A + 1000\ ^\circ C\ /24\ h \\
B + HT & : & B + 1000\ ^\circ C\ /24\ h
\end{array}
$$

Prior to the heat treatment specimen A was hot isostatically pressed. Compression creep tests were performed for 100 h at 700 °C and 800 °C in air with an initial load of 225 MPa. All specimens were cut and prepared for examination by optical microscopy and transmission electron microscopy (TEM).

3 RESULTS AND DISCUSSION

3.1 Adjustment of Microstructures

Both specimens were annealed at 1350 °C for 2 hours. The subsequent air cooling of specimen A resulted in the formation of a fully lamellar microstructure as it can be seen in Fig. 1 (a).

No visible change of the microstructure could be observed after the following annealing at 700 °C for 1 hour. From TEM observations (Fig. 1 (c)) the average lamellar spacing was determined to be approximately 200 nm. For specimen B the oil quenching supressed the phase formation, i.e. the microstructure consisted of supercooled alpha phase as it is shown in Fig. 2. Annealing at 700 °C for 1 hour led to the formation of a fully lamellar microstructure (Fig.1 (b)) which has some areas with Widmanstätten-type structures. The average lamellar spacing was determined to be in the range of 25 to 50 nm. Both specimens exhibit remarkably small lamellar colony sizes which can be attributed to the presence of Mo and Cr stabilized β /B2 phase during annealing [4]. The β /B2 phase acts as a grain-size controlling agent which restricts the growth of primary α grains as well as it presumably prevents the microstructure from cracking during oil quenching. A detailed discussion about the influence of the β /B2 phase would be beyond the scope of this article, but some more information about this topic can be found in [4, 5].

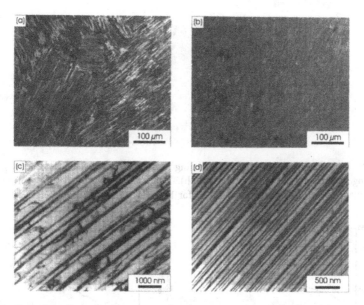

Figure 1: Fully lamellar microstructures after heat treatment A (left) and heat treatment B (right): (a) (b) polarized light optical micrographs; (c) (d) TEM images. The average lamellar spacing is approximately 200 nm for A and 25-50 nm for B.

3.2 Creep Behavior

The results of 100 hours compression creep tests at 700 °C with an initial load of 225 MPa are shown in Fig. 3. A severe primary creep is observed for both specimens. Specimen B shows a higher secondary creep rate even though the lamellar spacing of the microstructure was much smaller compared to specimen A. It can be assumed that the rapid formation of the lamellar microstructure comes along with high internal stresses and high dislocation densities which result in the observed strong primary creep. The heat treatment HT was expected to be effective as a "recovery" annealing to reduce internal stresses and dislocation densities and therefore to improve the primary creep behavior. The results from creep tests at 700 °C with an initial load of 225 MPa after this annealing is shown in Fig. 3 (b). The primary creep for both materials is significantly reduced, but the secondary creep of B + HT is much higher, whereas the secondary creep of A+HT is nearly unchanged. This behavior can be explained with a change of the microstructure as it will be discussed in the following section.

Figure 2: Supercooled α phase after annealing at 1350 °C for 2 hours and subsequent oil quenching (polarized light optical micrograph). The lamellar microstructure evolved during a subsequent annealing treatment at 700°C for 1 hour.

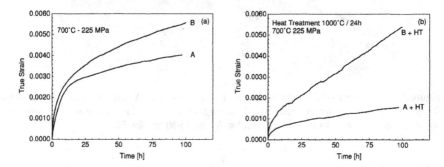

Figure 3: Creep behavior testet at 700 °C and 225 MPa in air. A distinct primary creep regime is observed (a). After annealing at 1000 °C for 24 hours the primary creep is significantly decreased, but now material B shows a low creep resistance (b).

3.3 Thermal Stability

In order to study the thermal stability all microstructures were examined after the creep tests and after annealing for 24 hours at 1000°C. No obvious change of the microstructure could be found after the creep tests at 700°C. After annealing for 24 h at 1000 °C the microstructure of specimen A contained some equiaxed grains along colony boundaries but still consisted mainly of lamellar grains (Fig. 4 (a)). The previously lamellar microstructure of material B, however, was completely recristallized to an equiaxed microstructure (Fig. 4(b)), i.e. in the creep test a fully lamellar microstructure (sample A+HT) was compared to an equiaxed microstructure (sample B+HT) with a much lower creep resistance. From these results it might be speculated that the lower creep resistance of specimen B with the smaller lamellar spacing (Fig. 3(a)) is related to a degradation of the

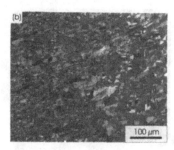

Figure 4: Microstructure after annealing for 24 hours at 1000 °C. Some equiaxed γ-grains occure along colony boundaries in specimen A+HT (a), Specimen B+HT is almost completely recrystallized (b).

microstructure during the creep test, even though no visible change could be found after creep deformation (which was less than 1%, see Fig. 3). Therefore, additional creep tests with an initial load of 225 MPa were performed at 800°C on material A and B to study the creep behavior and the microstructures after a significantly higher creep strain.

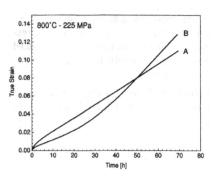

Figure 5: Creep behavior testet at 800 °C and 225 MPa.

Specimen A exhibits a constant secondary creep rate during the whole test (Fig. 5). At the beginning of the test, specimen B shows a higher creep resistance, but after approximately 10 hours the creep rate increases rapidly. After the tests the microstructure of specimen A was unchanged, whereas specimen B contained some lamellar areas as well as equiaxed γ grains as shown in Fig. 6. The micrograph of sample B indicates that the Widmanstätten-type areas within the microstructure are the first regions which change by recrystallizing to equiaxed grains.

Thus, it can be assumed that at the beginning of the creep test B the small lamellar spacing of specimen causes the superior creep behavior. However, the later observed increase of the creep rate is then caused by the degradation of the lamellar microstructure which becomes the dominating effect.

4 SUMMARY AND CONCLUSIONS

Fully lamellar microstructures in γ-TiAl (Cr, Mo, Si, B) based alloys with very small lamellar spacing were produced by two-stage heat treatments. A distinct primary creep regime was observed during compression creep tests at 700°C. The high primary creep was attributed to

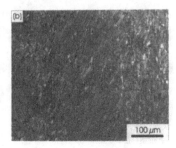

Figure 6: Microstructure after creep at 800°C. Specimen A is nearly unchanged (a), recrystallization occured in specimen B (b).

internal stresses and could be significantly reduced by a "recovery" annealing at 1000°C. The microstructure with the extremely fine lamellar spacing turned out to be thermodynamically unstable during recovery annealing as well as during creep tests. The degradation of the microstructure was dominating over any beneficial effect of the small lamellar spacing, i.e. the samples exhibited poor creep resistance. The degradation initiated from Widmanstätten-type areas within the microstructure.

To optimize the creep resistance of fully lamellar microstructures, a balance between small lamellar spacing and thermal stability has to be found and in addition Widmanstätten-type areas have to be avoided. There is no advantage in adjusting an extremely fine-spaced lamellar microstructure by annealing supercooled α-phase, as this results in a highly unstable microstructure with a high amount of Widmanstätten-type areas.

It can be assumed that β/B2 phase, which was stabilized by Mo and Cr, plays an important role for the microstructural stability and creep behavior. This will be adressed more detailed in a future publication.

REFERENCES

[1] T. A. Parthasarathy, M. Keller and M. G. Mendiratta, Scripta Mat. **37** (7), 1025-1031 (1998).

[2] K. Maruyama, R. Yamamoto, H. Nakakuki, and N. Fujitsuna, Mat. Sci. And Eng. **A239-240**, 419-428 (1997).

[3] A. Chatterjee, U. Bolay, U. Sattler, and H. Clemens in *Intermetallics and Superalloys*, edited by D.G. Morris, S. Naka and P. Caron, (Proceedings of Euromat 1999 **6**, Wiley-VCH, 2000), pp. 233-239.

[4] D. Zhang, P. Kopold, V. Güther, and H. Clemens, Z. Metallkd. **91** (3), 206-210, (2000).

[5] D. Zhang, E. Arzt and H. Clemens, Intermetallics **7**, 1081-1087, (1999).

Mat. Res. Soc. Symp. Proc. Vol. 646 © 2001 Materials Research Society

The Influence of the Texture on the Creep behavior of γ-TiAl Sheet Material

Arno Bartels[1], Wolfram Schillinger[1], Anita Chatterjee[2] and Helmut Clemens[3],
[1]Dep. Materials Science and Technology, Technical University of Hamburg-Harburg, D-21071 Hamburg, Germany ,
[2]Max-Planck-Institut für Metallforschung, D-70174 Stuttgart, Germany
[3]GKSS-Research Centre Geesthacht, D-21502 Geesthacht, Germany

ABSTRACT

In hot rolled Ti-46.5at%Al-4at%(Cr,Nb,Ta,B) sheets a strong modified cube texture is found. The c-axes of the tetragonal unit cells in the grains are aligned with the transverse direction of the sheets. This texture causes an anisotropy of the creep resistance which is improved in transverse direction. Heat treatments with different subsequent cooling rates were performed in order to obtain lamellar microstructures with a different spacing of lamellae. Creep experiments exhibit an increase of the creep resistance which is highest after fast cooling. The texture measurements show no longer an alignment of c-axes after the heat treatment in the α-phase field, but a weak {110}-fiber texture in rolling direction occurs which causes a small improvement of the creep resistance in rolling direction. However, the creep resistance of the lamellar microstructure is more determined by the morphology than by the texture.

INTRODUCTION

In recent years Plansee AG, Reutte/Austria, has established the rolling process of γ-TiAl based alloys on industrial scale [1,2]. After a final heat treatment ('primary annealing'), which is conducted to flatten the sheets, a strong modified cube texture is found by x-ray diffraction measurements [3]. The c-axes of the tetragonal unit cells in the grains are aligned with the transverse direction of the sheets. This texture causes anisotropies of the yield stress which depends on the temperature. Especially in the range 700°C to 800°C the yield stress is higher in transverse direction than in rolling direction [3,4]. These anisotropies occur also during dislocation creep because the texture influences the Taylor factors. This paper deals with textures of sheets and the resulting creep anisotropy. We will describe heat-treatments which are performed to change the microstructure and the texture with the aim to improve the creep resistance of the sheets.

EXPERIMENTAL AND CREEP MEASUREMENTS

The Ti-46.5Al-4(Cr,Nb,Ta,B) sheet material was provided by Plansee AG. It was hot rolled from prealloyed and hot isostatically pressed powder compacts. After primary annealing (PA) at 1000°C for 2 hours the microstructure is fine grained and nearly equiaxed. The primary annealed microstructure is shown in Figure 1. It consists of equiaxed γ-TiAl grains and α_2-Ti$_3$Al particles at γ-grain boundaries and triple points. In addition, a small fraction Cr-stabilized β/B2-phase and (Ti,Ta)-borides is present [2].

Three types of specimens were cut from the sheet so that the tensile stress was applied in rolling direction (RD), in transverse direction (TD) and in diagonal direction (under 45° between

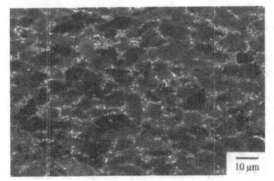

Figure 1. Microstructure of a sheet of Ti-46.5A-4(Cr,Nb,Ta,B) after primary annealing. The rolling direction is horizontal. The SEM-BSE picture shows α₂-Ti₃Al as gray spots and the β/B2-phase as white spots at grain boundaries.

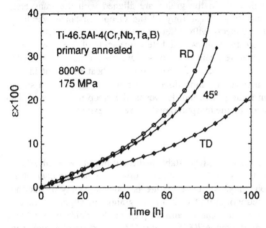

Figure 2. Creep tests conducted on the PA sheet at 800°C and 175 MPa. The tensile specimen were cut by spark erosion in rolling direction (RD), in transverse direction (TD) and under 45° to these directions.

RD and TD). The specimens had an overall length of 50 mm and a gauge area of 30 mm × 3 mm. The sheet thickness was 1mm. Tensile Creep tests were performed at 800°C in air under constant load. The initial stress was 175 MPa. As shown in Figure 2 the creep in RD was faster than in diagonal direction. The slowest creep is observed in TD. The observed acceleration of creep is not only tertiary creep. Due to the large creep strains under constant load the stress increases during creep and, therefore, also the creep rate.

The observed creep rates are larger than 2×10^{-7} s^{-1} and thus too large for most applications. In order to improve the creep resistance the microstructure was transformed into a designed fully lamellar microstructure (DFL). With a two step heat-treatment a variation of the lamellar spacing but with nearly constant colony size was possible [2]. The temperature of the heat-treatment within the α-phase field and the holding time were 1350°C and 12 min, respectively. With controlled cooling rates between 1 K/min and 200 K/min a variation of the mean interface spacing between 1.2 µm (Figure 3) and 140 nm (Figure 4) was possible. The colony size is about 130µm. At 800°C tensile creep tests under constant load with an initial stress of 175 MPa were performed in rolling direction as well as in transverse direction.

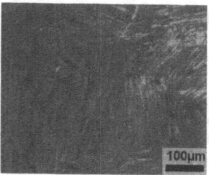

Figure 3. *Microstructure after annealing at 1350 °C for 12 min and subsequent cooling at 1K/min. The mean interface spacing is 1.2μm.*

Figure 4. *As figure 3 but with a cooling rate of 200 K/min. Mean interface spacing: 140 nm.*

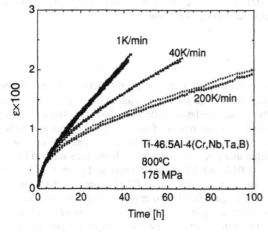

Figure 5.
Creep test after annealing at 1350 °C for 12 min and subsequent controlled cooling with indicated cooling rates. Open symbols: rolling direction; closed symbols: transverse direction.
No significant anisotropy is observed.
Mean interface spacings:
1.2 μm (1 K/min);
290 nm (40 K/min);
140 nm (200 K/min).

The tests are plotted in Figure 5. The results of the tests in rolling and transverse direction a nearly identical, but the creep resistance in rolling direction seems to be slightly improved. The difference in creep between rolling and transverse does not exceed significantly the possible error of measurements. However, strong dependence on interface spacings was found. In a recent paper [5] it could be shown, that both, the dislocation forest build by dislocations emitted from interfaces, as well as the interfaces themselves are effective barriers to dislocation motion, whereas the mean interface spacing represents a theoretical limit for the free path of the dislocations. Modeling with these assumptions has led to a quantitative description of the influence of interface spacing on the creep rate.

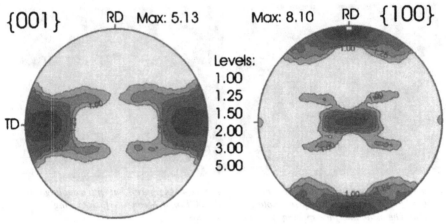

Figure 6. *Recalculated pole figures {001} and {100} of a Ti-46.5Al-4(Cr,Nb,Ta,B) sheet after rolling and subsequent primary annealing at 1000°C for 2 hours.*

TEXTURE MEASUREMENTS

The textures of the sheets were determined using x-ray diffraction. A set of four incomplete pole figures ($\varphi \leq 75°$) was measured. Due to a new parallel beam geometry it was possible to measure the pole figure of the weak superlattice reflections {001}, {110} and {201}, which contain the information of the orientation differences of the tetragonal c-axes and a-axes. From these three pole figures, together with the {111}-pole figure, the complete orientation distribution function (ODF) was calculated using an extension of the harmonics method [6]. As an example, the complete pole figures {001} and {100} are plotted in Figure 6. The main component is a modified cube texture with the tetragonal c-axis [001] aligned in the transverse direction TD and the a-axes [100] and [010] aligned in RD and in normal direction (the center of the pole figures). A detailed analysis of other components in the ODF is beyond scope of this paper, but in addition to the cube component small volume fractions of copper, brass and s-components can be detected (if we use the fcc nomenclature). However, these components play a minor role for the mechanical anisotropy, which is mainly caused by the modified cube texture.

The textures of the DFL sheet were determined in the same way. The relatively large size of the lamellar colonies leads to the problem that only a limited number of colonies can be detected by the x-ray beam. In order to obtain a better statistic the specimens were oscillated by 6mm to measure a larger area of the sheet. The ODF of all sheets with DFL microstructure show no more significant differences in the alignment of the a- and c-axes, but a new component is observed with <110> and <101> in RD. Therefore, in Figure 7 the recalculated {110}- and {111}-pole figures with view in RD are plotted, i.e. the centers of the pole figures represent the RD. The {110}-pole figure shows a {110}-fiber texture component, but the ring at $\varphi=45°$ exhibits structures which could be caused by a preferred orientation of the lamellae. Therefore, we connected in the {111}-pole figure all maxima which belongs to the {111}-maximum near ND (normal direction) with a solid line. If the maximum near ND is due to the normal direction of the lamellae, then there must exist the corresponding orientation maxima of the twinned

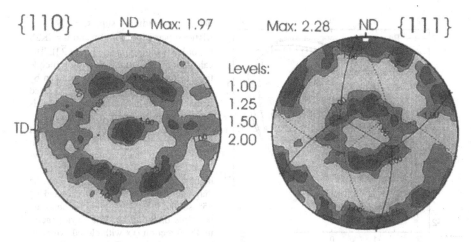

Figure 7. *Recalculated pole figures {110} and {111} of a sheet with DFL microstructure. The solid and dotted lines in the {111} pole figure connect the {111} orientations, which would occur, if the maximum near ND would be caused be preferred oriented lamellar colonies.*

lamellae in those orientations, which are connected in Figure 7 by dotted lines. However, no maxima at these orientations can be seen and, therefore, it is concluded that the maximum near ND does not represent a preferred orientation of lamellae.

In DFL microstructures the cooling rate has a pronounced influence on the texture. A slow cooling rate (e.g. 1 K/min) results in a large interface spacing and relatively sharp {110}+{101} fiber component. After fast cooling the texture of the fine lamellar DFL microstructure is weaker. In addition, other orientations like the <100> orientation occur. It is assumed, that these orientations belong to a small volume fraction with a cube component, which apparently survived the transformation process.

DISCUSSION

The strong anisotropy of creep in sheets with PA microstructure is caused by the texture in combination with a higher critical resolved shear stress (CRSS) for the eight superdislocations of the <101> type than for the four ordinary ½<110> dislocations (both types have {111} slip planes). In TD the ordinary dislocation can hardly be activated, because the <001> directions possess a Schmid-factor of Zero and they are concentrated in TD due to the modified cube texture. <001> superdislocations exhibit Schmid-factors of ±0.408, but they have a high CRSS at 800° [3]. In <100>, i.e. RD, ordinary slip is possible on all four slip systems and in contrast to TD twinnig is possible under tension stress. Consequently, in RD the creep is faster than in TD.

In a more realistic description we have to take into account that we are dealing with polycrystalline materials and that the texture does not reveal a single orientation. Therefore, we are using the single crystal yield surfaces of γ-TiAl [7] and calculate the yield loci with the Los Alamos polycrystal plasticity code (LAPP) [8] according to the measured texture. In these simulations we made the assumption that the ratios of the CRSS are $\tau_0/\tau_s = \tau_t/\tau_s = 0.4$ (τ_o = CRSS

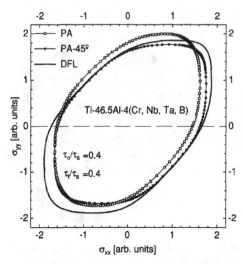

Figure 8. Yield loci calculated with the measured textures. For PA and DFL xx is RD and yy is TD. For PA-45° xx is the diagonal direction of 45° between RD and TD and yy is 135°.

of ½<110> ordinary slip, τ_s = CRSS of <101> superdislocation slip and τ_t =CRSS for order twinning on {111}<11$\bar{2}$]). The calculations in Figure 8 show in principle the Taylor-factors of yielding and can by used in the discussion of the obtained creep results, if the creep process is determined by dislocation motion only.

The yield loci of the PA sheet (open symbols) show in xx-direction (RD) a much lower value than in yy-direction (TD). This explains the much higher creep resistance in TD. For the diagonal direction the ODF is rotated by 45°, so that the 45°-direction becomes the principle direction xx. The Taylor factor in 45°-direction (xx with closed symbols) is higher than in RD (xx with open symbols), but is smaller than in TD (yy with open symbols). We must conclude that the creep resistance lies between those of RD and TD, which confirm the experimental results in Figure 2. In the 45°-direction the preferred orientation is <101>, which enables the slip on only two ordinary slip systems but disables twinning under tension. Therefore, this direction is harder than the rolling direction with the activation of four ordinary slip systems assisted by twinning.

The yield loci calculation was also performed with the texture of a DFL microstructure (solid line in Figure 8). The yield loci curve is nearly symmetric. The difference between xx-(RD) and yy-direction (TD) is small. If the {110}-fiber is strong, then the xx-direction exhibits the higher Taylor-factor. In contrast, if the cube component becomes comparable, the yy-direction exhibits the higher Taylor-factor. From Figure 8 it is evident, that for DFL sheet the calculation in TD show smaller values than for the PA sheet, although the creep resistance of the DFL sheet is much higher. Therefore, it is concluded, that the creep resistance of the DFL sheet is much more determined by the morphology of the microstructure than by the prevailing texture.

REFERENCES
1. H. Clemens, Z. Metallkd. **86,** 814 (1995)
2. H. Clemens, H. Kestler, Adv. Eng. Mat. **2,** 551 (2000)
3. A. Bartels, H. Clemens, Ch. Hartig, H. Mecking, in *High-Temperature Intermetallic Alloys VII*, ed. by C. Koch, C.T. Liu, N. Stoloff, A. Wanner, MRS Vol. 460, 1997, p. 141
4. A. Bartels, Ch. Hartig, St. Willems, H. Uhlenhut, Mater. Sci. Eng. **A239-240,** 14 (1997)
5. A. Chatterjee, H. Mecking, E. Arzt, H. Clemens, Mater. Sci. Eng. accepted for publication
6. Ch. Hartig, X.F. Fang, H. Mecking, M. Dahms, Acta metall. mater. **40,** 1883 (1992)
7. H.. Mecking, Ch. Hartig, U. F. Kocks, Acta mater. **44,** 1309 (1996)
8. Los Alamos Polycrystal Plasticity Code, Los Alamos Nat. Lab. LA-CC-88-6, NM, USA

Mat. Res. Soc. Symp. Proc. Vol. 646 © 2001 Materials Research Society

Creep Mechanisms in Equiaxed and Lamellar Ti-48Al

G. B. Viswanathan [1], S. Karthikeyan [1], V. K. Vasudevan [2] and M. J. Mills [1]

[1]Department of Materials Science and Engineering, The Ohio State University, Columbus, OH 43210, [2]Department of Materials Science and Engineering, University Of Cincinnati, Cincinnati, OH 45221

Abstract

Minimum creep rates as a function of stress have been obtained for Ti-48Al binary alloy with a near gamma and a fully lamellar microstructure. TEM investigation reveals that deformation structures in both microstructures are dominated by jogged screw 1/2[110] dislocations in γ phase. A modified jogged screw model is adopted to predict minimum creep rates where the rate controlling step is assumed to be the non-conservative motion of 1/2[110] unit dislocations. In the case of equiaxed microstructure where the deformation was mostly uniform, the creep rates predicted by this model were in agreement with experimental values. Conversely, deformation in lamellar microstructures were highly inhomogeneous where the density of jogged 1/2[110] unit dislocations were seen in varying proportions depending on the width of the γ laths. The creep rates and stress exponents in these microstructures is explained in terms of active volume fractions of γ laths participating in deformation for a given applied stress.

1.0 Introduction

The most promising Ti-48Al (at. %) based alloys can have broad categories of microstructures such as equiaxed, duplex and fully lamellar and understanding of creep behavior of these microstructures is very critical to applications [1,2]. Activation energies similar to those for self diffusion, and stress exponents in the range 4-6, reported in these microstructures [2-5] suggest a creep mechanism involving recovery by dislocation climb processes. However, the absence of subgrain formation during minimum creep regime suggest that the power law behavior is different from that of a pure metal. Accordingly, a model based on the mobility of jogged screw dislocations was previously proposed and tested for equiaxed microstructure [6]. In the case of lamellar microstructures, several explanations such as interlocked boundaries and decreased slip distances due to numerous lath boundaries [2], twinning , dynamic recrystallization [7] and sliding [8] have been suggested to be the cause of superior creep resistance. In this paper, an attempt has been made to understand the creep behavior in lamellar microstructures. Constant load creep experiments were performed at various stress levels to obtain stress exponent values. Based on TEM evidence, it is shown that the deformation in lamellar microstructures is similar to that of equiaxed microstructure. Suitable modifications to the original jogged screw model have been proposed to explain creep in lamellar microstructures.

2.0 Experimental Methods

The creep study here is based on an alloy of composition: 47.86 at.%, O-0.116 wt.%, N-0.016 at.%, C-0.041 at.%, H-0.076 at.% and the balance titanium. To briefly summarize the experimental procedures, cylindrical blanks of the extruded alloy were suitably heat treated to produce a "near-gamma" equiaxed microstructure (g.s.~50 μm) and a "fully-lamellar" microstructure (g.s.~200 μm). Stress-increment tests were used to obtain the stress dependence of the creep rate and select monotonic tests were also performed to confirm the results of the stress-increment tests Details regarding processing, heat treatment and testing can be found elsewhere [3,9]. Thin foils were prepared and deformation structures were examined in a Philips CM200 TEM operated at 200 kV.

3.0 Results

3.1 Creep Behavior.

Figure 1 shows the log strain rate versus log applied stress curve for equiaxed microstructure (EQ) tested at T=1041K and fully lamellar (FL) microstructure tested at T=1088K. It is evident that the Ti-48Al(FL) is more creep resistant than the Ti-48Al(EQ) structure despite the fact that FL microstructure

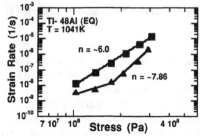

Figure 1 Plot showing log strain rate versus log applied stress

was tested at a higher temperature. For limited stress ranges, the creep response in γ-TiAl, both in the equiaxed and fully lamellar conditions, seems to obey a power law type Dorn equation relating the strain rate to the stress as

$$\dot{\varepsilon} = A \cdot \sigma^n \cdot \exp(-Q/RT) \tag{1}$$

where A is a constant, Q is the creep activation energy, R is the gas constant and T, the temperature. It is seen that the equiaxed structures exhibit a stress exponent n of around 6. Similar results have been widely reported [2-5]. The stress exponents for fully lamellar structures however seem to be clearly lower than five (around 3) at lower stresses and significantly larger than five (~8) at higher stresses. Higher stress exponent values for lamellar structures have been frequently reported [2,3].

3.2 Deformation structures

Figure 2 shows that deformation microstructures in the equiaxed Ti-48Al alloy, at creep strains corresponding with the minimum creep rate, are dominated by 1/2<110] type dislocations. There is little tendency for subgrain formation as might be expected for an n~5 behavior. Similar dislocation structures have been frequently reported to dominate creep microstructures [5,6]. The dislocations tend to be elongated in the screw orientation and appear to be frequently pinned along their lengths, as can be seen in Figure 2. The segments on either side of the pinning points are seen to be bowed-out, forming local cusps along the length of the dislocations. The average spacing between the apparent pinning points is around 200nm. Tilting experiments in the TEM have confirmed that these pinning points are jogs on the screw dislocations [7]. Figure 3 shows that the deformation in lamellar microstructures is mostly contained within γ laths and also dominated by 1/2[110] type dislocations with minimal twinning. The deformation is highly inhomogeneous; a higher density of 1/2<110] dislocations was seen in thicker laths compared to thin ones. Interestingly, the morphology of 1/2[110] dislocations was very similar to those observed in equiaxed microstructure, with jogs pinning the screw segments. However, jogs were less frequently observed within finer γ laths. Figure 4a show a set of hard mode 1/2[110] dislocations (marked by arrow) bowing across the lamellar boundaries. When imaged with the lamellar plane steeply inclined to the beam direction in as shown in Figure 4b, these dislocations can be seen channeling between the interfaces, thus leaving behind long segments of dislocations along the interface.

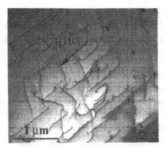

Figure 2 TEM micrograph showing deformation structures in equiaxed (EQ) structures

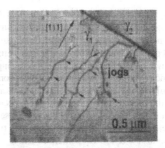

Figure 3 TEM micrograph showing deformation structures within wider γ lath in fully lamellar (FL) microstructure

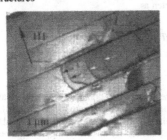

Figure 4 TEM micrographs showing 1/2[110] dislocations channelling within thinner γ lath. Dislocations imaged when lamellar interfaces are (a) parellel (b) inclined to beam

4.0 Discussion

4.1 Equiaxed Microstructures

The observation of cusped screw segments in the equiaxed structures in this alloy, and the general absence of subgrains, has led to the conclusion that the creep rate is controlled by the non-conservative motion of jogs along the length of the screw dislocations. The basic premise of this approach has been discussed in detail previously [7]. Briefly, it is presumed that the jogs form due to frequent cross-slip of adjacent segments along the near screw line direction. When the jogs are relatively short, non-conservative movement of the jog can occur, as assumed in the conventional model [10]. As the jog height h increases, a critical height h_d is reached above which jogs will no longer be dragged non-conservatively, but the near edge segments will by pass each other and act as new dislocation source. This critical jog height, h_d, is given by the dipole breakaway expression

$$h_d = \left(\frac{Gb}{\{8\pi(1-\nu)\tau\}} \right) \qquad (2)$$

where G is the shear modulus, b is the Burgers vector and ν, Poisson's ratio. Taking into account this stress dependence of jog heights, the modified strain rate expression for the jogged-screw model is expressed as follows

$$\dot{\gamma} = \left(\frac{\pi D_s}{\beta b h_d} \right) \left(\frac{\tau}{\alpha G} \right)^2 \left[\exp\left(\frac{\tau \Omega l}{4\beta h_d kT} \right) - 1 \right] \qquad (3)$$

where D_s is the self diffusion coefficient, h_d is the critical jog height, β is a parameter that characterizes the average jog height. τ is the applied shear stress, α is the Taylor factor, Ω is the atomic volume, l is the spacing between the jogs, k is the Boltzman's constant and T, the temperature. We previously assumed that only vacancy producing jogs were present on the screw segments [6]. However, there is no reason to expect that only vacancy producing jogs will be present because these jogs are intrinsic in nature and are not produced through dislocation intersection events. Our TEM tilting experiments have shown [6] that jogs of opposing signs are present on the screw segments. When both types of jogs are present, the vacancy absorbing jogs will lag behind the vacancy producing jogs. This will lead to variation in line tension values along various segments of the dislocation. In this model, it is assumed that a steady state will be reached where both types of jogs will move with an average velocity according to $v = \alpha_p v_p + \alpha_a v_a$ where α_p and α_a are the fractions of vacancy producing and vacancy absorbing jogs respectively. Barrett and Nix [10] have treated this problem for $\alpha_p = \alpha_a = 0.5$ and obtained a different velocity law in which the exponential term in Equation 3 was replaced by a sinh function. Adopting this form and assuming the vacancy producing and absorbing jogs are equal in number, then the strain rate equation (3) could be rewritten as

$$\dot{\gamma} = \left(\frac{\pi D_s}{\beta b h_d}\right)\left(\frac{\tau}{\alpha G}\right)^2 \sinh\left(\frac{\tau \Omega l}{4\beta h_d kT}\right) \qquad (4)$$

Figure 1 shows the reasonable agreement between creep rates predicted from equation (4) and experimental data for equiaxed microstructure.

4.2 Lamellar Microstructures

It has been argued [11] that grains oriented for soft mode deformation dominate the creep behavior of polycrystals. The increase in creep rates in the soft mode is presumably due to increased 1/2<110] dislocation mobility parallel to and/or on the interface plane, with larger slip lengths compared to the hard slip modes where lamellar interfaces act as major obstacles. By this reasoning, it might be expected that the creep rates would be increased by lamellar spacing refinement. This is clearly contradictory to the recent findings [4,12] where finer lamellar structures have yielded lower creep rates. This discrepancy brings our attention primarily to hard mode 1/2<110] dislocations (Fig 4) whose mobility is restricted by the lamellar spacing. In fact, the fully lamellar structure often has a distribution of lamellar thickness. Thinner lamellae have hard-mode dislocations channeling through them, while the thicker lamellae have near-screw dislocations that are often cusped in configurations similar to those in equiaxed structures. The stress to move these channeling dislocations through a thin, capped film is given by [13]:

$$\tau_b = \left(\frac{Gb}{4\pi\lambda}\right)\ln\left(\frac{\lambda}{b}\right) \qquad (5)$$

where λ is the lamellar spacing. The effective stress on the channeling dislocations would be given by $\tau_{applied} - \tau_b$. As the lamellar thickness decreases, the effective stress available to propagate dislocations decreases. So for a given applied stress, lamellae thinner than a critical cut off λ_c would not participate in the deformation process. As the applied stress increases, the critical limit λ_c decreases and a greater volume fraction of the material is able to participate in the deformation process This is the basic premise for an approach to modeling creep rates in fully lamellar structures. In the present model, the lower cutoff λ_c is assumed to be the minimum lamellar spacing (λ_{js}) required for jogs to be present

To evaluate the creep rate it is required that the strain rate be averaged over the distribution of lamellar spacings

$$\dot{\gamma} = \frac{\int_{\lambda_{jt}}^{\lambda_{max}} F(\lambda) \cdot v \cdot \rho \cdot b \cdot d\lambda}{\int_{0}^{\lambda_{max}} F(\lambda) \cdot d\lambda} \tag{6}$$

where $F(\lambda)$ is the distribution of the volume fraction of gamma lamellae as a function of λ and is usually a log-normal distribution $N(\lambda)$, v is the velocity of dislocations under the effective stress, ρ is the dislocation density and b, the Burgers vector. TEM observations have been used to compile histograms of γ lath spacing from which $F(\lambda)$ can be obtained. The measure of distribution for the Ti-48Al lamellar microstructure is shown in Figure 5.

Now, we proceed to develop a velocity law as a function of lamellar spacing. As the jogged dislocation moves forward by the non-conservative dragging of the jogs, it lays out dislocation edge segments along the interface. This is schematically shown in Figure 6. So at steady state, work done,

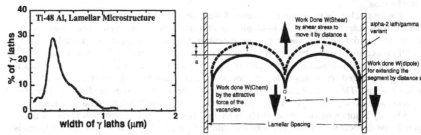

Figure 5. Distribution of γ lath widths in a fully lamellar microstructure

Figure 6 Schematic of jogged dislocation channeling within γ laths under applied stress

W_{Shear} by the applied shear stress in moving the dislocation forward by a, is balanced by the work done, W_{Chem} by the attractive forces due to the creation of vacancies (due to the non-conservative motion of jogs) and the work done, W_{dipole} for extending the dislocation segments laid out along the interface. Assuming once again that the vacancy producing and vacancy absorbing jogs are equal in number, we can arrive at the following expression for the velocity of the non-conservative motion of screw dislocations :

$$v(\lambda) = \frac{4 \cdot \pi \cdot D_{L}}{h} \cdot \left\{ \sinh\left[\frac{\Omega \cdot \lambda}{h \cdot n(\lambda) \cdot k \cdot T} \cdot (\tau - \frac{E_{dipole}}{b \cdot \lambda}) \right] \right\} \tag{7}$$

and $\quad E_{dipole} = \frac{G \cdot b^2}{2 \cdot \pi (1-v)} \cdot \ln(\frac{\lambda}{b}) + \frac{2 E_{core}}{b} \tag{8}$

where D is the diffusion coefficient, h jog height, $n(\lambda)$ is the number of jogs, E_{core} is the core energy of the dislocation being laid-out at the interface. Once again, assuming that the jog height h, is stress dependent (through Equation 2)), and β to be 0.5 representative of an average jog height [ref. 7], we can use Equation (6) to determine the strain rate as a function of stress. Figure 7 shows a comparison of the predicted creep rates and the experimental values. Assuming a stress dependent jog height, this model seem to over predict the creep rates. A possible source of this over-prediction is that the dislocation density is assumed to vary with the Taylor expression, and to be independent of lamellar spacing. TEM results suggest that narrow lamellea which are nevertheless active seem to exhibit lower dislocation densities than do the wider lamellea. However, this approach may provide an explanation for both the reduced strain rates and higher stress exponents observed for the lamellar structures. Note that this model does not provide an explanation for the significantly lower stress exponents observed at lower stresses. This regime has previously been attributed to interface dislocation motion by Wang and Nieh

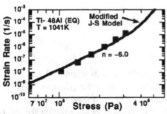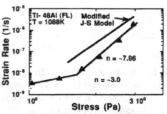

Figure 7 Comparison between the minimum creep rates predicted by modified jogged screw model and the experimental values for (a) equiaxed and (b) lamellar

[8]. Additional TEM study is needed to determine the change, if any, in the deformation mechanisms that may be responsible for the markedly different stress dependence observed in this regime.

5. Conclusions

Stress exponent values in the range of ~6 and ~8 were obtained for equiaxed and. lamellar microstructures respectively for Ti-48Al alloy at a temperature of T=1041K. TEM analysis indicates that the deformation microstructure in both microstructures were dominated by unit 1/2[110] dislocations pinned by jogs. It is suggested that that the creep in both microstructures is controlled by the non conservative dragging of jogs by 1/2110] screw dislocations. A modification of the original jogged-screw model is proposed in which the average jog height is assumed to depend on stress. This modified model results in good agreement between predicted and measured creep rates while using reasonable model parameters. On the other hand, lamellar structures showed highly inhomogeneous deformation behavior. There seems to be extensive activity of 1/2[110] type dislocations in lamellae above a critical thickness. This cutoff lamellar thickness is related to the minimum stress required to cause channeling of dislocations: lamellae thinner than the cutoff thickness experience no effective stress. The observation of jogged segments in the thicker lamellae suggests that a modification of the jogged-screw velocity law could be used by incorporating the effective stress, instead of the applied stress. To evaluate the creep rate it is required that the strain rate is averaged over the range of lamellar spacings.

Acknowledgements

G.B.V., S.K., and M.J.M. acknowledge the support of the National Science Foundation under grant DMR- 9709029 with Bruce MacDonald as program manager.

References

1. Y. W. Kim, JOM, 41, 24 (1989).
2. J. Beddoes, W. Wallace and L. Zhao, Int. Mater. Rev., 40, p. 197 (10995)
3. G. B. Viswanathan and V. K. Vasudevan *Gamma Titanium Aluminides*, ed. Y-W Kim, R. Wagner and M. Yamakuchi (eds.), TMS, Warrendale, Pennsylvania, p.967(1995).
4. T.A. Parthasarathy, M.G. Mendiratta and D.M. Dimiduk, Scripta Matel, 37 p.315 (1997)
5. M. Lu and K.J. Hemker, Acta. Mater., 45 (1997) 3573
6. G.B. Viswanathan, V.K. Vasudevan and M.J. Mills, Acta Mater., 47 p.1399 (1999)
7. B Skrotzki, T. Rudolf, A. Dlouhy and G. Eggeler, Scripta Materialia, 39 p.1545 (1998)
8. J.N. Wang and T. G. Nieh, Acta. Mater., 46 (1998) 1887
9. G. B. Viswanathan, Ph.D. Thesis (1998)
10. C. R. Barrett and W. D. Nix, Acta Metall., 13 p. 1247 (1965),
11. T.A Parthasarathy, P.R. Subramanian, and D.M. Dimiduk, Acta. Mater., 48, p. 541 (2000).
12. T.A. Parthasarathy, M. Keller and M.G. Mendiratta, Scripta Matl, 38 p.1025 (1998).
13. W.D.Nix, Scripta Materialia, 39(4/5) (1998) 545

Mat. Res. Soc. Symp. Proc. Vol. 646 © 2001 Materials Research Society

An Investigation of The Effects of Temperature on Fatigue Crack Growth Behavior of a Cast Nearly Lamellar Ti-47Al-2Cr-2Mn + 0.8 Vol. %TiB$_2$ Gamma Titanium Alloy

J. Lou, C. Mercer and W.O. Soboyejo
The Princeton Materials Institute and The Department of Mechanical and Aerospace Engineering, Princeton University, Olden Street, Princeton, NJ 08544

ABSTRACT

This paper presents the results of a study of the effects of temperature on fatigue crack growth in Ti-47Al-2Cr-2Mn + 0.8 Vol. %TiB$_2$ gamma titanium aluminide intermetallics. Fatigue crack growth rate data are presented for the cast lamellar microstructure at 25,450 and 750°C. The trends in the fatigue crack growth rate data are explained by considering the combined effects of crack-tip deformation mechanisms and oxide-induced crack closure. Faster fatigue crack growth rates at 450°C are attributed to the high incidence of irreversible deformation-induced twinning, while slower crack growth rates at 700°C are due to increased deformation by slip and the effects of crack-tip shielding provided by oxide-induced wedging, which is analyzed using a modified Dugdale-Barenblatt model.

INTRODUCTION

Since the development of XD™ gamma titanium aluminides at Lockheed-Martin and Howmet [1] over a decade ago, there have been considerable efforts to refine the microstructures of cast gamma alloys by alloying with B to promote the formation of TiB$_2$ particles/flakes [2,3]. The resulting refined lamellar microstructures (colony size of ~200-300 μm) have been found to exhibit relatively high tensile strengths (~700 MPa) [4], and increased fatigue endurance limits in smooth bar fatigue strengths that have been found to increase with increasing tensile yield strength [4]. However, there have been only a few studies [5,6] of the fatigue crack growth behavior of near-commercial XD™ gamma alloys in the potential service-temperature regime between 25 and 750°C.

This paper presents the results of a recent study of the effects of temperature on the fatigue crack growth behavior of Ti-47Al-2Mn-2Nb+0.8 vol.%TiB$_2$ alloy. Fatigue crack growth rate data are presented for long fatigue cracks that were tested in lab air (approximately 40% relative humidity) at 25, 450 and 700°C. As in earlier studies [5,6], the fastest fatigue crack growth rates were obtained at the intermediate temperature of 450°C. These are attributed to the effects of crack-tip deformation by twinning process. In contrast, slower fatigue crack growth rates at 700°C (compared to those at 25 and 450°C) are associated with the combined effects of oxide-induced crack closure and increased deformation by slip. The possible shielding contributions from oxide-induced crack closure are quantified using a modified Dugdale-Barenblatt model [7]. The possible causes of the observed effects of temperature on fatigue crack growth rates in Ti-47Al-2Mn-2Nb+0.8 vol.%TiB$_2$ gamma alloy are then elucidated.

MATERIALS AND EXPERIMENTAL PROCEDURES

The cast material that was used in this study was supplied by Howmet, Whitehall, MI. The alloy was supplied in the form of 152 mm long plates with rectangular (102 mm x 19 mm) cross-sections. The casting conditions are summarized in Refs. 8 and 9. The ingots were HIPed at 1260°C and 175

MPa for 4 hours. The resulting nearly lamellar microstructure of the Ti-47Al-2Mn-2Nb+0.8 vol.%TiB$_2$ alloy is shown in Fig. 1. This shows the scanning electron microscopy image of nearly lamellar colony microstructure, which was found to have an average lamellar volume fraction of approximately 92% and an average colony size of ~100 ± 32 µm. The average equiaxed gamma grain size was approximately 28 ± 4 µm and the average γ lath spacing was ~0.7 µm. The volume fraction of α$_2$ phase in the alloy was ~21%, and the average α$_2$ lath width was ~0.55 ± 0.04 µm.

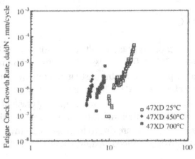

Stress Intensity Factor Range, ΔK, MPa√m

Fig. 1 Microstructure of Ti-47Al-2Mn-2Nb+0.8 vol.%TiB$_2$ Alloy

Fig. 2 Effects of Temperature on Fatigue Crack Growth Rates

Fatigue crack growth experiments were carried out on 50 mm long single edge notch bend (SENB) specimens, with rectangular (12.7 mm x 6.35 mm) cross-sections. The specimens were fabricated using electro-discharge machining (EDM) techniques. Their initial notch-to-width ratios were ~0.25. The SENB specimens were pre-cracked under far-field compression loading [10] prior to fatigue crack growth testing under three-point bend loading. A cyclic frequency of 10 Hz was used in the fatigue crack growth experiments, which were conducted at 25, 450 and 700°C. The fatigue crack growth tests were performed in laboratory air at a stress ratio, R = K$_{min}$/K$_{max}$, of 0.1. Crack growth was measured using a direct current potential drop technique. Low (below the fatigue thresholds) initial stress intensity factor ranges, ΔK, were applied initially. The ΔK levels were then increased in increments of 5-10% until crack growth was detected within ~10^5 cycles. The cracks were then allowed to grow under constant stress ranges (increasing ΔK) until they reached the high ΔK regime where the growth rates began to accelerate towards catastrophic failure. The tests were stopped prior to specimen fracture to allow subsequent examination of the crack-tip regions of the specimens using transmission electron microscopy (TEM). Before removal of slices for TEM observation, the sides of the specimen surfaces were examined in an SEM that was operated in the back-scattered electron (compositional contrast) mode. This was used to determine the presence and thickness of any oxide layer which may have formed between the crack surfaces during fatigue testing, particularly at elevated temperature. X-ray diffraction (XRD) and scanning Auger microscopy techniques [11] were then used to determine the compositions of oxide layers detected by back-scattered electron imaging of the specimen surfaces. The remaining portions of the SENB specimens (after removal of slices for TEM observation) were fractured under monotonic loading. Finally, the fracture surfaces of the specimens were examined using the SEM, which was used to assess the effects of test temperature on the micromechanisms of fracture under cyclic loading.

RESULTS AND DISCUSSION

Fatigue Crack Growth Rates:
The fatigue crack growth curves obtained for Ti-47Al-2Mn-2Nb+0.8 vol.%TiB$_2$ alloy at three temperatures are presented in Fig. 2. Fatigue threshold values at 25, 450 and 700°C are 9.2, 5.1 and 6.8 MPa√m. Respective Paris exponents are 6.3, 10.1 and 7.5, while the Paris coefficients are 1.6×10^{-13}, 2.1×10^{-14}, and 2.6×10^{-13}, respectively. The relatively low values of C and m at 25°C are associated with the slowest fatigue crack growth rates, while the relatively high m and low C values at 450°C are generally associated with the fastest fatigue crack growth rates. The low m value at 700°C is associated with a slow fatigue crack growth near threshold regime (~ 6.8 MPa√m).

Oxide-Induced Crack Closure Analysis:
X-ray diffraction (XRD) was used to determine the compositional nature of the oxide layer formed on the fracture surfaces during cyclic loading at 700°C. The analyses revealed that the oxide layer was predominantly alumina (Al$_2$O$_3$). No discernible X-ray peaks corresponding to TiO$_2$ were detected, as in previous studies of thermally-induced oxide layers in other gamma-based titanium aluminide alloys [11-13]. Scanning Auger spectroscopy was also employed to give further insight into the compositional nature of the oxide layers formed at this temperature. The results were consistent with the XRD analysis, in that Al$_2$O$_3$ was the major constituent of the oxide layer. The Auger spectroscopy also indicated the presence of low levels of rutile (TiO$_2$) in the oxide.

The possibility of crack-tip shielding contributions from oxide-induced crack closure was investigated by examining the sides of the fatigue cracks in the SENB specimens in the SEM using backscattered (compositional contrast) imaging conditions. No discernible oxide layers were detected on the fracture surfaces of the specimens tested at 25 and 450°C. However, a relatively thick oxide layer (~ 3 μm of excess oxide thickness) was detected on the sides of the specimen tested at 700°C, as shown in Fig. 3.

The excess thickness of this oxide layer was, therefore, comparable to the maximum crack-tip opening displacements, δ, during cyclic loading, which may be estimated from δ = $K_{max}^2/2E\sigma_y$ (where K_{max} represents the maximum stress intensity factor, E is the Young's modulus and σ_y is the cyclic yield stress, which is approximately equal to twice the monotonic yield stress). Values of δ for this material are found to vary between ~ 1.5 – 2.5 μm for the range of ΔK over which fatigue crack growth was monitored, (assuming a Young's Modulus of 150 GPa for TiAl). This indicates that oxide-induced crack closure is likely to occur at 700°C, since the estimated values of δ are less than the measured oxide thickness at 700°C, i.e. the crack is wedged open by the oxide layer. Further evidence of the possible role of oxide-induced crack closure can be obtained by estimating the oxide-induced closure stress intensity factor, K_{cl}, with a modified Dugdale-Barrenblatt model that was first proposed by Suresh and Ritchie [7] in their previous work on steels. For an oxide wedge of thickness, d, and a wedge length of 2l, the model yields:

$$K_{cl} = \frac{dE}{4(\pi l)^{1/2} (1-v^2)} \tag{1}$$

where E is the Young's modulus and ν is Poisson's ratio. For E = 150 GPa, ν = 0.3, d = 3 μm (the thickness of the oxide layer), values of K_{cl} may be estimated for different values of 1, where 2l is assumed to be equal to the length of the fatigue crack (minus the notch length). Values for effective stress intensity factor range (taking crack closure into consideration), ΔK_{eff}, can then be calculated from $\Delta K_{eff} = K_{max} - K_{cl}$.

Plot of da/dN versus ΔK_{eff} is presented in Fig. 4 for the specimen tested at 700°C. A plot of da/dN versus the applied stress intensity factor range, ΔK_{app}, is also shown presented on the same axes in Fig. 4 for comparison. This shows that, for a particular fatigue crack growth rate, the effective driving force for crack growth is less than the applied stress intensity factor range.

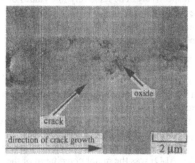

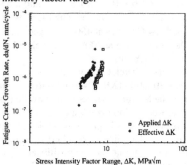

Fig. 3 SEM Micrographs Showing Unbroken surface of Fatigue Specimens Tested at 700°C for Ti-47Al-2Mn-2Nb + 0.8 vol.%TiB₂

Fig. 4 Plots of fatigue crack growths rates at 700°C versus effective and applied stress intensity factor ranges for Ti-47Al-2Mn-2Nb + 0.8%TiB₂

It is important to note here that the effectiveness of oxide-induced closure is most significant in the near-threshold regime where the oxide thickness is comparable to the minimum crack-tip opening displacements. Furthermore, fatigue crack growth is relatively fast beyond the near-threshold regime (Fig. 4), where the crack-tip opening displacements are greater than the excess oxide thickness. The beneficial effects of oxide-induced crack closure are, therefore, limited to the near-threshold regime where the crack-tip opening displacements are comparable to the excess oxide thickness.

Crack-Tip Deformation Mechanisms

Transmission electron micrographs showing the undeformed and deformed substructures of the three alloys are presented in Fig. 5. The undeformed substructure of the Ti-47Al-2Cr-2Nb+0.8%TiB₂ alloy is presented in Fig. 5 (a). This material was relatively free of deformation, but exhibited a fairly high incidence of microstructural twinning, in the form of micro-twinning across lamellae (Fig. 5 (a)). It is not clear why microstructural micro-twinning occurs in this material, but it may be associated with the effects of Mn and Cr on the twinning crystallography. Alternatively, the presence of grain size controlling TiB₂ particles may have an effect on microstructural twinning.

The deformation features observed in the crack-tip regions of the fatigue crack growth specimens tested at 25 and 450°C are presented in Figs. 5 (b) and (c), respectively. The specimen tested at 25°C exhibits a fairly high density of dislocations and stacking faults within the lamellae (Fig. 5 (b)), indicating that slip is the predominant mechanism of crack-tip deformation at room temperature. At 450°C, a mixed mode of deformation consisting of deformation-induced twinning and slip was observed (Fig. 5 (c)). So, the irreversibility of the deformation-induced twinning process is considered to be the main reason for the faster fatigue crack growth rates observed at 450°C.

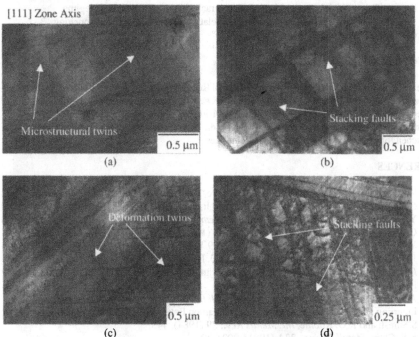

Fig. 5 Transmission electron micrographs showing (a) undeformed substructure, and crack-tip regions for fatigue specimens tested at (b) 25˚C, (c) 450˚C and (d) 700˚C, for Ti-47Al-2Mn-2Nb+0.8%TiB$_2$.

The crack-tip deformation mechanisms observed in the Ti-47Al-2Mn-2Nb +0.8% TiB$_2$ alloy at 700°C shows very little evidence of deformation-induced twinning. Instead, dislocation networks and a re-appearance of the stacking faults observed at room-temperature, were evident (Fig. 5 (d)). This suggests that the predominant mechanism of crack-tip deformation at elevated temperature has reverted back to that active at room temperature (i.e. conventional slip). Again, the partially reversible nature of slip (discussed above) will lead to slower fatigue crack growth rates at 700°C, as compared with those at 450°C.

CONCLUSIONS

1. The faster fatigue crack growth rates at 450°C are attributed largely to the higher incidence of crack-tip deformation by deformation-induced twinning at this temperature. However, the slower fatigue crack growth rates at 700°C (compared to those at 450°C) are partly associated with increased crack-tip deformation by slip processes.

2. The slower fatigue crack growth rates at 700°C are also due to the effects of oxide-induced crack closure. This occurs primarily as a result of thermally-grown oxides that have sufficient thickness to wedge open the cracks in the near-threshold regime. However, the fatigue crack growth rates are relatively fast once the crack opening displacements exceed the oxide thickness.

ACKNOWLEDGMENTS

The research was supported by The Office of Naval Research with Dr. George Yoder as Program Monitor. The authors are grateful to Dr. Yoder for his encouragement and support. Appreciation is extended to Mr. Paul McQuay of Howmet for the provision of the cast gamma alloys that were used in this study. Finally, the authors would like to thank Mr. Henk Colijn, Mr. Lloyd Barnhardt and Mr. Jon Mitchell for assistance with experimental techniques.

REFERENCES

1. L. Christodoulou, P.A. Parrish and C.R. Crowe, *High Temperature/High Performance Composites*, eds. F.D. Lemkey et al. (Pittsburgh, PA: MRS, 1988), 29-34.
2. Y-W. Kim and D.M. Dimiduk, *Journal of Metals*, 43 (1991), 40-47.
3. C. Mercer and W.O. Soboyejo, *Acta Mater.*, 45 (10) (1997), 4385-96, 1997.
4. Y-W. Kim, *Journal of Metals*, 46 (1994), 30-39.
5. J.P. Campbell, K.T. Venkateswara Rao and R.O. Ritchie, *Proceedings of the Sixth International Congress on Fatigue*, ed. G. Lüterjing and J. Nowack, (Oxford, U.K.: Pergamon Press, vol III, 1996), 1779-84.
6. A.L. Mckelvey et. al., *Proceedings of the Sixth International Congress on Fatigue*, ed. G. Lüterjing and J. Nowack, (Oxford, U.K.: Pergamon Press, Vol. III, 1996), 1743-78.
7. S.Suresh and R.O. Ritchie, *Int. Metals Rev.*, 29 (1984), 445.
8. K.S. Chan, *Metall. Trans.*, 22A (1991), 2021-29.
9. D.E. Larsen, Jr. and M. Behrendt, *Proceedings of The Symposium on Fatigue and Fracture of Ordered Intermetallic Materials I*, ed. W.O.Soboyejo, T.S. Srivatsan and D.L. Davidson (Warrendale, PA: The Minerals, Metals & Materials Society, 1993), 27-37.
10. S. Suresh and J.R. Brockenbrough, *Acta Metall.*, 36 (1988), 1444-70.
11. W.O. Soboyejo, J.E. Deffeyes and P.B. Aswath, *Materials Science and Engineering*, A138 (1991), 95-101.
12. Aswath, P. B., Soboyejo, W. O. and Suresh, S., in *FATIGUE '90, Proceedings of an International Conference on Fatigue*, 1990, Kitagawa, H. and Tanaka, T., Editors, EMAS, Warley, U. K., pp. 1941-46.
13. McKelvey, A. L., Venkateswara Rao, K. T. and Ritchie, R. O., *Scripta Mater.*, 37 (1997), pp. 1797-1803.

Mat. Res. Soc. Symp. Proc. Vol. 646 © 2001 Materials Research Society

Twinning in Crack Tip Plasticity of Two-Phase Titanium Aluminides

Fritz Appel
Institute for Materials Research, GKSS Research Centre,
Max-Planck-Straße, D-21502 Geesthacht, GERMANY

ABSTRACT

Intermetallic titanium aluminides based on γ(TiAl) are prone to cleavage fracture on low index lattice planes. Unfavourably oriented grains may therefore provide easy crack paths so that the cracks can rapidly grow to a length which is critical for failure. The effect of crack tip plasticity on crack propagation in γ(TiAl) was investigated by conventional and high-resolution electron microscopy. Crack tip shielding due to mechanical twinning was recognized as toughening mechanism, which occur at the atomic scale and apparently is capable to stabilize fastly growing cracks. The potential of the mechanism will be discussed in the context of novel design concepts for improving the strength properties of γ-base titanium aluminide alloys.

1. INTRODUCTION

Titanium aluminides based on the intermetallic phases γ(TiAl) and α_2(Ti$_3$Al) meet many demands for high temperature technology. However, as with many other intermetallics an inherent drawback for practical use is their tendency to undergo brittle transgranular fracture at low and ambient temperatures [1-3]. It is now recognized that there are at least three causes of brittleness in such alloys:

- the occurence of cleavage on low index planes
- insufficient deformation modes which can simultaneously operate at given stress
- low dislocation mobilities.

Extensive fractographic studies have demonstrated that the microstructure exerts a significant effect on crack propagation [4-6]. For toughening the material a lamellar morphology of the γ(TiAl) and α_2(Ti$_3$Al) phases is beneficial, because shear ligament bridging, crack deflection and microcrack shielding occur. These processes take place on a scale ranging from ten to some hundred microns; thus, continuum fracture mechanics is capable of capturing these prominent features of crack propagation [4]. However, plastic dissipation at the crack tip may also occur at the atomic level and constitute a major fraction of the total work of fracture. Rice and Thomson [7] have proposed that the intrinsic brittleness or ductility of a material is closely related to the competition between cleavage decohesion and crack tip shielding or blunting due to dislocation emission. Stable crack growth requires the plastic zone to keep up with the cleavage crack, which is difficult if the mobility and multiplication rate of the dislocations are low. In this respect mechanical twinning might be an important mechanism, because the growth rate of twins is often an appreciable fraction of the elastic wave velocity [8]. There have been relatively few studies on the contribution of twinning to crack tip plasticity, although twinning is a prominent deformation mode in γ(TiAl). This results probably from the fact that brittle fracture is a rapid process and most analytical techniques are not capable of imaging the details on the appropriate size scale. The present paper describes an electron microscope study of crack tip plasticity in two-phase titanium aluminides. The main objectives of the paper are to ascertain the effects of mechanical twinning on crack tip shielding and to put the results into perspective of novel alloy design concepts.

2. DISLOCATION GLIDE AND MECHANICAL TWINNING IN γ(TiAL)

The deformation behaviour of γ(TiAl) is closely related to its $L1_0$ structure. Dislocations with the Burgers vectors $1/2<110]$, $1/2<11\bar{2}]$ and $<011]$ are liable to glide on close-packed $\{111\}$ planes and contribute to deformation at low and intermediate temperatures [10, 11]. Mechanical twinning along $1/6<11\bar{2}]\{111\}$ can also be an important deformation mode at low and ambient temperature. However, there is only one distinct true twinning $b_1 = 1/6<11\bar{2}]$ on (111) that does not alter the ordered $L1_0$ structure of γ(TiAl) [1, 3, 8]. Shear in the reversed direction $-2b_1$ is the so-called anti-twinning. The two other partials with the Burgers vectors $b_2 = 1/6[\bar{2}11]$ and $b_3 = 1/6[1\bar{2}1]$ lead to pseudo-twinning. The relative contributions of the individual mechanisms to deformation depend mainly on the aluminium concentration, the content of ternary elements and the deformation temperature [11]. There is growing evidence that the activation of ordinary dislocations and superdislocations requires significantly different critical shear stresses so that these glide systems often cannot simultaneously operate [11]. Thus, there are many more restrictions upon dislocation glide and crack tip plasticity in γ(TiAl) than for disordered f.c.c. metals. In this respect mechanical twinning may compensate for the lack of independent slip systems which can operate at comparable stresses.

3. STRESS ACCOMMODATION AT CRACK TIPS DUE TO MECHANICAL TWINNING

The experimental investigations were performed on different two-phase alloys with the base line composition (in at.%) Ti-(45-47)Al-(1-10)Nb-(3-4)Cr, Mn, Ta-(0.2-0.5)B,C. More details of alloy preparation and thermal treatments were given elsewhere [9]; the alloys will also be specified at the respective examples. From undeformed samples slices 0.5 mm thick were cut by spark erosion. Electron transparent foils were prepared by mechanical grinding the slices down to 150 μm and subsequent twin-jet electrolytic polishing. The thin foils usually contained cracks, which were probably generated during preparation and handling. The crack tips were examined for signs of stress induced plasticity utilizing conventional and high resolution electron microscopy. These microscope techniques provide direct evidence but suffer from the disadvantage that thin foils are not generally representative of bulk behaviour. This is because the cracked sample geometry and loading conditions were not well defined. Furthermore, due to image forces dislocations can easily escape to the surface of a thin foil and alter the defect content. However, this may not be a serious drawback if the goal of the investigation is not the characterization of gross structures but the elucidation of the atomic details of defect emission at crack tips.

In terms of dislocation dynamics, the question of whether an atomically sharp crack can or cannot propagate in a brittle manner depends on the rates at which the dislocations can propagate and multiply. In γ(TiAl) alloys the mobility of both ordinary and superdislocations is impeded by high Peierls stresses, localized obstacles and jogs [11, 12]. This gives rise to rate dependent drag forces acting on the dislocations and certainly reduces the tendency for slip to accommodate the strain at the high strain rates encountered at the crack tip. Furthermore, dislocation multiplication within the plastic zone of crack tips was found to be closely related to the operation of cross glide sources [2] and thus needs some glide or time. Plastic zones consisting of dislocation assemblies may therefore easily be outrun by fast cracks. The enhanced stress concentration ahead of fast cracks and the limited capability of plastic relaxation by dislocation glide, on the other hand, make twinning feasible.

An interesting interactive case of fracture and twinning is demonstrated in figure 1. The crack originated from the perforation of the thin foil and was probably subjected to mixed mode loading, which apparently involve a significant mode II shear component. The crack propagated across the lamellae along {111} planes with a common <110> direction. Thus, the crack was deflected at the interfaces according to the orientation relationship of the adjacent lamellae [2], (figure 1a). Within the individual lamellae the crack follows the {111} planes almost along the atomic planes, which is consistent with their low cohesion energy and stress intensity factor predicted by Yoo and Fu [1, 3]. In response to the stress field associated with the crack tip a narrow twin is formed ahead of the crack tip (figure 1b). The width of the twin band gradually becomes narrow with increasing distance from the crack tip as expected from the stress field of the crack. This leads to the impression that the nucleation of the twin takes place sequentially as the crack advances. Twinning seems always to precede crack propagation as indicated by the heavily twinned crack wakes (figure 1c). This is probably the reason for the formation of a high density of cleavage facets, which may be another factor of energy absorption in the propagation of the crack.

However, as mentioned in Sect. 3, an important difference between twinning and dislocation glide is that twining is polarized; i.e. reversal of the twinning shear direction will not produce a twin [1, 3, 8]. Furthermore, in the $L1_0$ structure only one twinning system per {111} plane is available. Thus, for a given shear direction certain γ grains or lamellae should not twin. Due to this lack of twinning systems, crack tip shielding by mechanical twinning is certainly limited to favourably oriented grains. Nevertheless, concerning a larger assembly of grains, the mechanism certainly contributes to some energy dissipation at the crack tip and consequently, to some resistance against unstable crack propagation.

4. IMPLICATIONS FOR ALLOY DESIGN

Several strategies have been evolved in order to improve the high strength and creep resistance of titanium aluminides by solid solution and precipitation hardening [12]. The challenge is to establish these mechanism without compromising desirable low-temperature properties such as ductility and toughness. Thus, for the alloy design a detailed understanding of the effects of strengthening agents on the operation of deformation modes is quite important.

A significant strengthening effect has been reported for Nb when added with a larger amount to two-phase alloys [13]. Alloys of the base line composition Ti-45Al-(5-10)Nb at room temperature typically exhibit flow stresses in excess of 1000 MPa and an appreciable tensile elongation of ε_f = 1-2 %. Detailed TEM observations have revealed that $1/6<11\overline{2}]\{111\}$ order twinning is a prominent deformation mechanism in the Nb containing alloys [13]. In these

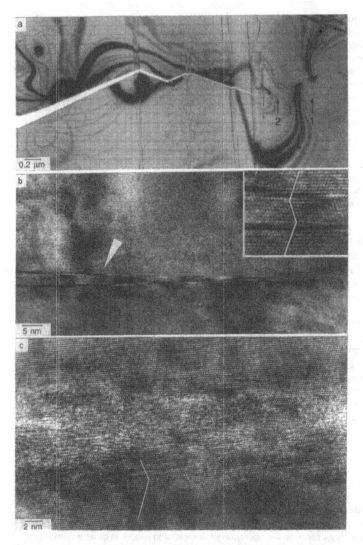

Figure 1. *Association of twinning and fracture in* γ*(TiAl). (a) Crack propagation along {111} planes and deflection at lamellar interfaces. (b) High resolution image of detail (1) in figure 1a showing twin formation ahead the crack tip. (c) Detail (2) in figure 1a showing twinning on either side of the crack, Ti-47Al-4(Cr, Nb, Ta, B) with nearly-lamellar microstructure.*

materials the superdislocations were found to be widely dissociated, which has seldom been observed in "conventional" alloys. As dissociated superdislocations, planar faults and twins often coexist in the same grain or lamella (figure 2), it is speculated that the twins originate from the superposition of extended stacking faults on alternate {111} planes [13]. The preponderance of faults in the Nb containing alloys as compared to conventional alloys may be taken as an indication that Nb lowers the stacking fault energy of TiAl.

Appreciable improvements in strength and creep resistance have also been achieved by carbon additions when introduced as a fine dispersion of Ti_3AlC precipitates of perovskite type [14]. The precipitates are elongated along their [001] axis and exhibit significant coherency stresses due to their lattice mismatch with the γ matrix. These structural features lead to a strong glide resistance for all types and characters of dislocations, which in view of the considerations made in the previous section might be harmful for ductility. However, as with the TiAl-Nb alloys, mechanical twinning seems to be relatively easy. Twin nucleations were often found together with widely separated superdislocations (mostly of type <011]. Thus, it is tempting to speculate that the twins originate from overlapping stacking faults trailed by the superdislocations. The process is certainly supported by the high coherency stresses occurring at the precipitates. In this way a fine dispersion of deformation twins can be generated, which apparently is beneficial for the deformability of the material.

0.2 μm

Figure 2. *Defect structure in a Ti-45Al-10Nb alloy with duplex structure observed after tensile deformation at room temperature to a plastic strain $\varepsilon_f = 1$ % and stress $\sigma = 1050$ MPa. Widely dissociated superdislocations, stacking faults and twins coexisting in a γ grain, that is subjected to constraint stresses due to twinning in the adjacent grain. Weak-beam dark field image.*

5. CONCLUSIONS

In closing it may be noted that mechanical twinning plays a major and complex role in stabilizing brittle fracture in γ(TiAl), because twins can rapidly propagate. Mechanical twinning also compensates for the lack of independent slip systems that at a given stress can simultaneously operate. In γ(TiAl) twinning apparently can be supported by alloying with Nb and the implementation of precipitation hardening. These characteristics are particularly important for novel high strength γ base alloys, which exhibit a relatively good ductility at low and ambient temperatures.

The {111} planes of γ(TiAl) serve at the same time as slip planes, twin habit planes and cleavage planes. Thus, microcracks may easily be generated by blocked slip in front of dislocation pile-ups or at intersecting twin bands. These synergistic effects of dislocation glide, twinning and fracture need further investigation.

ACKNOWLEDGEMENTS

The author acknowledge helpful discussions and continuous support by Ms E. Tretau, U. Christoph, St. Eggert, U. Lorenz, J. Müllauer, M. Oehring and J. Paul.

The financial support of this research by the Helmholtz-Gemeinschaft (HGF-Strategiefonds) is gratefully acknowledged.

REFERENCES

1. M.H. Yoo, C.L. Fu, J.K. Lee, *Twinning in Advanced Materials*, M.H. Yoo, M. Wuttig (Eds.), TMS, Warrendale, PA, 1994, p. 97.
2. F. Appel, U. Christoph, R. Wagner, Philos. Mag. A 72 (1995) 341.
3. M.H. Yoo and C.L. Fu, Metall. Trans. A29 (1998) 49.
4. K.S. Chan, Metall. Trans. A24 (1993) 569.
5. D.M. Dimiduk, *Gamma Titanium Aluminides*, Y-W. Kim, R. Wagner, M. Yamaguchi (Eds.), TMS, Warrendale, PA, 1995, 3.
6. Y-W. Kim, D.M. Dimiduk, *Structural Intermetallics*, R. Darolia, C.T. Liu, P.L. Martin, D.B. Miracle, R. Wagner, M. Yamaguchi (Eds.), TMS, Warrendale, PA, 1997, p. 531.
7. J.R. Rice, R. Thomson, Philos. Mag. 29 (1974) 73.
8. J.W. Christian, S. Mahajan, Progress in Materials Science 39 (1995) 1.
9. F. Appel, U. Lorenz, M. Oehring, U. Sparka, R. Wagner, Mater. Sci. Eng. A233 (1997) 1.
10. M. Yamaguchi, Y. Umakoshi, Progress in Materials Science, 34 (1990) 1.
11. F. Appel, R. Wagner, Mater. Sci. Eng. R22 (1998) 187.
12. F. Appel, M. Oehring, R. Wagner, Novel Design Concepts for Gamma-Base Titanium Aluminides, Intermetallics, 2000, in print.
13. F. Appel, U. Lorenz, J.D.H. Paul, M. Oehring, *Gamma Titanium Aluminides 1999*, Y-W. Kim, D.M. Dimiduk, M.H. Loretto (Eds.), TMS, Warrendale, PA, 1999, p. 381.
14. U. Christoph, F. Appel, R. Wagner, Mater. Sci. Eng. A 239-240 (1997) 39.

Mat. Res. Soc. Symp. Proc. Vol. 646 © 2001 Materials Research Society

An In-Situ Study of Crack Propagation in Binary Lamellar TiAl

P. Wang, N. Bhate, K.S. Chan* and K.S. Kumar

Division of Engineering, Brown University, Providence, RI 02912
* Southwest Research Institute, San Antonio, TX 78238

Abstract

Single-colony thick compact tension specimens of binary lamellar Ti-46.5%Al were tested within a scanning electron microscope to examine the contribution of colony boundaries to crack growth resistance under monotonic loads. These specimens were obtained by machining slices from bulk material that had been heat treated to grow the colony size. Thus, the lamellae in adjacent colonies exhibit significant misorientation across the boundary. The orientation of the lamellae within a colony has been characterized in terms of two angles defined with respect to the notch orientation: an in-plane angle α and a through thickness angle β. The change in these two angles across the colony boundary quantifies the misorientation. In addition a third angle, θ, defines the colony boundary tilt to the vertical plane. These parameters were measured in several specimens and the crack growth resistance across the boundary was qualitatively and quantitatively characterized. The importance of the through-thickness angle β in providing resistance to crack growth is illustrated.

Introduction

Crack propagation studies have been conducted on polycrystalline and single-crystalline (PST crystals) lamellar two-phase $TiAl/Ti_3Al$ alloys with and without further alloying [1-4]. These studies have confirmed the presence of intrinsic (matrix slip and ductile phase toughening) and extrinsic (crack deflection, ductile-phase bridging, shear ligament toughening, microcrack shielding) toughening mechanisms in these alloys. Recently, Chan et al. [5] examined cracking of lamellar Ti-46.5%Al polycrystalline compact tension specimens in-situ in the SEM. They reported a low initiation K, predominantly interlamellar cracking within the colony with crack advancing by the linking up of microcracks, minimal number of microcracks ahead of the main crack, negligible resistance to crack growth within the colony, *and noticeable resistance to crack growth across colony boundaries.* They further showed that the magnitude of the resistance offered by the boundaries to crack growth was dependent on the lamellar misorientation (as measured on the surface) across the boundary. Frequently, crack renucleated across the boundary and advanced forward and backward to link up with the crack in the previous colony. Multiple cracking was also reported to occur within individual colonies. It is however unclear how much of this resistance as well as observed damage on the surface was a consequence of subsurface colonies and associated boundaries since the average colony size reported in [5] was 640 μm and the specimen thickness was 4.75 mm. It was therefore recognized that it would be beneficial to conduct such tests on specimens that were "single-colony thick" (but polycolony in-plane) to more clearly establish the role of colony boundaries (or colony misorientation) in providing resistance to crack growth.

Computational efforts to model deformation and fracture of such microstructures [6] have been based on two-dimensional calculations. Extending such models into three dimensions (3D) is at its infancy; in the early stages of developing 3D models, specimens modeled as being single-colony thick is a more likely scenario than one where multiple colonies are modeled in the thickness dimension. In such a situation, experimental results generated from single-colony thick specimens could provide direct contact with the model. With respect to crack propagation, both within a colony as well as across a boundary in these lamellar structures, it is necessary to conduct calculations in 3D as the lamellae can not only have an in-plane "kink" angle α with respect to the notch but also a through-thickness inclination β (Figure 1). The effect of the through-thickness inclination on fracture is not represented in a 2D computation.

To address these issues, we created (by heat treatment and sectioning), compact tension specimens containing several colonies that extended from one surface of the specimen through the thickness to the other surface. Such specimens were tested within an SEM and load versus crack extension data were obtained. The in-plane microstructure along the crack path was recorded during the test. These results are presented and the effect of microstructure across colony boundaries on crack growth resistance is discussed.

Experimental Procedure

The alloy examined in this study was binary Ti-46.5 atomic percent Al with a duplex microstructure in the as-forged condition. Processing details were provided in [5]. Blocks of material from the as-forged disk were heat-treated at 1450°C for 6-10h to grow the grains and furnace-cooled to obtain a fully lamellar structure. Slices (19.8 mm x 19.05 mm x 1.2-1.3 mm) were obtained by electrodischarge machining and both surfaces were ground, polished and examined to verify that several of the colonies had the same lamellar orientation on both sides. The holes for the loading pin and the notch were then electrodischarge machined into the selected specimens. Using a specially designed fixture, these specimens were then fatigue precracked under a compression-compression stress state. The fatigue precracked specimens were incrementally loaded to fracture within an SEM. Crack propagation was monitored as a function of load increase and the interaction of the propagating crack with surface microstructure

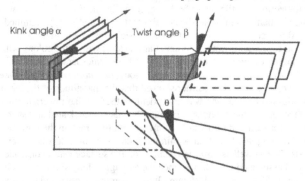

Figure 1: A schematic illustration describing the orientation of the lamellar plates with respect to the notch through an in-plane kink angle α and a through-thickness twist angle β. In addition, the colony boundary plane can be inclined with respect to the vertical by θ.

was recorded by taking still photographs or using a video camera. From such data, strain energy release rate as a function of crack extension was calculated and the details of the calculations are provided in the next section.

Results and Discussion

As a general observation, it is noted that cracks propagated along the lamellae within a colony although it is not clear whether they run along the interlamellar interface or through the lamellae. Crack advance within the colony typically proceeded rapidly with minimum resistance. Occasionally, secondary parallel cracks nucleated and propagated within the colony but more often, the original crack simply ran to the boundary with little indication of damage on the surface in the process zone. The orientation of the lamellae in each of the colonies traversed by the crack is described through the kink angle, α and the twist angle β; a clockwise and counterclockwise rotations with respect to the notch axes are distinguished using positive and negative notations. The lamellar misorientation across a colony boundary is then described through $\Delta\alpha$ and $\Delta\beta$ using the α and β for the two adjacent colonies under consideration. In addition, a grain boundary inclination angle θ (Figure 1) is used to quantify the inclination of the boundary to the vertical plane. Using low magnification images from the top and bottom surfaces of the fractured specimens that had been transferred on to transparent sheets and by overlaying them carefully, the relative displacements of crack paths as well as colony boundaries can be obtained. Knowing the specimen thickness, the angles α, β, and θ can be obtained. A compilation of these angles is provided in Table I for seven different specimens.

We first examine crack propagation from colony 2 to colony 3 in specimen # 3. In this case, $\Delta\beta$ is minimal but $\Delta\alpha$ is 15 degrees. In addition, the colony boundary is almost vertical (Table I; specimen #3, θ_{2-3}). The associated micrograph is shown in Figure 2a. The crack traverses the boundary with evidence of minimal damage in the region adjacent to the boundary. In contrast, the $\Delta\alpha$ between colonies 3 and 4 in the same specimen is 20 degrees (about the same as the difference between colonies 2 and 3) but $\Delta\beta$ is 44 degrees. In addition, the colony boundary is also inclined (θ = 19 degrees). When the crack traverses this boundary, there is a significant number of secondary cracks that originate at the boundary and run back into colony 3 (Figure 2b). The ligaments between these cracks exhibit plastic deformation. The presence of all this damage at this boundary in contrast to the previous case (colonies 2/3) is indicative of a higher level of difficulty associated with traversing this boundary.

Table I. Lamellar Orientation Description in Terms of Specimen Axes

#	Colony 1		Colony 2		Colony 3		Colony 4		θ		
	α	β	α	β	α	β	α	β	1-2*	2-3	3-4
1	-3	-5									
2	+17	+9	+4	+4	-14	+49			0	4	
3	+29	-25	-17	+7	-2	+4	+18	+48	32	4	19
4	-13	+30	-67	+5							
5	-30	+32	-40	+3	-23	-32			8		
6	+15	+47	+10	-17	+9	-29					
7	-16	+48	-9	-19							

* denotes boundary between colony 1 and colony 2.

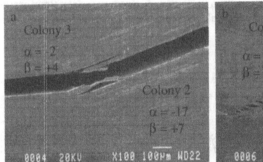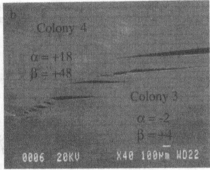

Figure 2. Crack propagation across colony boundaries in specimen #3 (a) across boundary between colonies 2 and 3, and (b) across boundary between colony 3 and 4.

A similar response (i.e. multiple cracking and plastic deformation of the partially constrained ligaments in the vicinity of the boundary) was also noted in specimen # 5 as a crack traversed the boundary between colonies 1 and 2 (Figure 3a,b). Here the colony boundary exhibits $\theta = 8$ degrees, $\Delta\alpha = 10$ degrees, but $\Delta\beta = 29$ degrees. These two examples illustrate that the through-thickness misorientation at a colony boundary provides significant crack growth resistance, perhaps more so than a comparable in-plane misorientation $\Delta\alpha$. There were no situations in the seven specimens examined where $\Delta\alpha$ was >30 degrees and $\Delta\beta$ was small. In specimen #4 where a large $\Delta\alpha$ is present between colonies 1/2 (Table I), the fatigue precrack was already in colony 2 when the fracture test commenced. Recent computational efforts [6] have shown that the contribution of the in-plane kink angle is significant only when the $\Delta\alpha$ is high (e.g. > 60°). Lastly, an additional interesting observation is that the damage pattern in the vicinity of the boundary on the two surfaces of the single colony specimen are distinctly different (Figure 3a,b) illustrating the complexity in crack propagation across such boundaries. This difference is possibly a consequence of θ.

Attempts to obtain a mode I K versus Δa curve from the load-crack length data were futile because the crack incorporates substantial levels of mode mixing. Instead, using the data from all of these specimens, it was felt that it would be prudent to obtain a strain energy release rate, G, versus crack length curve. An examination of the literature revealed that there is relatively little by the way of analytical solutions for a compact tension specimen geometry when the crack plane is inclined to the loading axis. In such a case, the mode of cracking comprises combinations of Mode I, II and III components. Specifically, with respect to Figure 1, when the crack plane in the colony containing the fatigue precrack is inclined to both the width and thickness of the specimen (i.e. non-zero α and β), the stress intensity factor solution based on a Mode I crack is not applicable. Chan and Cruse [7] have examined this problem using the boundary integral equation technique and have provided boundary correction factors for the K_I and K_{II} components of an inclined crack for a variety of crack orientations. In their calculations, they used an a/W = 0.2 (a = notch length from the loading axis and W = specimen width). In this study, the a/W = 0.45 and therefore these correction factors had to be recalculated for several orientations (0°, 30°, 45°, 60° and 70°) using the approach outlined in [7]. It is pertinent to

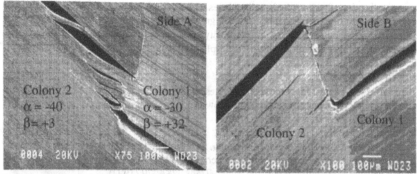

Figure 3. Damage resulting in the vicinity of the boundary between colonies 1 and 2 in specimen #5 when the crack traverses the boundary as seen on the two surfaces of the single-colony thick compact tension specimen.

recognize however that the boundary integral equation methodology and its implementation are limited to two-dimensional elastic fracture mechanics problems of a finite plate containing a single traction-free crack. Thus, this approach will only provide the Mode I and Mode II stress intensity factors. There is no analytical solution available for a compact tension specimen containing an inclined through-thickness crack. However, Yoshioka et al [8] have provided stress intensity factors for a three-dimensional slant through-crack at the center of a plate of width W that is subjected to a uniform tensile stress σ. The effect of slant angle β and of Poisson's ratio ν were examined. We have used this formulation as an approximation to obtain K_{III} and then simply summed the contributions of K_I, K_{II} and K_{III} appropriately to obtain G. Such a relationship (assuming isotropy) is given as $G.E = K_I^2 + K_{II}^2 + (1-\nu) K_{III}^2$, where E is the Young's modulus. The variation of G.E with actual crack length 'a' is shown in Figure 4 for the seven specimens. Data in Figure 4 for each specimen are for crack extension within a single colony. Data for specimen #4 are from colony 2 since the fatigue precrack had already traversed colony 1 and entered colony 2. In this case alone, an R curve behavior is observed; more data are needed to confirm such a characteristic. In all other specimens, data were obtained only from the first colony as we do not have a satisfactory approach at present to cope with a crack that kinks and twists across a boundary . The two open squares (at a ≈11.5 mm) represent a situation (specimen #2, Table I) where a crack was held up at a colony boundary such that on one surface it was in the first colony whereas on the second surface, it was in the new colony (Figure 5a). It is worth noting that in this case, there is a $\Delta\alpha$ of 18 degrees but a substantial $\Delta\beta$ of 45 degrees. The difference in G.E between the two open squares for minimal advance in crack (compare Figure 5a with 5b) is a measure of the resistance offered by the boundary as a consequence of the lamellar misorientation across the colony boundary.

Conclusions

This study illustrates that the various damage modes observed in multi-colony thick specimens are also observed in single-colony thick specimens. Within the colonies, crack propagation is primarily along the lamellar planes with minimum resistance to propagation. Colony boundaries can offer resistance to crack growth provided the misorientation of the lamellae across the boundary is large. Specifically, differences in the twist angle β appear more effective than a comparable difference in the in-plane kink angle α.

Acknowledgments

This effort was supported by the Materials Research Science and Engineering Center on Micro- and Nano-Mechanics of Materials at Brown University (NSF Grant DMR-9632524).

References

1. K.S. Chan and Y-W. Kim, Metall. and Mater. Trans., 25A, 1217 (1994).
2. K.S. Chan and Y-W. Kim, Acta Metall. Mater., 43, 439 (1995).
3. K.S. Chan and D.S. Shih, Metall. and Mater. Trans., 28A, 79 (1997).
4. S. Mitao, T. Isawa and S. Tsuyama, Scripta Metall. Mater., 26, 1405 (1992).
5. K.S. Chan, J. Onstott and K.S. Kumar, Metall. Mater. Trans., 31A, 71 (2000).
6. J.J.M. Arata, A. Needleman, K.S. Kumar and W.A. Curtin, Int. Jour. Frac., 105, 321 (2000)
7. K.S. Chan and T.A. Cruse, Eng. Frac. Mech., 23, 863-874 (1986).
8. S. Yoshioka, M. Miyazaki, K. Watanabe, H. Kitagawa and Y. Hirano: from the Stress Intensity Factors Handbook, Volume 2; Editor-in-Chief: Y. Murakami, The Society of Materials Science, Japan. Pergamon Press, New York, 1987, p. 833.

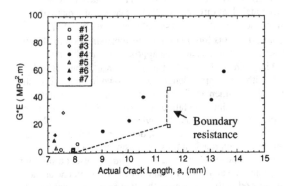

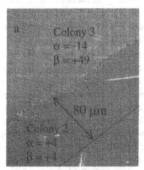

Figure 4. The variation in the product of elastic modulus and strain energy release rate, (G.E) with actual crack length.

Figure 5. Crack growth resistance at a colony boundary in specimen #2. The approximate location of the colony boundary on the reverse side of the specimen is indicated by the dashed line in (a). In (b) crack advance by a minimal amount is seen upon a significant increment in remote load. This translates into a discontinuous increase in G.E as evidenced by the open squares in Figure 4.

Mat. Res. Soc. Symp. Proc. Vol. 646 © 2001 Materials Research Society

STRAIN-RATE DEPENDENCE OF THE BRITTLE-TO-DUCTILE TRANSITION TEMPERATURE IN TiAl

M. Khantha, V. Vitek and D. P. Pope
Department of Materials Science and Engineering, University of Pennsylvania,
Philadelphia, PA 19104-6272

ABSTRACT

The brittle-to-ductile transition (BDT) and the strain-rate dependence of the brittle-to-ductile transition temperature (BDTT) have been recently investigated in single crystals of TiAl [1]. It was found that the activation energy associated with the BDTT is 1.4 eV when the slip is dominated by ordinary dislocations and 4.9 eV when it is dominated by superdislocations. Despite this difference in the activation energies, the BDTT, while varying with the strain-rate, remains in the same temperature range, viz., between 516-750C and 635-685C for ordinary and superdislocations, respectively. In this paper, we examine how the activation energy of the BDTT can vary with the type of dislocation activity and explain why it can attain values which are clearly much larger than the activation energy for dislocation motion. We describe a strain-rate dependent mechanism of cooperative dislocation generation in loaded solids above a critical temperature and use it to explain the characteristics of the BDT in TiAl. We show that the activation energy associated with the BDTT is a composite value determined by two or more inter-dependent thermally activated processes and its magnitude can be much larger than the activation energy for dislocation motion in certain materials. The predictions of the model are in good agreement with observations in TiAl.

INTRODUCTION

The brittle-to-ductile transition (BDT) exhibited by most crystalline materials is a strain-rate dependent phenomenon marked by a rapid increase of dislocation activity at the crack tip concomitant with a sharp increase of the fracture toughness [2, 3]. The brittle-to-ductile transition temperature (BDTT) increases with increasing strain-rate in all materials and an apparent activation energy can be associated with this strain-rate dependence. In silicon and other semiconductors [2, 3], this activation energy is almost exactly equal to the activation energy for dislocation motion. It is not known if such a correlation exists in other classes of materials.

The BDT in TiAl single crystals [1] was recently investigated for two different crack geometries. It was found that the activation energy associated with the BDTT is 1.4 eV when the slip is dominated by ordinary dislocations (Burgers vector 1/2<110]) and 4.9 eV when it is dominated by superdislocations (Burgers vector <011]). Despite this difference in the activation energies, the BDTT, while varying with the strain rate, is in the same temperature range, viz., between 516-750C and 635-685C for ordinary and superdislocations, respectively [1]. In addition, the variation of the fracture toughness with temperature was identical for both geometries over the entire temperature range for a fixed strain-rate. This suggests that the activation energy for motion of ordinary dislocations and superdislocations must be similar. The activation energy for dislocation motion is not known in TiAl but there is considerable evidence which suggests that the activation energy for the glide of ordinary dislocations is only slightly lower than that for superdislocations [4, 5]. In both cases, the activation energy is not expected to be as high as in Si (1.8-2.2 eV) because both types of dislocations are known to glide at low temperatures. This raises the question why the apparent activation energy of the BDTT is significantly different for ordinary and superdislocations when this difference cannot be sought in different mobilities of these two types of dislocations.

We describe a strain-rate dependent process of cooperative dislocation generation in loaded crystals that can lead to the formation and expansion of many dislocation loops without any energy barrier above a critical temperature [6-9]. In particular, the model shows that the apparent activation energy associated with the BDTT represents a composite phenomena of two or more inter-dependent thermally activated processes. Its value is determined not only by the activation energy for dislocation motion but also by the density of glissile dislocations. The latter includes pre-existing mobile dislocations and dislocations that are precursors to the onset of the cooperative instability, generated just below the BDTT, by thermal activations. In this paper we use this model to explain the BDT and predict the BDTT in TiAl for activity of both ordinary and superdislocations. We show that small changes in the density of pre-existing glissile dislocations (ordinary or superdislocations) can result in large changes in the apparent activation energy associated with the BDTT without affecting the magnitude of the BDTT significantly. The results are in good agreement with observations.

STRAIN-RATE DEPENDENT COOPERATIVE PROCESS OF DISLOCATION GENERATION

The cooperative dislocation generation proceeds by the concurrent nucleation and evolution of many atomic-size dislocation loops that form in a loaded crystal at finite temperatures. This process, which occurs owing to the combined effect of dislocation interactions and entropy [6-9], differs manifestly from the nucleation and subsequent expansion of an isolated dislocation loop [10]. At finite temperatures, in a loaded crystal, there is a small probability of forming dislocation loops of interatomic dimensions by thermal fluctuations since their energy ranges approximately from 1.0 - 2.0 eV. The density of such loops of radius r at temperature T is determined by the Boltzmann factor, $\exp[-H(r)/k_B T]$, where H(r) is the formation enthalpy of the loop and k_B, the Boltzmann constant. For an isolated shear loop [8], when the material is loaded by a shear stress σ in the direction of the Burgers vector,

$$H(r) = K_0 \left[r \ln(r/r_0) + cr \right] - \sigma b \pi r^2 \qquad (1)$$

where $K_0 = \left[\mu_0 (2 - v_0) b^2 / 4(1 - v_0) \right]$, b is the magnitude of the Burgers vector, μ_0 the elastic shear modulus in the slip plane, v_0 the Poisson ratio, $K_0 c$ the core energy of the dislocation and r_0 the elastic cut-off radius. Dislocation loops formed by thermal fluctuations mostly shrink and disappear because their radii are considerably smaller than the critical radius (~ ten lattice spacing) at which H(r) reaches a maximum (~ 10 eV) for applied stresses appreciably smaller than the ideal shear strength of the material. Hence, the homogeneous nucleation of dislocations has always been regarded as highly improbable [11]. The cooperative generation of dislocations is entirely different.

The fundamental principle underlying this process is the recognition that interactions between dislocation loops, even when they are of atomic-sizes, can lead to changes in the dislocation configuration whereby some loops expand while others shrink in order to minimize their total energy. Thus, when a dislocation loop is formed in a loaded crystal containing other dislocation loops, the re-arrangement of the existing loops under the stress field of the newly formed loop is associated with an incremental *net plastic strain*. A straightforward way to investigate the effect of dislocation interactions is to examine how the incremental plastic strain can influence the subsequent nucleation of dislocations in the loaded crystal [7]. Here, we are guided by the well-known result that dislocation glide resulting in net plastic strain is associated with a decrement of the effective modulus that relates stresses and total strains in the crystal [11]. The reduction of the effective moduli due to interactions between atomic-size dislocation loops is usually many orders of magnitude smaller than the typical 10-20% reduction observed for high densities of mobile dislocations. Nevertheless, even this small reduction of the effective moduli can have a spectacular effect on the nucleation and interaction of subsequent dislocations as a function of increasing temperature. This is a consequence of a complex feedback process which couples the modulus decrement to the density of thermally nucleated dislocations while their interactions determine the magnitude of the decrement. Thus, even a very small decrement in the effective modulus results initially in a small but exponential increase in the density of thermally nucleated dislocations. The enhanced density leads to increased interactions and hence a slightly larger decrement of the modulus along with a greater increase in the density of loops. This feedback process is accentuated as temperature increases at a fixed applied load until it leads to an instability as explained below.

The self energy of a newly formed dislocation loop in a crystal containing other loops is proportional to a combination of effective moduli just as the energy of an isolated dislocation loop is proportional to a combination of elastic moduli. Thus, the formation enthalpy of a loop in a medium containing other loops can be written as in equation (1) but with K_0 replaced by $K_{eff} = K_0/\varepsilon$, where ε, called a 'screening function', is akin to a 'dielectric function' and K_{eff} is the energy coefficient written in terms of the effective moduli. ε reflects the decrease of the self energy of a dislocation loop due to the interactions between other loops. At low temperatures, the difference between the effective and elastic moduli is very small and ε is nearly equal to unity which is its minimum value. As the temperature increases, the probability of formation of dislocation loops in a loaded crystal increases which in turn, increases the plastic strain in the medium. Consequently, the effective moduli decrease and ε increases slowly from its baseline value of unity. As a result, the formation enthalpy of a loop present amidst other dislocation loops is smaller than that of an isolated loop of the same size. This, in turn, promotes the formation of more dislocation loops in the crystal which further increases the plastic strain. A positive or 'cooperative' feedback is set up between the formation of additional sub-critical dislocation loops and the continued reduction of the effective moduli.

Concomitant with the reduction of the formation energy, the critical radius for expansion of the loop and the related activation enthalpy decrease progressively as the temperature increases [8, 12]. (In contrast, the barrier for the expansion of a single dislocation loop does not vary with temperature.) At the same time, the configuration entropy associated with the dislocation loops increases as their number increases with temperature. Ultimately, at a critical temperature, T_c, the free energy of the loops vanishes. The unstable expansion and glide of the loops above T_c implies that spontaneous nucleation and glide of many dislocation loops can occur in the stressed crystal. The ensuing massive dislocation activity makes the effective moduli approach zero or, equivalently, ε diverge to infinity above T_c while the value of ε at T_c remains finite. For large applied loads, of the order of $\mu_0/100$, the critical temperature is typically half of the melting temperature [8].

The static model, described above, treats a dislocation-free crystal and examines how collective generation of glissile dislocations becomes feasible above a certain temperature under large applied loads. We now consider two types of dislocation activity which can significantly affect the onset of this instability. First, the glide of pre-existing mobile dislocations below T_c contributes to the plastic strain similarly as the formation of sub-critical dislocation loops does in the static model [7]. Hence, this additional plastic strain is expected to aid the cooperative instability and bring about a lowering of the critical temperature. Second, with increasing temperature, dislocation interactions lower the activation barrier for the cooperative generation of dislocations significantly. The barrier falls below 2 eV typically in the range 100-200K below T_c, thus enabling the generation of glissile dislocations by thermal activation in this temperature regime [8, 12]. The glide of such 'thermally nucleated' dislocations contributes to the plastic strain similarly as the glide of pre-existing dislocations and leads to further lowering of the critical temperature.

For a given set of material parameters and fixed external loads, the cooperative instability commences at a critical temperature when ε attains a critical value, ε_c. (The magnitude of ε_c is typically in the range 1.0 - 1.5 for most materials.) The glide of pre-existing dislocations causes the initial modulus decrement which then influences the nucleation of sub-critical dislocation loops and lowers the activation barrier for collective expansion of the loops. In a 100-200K interval below the critical temperature, thermally nucleated dislocations can also glide macroscopically and this in turn affects the nucleation of sub-critical dislocation loops and the onset of the cooperative instability. It is ultimately the expansion of sub-critical loops which occurs when the free energy becomes zero that triggers the cooperative instability. The macroscopic glide of dislocations is always strain-rate dependent, and thus, the BDTT predicted by this model [9] is also strain-rate dependent.

In a previous paper, it was shown [13] that the dislocation dynamics in the vicinity of cracks is of "similarity" type when the velocity of dislocations is a power law function of the stress with exponent m and the motion is thermally activated with activation barrier U_m. At the BDTT the density of thermally nucleated dislocations, n_f is proportional to $\exp(-U_n/k_B T_c)$ where U_n is an apparent formation energy and T_c refers to the strain-rate dependent BDTT [9]. Using this analytical form it can be shown [9] that the apparent activation energy, U_{app}, associated with the strain-rate dependence of the BDTT is given by

$$U_{app} = U_m + ((m+2)/2(m+1))U_n \qquad (2)$$

The formation energy, U_n, is however, not a constant. It is temperature-dependent and its value depends on how effectively the glide of pre-existing dislocations lowers the formation energy for cooperative generation of dislocations. Depending on the density of pre-existing dislocations and their mobility, the apparent activation energy associated with the strain-rate dependence of the BDTT is predicted to be either equal to or larger than the activation energy for dislocation motion.

RESULTS AND DISCUSSION: APPLICATION TO TiAl

According to our model, the glide of pre-existing dislocations and dislocations thermally nucleated below the BDTT, gives rise to the strain-rate dependence of the BDTT. The model also shows that the apparent activation energy associated with the BDTT, U_{app}, is not connected with a well-defined activation process. Rather, it is determined by the combined interplay between the nucleation of dislocations at BDTT and the motion of pre-existing and newly formed dislocations. The density of pre-existing dislocations, n_0, can influence the apparent formation energy of thermally nucleated dislocations below the BDTT and this can cause U_{app} to be bigger than the activation energy for dislocation motion, U_m (see Equation 2). We now examine how U_{app} varies in TiAl with changes in n_0 and the type of dislocation activity.

Table 1. Model predictions for the strain-rate dependence of the BDTT in TiAl

Dislocation type	Initial dislocation density n_0 (m^{-2})	BDTT (K) for strain rates 10^{-5}, 10^{-4}, and 10^{-3} sec^{-1}	U_m (eV)	U_{app} (eV)
1/2<110]	10^8	763, 800, 820	1.0	4.9
1/2<110]	10^{10}	672, 746, 807	1.0	1.6
<011]	10^8	926, 962, 974	1.2	6.8
<011]	10^{10}	808, 896, 960	1.2	2.0
<011]	10^{12}	929, 976, 999	2.0	5.0
<011]	10^{14}	841, 903, 970	2.0	2.5

The parameters that determine the BDTT in our model include the shear modulus on the slip plane, Burgers vector, applied stress, dislocation core energy, elastic cut-off radius, velocity-stress exponent, m, the activation barrier for dislocation motion, U_m, and two constants related to the dislocation mobility and the similarity solution, respectively[1]. We set the activation energy for dislocation motion, U_m, for ordinary dislocations to a low value of 1 eV. In the case of superdislocations, we investigate how the BDTT varies for two different values of the activation energy for dislocation motion. In the first case, a value slightly higher than that for ordinary dislocations is assumed, namely $U_m = 1.2$ eV, consistent with observations. In the second case, a much higher value is chosen, namely, $U_m = 2.0$ eV. This value is close to the activation energy for motion of dislocations in Si. The velocity-stress exponent m is set equal to unity and the BDTT is calculated for strain rates in the range 10^{-5} to 10^{-3} sec^{-1}. The variation of the BDTT with the density and type of dislocation activity is shown in Table 1.

We find that the apparent activation energy associated with the BDTT can change dramatically for two orders of magnitude change in the density of pre-existing dislocations in TiAl, irrespective of the nature of dislocation activity. This is especially apparent when the activation energy for dislocation motion has a low value such as 1.0-1.2 eV. The reason for this is as follows: When the activation energy for dislocation motion is small, the modulus decrement from the glide of pre-existing dislocations makes a significant contribution to ε when the density, n_0, is large. This leads to a significant reduction of the apparent formation energy for thermally nucleated dislocations (U_n), and thus U_{app} (see Equation 2) which then results in lowering of the BDTT. Interestingly, the calculated values of the BDTT lie in the same range for the two densities despite the significant difference in U_{app} for ordinary and superdislocation activity. A higher value of U_{app} necessarily corresponds to a higher BDTT at low strain-rates but this trend becomes unnoticeable at higher strain-rates. This follows from the fact that at higher strain-rates, the time available for dislocation glide before fracture occurs is very small. Therefore, the contribution to ε from the glide of pre-existing and thermally nucleated dislocations becomes very small and the onset of the cooperative instability is almost entirely controlled by the nucleation of sub-critical dislocation loops albeit at higher temperatures. It is this feature which leads to considerable overlap in the range of the BDTT for two widely different values of U_{app}.

Let us consider the case where the activation energy for superdislocation motion is set to 2.0 eV, a value close to that for dislocation motion in Si. The model predicts that U_{app} again varies appreciably with the initial density of glissile dislocations. However, compared to smaller values of U_m, the value of U_{app} is only 25% larger than the activation energy for motion when the mobile dislocation density is high. It is interesting to compare this trend for TiAl with that for Si for a similar value of U_m. Experiments in Si [2, 14, 15] indicate that $U_{app} \sim U_m$. It is also known that the initial density of glissile dislocations is very low in most experiments. These observations, at first, seem to contradict the trend seen in TiAl for the case of $U_m = 2.0$ eV. However, in order to perform a meaningful comparison we calculated the strain-rate dependence of the BDTT in Si using the parameters appropriate to this material [2]. We varied the initial mobile density (n_0) from a low value of 10^6 m^{-2} to 10^{10} m^{-2}. We found, $U_{app} = 2.2$ eV for

---

[1] The Burgers vector is set as b = 2.83 Å for ordinary dislocations and b = 5.7 Å for superdislocations in TiAl. μ_0 = 70 GPa and ν_0 = 0.23 represent average values for polycrystalline TiAl; the cut-off radius, r_0, is set equal to the appropriate Burgers vector in the two cases; the core energy factor c = 0.25 for ordinary dislocations and c = 0.23 for superdislocations. We assume the material to be loaded by a constant large stress, σ = 3 GPa for both types of dislocation activity. Such stress levels are expected in the vicinity of a crack. In addition to these parameters, two new constants enter the analysis: (i) The time available for dislocation glide at a certain strain-rate before fracture takes place which is estimated from experimental data [1]; and (ii) the constant appearing in the 'similarity' solution for dislocation motion near crack tips [13] which is chosen such that the BDTT lies in a reasonable range.

[2] For Si b = 2.21 Å corresponding to 1/6<112> Shockley partial dislocation, μ_0 = 60.5 GPa (the shear modulus on the (111) plane), ν_0 = 0.22, r_0 = 3.83 Å (the nearest-neighbor spacing), c = 0.53 such that the formation enthalpy of an isolated loop of radius 5Å is approximately 1eV and m = 1. The stress was assumed to be the same as in TiAl, namely, σ = 3 GPa. The activation energy for dislocation motion, U_m, was set equal to 2.1 eV.

$n_0 = 10^{10}$ m^{-2} and $U_{app} = 2.6$ eV for $n_0 = 10^6$ m^{-2}. Thus, U_{app} is quite close to U_m (2.1 eV) and does not depend sensitively on the density of pre-existing dislocations. While this is in good agreement with experiments in Si [2, 14, 15], it is different from the trend obtained for TiAl. The reason U_{app} varies appreciably in TiAl even for large values of U_m unlike the case in Si is related to a number of factors. In addition to the density of pre-existing dislocations, other parameters which can also raise or lower the values of U_n and hence, U_{app}, are the Burgers vector, the shear modulus and dislocation core energy. All these quantities are different for superdislocations in TiAl compared with the values used for partial dislocations in Si. Thus, in addition to the activation energy for dislocation motion and the density of pre-existing mobile dislocations, other material parameters also influence U_n and hence, U_{app}. Summarizing the model predictions, we find that the most plausible explanation for the large difference in U_{app} between ordinary and superlattice dislocation activity in TiAl observed by Booth and Roberts [1] is that the initial mobile densities of the two types of dislocations are significantly different but their mobilities are not vastly different.

In conclusion, a cooperative mechanism of dislocation generation above a critical temperature can give rise to massive dislocation activity of the type associated with the BDT. The strain-rate dependence of the critical temperature arises from the motion of pre-existing dislocations and dislocations which are "thermally nucleated" below the critical temperature by the cooperative process. The corresponding activation energy is not associated with one unique thermally activated process (such as dislocation motion) but represents an apparent value similarly as the activation energy associated with diffusion coefficients. This dependence is more complex in the case of BDTT than in diffusion because two or more thermally activated processes related to dislocation nucleation and glide are coupled together. Depending on their contributions, the apparent activation energy associated with the BDTT is either equal to or larger than the activation energy for dislocation motion.

ACKNOWLEDGMENT

This research was supported by the US Air Force Office of Scientific Research grant F49620-98-1-0245.

REFERENCES

1. A. S. Booth and S. G. Roberts, *Acta Mat.* **45**, 1045 (1997).
2. C. St. John, *Philos. Mag.* **32**, 1193 (1975).
3. S. G. Roberts, A. S. Booth and P. B. Hirsch, *Mat. Sci. & Eng.* **A176**, 91 (1994).
4. K. Kishida, H. Inui and M. Yamaguchi, *Phil. Mag. A* **78**, 1 (1998).
5. M.-C. Kim, M. Nomura, V. Vitek and D. P. Pope, *High-Temperature Ordered Intermetallic Alloys VIII* (edited by E. George, M. Mills and M. Yamaguchi), Materials Research Society, Vol. 552, p. KK3.1.1 (1999).
6. M. Khantha, D. P. Pope and V. Vitek, *Phys. Rev. Lett.* **73**, 684 (1994).
7. M. Khantha and V. Vitek, *Acta Mater.* **45**, 4675 (1997).
8. M. Khantha and V. Vitek, *Mat. Sci. & Eng.* **A234-236**, 629 (1997).
9. M. Khantha, V. Vitek and D. P. Pope, *Mat. Sci. & Eng. A* to appear, (2000).
10. J. R. Rice and R. Thomson, *Philos. Mag.* **29**, 73 (1974).
11. F. R. N. Nabarro, *Theory of Crystal Dislocations*, Dover: New York (1967).
12. M. Khantha, M. Ling and V. Vitek, *Mat. Sci. Res. Int.* **5**, 234 (1999).
13. M. Khantha, *Scr. Metall. Mater.* **31**, 1355 (1994).
14. M. Brede and P. Haasen, *Acta Metall.* **36**, 2003 (1988).
15. J. Samuels and S. G. Roberts, *Proc. R. Soc. Lond.* **A 421**, 1 (1989).

Titanium Aluminides II
and Metal Silicides I

Mat. Res. Soc. Symp. Proc. Vol. 646 © 2001 Materials Research Society

Experimental and Theoretical Investigations of the Phase Transformation in Al-rich TiAl Intermetallic Compounds

T. Koyama, M. Doi and S.Naito[1]

Department of Materials Science and Engineering, Nagoya Institute of Technology,
Gokiso-cho, Showa-ku, Nagoya 466-8555, JAPAN
[1]Gifu Shotoku Gakuen University, Faculty of Economics and Information, Gifu, JAPAN.

ABSTRACT

The phase decomposition of the Al-rich γ TiAl intermetallic compound in the TiAl($L1_0$) and Ti_3Al_5(P4/mbm) two phases region is investigated experimentally. On the phase decomposition of the Ti-56at%Al alloy, the single precipitate(Ti_3Al_5) shape is an oblate spheroid at the early stage of precipitation and each particle is aligned along certain direction of the orientation about 20 degrees from [100]. During coarsening, the precipitates encounter each other, then, the shape of the particle becomes the slanted or bended plate. In the case of phase decomposition of the Ti-58at%Al alloy, the tweed-like structure is observed at the beginning of the aging. The precipitates are connected each other during coarsening, finally the microstructure becomes the large layered structure with a zigzag-shaped interface. These microstructure changes are simulated based on the phase field model. The morphology and the time development of the simulated microstructure are in good agreement with the experimental results.

INTRODUCTION

Recently, the Ti-Al system has been investigated as a next candidate of the high strength heat resistant materials having an excellent lightweight and corrosion resistance. However, the improvement of the mechanical property in this alloy system has mainly been focused on the research, so that there are many uncertain points remained concerning the morphological stability of the microstructure and the dynamics of the microstructure changes. In the present study, we investigated the isothermal phase transformations in the Al-rich TiAl($L1_0$) intermetallic compound experimentally, where a Ti_3Al_5(P4/mbm) phase precipitates[1-3]. The dynamics of the microstructure developments were analyzed based on the phase field model.

EXPERIMENTAL PROCEDURES

Ti-56 and 58at%Al alloys were prepared by an arc-melting in a high vacuum. The specimens were cut into plate-shape in thickness about 0.5~1mm, and solution treated at 1623K for 72ks in a vertical electric furnace in an Argon atmosphere, and then drop-quenched into ice brine. The initial state of the specimens was the single phase with a $L1_0$ structure. Aging was isothermally performed at 973K for various durations in vacuum. The microstructure changes were investigated with the transmission electron microscopy (JEM-2000FX,EXII operated at 200kV). The thin foil for the TEM were prepared by electropolishing in an electroyte of CH_3OH 70% and $HClO_4$ 30% kept at about 220K. The electron beam was incident vertically upon the {100} or {110} planes. The wavelength and orientation relationship of the tweed or lamella structure were measured by mean of the image analyzing system for TEM micrographs.

EXPERIMENTAL RESULTS

Figure 1 shows the dark field images of the Ti-56at%Al aged at 973K for a)604.8ks, b)1814.4ks, c)3024.0ks and d)2419.2ks. The time evolution of the phase decomposition is observed from (a) to (c), where the foil plane of the specimen is (010). The white part is a $Ti_3Al_5(P4/mbm)$ phase[1-3] and black part indicates an TiAl($L1_0$) matrix. The microstructures observed on another foil plane, i.e. (001), is represented in figure 1(d). The single particle shape will be an oblate spheroid on the (001) plane at the early stage of precipitation. Each particle is aligned along certain direction, and the orientation of which is about 20 degrees from [100] direction. During coarsening, the particles encounter each other, then, the shape of the particle becomes the slanted or bended plate (the angle of inclination is about 20 degrees from [100]).

Figure 2 shows the dark and bright field images of the Ti-58at%Al aged at 973K for a)259.2ks, b)604.8ks, c)1814.4ks, d)1814.4ks(bright field image) and e)3024.0ks. From (a) to (c), the foil plane is (010). The microstructures taken from another foil planes, i.e. (110) and (001), are represented in figure 2(d) and (e), respectively. The very fine tweed-like structure is observed at the beginning of the phase transformation, then, precipitates start to connect each other with progress of aging, finally the microstructure becomes the large layered structure with a zigzag-shaped interface. The global shape of the Ti_3Al_5 phase should be an irregular raft layer perpendicular to the c-axis of $L1_0$ crystal structure. The angle of the zigzag-shaped interface depends on the aging time, and finally converged into the certain orientation of about 20 degrees from the [100] direction.

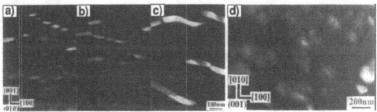

Figure 1. The dark field images of the Ti-56at%Al aged at 973K for a)604.8ks, b)1814.4ks, c)3024.0ks and d)2419.2ks.

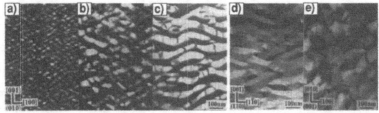

Figure 2. The dark and bright field images of the Ti-58at%Al aged at 973K for a)259.2ks, b)604.8ks, c)1814.4ks, d)1814.4ks(bright field image) and e)3024.0ks.

COMPUTER SIMULATION METHOD

In the phase-field model[4-7], a microstructure is represented by either the composition or structural order parameter fields. The precipitate, Ti_3Al_5, and matrix, TiAl, differ not only in composition but also in structure, both the composition $c(\mathbf{r},t)$ and the structural order parameter $s(\mathbf{r},t)$ at position $\mathbf{r}$ and time t are used to describe a morphology of two phase mixture. The value of $s(\mathbf{r},t) = 0$ corresponds to the $TiAl(L1_0)$ phase and $s(\mathbf{r},t) = \pm 1$ means $Ti_3Al_5(P4/mbm)$. The total free energy of the inhomogeneous system may be expressed as

$$G_{system} = \int_{\mathbf{r}} \left[G_c\{c(\mathbf{r},t), s(\mathbf{r},t)\} + \kappa_c\{\nabla c(\mathbf{r})\}^2 + \kappa_s \left| \nabla s(\mathbf{r}) \right|^2 + E_{str}\{c(\mathbf{r},t), s(\mathbf{r},t)\} \right] d\mathbf{r} , \qquad (1)$$

where κ_c and κ_s are gradient energy coefficients ($\kappa_c = 1.3 \times 10^{-13}$, $\kappa_s = 1.0 \times 10^{-15}$ ($J \cdot m^2 / mol$)). The chemical free energy density, $G_c\{c(\mathbf{r},t), s(\mathbf{r},t)\}$, is approximated as

$$G_c(c,s) = \frac{1}{2} A(c - c_1)^2 + \frac{1}{2} B(c - c_2)s^2 - \frac{1}{4} Cs^4 + \frac{1}{6} Ds^6 . \qquad (2)$$

The coefficients(unit is J/mol) in equation (2) are chosen as follows: $B = 20(c_{01} - c_{02})$, $c_2 = (c_{01} + c_{02})/2 - C/(2B) + D/(3B)$; $A = 10$, $C = 0.06$, $D = 0.1$, $c_1 = 0.5$, $c_{01} = 0.54$ and $c_{02} = 0.625$. The elastic strain energy density[5], $E_{str}\{c(\mathbf{r},t), s(\mathbf{r},t)\}$, is represented as

$$E_{str}(c,s) = \frac{1}{2} C_{ijkl} e_{ij}^T(\mathbf{r}) e_{kl}^T(\mathbf{r}) - C_{ijkl} e_{kl}^T(\mathbf{r}) e_{ij}^c(\mathbf{r}) + \frac{1}{2} C_{ijkl} e_{ij}^c(\mathbf{r}) e_{kl}^c(\mathbf{r}) , \qquad (3)$$

where C_{ijkl} is the elastic constant. The stress free strain[5], $e_{ij}^T(\mathbf{r})$, is expressed as $e_{ij}^T(c,s) = \eta_{ij}^c \delta_{ij}(c - c_0) + \eta_{ij}^s \delta_{ij} s^2$, where η_{ij}^c and η_{ij}^s are the lattice mismatch matrices with respect to deviation in the composition and the structural order parameter, respectively. c_0 is an alloy composition and δ_{ij} is a Kronecker's delta. The constrained strain[8], $e_{ij}^c(\mathbf{r})$, is given by

$$e_{ij}^c(\mathbf{r}) = \frac{1}{2} \sum_{\mathbf{k}} \begin{Bmatrix} [n_i \Omega_{mj}(\mathbf{n}) \sigma_{mn}^c n_n + n_j \Omega_{mi}(\mathbf{n}) \sigma_{mn}^c n_n] Q_c(\mathbf{k}) \\ + [n_i \Omega_{mj}(\mathbf{n}) \sigma_{mn}^s n_n + n_j \Omega_{mi}(\mathbf{n}) \sigma_{mn}^s n_n] Q_s(\mathbf{k}) \end{Bmatrix} \exp(i\mathbf{k}\mathbf{r}) , \qquad (4)$$

where $\Omega_{pl}(\mathbf{n})$ is the inverse matrix of $\Omega_{pl}^{-1}(\mathbf{n}) = C_{pqkl} n_q n_k$. $\sigma_{ij}^q = C_{ijkl} \eta_{kl}^q$ is the eigen stress ($q = c$ or s) with respect to the lattice mismatch η_{ij}^q . $\mathbf{k}$ is a vector in the reciprocal space and $\mathbf{n} \equiv \mathbf{k}/|\mathbf{k}|$ is unit vector. $Q_c(\mathbf{k})$ and $Q_s(\mathbf{k})$ are the Fourier transform of $c(\mathbf{r},t)$ and $s(\mathbf{r},t)$ fields, respectively. The values in η_{ij}^c and η_{ij}^s are determined from the experimental data of lattice parameters for the TiAl and Ti_3Al_5 phases[1,9], i.e. $\eta_{11}^c = \eta_{22}^c = -0.0865$, $\eta_{33}^c = 0.0296$ and $\eta_{11}^s = \eta_{22}^s = 0.00692$, $\eta_{33}^s = -0.0137$, respectively.

The temporal evolution of the microstructure is calculated by solving the time-dependent Landau equation for the non-conserved order parameter variable and the Cahn-Hilliard diffusion equation for the conserved composition variable, simultaneously[7,10]:

$$\frac{\partial c(\mathbf{r},t)}{\partial t} = M_a \left[\frac{\partial^2}{\partial x^2}\left\{\frac{\delta G_{system}}{\delta c(\mathbf{r},t)}\right\} + \frac{\partial^2}{\partial y^2}\left\{\frac{\delta G_{system}}{\delta c(\mathbf{r},t)}\right\} \right] + M_c \left[\frac{\partial^2}{\partial z^2}\left\{\frac{\delta G_{system}}{\delta c(\mathbf{r},t)}\right\} \right],$$

$$\frac{\partial s(\mathbf{r},t)}{\partial t} = -L\left[\frac{\delta G_{system}}{\delta s(\mathbf{r},t)} + \xi_s(\mathbf{r},t) \right]$$

(5)

where M_a and M_c are the atomic mobilities along the a- and c-axis of the TiAl(L1$_0$) ordered crystal structure[11]. L is a kinetic coefficient which characterizes the interface motion. We assumed $L = 200M_a$ in this study. $\xi_s(\mathbf{r},t)$ is a random white noise. The simulation was conducted in a two-dimensional space. The 128×128 or 256×256 uniform square grid was used to spatially discretize the field equations. The elastic constants[9] used for the calculation are listed in Table I. The value of the diffusion coefficient, D_a, along the a-axis was assumed as $D_a \equiv M_a RT = 1.22 \times 10^{-21} (\mathrm{m}^2 / \mathrm{s})$ [12].

SIMULATION RESULTS AND DISCUSSION

Figure 3 and 4 demonstrate the calculated microstructure developments of the Ti-56at%Al and 58at%Al, respectively. The calculation is performed on the 2-dimensional plane of (010) and the mobility ratio, M_c / M_a, is assumed as 0.33. The completely black part indicates the composition of 54at%Al and the white part corresponds to the 62.5at%Al, therefore, the white region is a Ti$_3$Al$_5$ phase. The morphological changes of the microstructure are in very good agreement with the experimental observation in figures 1 and 2.

The temporal changes of the wavelength of the modulated structure and the precipitate alignment angle deviated from the [100] in Ti-58at%Al are shown in figure 5(a) and (b), respectively. The solid square is the experimental result and open circle is the calculated one. It is noted that the present simulation reproduces the features of the microstructure quantitatively.

Table I. Elastic constants used for the calculation (GPa)[9]

	C_{11}	C_{22}	C_{33}	C_{44}	C_{55}	C_{66}	C_{12}	C_{13}	C_{23}
TiAl(L1$_0$)	144.4	144.4	139.5	66.4	66.4	51.3	67.3	70.5	70.5

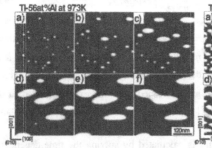 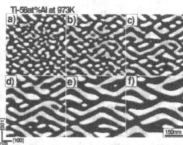

Figure 3. Calculated microstructure changes in Ti-56at%Al at 973K. a)48.0ks, b)480ks, c)960ks, d)1.95Ms, e)3.36Ms, f)4.32Ms

Figure 4. Calculated microstructure changes in Ti-58at%Al at 973K. a)75.1ks, b)150ks, c)376ks, d)751ks, e)1.50Ms, f)3.76Ms

The angle of the Ti_3Al_5 phase depends not only on the elastic anisotoropy but also the ratio of the mobility M_c/M_a. It is suggested that the diffusion coefficient in the Ll_0 based ordered structure depends on the orientation because of its atomic arrangement in the ordered structure. In TiAl phase, the diffusion coefficient along the c-axis is smaller than that along the a(or b)-axis[12]. Figure 6 shows the simulated microstructure changes with the various values of the mobility ratio M_c/M_a. With increasing the M_c/M_a, the angle of the Ti_3Al_5 phase from the [100] direction at the initial stage of phase decomposition becomes smaller. The angle converges gradually to the unique value during coarsening, which is determined from the condition of the minimum elastic strain energy. The normalized elastic energy function[5], $B(\mathbf{n})$, which characterize the orientation dependencies of the elastic strain energy, is calculated numerically as shown in the figure 7. It should be noted from figure 7(a) that the elastic field is almost symmetric inside the (001) plane, and the elastically soft direction is about the 70 degrees from the (001) plane(see figure 7(c)), which will lead the angle between plate-shaped precipitates and (001) plane to be 20 degrees. It is concluded that the morphological feature of the precipitated phase is determined depending on the balance of both of the diffusion coefficient anisotropy and elastic anisotropy in the intermetallic compound.

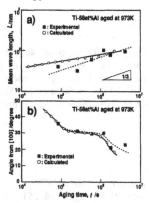

Figure 5. *The temporal changes of the wave length of the modulated structure in Ti-58at%Al(a), and its angle from the [100] direction(b).*

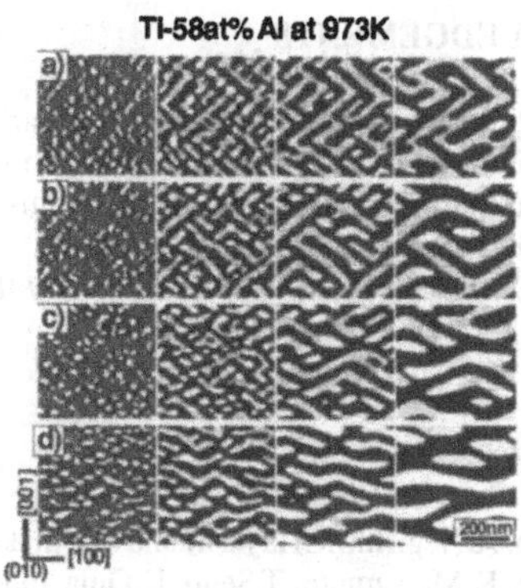

Figure 6. *Simulated temporal microstructure changes with the mobility ratio M_c/M_a. M_c/M_a= a)0.1, b)0.25, c)0.33, d)1.*

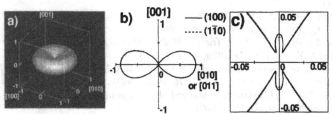

Figure 7. *a) Three dimensional illustration of the normalized elastic energy function. The distance from the origin to the surface(indicated by a white region in the figure) is $|B(\mathbf{n})|$, where the direction is indicated by $\mathbf{n}$. b) The solid and dotted curves are the (010) and (110) sections of (a), respectively. c) The magnified center part in (b).*

CONCLUSIONS

We investigated the phase decomposition of the Al-rich γ TiAl intermetallic compound in TiAl(L1$_0$) and Ti$_3$Al$_5$(P4/mbm) two phases region. The dynamical microstructure changes were analyzed based on the phase field model. Results obtained are as follows:

On the phase decomposition of the Ti-56at%Al alloy, the single precipitate particle shape is an oblate spheroid at the early stage of aging and each particle is aligned along certain direction of the orientation about 20 degrees from [100]. In the case of phase decomposition of the Ti-58at%Al alloy, the tweed-like structure is observed at the beginning of the aging. The precipitates encounter each other during coarsening, finally the microstructure becomes the large layered structure with a zigzag-shaped interface. The microstructure changes are simulated based on the phase field model. The morphology and the time development of the simulated microstructure are in good agreement with the experimental results. The simulation results indicate that the complex shape of the precipitate phase is mainly controlled by the anisotropy of the diffusion coefficient in the matrix phase with a L1$_0$ ordered structures and by the elastic anisotropy. Especially, the stable shape of the precipitates at later stage will be determined by the condition of the minimum elastic strain energy.

ACKNOWLEDGEMENTS

The authors are grateful to Ms. N.Ogiso and Mr. T.Taniguchi who were all students in the graduate school of Nagoya Institute of Technology for their experiments in part. The present research was partially supported by the Ministry of Education, Science, Sports and Culture, Grant-in-Aid for Encouragement of Young Scientists, 12750583, 2000 and for Scientific Research (C), 11650677, 1999, Japan, also by the research funds from the Research Promotion Association for Light Metals, Japan.

REFERENCES

1. T.Nakano, A.Negishi, K.Hayashi and Y.Umakoshi, *Acta mater.*, **47**, 1091 (1999).
2. T.Nakano, K.Matsumoto, T.Seno, K.Oma and Y.Umakoshi, *Phill. Mag. A*, **74**, 251 (1996).
3. R.Miida, S.Hachimoto and D.Watanabe, *Japanese J. of Applied Phys.*, **21**, L59 (1982).
4. R.Kobayashi, *Physica D*, **63**, 410 (1993).
5. A.G.Khachaturyan, *Theory of Structural Transformations in Solis*, Wiley and Sons, New York, (1983)
6. L-Q.Chen and Y.Wang, *JOM*, **48**, 13 (1996).
7. D.Y.Li and L.Q.Chen, *Acta mater.*, **46**, 639 (1998).
8. T.Koyama and T.Miyazaki ; *Mater. Trans. JIM*, **39**, 169 (1998).
9. K.Tanaka and M.Koiwa, *Intermetallics*, **4**, S29 (1996).
10. J.E.Hilliard, *Phase Transformation*, ed.by H.I.Aaronson,ASM,Metals Park,Ohio,497 (1970).
11. S.Nambu and A.Sato, *J. Am. Ceram. Soc.*, **76**, 1978 (1993).
12. H.Kadowaki, T.Ikeda,H.Nakajima, H.Inui and M.Yamaguchi, *Abstract of the Spring Meeting of Japan Institute of Metals*, 166 (2000).

Mat. Res. Soc. Symp. Proc. Vol. 646 © 2001 Materials Research Society

Anomalous Strengthening Mechanism in NbSi$_2$-Based Silicide Single Crystals

Yukichi Umakoshi, Takayoshi Nakano and Masafumi Azuma
Department of Materials Science and Engineering, Graduate School of Engineering,
Osaka University, 2-1, Yamada-oka, Suita, Osaka 565-0871, Japan

ABSTRACT

Anomalous strengthening behavior of NbSi$_2$-based silicide single crystals with C40 structure was examined focusing on the effect of substitutional alloying elements and the formation of a solute dragging atmosphere. After pre-straining to 1% at temperatures between 600 and 1600°C, single crystals were deformed at 400°C to examine the effects of dislocation sources and segregation of solute atoms around dislocations. The stress amplitude of yield drop was derived by the segregation of solute atoms at the superlattice intrinsic stacking fault (SISF) between two superpartials. Introduction of dislocation sources by pre-straining remarkably improved fracture strain, but the segregation of solute atoms around dislocations during pre-straining at higher temperatures induced a rapid drop of fracture strain. Change in the SISF energy by static and dynamic aging was observed to obtain evidence of segregation of solute atoms at the SISF between two superpartials.

INTRODUCTION

NbSi$_2$-based silicides with the C40 structure deform by the $(0001)<1\overline{2}10>$ slip and show anomalous strengthening with a peak at and above 1400°C [1,2]. The strength and peak temperature could be remarkably improved by addition of substitutional alloying elements [2]. Although the limited slip system is unfavorable for industrial application, high strength and high anomalous peak temperature make the NbSi$_2$-based phase attractive as a superior reinforcement phase in MoSi$_2$ matrix silicides for a new refractory structural material operating at ultra high temperatures [3-5].

The critical resolved shear stress (CRSS) for the basal slip and the anomalous peak temperature showed no orientation dependence but they varied depending on the strain rate [1,2]. In addition, serrations in the stress-strain curves appeared at a certain temperature and strain rate. From these results, we have proposed an anomalous strengthening mechanism for NbSi$_2$-based silicides based on the formation of dragging atmosphere around moving dislocations [2].

This paper describes evidence of the dragging atmosphere around dislocations and effect of this atmosphere on the SISF energy and plastic deformation behavior in NbSi$_2$-based silicides by pre-straining and aging at high temperatures.

EXPERIMENTAL PROCEDURE

Binary NbSi$_2$, ternary (Nb$_{0.97}$Mo$_{0.03}$)Si$_2$, (Nb$_{0.9}$Mo$_{0.1}$)Si$_2$, (Nb$_{0.9}$W$_{0.1}$)Si$_2$, (Nb$_{0.97}$Ti$_{0.03}$)Si$_2$ and Nb(Si$_{0.97}$Al$_{0.03}$)$_2$ silicide single crystals were grown by a floating zone method at a rate of 5mmh^{-1} in a flowing argon atmosphere. Compression specimens with 2x2mm^{2}x5mm and a loading axis shown in figure 1 were cut from the crystals by spark machining and were mechanically polished with a diamond paste. Compression tests were carried out at a nominal strain rate of $1.7\times10^{-4}\text{s}^{-1}$ at temperatures between 1000 and 1700°C to examine the temperature dependence of yield stress. The specimens were deformed at 400°C after pre-deformation to 1% in the temperature range between 600 and 1600°C to examine the effect of dislocation sources on the plastic deformation behavior. After pre-deformation to 1% at 600°C and subsequent annealing for 1h at temperatures between 600 and 1600°C, the specimens were deformed again at 600°C. The slip plane was determined by two-face slip trace analysis using an optical microscope. The superlattice intrinsic stacking fault (SISF) energy was measured based on the separation distance between two superpartials by transmission electron microscope observation.

RESULTS

Deformation of all specimens at the tested temperature was controlled by the (0001) $<1\overline{2}10>$ slip. Anomalous strengthening depended strongly on the substitutional alloying elements [2]. Figure 1 shows the temperature dependence of the CRSS for the $(0001)[2\overline{1}\,\overline{1}0]$ slip for binary and ternary NbSi$_2$-based silicide single crystals as reported in our previous papers [1,2]. The CRSS for binary NbSi$_2$ increases anomalously above 1000°C and reaches a maximum around 1400°C. The CRSS for ternary silicides containing Al or Ti which crystallizes in C54 structure as a disilicide is higher than that for binary silicide but there is no significant change in the anomalous strengthening behavior between the binary and ternary silicides. In contrast, addition of Mo or W which crystallizes in the C11$_b$ structure as a disilicide enhances the anomalous strengthening; the strength rises and the peak is shifted to a higher temperature of around 1600°C.

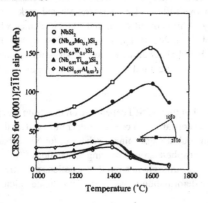

Figure 1
Temperature dependence of CRSS in NbSi$_2$ and ternary NbSi$_2$-based silicides single crystals [1,2].

Figure 2 shows the stress-strain curves of $(Nb_{0.97}Mo_{0.03})Si_2$ single crystals deformed at 400°C with and without pre-straining at higher temperatures. The specimen deformed little at 400°C without pre-deformation and it fractured just after yielding. After pre-deformation the samples became deformable even at 400°C. A discontinuously serrated stress-strain relation appeared after displaying the yielding phenomenon. The serrated flow suggests the discontinuous motion

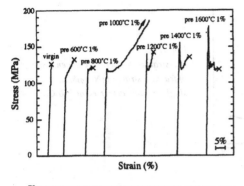

Figure 2

Stress-strain curves of $(Nb_{0.97}Mo_{0.03})Si_2$ single crystals deformed at 400°C with and without pre-straining at higher temperatures.

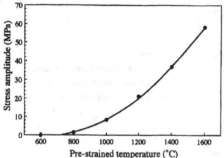

Figure 3

Pre-strained temperature dependence of the stress amplitude of yield drop in $(Nb_{0.97}Mo_{0.03})Si_2$ single crystals deformed at 400°C.

of dislocations under applied stress. The amplitude and frequency of serrations increased with increasing pre-deformation temperature. The yielding phenomenon also changes depending on the pre-straining temperature.

Figure 3 shows the stress amplitude of the yield drop as a function of pre-strained temperature. The yield drop monotonously increases with the pre-straining temperature, and yield drop may correspond to the release of the dislocations introduced during pre-deformation from obstacles and the reactivation of dislocation sources. The increase in yield drop at higher pre-straining temperature suggests that dislocations created during pre-straining are more effectively trapped at higher temperature.

Figure 4 shows a variation in the fracture strain of $(Nb_{0.97}Mo_{0.03})Si_2$ single crystals with the pre-strained temperature. The fracture strain rapidly increases and then suddenly drops at higher temperatures. Mobile dislocations and dislocation sources introduced by pre-straining around 1000°C ease plastic deformation and improve the fracture strain at 400°C. However, at high pre-straining temperatures above 1000°C, the dragging atmosphere may be formed around the created dislocations on heterogeneous slip bands and these dislocations are then strongly trapped, resulting in a sudden drop in the fracture strain.

After pre-straining to 1% at 600°C and subsequent annealing for 1h at various temperatures, specimens were deformed again at 600°C. Since segregation of solute atoms occurred during annealing around the dislocations created by pre-deformation, remarkable yield drop and

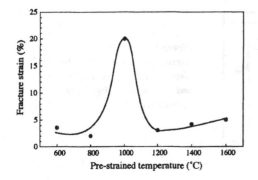

Figure 4
Pre-strained temperature dependence
of fracture strain in $(Nb_{0.97}Mo_{0.03})Si_2$
single crystals deformed at 400°C.

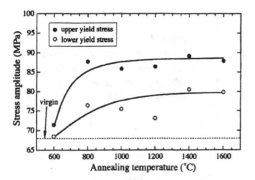

Figure 5
Annealing temperature dependence
of yield stress in $(Nb_{0.97}Mo_{0.03})Si_2$
single crystals deformed at 600°C.

serrations appeared in the stress-strain curves depending on the annealing temperature. The upper and lower stresses are shown as a function of annealing temperature in figure 5. As the annealing temperature rises, the upper and lower yield stresses increase and reach saturation. The increase in the upper yield stress corresponds to the formation of a dragging atmosphere around dislocations during annealing. The increase in the lower yield stress suggests that the frictional stress for the motion of the multiplied dislocations increases with increasing annealing temperature.

DISCUSSION

One of the most accepted mechanisms for anomalous strengthening in NbSi$_2$-based silicides is due to the dragging atmosphere around moving dislocations [2]. In the stacking sequence of atoms on the closely packed planes of (110) for the C11$_b$ structure, (0001) for the C40 structure and (001) for the C54 structure, there are two fold periodicity of ABAB, three fold periodicity of ABDABD and four fold periodicity of ADCBADCB, respectively (See the reference [2]). A $1/3 < 1\overline{2}10 >$ superlattice dislocation whose motion controls the deformation of NbSi$_2$-based silicides is dissociated into paired $1/6 < 1\overline{2}10 >$ superpartials bounded by an SISF. After introduction of the SISF in the C40 structure on the (0001) plane, the stacking sequence of atoms around the SISF changes from three-fold to two-fold periodicity corresponding to that on (110) in the $C11_b$ structure. This means that the $C11_b$ structure is locally created in the C40 structure at the SISF after the motion of the superpartial, resulting in an increase in chemical potential at the SISF compared with that in the matrix. Solute atoms which crystallize in the $C11_b$ structure as a disilicide may gather near the SISF to maintain the same chemical potential in the matrix and at the SISF. The fact that addition of Mo or W enhanced the anomalous strengthening in $NbSi_2$-based silicides accompanied by an increase in peak temperature and strength, but that Ti or Al addition was not effective in the strengthening suggests that the anomalous strengthening is closely related to the atomic stacking sequence and the segregation of solute atoms at the SISF between the superpartials [2].

Mobile dislocations and dislocation sources created by pre-deformation are known to induce improvement in ductility because motion and multiplication of dislocations occur more easily than nucleation [6]. In fact, the pre-straining improves the ductility of $(Nb_{0.97}Mo_{0.03})Si_2$ deformed at 400°C. If the dislocations and dislocation sources created during pre-straining at higher temperature do not change, deformation behavior of yield stress and fracture strain at 400°C shows no significant annealing temperature dependence. Monotonous increase in yield drop with increasing annealing temperature suggests that dislocations created at higher temperatures were more effectively trapped by the dragging atmosphere of Mo atoms and/or vacancies. The initial increase in the fracture strain in Fig.4 is due to the more homogeneous distribution of dislocations created during pre-straining at high temperatures. But the sudden drop of the fracture strain at the higher temperatures is caused by the created dislocations with the dragging atmosphere being immobilized at 400°C. Mo atoms and/or vacancies may gather around dislocations to decrease the chemical potential at the SISF during annealing at high temperatures after deformation at 600°C. The higher the annealing temperature, the more easily Mo atoms and/or vacancies gather at the SISF and form the dragging atmosphere around dislocations. Formation of the dragging atmosphere causes the strengthening. Annealing in the temperature range where the anomalous strengthening occurs induces an easy formation of the dragging atmosphere.

If Mo atoms preferentially segregate at the SISF at 1600°C, the SISF energy will decrease. The SISF energy was measured from the separation distance between two superpartials bound by a SISF and the results are shown in Table I. Since the SISF energy in ternary silicide is lower than that in binary $NbSi_2$, addition of Mo decreases this energy. When $(Nb_{0.97}Mo_{0.03})Si_2$ single crystal is annealed at 1600°C for 1h after pre-deformation at 600°C, the SISF energy decreases remarkably. This means that Mo atoms preferentially gather at the SISF by diffusion and forms the dragging atmosphere during annealing. The yield drop and serrated flow after pre-straining and/or annealing are due to the formation of dragging atmosphere around dislocations.

CONCLUSIONS

(1) Addition of Mo which stabilizes the $C11_b$ structure as a disilicide causes strong anomalous strengthening in $NbSi_2$-based silicides.

Table I *SISF energy for NbSi$_2$ and (Nb$_{0.97}$Mo$_{0.03}$)Si$_2$ single crystals.*

	Pre-deformation at 600°C (NbSi$_2$)	Pre-deformation at 600°C (Mo addition)	Annealing at 1600°C (Mo addition)
E$_{SISF}$ (mJ/m^2)	231	195	110

(2) Anomalous strengthening in NbSi$_2$-based silicides is due to the formation of a dragging atmosphere of substitutional Mo atoms around moving dislocation containing an SISF.
(3) Segregation of Mo atoms around dislocations reduces the SISF energy.
(4) Fracture of NbSi$_2$-based silicides depends on nucleation and multiplication of dislocation sources. Formation of a dragging atmosphere around dislocations decreases both the dislocation mobility and the fracture strain.

ACKNOWLEDGMENT

This work was supported by a Grant-in-Aid for Scientific Research Development from the Ministry of Education, Science, Sports and Culture of Japan.

REFERENCES

[1] Y. Umakoshi, T. Nakashima, T. Nakano and E. Yanagisawa, *High Temperatures Silicides and Refractory Alloys*, ed. C.L. Briant, J.J. Petrovic, B.P. Bewlay, A.K. Vasudevan, H.A. Lipsitt, (MRS, Pittsburgh, 1994), **322**, pp. 9.
[2] T. Nakano, M. Kishimoto, D. Furuta and Y. Umakoshi, Acta Mater., **48**, 3465 (2000).
[3] W.J. Boettinger, J.H. Perepezko and P.S. Frankwicz, Mater. Sci. Engng., **A155**, 33 (1992).
[4] T. Nakano, M, Azuma and Y. Umakoshi, Intermetallics, **6**, 715 (1998).
[5] J.J. Petrovic and A.K. Vasudevan, Mater. Sci. Engng., **A261**, 1 (1999).
[6] M. Moriwaki, K. Ito, H. Inui and M. Yamaguchi, Mater. Sci. Engng., **A239-240**, 69 (1997).

Mat. Res. Soc. Symp. Proc. Vol. 646 © 2001 Materials Research Society

Creep Studies of Monolithic Phases in Nb-Silicide Based In-Situ Composites

B.P. Bewlay[1], C.L. Briant[2], E.T. Sylven[2], M.R. Jackson[1] and G. Xiao[2]
[1]GE Corporate Research and Development, Schenectady, NY 12301, USA
[2]Division of Engineering, Brown University, Providence, RI 02912, USA

ABSTRACT

Nb-silicide composites combine a ductile Nb-based solid solution with high-strength silicides, and they show great promise for aircraft engine applications. Previous work has shown that the silicide composition has an important effect on the creep rate. If the Nb:(Hf+Ti) ratio is reduced below ~1.5, the creep rate increases significantly. This observation could be related to the type of silicide present in the material. To understand the effect of each phase on the composite creep resistance, the creep rates of selected monolithic phases were determined. To pursue this goal, monolithic alloys with compositions similar to the Nb-based solid solution and to the silicide phases, Laves, and T2 phases, were prepared. The creep rates were measured under compression at 1100 and 1200°C. The stress sensitivities of the creep rates of the monolithic phases were also determined. These results allow quantification of the load bearing capability of the individual phases in the Nb-silicide based in-situ composites.

INTRODUCTION

Nb-silicide based in-situ composites are being explored for structural applications at very high temperatures [1-4]. These composites consist of Nb_5Si_3 and Nb_3Si type silicides toughened with a Nb solid solution (abbreviated by (Nb) in the present paper). More recent Nb-silicide based in-situ composites are highly alloyed with elements such as Cr, Ti, Hf, B and Al. These in-situ composites have demonstrated a promising combination of high-temperature strength, creep resistance, and room-temperature fracture toughness [1-3]. With the appropriate combination of alloying elements it is possible to achieve the required balance of room temperature toughness and high temperature creep resistance. Alloying elements such as Cr and B have beneficial effects on oxidation resistance, stabilizing Laves phases and T2 niobium borosilicide phases, respectively. The Nb_5Si_3 and Nb_3Si have the tI32 and tP32 ordered tetragonal structures with 32 atoms per unit cell. The unit cells also possess large lattice parameters; the large Burgers vectors and complex dislocation cores associated with these structures would suggest that dislocation creep makes only a small contribution to creep deformation in these silicides. When Nb_5Si_3 is alloyed with Ti and Hf, the less complex hP16 structure can also be stabilized [1, 6]. The Laves phases typically have C14, C15, or more complex structures of a hexagonal form [3].

The present study was performed to determine the creep rates of monolithic intermetallic phases and to develop the constitutive creep laws for these phases. These creep laws are also required to perform predictive modeling of more complex two-phase and multi-phase systems [7]. The monolithic phases described in this paper were produced by directional solidification, or arc melting, followed by homogenization heat treatments. Previous work indicated that creep deformation in binary Nb_5Si_3 is controlled by bulk diffusion of Nb in the Nb_5Si_3 [4]. The aim of the present paper is to describe high-temperature creep behavior of the monolithic intermetallic phases that exist in Nb-silicide in-situ composites.

EXPERIMENTAL

The samples were prepared using directional solidification [1, 3] or multiple arc melting. The starting charges were prepared from high purity elements (>99.99%). The samples were examined using scanning electron microscopy and Electron Back-Scatter Diffraction in the SEM (EBSD).

The compositions of the monolithic phases that were investigated are shown in Table I. The compositions of these monolithic alloys were selected on the basis of electron microprobe analyses (EMPA) of the phases in multi-phase composites [1, 3]. The monolithic phases that were generated from ternary alloys were given the post-script 3, for example, the Nb_5Si_3 modified with 10% Ti was described by silicide-3. The monolithic phases that were generated from quaternary and higher-order alloys were given the post-script C, such as silicide-C. The hP16-3 composition was based on EMPA of the hP-16 in previous studies [8]. The hP16-C composition was based on analyses of the hP16 phase in complex composites [1]. The (Nb) compositions were based on EMPA data from binary Nb-Si, ternary Nb-Ti-Si [8] and higher order alloys [1].

The microstructure of hP16-3 consisted principally of the hP16 phase, as confirmed using EBSD, and a volume fraction of ~0.05 of residual bcc (Nb). The homogenization treatment was not completely effective in removing the eutectic (Nb) associated with solidification. The microstructure of the hP16-C was similar to that of the hP16-3. The grain size of the hP16 phase for both alloys was ~100μm. The other monolithic alloys were essentially single phase with only minor volume fractions of phases emanating from solidification segregation.

Compression creep tests were performed at a temperatures of 1100 and 1200°C, and at stress levels in the range 70-280 MPa. The cylindrical specimens that were used were 7.6 mm in diameter and up to 30 mm in length. The samples were machined by EDM to final dimensions. In each test the sample was placed between two 18.7 mm diameter silicon nitride platens to prevent breakage of the graphite rams. Pure niobium foil was placed at the interface between the platens and the sample to prevent any contamination of the sample or reaction with the platens. Creep testing was performed in a vacuum of ~4x10^{-5} Torr.

Table I : *Compositions of the monolithic phases that were investigated (the compositions are given in atom per cent). The silicide phases are both Nb_5Si_3-based.*

PHASE	Nb	Ti	Hf	Si	Cr	Al	B
Laves-C	21.0	11.0	5.5	8.5	53.0	1.0	
Laves-3	30.0			15.0	55.0		
Silicide-C	38.5	16.0	6.0	37.0	1.0	1.0	0.5
Silicide-3	53.0	10.0		37.0			
T2-C	41.5	13.0	3.0	12.5	4.0	0.5	25.5
T2-3	62.5			12.5			25.0
hP16-3	20	44		36.0			
hP16-C	25.5	25.5	13	36.0			
(Nb)₃SI-C	49.0	18.2	7.8	25.0			
(Nb)-C	63.1	27	5	0.9	2	2	
Nb-1Si	99			1			
Nb-46Ti-1Si	53	46		1			

In order to perform a creep experiment, a sample was loaded at temperature to the first stress level of 70 MPa and held at that level for 24 hours. The creep rate was determined by analysis of the creep curves. After the 24-hour test, the sample dimensions were determined and the load was increased. This process was continued until the test was terminated or the sample failed.

RESULTS AND DISCUSSION

Creep Behavior at 1100°C

The data for the creep tests at 1100°C are shown in Figure 1. The data are not complete for all compositions, as will be discussed below. The data are well behaved for the ternary Nb_5Si_3 silicide-3 and the Nb_3Si-C. There is some scatter in the Nb_5Si_3–C data; the highest creep rate that was measured was $2x10^{-8}s^{-1}$ at 210 MPa, and all other measurements were below this value. At creep rates $<10^{-8}s^{-1}$ there is generally more scatter in the data because these rates are very close to the measurement limit for the creep system that was employed ($\sim5x10^{-9}s^{-1}$); this limit was governed by the dilatometer resolution, mechanical stability, and electrical noise. The silicide-3 and silicide-C have creep rates that are close to those of the binary Nb_5Si_3 at 1200°C, but they are slightly higher. The ternary Nb_5Si_3 with Ti substituted for Nb has a lower creep rate than the Nb_5Si_3-C. The Nb_3Si-C has the highest creep rates. At 1100°C the creep rate of the T2 phase was $\sim3x10^{-8}s^{-1}$, but there was little sensitivity of the creep rate to stress in the range studied.

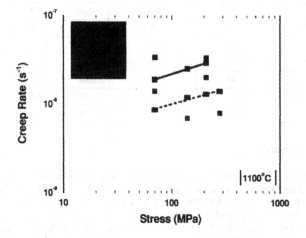

Figure 1 : *Secondary creep rates at 1100°C for the monolithic phases that were investigated.*

Due to limited ductility of some of the phases at 1100°C, cracking occurred in some cases during the creep test, and this led to some difficulties/inconsistencies in some creep strain measurements. For some of the alloys, such as Laves-3 and Laves-C, the creep rates were very low

($< 5\times10^{-9}$ s^{-1}) at low stresses (70 MPa), and increasing the stress led to cracking prior to creep. This behavior was observed for the Laves-C at both 1100°C and 1200°C. Thus, there are no data in Figure 1 for the Laves phases and the T2-3 phase. The microstructures of the creep samples were analyzed after the creep test in order to investigate the type of cracking that occurred.

Creep data for the binary Nb$_5$Si$_3$ at 1200°C, as reported by Subramanian et al. [4], indicate that even at 1200°C the creep rate of the binary Nb$_5$Si$_3$ is less than that of any of the other phases at 1100°C. In order to determine the stress sensitivity of the creep rate, the creep rate ($\dot{\varepsilon}$) and stress (σ) were related using a power law equation, $\dot{\varepsilon}= B\ \sigma^n$; where n is the stress exponent and B is a constant at any specific temperature. The grain size of all the materials was large and similar in each case (~100μm), and in this analysis no attempt was made to incorporate any grain size dependence. The constants obtained from these analyses are shown in Table II.

The monolithic (Nb) alloys have exponents of ~3 and in these systems deformation is probably controlled by dislocation creep. These exponents are similar to those reported previously for Nb-1.25Si [7]. At 1100°C further analysis of the exponents for the intermetallics is limited by the small data set. Exponents for the silicide-3 and Nb$_3$Si-C were determined, but they had values less than unity. This would suggest that there is some creep threshold, or other complicating factor, controlling creep deformation in these phases.

Table II : Power law constants for creep of the monolithic phases that were investigated.

Phase	Stress Range (MPa)	Temperature (°C)	Constant	Exponent
Nb$_5$Si$_3$	100-280	1200	6.16×10^{-11}	1.0
Nb-10Si	70-140	1200	6.57×10^{-12}	1.9
Silicide-C	70-140	1200	3.84×10^{-11}	1.5
Nb$_3$Si-C	70-140	1200	1.07×10^{-14}	3.6
Laves-3	70-140	1200	1.11×10^{-9}	1
T2-C	70-140	1200	2.15×10^{-9}	0.9
(Nb)-3	3-80	1100	1.9×10^{-10}	3.3
(Nb)	3-80	1100	4.7×10^{-14}	2.9

Creep Behavior at 1200°C

The data for the creep tests at 1200°C are shown in Figure 2. The data are well behaved for all alloys. Figure 2 also shows data for the binary monolithic Nb$_5$Si$_3$ and the Nb$_5$Si$_3$-Nb composite prepared from the binary Nb-10Si alloy [7]. The binary Nb$_5$Si$_3$ possessed the lowest creep rates and the Nb$_3$Si-C displayed the highest creep rates of the tetragonal phases. The hP16 phases have creep rates similar to the T2-C phase at stresses up to 140 MPa. However, on increasing the stress above 140MPa the creep rate of the hP16-C increases dramatically to a level of 3×10^{-5}s^{-1} at 210MPa, as shown in Figure 2. This behavior suggests a change in the creep mechanism with increasing stress. The hP16 phases have the worst performance, and at high stresses are beyond the scale of Figure 2.

The silicide-3 and silicide-C have creep rates that are slightly higher than those of the binary Nb$_5$Si$_3$. The ternary Nb$_5$Si$_3$ with Ti has a lower creep rate than the Nb$_5$Si$_3$-C. The T2-C creep curve was higher than those of the Nb$_5$Si$_3$ type silicides, although it is lower than that of the Nb$_3$Si-C. The creep rate of the T2-3 was ~1x10^{-8}s^{-1}, but there was little sensitivity of the creep rate to stress. The T2-3 also has a lower creep rate than the T2-C; the addition of Ti, Hf, Cr and Al led to an increase in the creep rate of the T2. The Laves-3 possessed creep rates similar to those of the silicide-3.

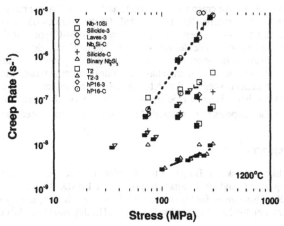

Figure 2 : *Secondary creep rates at 1200°C for the monolithic silicides, Laves phases, hP16 and T2 phases that were investigated.*

Of the phases studied, the Nb$_3$Si-C has the highest creep rate. Unfortunately, data for binary Nb$_3$Si was not obtained because this phase is generated by a peritectic reaction during solidification of hypereutectic alloys and it is extremely difficult to prepare in the bulk form as a metastable single phase. At 1200°C several of the monolithic phases, including the Laves phase and the Nb$_3$Si -C, have creep performance only equal to, or inferior to, that of the Nb$_5$Si$_3$-Nb composite from the binary Nb-10Si alloy. The Nb$_3$Si-C has creep rates almost an order of magnitude higher than the Nb-10Si, and the stress exponent is almost two times higher. The hP16-C creep rates were also significantly higher than in the Nb-10Si composite.

The stress exponents were obtained and are shown in Table II. In the case of the Nb$_5$Si$_3$, the stress exponent was almost one and the mechanism for creep deformation was reported to be Nabarro-Herring creep, the creep deformation being limited by Nb diffusion [4]. In the cast and heat treated conditions the dislocation densities in the monolithic phases investigated in the present study were very low and Harper–Dorn creep probably did not make a significant contribution to creep. The potential creep mechanisms are Nabarro-Herring, grain boundary sliding, or power law creep. Examination of the creep exponents in Table II indicates that the creep deformation of the monolithic phases is controlled by a range of mechanisms. For example, the T2-C, Laves-3, and silicide-C have exponents close to unity, as is the case for binary Nb_5Si_3. At present there are insufficient data to establish a stress exponent for the silicide-3. Deformation of the silicide-C is probably also controlled by Nabarro-Herring type creep, but the diffusing species that control deformation are still being investigated. The Nb_3Si-C displays a high stress sensitivity with a creep exponent of ~3.6. This high stress exponent suggests that creep deformation is controlled by dislocation glide, as is the case for pure metals, despite the fact that the dislocation structures in this intermetallic are complicated.

CONCLUSIONS

The (Nb) solid solutions have creep rates at 1100°C and 1200°C that are more than an order of magnitude higher than the intermetallics investigated. Of the intermetallics investigated, the Nb_5Si_3 type silicides had the lowest creep rates. The complex Nb_3Si-C type silicide had the highest creep rate of the tetragonal silicides investigated. The hP16 silicide phases have higher secondary creep rates than any of the tetragonal silicides, or the T2 phases, at stresses greater than 140 MPa. At lower stresses, the hP16 silicide has creep rates similar to those of the T2 phase. Analysis of the creep exponents suggests that deformation of the complex Nb_5Si_3 type silicide is probably controlled by Nabarro-Herring type creep, as is the binary Nb_5Si_3 silicide. In contrast, creep of the Nb_3Si silicide appears to be controlled by a dislocation controlled mechanism.

ACKNOWLEDGMENTS

The authors would like to thank D.J. Dalpe for the directional solidification, C. Bull for assistance with creep testing, and Dr. V. Smentkowski for the EBSD. The authors would also like to thank Dr. P.R. Subramanian for his comments. This research was partially sponsored by AFOSR under contract #F49620-00-C-0014 with Dr. C.S. Hartley and Dr. S. Wu as Program Managers.

REFERENCES

[1]   B.P. Bewlay, M.R. Jackson and H.A. Lipsitt, *Metall. and Mater. Trans.*, 1996, Vol 279, pp. 3801-3808.
[2]   M.G. Mendiratta, J.J. Lewandowski and D.M. Dimiduk, *Metall. Trans. 22A (1991)*, pp. 1573-1581.
[3]   P.R. Subramanian, M.G. Mendiratta, D.M. Dimiduk and M.A. Stucke, *Mater. Sci. Eng., A239-240*, 1997, pp. 1-13.
[4]   P.R. Subramanian, T.A. Parthasarathy, M.G. Mendiratta and D.M. Dimiduk, *Scripta Met. and Mater.*, Vol. 32(8), 1995, pp. 1227-1232.
[5]   B.P. Bewlay, M.R. Jackson and H.A. Lipsitt, *Journal of Phase Equilibria*, Vol 18(3), 1997, pp. 264-278.
[6]   B.P. Bewlay, R.R. Bishop and M.R. Jackson, *Z. Metallkunde*, 1999, Vol 90, pp. 413-422.
[7]   G.A. Henshall, M.J. Strum, P.R. Subramanian, and M.G. Mendiratta, *Mat. Res. Soc. Symp. Proc.* 364, 1995, pp. 937-942.
[8]   B.P. Bewlay, R.R. Bishop and M.R. Jackson, *Journal of Phase Equilibria*, Vol 19(6), 1998, pp. 577-586.

Mat. Res. Soc. Symp. Proc. Vol. 646 © 2001 Materials Research Society

The Effect of Silicide Volume Fraction on the Creep Behavior
of Nb-Silicide Based In-Situ Composites

B.P. Bewlay[1], C.L. Briant[2], A.W. Davis[2] and M.R. Jackson[1]
[1]GE Corporate Research and Development, Schenectady, NY 12301, USA
[2]Division of Engineering, Brown University, Providence, RI 02912, USA

ABSTRACT

This paper will describe the creep behavior of high-temperature Nb-silicide in-situ composites based on quaternary Nb-Hf-Ti-Si alloys. The effect of volume fraction of silicide on creep behavior, and the effects of Hf and Ti additions, will be described. The composites were tested in compression at temperatures up to 1200°C and stress levels in the range 70 to 280 MPa. At high (Nb) phase volume fractions the creep behavior is controlled by deformation of the (Nb) and, as the volume fraction of silicide is increased, the creep rate is reduced. However, at large silicide volume fractions (>0.7) damage in the silicide begins to degrade the creep performance. The creep rate has a minimum at a volume fraction of ~0.6 silicide. The creep performance of the monolithic and silicide phases will also be discussed.

INTRODUCTION

Nb-silicide based in-situ composites are potential candidates for use in high temperature structural applications. These composites consist of a Nb-based solid solution, denoted (Nb), strengthened by Nb_3Si and/or Nb_5Si_3 based silicides. Previous work has shown that these materials exhibit a good combination of room temperature fracture toughness and high temperature strength [1-3]. However, the high-temperature creep performance has been found to be highly dependent on the alloy composition and constituent phases. Two of the most common alloying additions are Ti and Hf, and these elements can adversely increase the creep rate to $>10^{-7}$ s^{-1} at stress levels greater than 100 MPa, if they are present in concentrations greater than ~30 and 8 percent, respectively [4]. The composite creep rate may also be increased because increasing Ti and Hf stabilizes a hexagonal silicide [5]. In contrast, Nb_3Si and tetragonal Nb_5Si_3 are beneficial to creep behavior provided their volume fraction and distribution within the composite are controlled.

The aim of the present paper is to describe the effect of volume fraction of silicide on the creep rate of Nb-silicide based in-situ composites. Alloys were prepared with Si concentrations from 12 to 22 atomic percent (all compositions are given in atom percent in the present paper). These alloys provided a broad range of silicide volume fractions. The Ti and Hf concentrations were held constant at 25 and 8 atomic percent, respectively.

EXPERIMENTAL

All samples were prepared using Czochralski cold crucible growth [3]. Composites were directionally solidified from quaternary alloys with compositions of Nb-8Hf-25Ti-XSi, where X was adjusted from 12 to 22%. The directional solidification procedure has been described in more detail previously [3]. Monolithic Nb alloys were also prepared with compositions of Nb-1Si, Nb-46Ti-1Si, and Nb-27Ti-5Hf-2Al-2Cr-0.9Si (denoted as Nb-C in the present paper), in order to determine the creep performance of the Nb solid solution in the in-situ composites at elevated temperatures.

All the creep data reported in the present paper were obtained in compression. Compression creep tests were performed at a temperature of 1200°C, and at stress levels in the range 70-280 MPa. The cylindrical specimens that were used were 7.6 mm in diameter and 15 mm long, and they were machined from DS samples such that the loading axis of the creep sample was parallel to the growth direction. In order to perform the creep test, the sample was first heated to the test temperature. A load was then applied to the sample, and it was maintained at a constant value for 24 hours. During this 24 hour period the strain was monitored and the creep rate was obtained through analysis of these data. A vacuum of approximately 5×10^{-5} Torr was maintained throughout the test. The sample was measured after each test and then used for the tests at higher stresses.

Detailed metallographic information was obtained using both optical microscopy and scanning electron microscopy before and after the creep test. Electron back scatter diffraction pattern analysis (EBSD) and orientation imaging in the scanning electron microscope were used to identify the phases and their crystallographic orientations.

RESULTS AND DISCUSSION

The typical microstructure of the composites with Si concentrations of 20% and less is shown in Figure 1. The microstructure consisted of large-scale $(Nb)_3Si$ tP32 phase (the grey phase) with large-scale (Nb) dendrites (the light phase). Both the Nb_3Si and the (Nb) possess substantial amounts of Hf and Ti in solid solution [5, 6]. There was some segregation in the (Nb) which led to varying BSE contrast at the interface between the (Nb) and the $(Nb)_3Si$. EPMA and EBSD data indicated that this was a result of Ti segregation and Hf/Nb depletion in these regions.

A typical microstructure of the composites with Si concentrations greater than 20% is shown in Figure 2. In addition to the faceted $(Nb)_3Si$ and (Nb) phases observed in the composites from lower Si concentrations, the Nb_5Si_3 tI32 phase was also observed as the primary solidification phase. The $(Nb)_5Si_3$ was the large-scale, dark, faceted phase in the micrograph of Figure 2.

Figure 3 shows the creep rate at 1200°C as a function of Si concentration for the Nb-25Ti-8Hf-XSi composites, where X was varied from 12 to 22 atomic percent. Data are shown for stresses of 70 to 280 MPa. As the Si concentration was increased from 12% to 18%, the volume fraction of $(Nb)_3Si$ increased from 0.25 to 0.62. There is a broad range of compositions for which the creep rate is less than $3 \times 10^{-8} s^{-1}$.

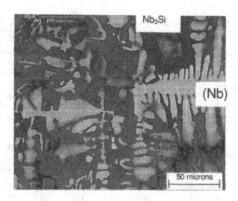

Figure 1. *Typical microstructure (BSE image) of the transverse section of a DS composite generated from a quaternary Nb-25Ti-8Hf-16Si alloy. The (Nb) is the light phase and the (Nb)₃Si is the grey faceted phase.*

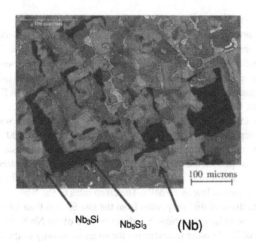

Figure 2. *Typical microstructure (BSE image) of the transverse section of a DS composite generated from a quaternary Nb-25Ti-8Hf-22Si alloy. The (Nb) is the light phase, the (Nb)₃Si is the grey faceted phase, and the (Nb)₅Si₃ the dark phase.*

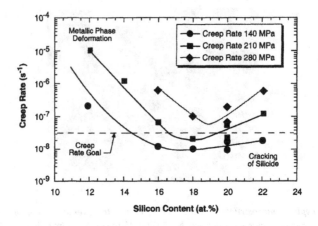

Figure 3. *The effect of Si concentration and volume fraction of (Nb) and silicide on the secondary creep rate at 1200 °C for stresses of 140-280 MPa.*

These data reveal several important points. First, for a constant Si concentration, the creep rate increased with increasing applied stress. Second, the creep rate possessed a minimum value at approximately 18 %Si. At higher Si concentrations the creep rate increased again. Microstructural analysis of the samples after the creep test indicated that at low Si concentrations, deformation was controlled by creep of the (Nb) and at high Si concentrations, composite deformation was controlled by cracking of the silicide.

In order to interpret these results further, the creep behavior of the (Nb) was investigated. Figure 4 shows the creep rate of the Nb-1Si, Nb-46Ti-1Si, and Nb-C at 1100°C and 1200°C as a function of stress. The compositions of these alloys cover the compositions of the Nb-based solid solutions in the composites that have been generated previously from ternary, quaternary, and higher-order alloys. The results at 1200°C show that the creep rate of the (Nb) is greater than 10^{-7} s^{-1} even at stresses as low as 70 MPa. The creep rates of the Nb-C, Nb-1Si and Nb-46Ti-1Si are similar to those of the composites from the low Si (less than 14%) quaternary alloys shown in Figure 3. These results also show that the creep rate of the Nb-Si solid solution is very sensitive to additions of Ti. In order to determine the stress sensitivity of the creep rate, the creep rate ($\dot{\varepsilon}$) and stress (σ) for the three alloys at 1100°C and 1200°C were fitted to a power law creep expression, $\dot{\varepsilon} = B \, \sigma^n$; where n is the stress exponent and B is a constant at any specific temperature.

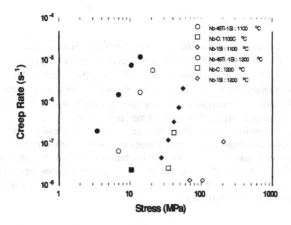

Figure 4. *Secondary creep rates as a function of stress at 1100°C and 1200°C for binary Nb-Si, ternary Nb-Ti-Si, and complex Nb-based monolithic solid solutions. The effect of alloying additions on creep rate is shown.*

The second set of creep data that must be considered are those for the Nb_5Si_3 silicide [7]. The Nb_5Si_3 creep rates are much lower than those for the Nb-based solid solutions and are similar to those for the high-Si quaternary alloys. In order to develop a better understanding of the response of the composite to increasing silicide volume fraction and increasing stress, the constitutive behavior of the composite was simulated using the expressions provided by Henshall et al. [8, 9], where σ_A is the applied stress, n is the stress exponent for the silicide (1.0), and m is the stress exponent for the Nb (2.9). V_s and V_w are the volume fractions of the silicide and Nb, respectively. A_s and A_w are the pre-exponents in the power law creep expressions for the silicide and Nb, respectively.

$$\sigma_A = V_s \left[\frac{1}{A_s^{1/n}}\right] \dot{\varepsilon}^{1/n} + V_w \left[\frac{1}{A_w^{1/m}}\right] \dot{\varepsilon}^{1/m}$$

The creep data for the Nb-1Si solid solution were used together with those for the binary monolithic Nb_5Si_3 [8]. This equation can predict the increase in the creep rate with increasing stress for a given Si concentration, as will be published subsequently [10]. However, this equation does not predict as strong a dependence on the silicide volume fraction as was observed experimentally. This difference may arise because the power law creep parameters used were those for Nb-1Si, rather than those for the actual quaternary (Nb) alloys of the composites tested. In addition, the damage accumulation that occurs at Si concentrations greater than 18% is at present not modeled. Thus, the results from the monolithic silicide and the (Nb) studies suggest that at low Si concentrations the creep rate is dominated by the (Nb), but that as the Si concentration is increased, and the silicide volume fraction is increased, the creep rate decreases.

CONCLUSIONS

The creep rates of the composites from the quaternary Nb-Hf-Ti-Si alloys decreased with increasing Si concentration from 12 to 18%. At higher Si concentrations, the creep rate increased as a result of crack linking and damage accumulation in the silicides. The composite creep rate increased with increasing stress, and at each stress level there was a minimum in the creep rate at ~18Si. The Nb solid solutions have creep rates at $1100°C$ and $1200°C$ that can be more than an order of magnitude higher than those of the composites.

Modeling results and experimental data from the monolithic silicides and Nb solid solutions indicate that at low Si concentrations the creep deformation is dominated by the (Nb), but as the Si concentration is increased, and the silicide volume fraction is increased, the composite creep performance is controlled by the silicide. However, the model underestimates the effect of increasing volume fraction of the silicide on the creep rate. The model is currently being extended to improve the predictive capability.

ACKNOWLEDGMENTS

The authors would like to thank D.J. Dalpe for the directional solidification, and C. Bull for assistance with creep testing. This research was partially sponsored by AFOSR under contract #F49620-00-C-0014 with Dr. C.S. Hartley and Dr. S. Wu as Program Managers.

REFERENCES

[1] P.R. Subramanian, M.G. Mendiratta, D.M. Dimiduk and M.A. Stucke, *Mater. Sci. Eng.,* A239-240, 1997, pp. 1-13

[2] P.R. Subramanian, M.G. Mendiratta and D.M. Dimiduk, *Mat. Res. Soc. Symp. Proc.,* 322 (1994), pp. 491-502.

[3] B.P. Bewlay, M.R. Jackson and H.A. Lipsitt, *Metall. and Mater. Trans.,* 1996, Vol 279, pp. 3801-3808.

[4] B.P. Bewlay, P. Whiting and C.L. Briant, *MRS Proceedings on High Temperature Ordered Intermetallic Alloys VIII,* 1999, pp. KK6.11.1- KK6.11.5.

[5] B.P. Bewlay, M.R. Jackson and H.A. Lipsitt, *Journal of Phase Equilibria,* Vol 18(3), 1997, pp. 264-278.

[6] B.P. Bewlay, R.R. Bishop and M.R. Jackson, *Journal of Phase Equilibria,* Vol 19(6), 1998, pp. 577-586.

[7] P.R. Subramanian, T.A. Parthasarathy and M.G. Mendiratta and D.M. Dimiduk, *Scripta Met. and Mater.,* Vol. 32(8), 1995, pp. 1227-1232.

[8] G.A Henshall and M.J. Strum, *Scripta Metall.,* Vol 30(7), 1994, pp. 845-850

[9] G.A. Henshall, M.J. Strum, P.R. Subramanian, and M.G. Mendiratta, *Mat. Res. Soc. Symp. Proc.* 364, 1995, pp. 937-942.

[10] C.L. Briant and B.P. Bewlay, to be published.

Mat. Res. Soc. Symp. Proc. Vol. 646 © 2001 Materials Research Society

Microstructures and thermoelectric power of the higher manganese silicide alloys

Kimiko Kakubo*, Yoshisato Kimura and Yoshinao Mishima
Tokyo Institute of Technology, Department of Materials Science and Engineering, 4259
Nagatsuta, Midori-ku, Yokohama 226-8502, Japan,
kakubo@materia.titech.ac.jp,*Graduate Student.

ABSTRACT

In the present work, we investigate microstructures and thermoelectric power of the binary higher manganese silicide (hereafter denoted as HMS) base alloys prepared by ingot metallurgy, i.e., arc-melting in an Ar gas atmosphere. Alloys with 63.5, 64.0 and 64.5 at% Si have a small amount of second phase, either the primary metallic MnSi phase or Si solid solution (hereafter denoted as (Si)). A slight revision is thereby made on the HMS single phase region of the Mn-Si binary phase diagram. The presence of MnSi seems to reduce the thermoelectric power of the HMS, while that of (Si) enhances it. An alloy having the HMS/(Si) eutectic microstructure, 67.9at%Si, shows the highest thermoelectric power in the Mn-Si binary alloys examined. Moreover, effect of doping the p-type elements, Fe and Cr, on the thermoelectric power and microstructural features is also investigated for a 63.5at%Si alloy which consists mostly of HMS phase with a small amount of (Si) phase. It is shown that both elements, especially 0.5 at% Cr addition, enhance the thermoelectric power of the alloys.

INTRODUCTION

Several transition-metal silicides exhibit relatively high thermoelectric power and are candidate materials for thermoelectric generation at high temperatures because of their excellent oxidation resistance. A compound in the binary Mn-Si system called higher manganese silicide (hereafter denoted as HMS) is known as a p-type thermoelectric material that can be used at high temperatures such as 1000K without any special protection against oxidation[1]. Low cost of constituent elements would be an advantage in commercial use since we are interested in the application of this material as a component of micro gas turbine engine based co-generators. The Mn-Si binary phase diagram in the literature is shown in Fig.1[2], according to which the HMS is a line compound having a complicated ordered crystal structure and its chemical composition as expressed as Mn:Si is reported to range from 1:1.71 to 1:1.75 [3]. It is slightly Mn-rich side of the composition for $MnSi_2$ by which reason it is called as HMS. At Si-poor side of the HMS, neighboring two phase region involves a metallic MnSi phase, while at Si-rich side, it involves the primary solid solution of Si phase, hereafter denoted as (Si).

In this work, microstructure and thermoelectric power of the binary Mn-Si alloys are first investigated with various silicon concentration of both Si-poor and –rich sides of the HMS. The major interests include whether the single phase HMS can be obtained or not, effects of microstructure and Si concentration on the thermoelectric power of two-phase alloys based on HMS on both sides of it, and the effect of doping ternary elements on the thermoelectric power of the HMS. As doping elements, Fe and Cr are chosen to the amount up to 1.0at% because both $FeSi_2$ and $CrSi_2$ are known to exhibit high thermoelectric power [4].

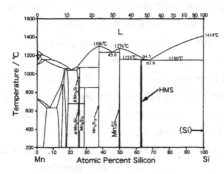

Fig.1. The binary Mn-Si equilibrium phase diagram[2].

EXPERIMENTAL PROCEDURES

Alloys were prepared by arc melting in an argon gas atomsphere using high purity raw materials such as 99.9%Mn, 99.9Si, 99.9%Fe and 99.9%Cr. The binary alloys prepared were with compositions between 63.5 to 67.9at%Si and ternary doping was made to a 63.5at%Si alloy by 0.5 to 1at% with Fe or Cr. As-cast specimens were then annealed in an evacuated quartz tube at 1273K for 48h.

Microstructural observation and phase identification were conducted by scanning electron microscopy (SEM) and energy dispersive X-ray spectroscopy. Phase relations involving MnSi, (Si) and HMS were investigated using scanning electron microscopy and the differential thermal analysis (DTA). The DTA measurements were performed in an argon atmosphere at heating and cooling rates of 2 to 10K/min.

The thermoelectric power is measured using the apparatus we have designed in a vacuum (about 10^{-1}Pa) and in a temperature range from room temperature to 1200K. Measurements of electromotive force due to temperature difference are made through spot-welded Pt-legs.

RESULTS AND DISCUSSION

Microstructure of the alloys based on HMS with different Si concentration on the binary Mn-Si system.

Microstructure observed by SEM-BEI (Back Electron Scattered Image) of as-cast alloys are shown in Fig.2 for a 63.5at%Si alloy, which may fall on the single phase HMS according to the phase diagram shown in Fig.1.The typical microstructure, top left, consists mainly of HMS phase and a second phase identified as MnSi having a straight line feature. Micrographs shown on the right and bottom left of the figure show features observed only locally. The matrix in both micrographs is the HMS but blocky MnSi, which is presumably the primary solidification phase, and a small region having a feature characteristic to a eutectic product are observed. The latter should be a HMS/(Si) eutectic reaction product formed locally during solidification according to Fig.1. The second phase involved in a micrograph on the bottom left of Fig.2 is (Si), which also could have been formed as nonequilibrium phase during solidification. By these observations, together with a fact that these local microstructural features disappears after annealing at 1273K, we could slightly modify the binary Mn-Si phase diagram in the vicinity of HMS as shown in Fig.3.

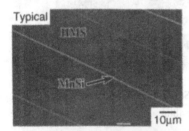

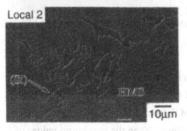

Fig.2. Back scattered electron images of as-cast 63.5at%Si alloy. The top left is a typical microstructure consisting mostly of HMS with a small amount of the second phase, MnSi, having a straight line feature. The right and the bottom left are the microstructure observed only locally.

Microstructures after annealing at 1273K are shown in Fig.4 for 64.5 and 67.9at%Si alloys. As expected by Fig.1 and 3, the microstructure is two phase consisting of HMS and (Si). The amount of (Si) phase increases with increasing Si concentration of the alloy and the alloy with 67.9at%Si exhibits a typical eutectic microstructure.

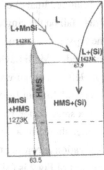

Si Concentration / at%

Fig.3. A slight revision on the Mn-Si binary phase diagram in the vicinity of HMS after the present work..

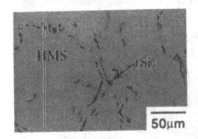 

Fig.4. A back scattered electron image of a 64.5Si (on the left) and a 67.9Si (on the right) alloys annealed at 1273K for 48h.

Thermoelectric power of the two-phase alloys based on HMS with different Si concentrations

Measurement of thermoelectric power as a function of temperature is carried out for 63.5, 64.5, 65.7 and 67.9at%Si alloys and the results are shown in Fig.5. Each alloy exhibits different temperature dependence of thermoelectric power, although the behavior is similar between 63.5 and 64.5at%Si alloys. When the maximum values of their temperature dependences are compared, it is first noted that it is lowest in the alloy with 63.5at% having MnSi as a second phase, it is higher with increasing Si concentration, namely with increasing amount of (Si) phase in the HMS in the alloys with more than 64.5at%Si. It is interesting to find that among all the binary alloys, highest thermoelectric power is obtained in a 67.9at% alloy, which has a eutectic microstructure consisting of HMS and (Si).

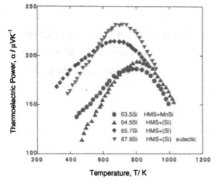

Fig.5. The temperature dependence of the thermoelectric power for 63.5, 64.5, 65.7 and 67.9at%Si alloys annealed at 1273K for 48h.

Effect of doping of ternary elements on the thermoelectric power of HMS

Fig . 6 is the temperature dependence of the thermoelectric power of 63.5at%Si alloys with or without Fe or Cr dopings. The base 63.5at%Si alloy consists mainly of HMS with a small amount of (Si) phase as described before. Therefore the results may as well indicate the effect of ternary addition on the thermoelectric power of the compound HMS itself.

The effect of doping is found to be a little complicated. For Fe dopings, the effect of both 0.5at% and 1.0at% additions improves the maximum value of thermoelectric power in its temperature dependence to a similar degree. However, the temperature to give rise the maximum shifts toward lower temperature as the amount of doping increases. For Cr dopings, the results seem to be less systematic because the addition by 0.5at% substantially enhances the maximum in thermoelectric power, while the completely reverse effect is found by 1.0at% doping. It is noted that the enhancing effect by 0.5at%Cr addition also accompanies a substantial shift in the temperature for the peak value of thermoelectric power.

In the forthcoming investigations, the followings should be considered and discussed in order to provide rigorous explanations on the results obtained hereby. They include the effects of second phase, regardless of the second phase being a semiconductor or not, and of morphology of second phase with a particulare emphasis of the role of eutectic microstructure in the thermoelectric power of the alloys based on

HMS. Also the origin to determine the peak in the temperature dependence of thermoelectric power as effected by ternary doping is of interest.

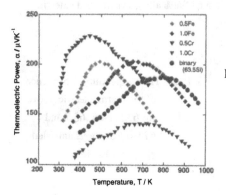

Fig.6. Effect of doping Fe or Cr on the temperature dependence of the thermoelectric power.

CONCLUSION

The following conclusions are drawn from this work:

1. Microstructure of the binary 63.5at%Si alloy, presumably the stoichiometric composition for HMS, is actually two-phase, containing some MnSi phase.

2. In the binary alloys with more than 64.5at% Si, the alloys are two-phase consisting of HMS and (Si) phase. With increasing (Si) phase, maximum thermoelectric power in its temperature dependence increases.

3. The highest thremoelectric power is obtained in an HMS/(Si) eutectic alloy at 67.9at%.

4. Doping 0.5at%Cr to the HMS alloys at 63.5at%Si is found to enhance thermoelectric power of the binary alloy most effectively.

REFERENCES

[1] V.A.KORSHUNOV and P.D.GEL'D, Fiz.Metal. matalloved 11,vol.6 (1961)

[2] T.B.Massalski, Binary Alloys Phase Diagrams, Second Edition, 2 (1990)

[3] M.RIFFEL, E.GROβ, U.RTOHRER, J. Mater. Sci. Mater. in ELECTRON.6 (1995)

[4] Y.Kimura, K.Shindo, Y.Ohta and Y.Mishima, Proc. On the Intn'l Federation of Heat Treatment and Surface Engineering Congress, vol.2, Institute of Materials Engineering Australasia Ltd, (2000), 169-176.

Iron Aluminide, Iridium and Other Ordered Intermetallic Alloys

Mat. Res. Soc. Symp. Proc. Vol. 646 © 2001 Materials Research Society

The Role of Carbon and Vacancies in Determining the Hardness of FeAl Intermetallic in the Quenched and the Aged States

C. García-Oca[1], D.G. Morris[1] M.A. Muñoz-Morris[1] and S.C. Deevi[2]
[1]Department of Physical Metallurgy, CENIM, CSIC, Avenida Gregorio del Amo 8, 28040 Madrid, SPAIN; [2]Chrysalis Technologies Incorporated, Richmond, Virginia, USA

ABSTRACT

The role of quenched-in vacancies in FeAl intermetallics on producing considerable hardening is well known, as is the softening on annealing as vacancies are annihilated. The present study examines quench hardening and anneal softening by quenched-in vacancies and interstitial carbon solute in Fe-40Al-C. Interstitial carbon is seen to be a more potent hardening agent than the vacancy, while the co-annihilation of vacancies and carbon atoms from solution during annealing leads to dislocation loop debris, and equiaxed or plate-like carbide precipitation, according to the annealing conditions. The processes occurring have been followed by detailed TEM studies, and are discussed in terms of the relative solubilities and diffusion rates of vacancies and carbon. The relevance of such interstitial solute hardening to the behaviour of other FeAl intermetallics is also briefly considered.

INTRODUCTION

Hardening by vacancies in FeAl alloys is well documented [1], as well as the subsequent softening as the vacancies anneal out to produce complex dislocation loop arrangements and APB faults [2,3]. There is also considerable work examining the influence of ternary additions, such as Ni or B, on hardening and softening and on the defects produced [4-6]. The effect of carbon on such quench and ageing has received little attention. Pang and Kumar [7] carried out an extensive study of the deformation and fracture behaviour of Fe-40Al-0.6C, including studies of the formation of perovskite precipitates (Fe$_3$AlC$_{-0.5}$, or κ phase), but did not examine vacancy effects. The present study examines high-temperature quench hardening and subsequent age softening in a Fe-40Al-0.4C alloy [8-10]. The full chemical composition of this material is Fe-39.2Al-0.4C-0.02B-0.19Mo-0.86O-0.05Zr (atomic %), with Zr and B present as large boride particles, Mo in solid solution, and O present as coarse oxide particles [10]. The insoluble dispersoids are few and widely separated and are not expected to influence carbon and vacancy solution and precipitation and, neglecting the small amount of Mo solute, the present material is nominally Fe-40Al-0.4C.

EXPERIMENTAL DETAILS

Small samples were annealed at temperatures between 500ºC and 1150ºC for 1 h and water quenched to examine hardening. The microstructure (grain size about 20 µm, boundaries pinned by oxides) does not change during such heat treatment. Samples quenched from 1000ºC were then annealed at 320-430ºC for times of 10 min to 1 month for age softening. Temperature precision was better than ± 5ºC during ageing. Vickers microhardness testing was carried out on polished surfaces using a 200 g load, 12-15 indents were measured giving values correct to within a few percent, or ± 5 Kg/mm^2 hardness. The initial structure was characterised by SEM and detailed studies of microstructure carried out by TEM on thin foils using a JEOL 2010 FX microscope.

RESULTS AND DISCUSSION

Hardness data after high temperature quenching are shown in Fig. 1. Hardness increases monotonically with temperature, showing a steady increase over the whole temperature range studied, 500-1000ºC. This hardness increase with temperature can be related [1,3] to the exponential increase in solute content, usually considered to be vacancies, retained by the rapid quench. Using a solution hardening model [1,3], the hardness increase (ΔH_v) can be written:

$$\Delta H_v \propto k \sqrt{C} = k \exp \{ - Q_f / 2RT \} \qquad (i)$$

where k is a constant, C the solute concentration, Q_f the activation energy of formation of solute and R and T the gas constant and the absolute temperature. The hardness increase over the lower temperature range in Fig. 1 leads to an activation energy for solution of about 55 kJ/mol, smaller than expected for vacancy formation in a Fe-40Al alloy (90-100 kJ/mol) and closer to the value for stoichiometric FeAl [1,3].

There are two possible origins of the hardening seen - the solute involved may be either vacancies or carbon. The activation energy deduced is smaller than for vacancy formation in Fe-40Al alloys [1,3]. The activation energy for solution of carbon in interstitial solution in FeAl is not well known, but can be estimated assuming an ideal solution, with the solubility at any temperature given by $C = \exp \{-Q_s/RT\}$. Taking the carbon solubility in FeAl as 0.5-1 % at 1000°C [11] we deduce an activation energy for interstitial solution of 52 ± 5 kJ/mol, consistent with the present data. This analysis assumes that sufficient carbon solute movement takes place during annealing that an equilibrium solute concentration is reached, i.e. that dissolution of any carbide precipitates at high temperature or precipitation of excess solute at lower temperatures takes place and any remnant particles do not contribute to hardness. This assumption may not be correct for the lowest temperatures but seems reasonable for high quench temperatures. Further support for the idea that quench hardening is predominantly due to carbon solution comes from the numerical value of the parameter k in equation (i). This parameter [1] can be written $6\gamma\mu$, where μ is the shear modulus and γ a coefficient representing the strength of the solute-dislocation interaction. γ has been shown to be larger for carbon interstitial atoms than for vacancies in FeAl [1]. Thus, even if similar numbers of vacancies and carbon atoms are present, the carbon atoms will dominate hardening. A further question concerns the possibility of carbon-vacancy association, or complex formation, modifying hardening. Based on studies of the formation of complexes in Fe-C alloys [12], where stable complexes are found only below 290°C, dissociating at higher temperature, the role of such complexes on solution hardening behaviour seems limited. We thus conclude that the high temperature quench hardening observed is mostly due to carbon solute and not to vacancies.

Hardness evolution on annealing quenched material is shown in Figs. 2 and 3. For the highest ageing temperature, 430°C, hardness falls rapidly from the shortest times. At a slightly lower temperature, 400°C, the initial rapid fall of hardness is followed by a period of slower fall to a plateau hardness. The period of slower hardness fall is seen as a shoulder on the curve from the rapid fall to the plateau. At lower temperatures a well-defined age hardening peak is seen, at 600 min - 1500 min for 380-320°C. Also, on ageing at 360 to 320°C an additional weak intermediate hardness peak is observed: at 360°C there is a rapid hardness fall from the quenched state to 60 min ageing, a weak hardness maximum at 120 min, and further softening before the main hardening peak at 1500 min; at 340°C the first hardness peak is found at 60 min before re-softening and the main hardening peak at 600 min; at 320°C the first weak hardening is simply a shoulder on the initial rapid softening region. It is emphasized that even though the hardness changes are small (e.g. at 360°C, first hardness minimum of 360 Kg/mm^2, first maximum of 390 Kg/mm^2, second minimum of 370 Kg/mm^2, and main maximum of 390 Kg/mm^2) they are repeatedly and consistently found on different treated samples. Several microstructural changes are taking place, including the initial loss of solution species and the subsequent formation of hardening precipitates.

Further analysis of the processes taking place can be made by examining the times for softening and for subsequent hardening. Fig. 4 shows the time to half softening, the time to the intermediate hardness peak, and the time to the main age hardening peak, for the range of ageing temperatures studied. A straight-line, Arrhenius relationship is respected at the highest temperatures, 430°C to about 370°C, and probably again at the lowest temperatures, 340°C to 320°C, but is not for the intermediate temperatures. It is also in this intermediate range that the additional weak hardness peak is observed. From the high temperature data, an activation energy for the initial softening

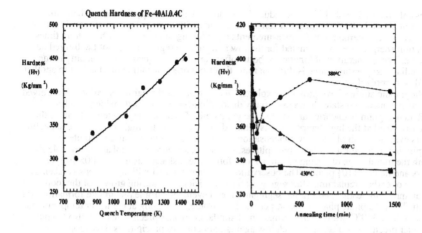

Fig. 1: Variation of hardness of quenched Fe-40Al-0.4C with quench temperature.

Fig. 2: Hardness variation of quenched Fe-40Al-0.4C on ageing at 430, 400, 380°C.

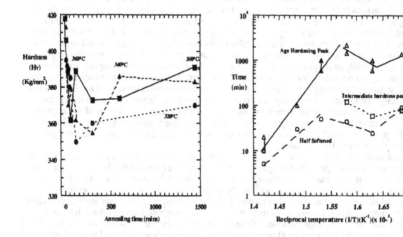

Fig. 3: Hardness variation of quenched Fe-40Al-0.4C on ageing at 360, 340, and 320°C.

Fig. 4: Arrhenius plot summarising times to half-softening, intermediate hardness peak, and main age-hardening peak.

process of about 130 ± 30 kJ/mol is deduced, and an activation energy for the age hardening process of about 340 ± 50 kJ/mol. The errors associated with these values are obviously rather large, given the uncertainty in the measurements of softening times and peak hardening times. Activation energies are not estimated for the low temperature region in view of the limited data available. The acceleration of kinetics of both softening and subsequent age hardening in the intermediate temperature range is due to a transition from one controlling mechanism to another, as will be described later.

Metallographic studies have provided explanations for many of the hardness variations observed. The starting quenched state shows essentially dislocation-free grains with oxide particles distributed at grain boundaries and within the grains, Fig. 5. At this stage essentially all of the solute dissolved at the high temperature is retained in solution. After high temperature annealing, there is evidence both of vacancy condensation as dislocation loops and also of solute precipitation in the form of planar precipitates, see Fig. 6, showing material annealed at 430°C. During the initial stage of ageing vacancy loops form. The dislocations have <100> Burgers vectors and lie on {110} planes. The observation of fringe contrast within the loops is taken as evidence of carbon remaining after segregation to the dislocation during growth of the loop. Longer annealing leads to additional segregation, and eventual carbide precipitate formation and growth. The precipitate phase is the κ, perovskite, carbide. Annealing at the lower temperatures, for example at 340°C, leads to the formation of similar co-planar precipitates and also to equi-axed finer precipitates, Fig. 7. At such low ageing temperatures precipitates thus adopt two morphologies: planar precipitates, especially at short ageing times, and small star-shaped precipitate groups with two or more arms, especially at longer ageing times. The star-shaped objects appear to be composed of fine equiaxed precipitates and attached segregated faults. The precipitates have been identified as κ carbide, lying in a well-defined orientation relationship in the FeAl matrix.

These studies of dislocation loop formation and precipitation during ageing make it possible to provide explanations for the hardness changes observed and the different response according to the ageing temperature. There is thus evidence to suggest rapid vacancy annihilation and the subsequent formation of planar precipitates at high ageing temperatures, but no other precipitate nucleation. At low temperatures there are similar dislocation loops, with associated segregation to the loop plane, found from the shortest ageing times, and later many fine star-morphology precipitates, especially at longer ageing times. On ageing at relatively low temperatures, there are then three processes occurring leading to the softening and the ageing peaks. At short times there is a loss of solution hardening - this occurs as vacancies and carbon interstitials are lost from the lattice. At long ageing times hardening occurs as many fine carbide particles form inside the grains. The weak intermediate peak may be associated with the rapid formation of the segregated loops and subsequently their planar precipitates. At high ageing temperatures such ageing occurs quickly, the processes overlap, and the loss of solution hardening and overage-softening take place within the shortest times examined.

The origin of relatively fast softening and age hardening at the lower temperatures is the change of ageing sequence and especially the appearance of many, finely-distributed carbide precipitates not found for high temperatures. The change of mechanism can be associated with the greater driving force for carbide nucleation, as well as the much lower vacancy mobility (especially relative to carbon mobility) at low temperatures [13]. The ageing temperatures are still probably too high for carbon-vacancy complexes to form [12]. The many sinks in the lattice for both carbon and vacancies mean that diffusion distances for solute annihilation are much reduced and hence the kinetics of both processes (loss of vacancies and of carbon) are accelerated. Faster kinetics of softening and age hardening are found (e.g. at 340°C than at 360°C) because of the more extensive precipitation at the lower temperature and the shorter average diffusion distance of solute from the matrix to the nearest sink. This may be visualised, semi-quantitatively, by comparing the distances between sinks at high temperature (Fig. 6a) and at low temperature (Fig. 7b): the reduction of sink separation from about 500nm to below 50nm, more than a factor

Fig. 5: Scanning electron micrograph showing the fine grain size and many oxide particles in the starting Fe-40Al-0.4C material.

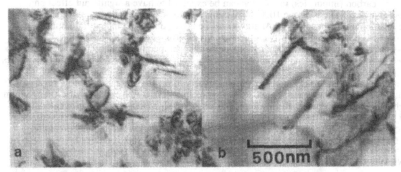

Fig. 6: Fe-40Al-0.4C, initially quenched from 1000°C and subsequently aged for:
(a) 10 min and (b) 120 min at 430°C. Dislocation loops resulting from vacancy annihilation are seen, with carbon segregating to the loop and forming κ carbide phase.

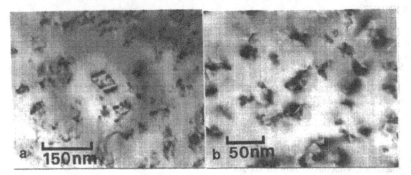

Fig. 7: Fe-40Al-0.4C, initially quenched from 1000°C and subsequently aged for:
(a) 60 min and (b) 600 min at 340°C. Plate-like and star-shaped carbide precipitates are seen to have formed, with no evidence of distinct vacancy debris.

of 10, would lead to a significant increase of kinetics if the structural refinement factor were considered alone.

Based on these arguments, the activation energy for softening in the high temperature regime, 130 kJ/mol, may be related to vacancy movement, required for vacancy loops to form (these act as sinks for carbon), or to the movement of carbon interstitials to the dislocation sinks. The activation energy for vacancy migration is about 160-180 kJ/mol for Fe-40Al [3], while that for carbon interstitial movement is 120 kJ/mol [13], consistent with the value deduced for softening. The activation energy for age hardening in the high temperature regime (340 kJ/mol) relates to the processes controlling particle nucleation and growth and cannot be easily associated with any diffusional kinetics.

CONCLUSIONS

Quench hardening of Fe-40Al-0.4C is controlled by the amount of carbon in interstitial solution. This occurs both because there is more carbon than vacancies in solution, and because the interstitial atom is a stronger hardener. From estimates of the concentration of thermal vacancies in Fe-40Al, carbon (impurities) in solution can be expected to have a significant effect on hardness at concentrations above 1000-1500 ppm. It is unclear whether previous reports of vacancy hardening in quenched Fe-Al alloys may be affected by such interstitial impurity hardening (this depends on impurity contents).

The way softening occurs during ageing depends on the temperature, and seems to depend on the relative kinetics of vacancy and carbon mobility. For high temperatures, highly mobile vacancies form dislocation loops where carbon segregates and later forms carbide precipitates. For low temperatures vacancies are relatively less mobile than carbon and similar carbon-segregated dislocation loops form but there is also extensive precipitation of fine carbides. The change from the high temperature regime to the low temperature regime is associated with an acceleration of overall kinetics as the diffusion distance required for loss of solute decreases: kinetics of diffusion continue to decrease at lower temperatures, but the nucleation of many precipitates means that the average diffusion distance of solute to the nearest sink is reduced.

ACKNOWLEDGEMENTS

We thank Chrysalis Technologies Incorporated, Richmond, USA, for their interest, material supply, and financial support under Contract Number CT 2699 to CENIM.

REFERENCES

1: Y.A. Chang, L.M. Pike, C.T. Liu, A.R. Bilbrey and D.S. Stone, *Intermetallics* 1, 107 (1993).
2: D. Weber, M. Meurin, D. Paris, A. Fourdeux and P. Lesbats, *J. Phys.* **38**, 332 (1977).
3: M.A. Morris, O. George and D.G. Morris, *Mater. Sci. and Eng.* **A258**, 99 (1998).
4: J.H. Schneibel, P.R. Munroe and L.M. Pike, Mater. Res. Soc. Symp. Proc. Vol. 460, Materials Research Society, Pittsburgh, 1997, p. 379.
5: C.H. Kong and P.R. Munroe, *Scripta Met. et Mater.* **30**, 1079 (1994).
6: K. Yoshimi, S. Hanada, T. Onuma and M.H. Yoo, *Phil. Mag.* **A73**, 443 (1996).
7: L. Pang and K.S. Kumar, *Acta Mater.* **46**, 4017 (1998).
8: M.R. Hajaligol, S.C. Deevi, V.K. Sikka and C.R. Scorey, *Mater. Sci. and Eng.* **A258**, 249 (1998).
9: S.C. Deevi and R.W. Swindeman, *Mater. Sci. and Eng.* **A258**, 203 (1998).
10: S.C. Deevi, D.G. Morris and V.K. Sikka, Proc. ASM Conf. on Structural and Functional Intermetallics, Vancouver, July 16-20, 2000.
11: M. Palm and G. Inden, *Intermetallics* 3, 443 (1995).
12: S. Takaki, J. Fuss, H. Kugler, U. Dedek and H. Schultz, Rad. Effects. 79, 87 (1983).
13: Smithells Metal Reference Book, Eds. E.A. Brandes and G.B. Brook, Butterworth-Heinemann, Oxford, 7th Edition (1992).

Mat. Res. Soc. Symp. Proc. Vol. 646 © 2001 Materials Research Society

Strain-Induced Ferromagnetism in Fe-40Al Single Crystals

D. Wu and I. Baker
Thayer School of Engineering, Dartmouth College, Hanover, NH 03755-8000, U.S.A.

ABSTRACT

Recent research [1-3] suggests that strain-induced ferromagnetism in lightly-strained FeAl arises chiefly from anti-phase boundary tubes. Magnetic and calorimetric measurements have been performed on two different orientations of B2-structured Fe-40Al single crystals that had been cold rolled to a variety of strains. The saturation magnetization and magnetic susceptibility are related to the enthalpy associated with the annealing out of antiphase boundary tubes and, hence, to the degree of deformation and crystal orientation.

INTRODUCTION

Many intermetallics which are paramagnetic when well-annealed become ferromagnetic upon plastic straining, e.g. Fe_3Al [4-8], FeAl [1-3, 9-16], CoAl [17], CoGa [18], Ni_3Sn_2 [19], Fe_3Ge_2 [19], Pt_3Fe [20-23]. Deformation induces disorders so that atoms of the ferromagnetic element are no longer isolated from each other but interact. After severe deformation, e.g. by ball-milling or prolonged crushing [5, 9, 10], the whole lattice is disordered, and the resulting ferromagnetism is well explained by the *local environment model* [24] in which the magnetic moment of an element depends on the number of like nearest neighbors.

After less severe deformation, e.g. tensile testing or rolling, strain-induced ferromagnetism can still be observed [4, 6, 7, 11-16]. In this case, disorder is present not throughout the whole lattice, but largely in antiphase boundaries (APBs). Huffman and Fisher [4] first suggested that these are the source of strain-induced ferromagnetism. Hence, Takahashi et al. [11-16] modeled the magnetic behavior of iron-rich FeAl using the *local environment model* applied to the APBs between gliding dislocations. However, Takahashi's approach cannot correctly predict the saturation magnetization of lightly-strained FeAl unless physically unrealistic assumptions are made concerning the width and influence of the APBs. Further, a 500 K anneal of cold-rolled Fe-40Al single crystals which are ferromagnetic at room temperature returns the crystals to their paramagnetic state, but changes neither the density nor configuration of dislocations [2]. Hence, APBs between gliding dislocations cannot account for strain-induced ferromagnetism.

Yang and Baker [1] first proposed that strain-induced ferromagnetism in lightly-strained intermetallics arises not from APBs between gliding dislocations but from APB tubes [25]. APB tubes have been observed in a number of lightly-strained intermetallics using transmission electron microscopy (TEM) [3, 26-38]. Further, Yamashita et al. [3] showed that polycrystals of B2-structured Fe-35Al strained under compression at 77 and 300 K contained APB tubes and exhibited spontaneous magnetization, whereas the same material strained at 650 K exhibited neither. They concluded that APBs coupling slip dislocations could not account for the strain-induced ferromagnetism and showed that a calculation of the saturation magnetization based on APB tubes gave a result of the correct order.

Yang et al. [2] came to the same conclusion via a different route. They cold rolled single crystals of Fe-40Al and then both heated them in a differential scanning calorimeter (DSC) and measured their magnetic properties. The cold rolled crystals produced three exothermic peaks in the DSC. The spontaneous magnetization disappeared upon annealing above the lowest temperature exothermic peak. They concluded that the lowest temperature exothermic peak at ~500 K was associated with the removal of APB tubes [1]. (That APB tubes are easily annealed out has been shown by a number of workers [30, 33].)

It seems fairly certain that strain-induced ferromagnetism in lightly-strained FeAl arises mostly at APB tubes. However, the phenomenology of this behavior is not well understood. Here, we present the results of experiments on cold-rolled Fe-40Al single crystals in which the saturation magnetization and magnetic susceptibility are correlated with the enthalpy associated with the annealing out of APB tubes and, hence, with the degree of deformation.

EXPERIMENTAL

Single crystals of Fe-40Al were grown in alumina crucibles under argon using a modified Bridgman technique. The crystals' orientations were determined by X-ray Laue back reflection. They were annealed for 10 hrs at 1473 K, followed by slow furnace-cooling, and further annealed at 673 K for 5 days to minimize the thermal vacancies [39]. Two sets of single crystal sheets were used: some were cut with [10 5 7] along the rolling direction and [$\bar{4}$ 1 5] normal to the sheet plane (A). The other set were cut with [8 1 1] along the rolling direction and [0 $\bar{1}$ 1] normal to the sheet plane (B). Thin sections (2.5 mm thick) of the crystals were cold rolled to a variety of strains up to ~38%, when severe cracking started to occur. The rolled crystals were both heated in a Perkin-Elmer DSC-7 at 20 K/min to 998 K and their magnetic behavior determined using a vibrating sample magnetometer (VSM), see references 1 and 2 for details. In the VSM, the magnetic field was applied perpendicular to the sheet thickness and the function generator was set to a sinusoidal output with a sweep rate of either 0.005 or 0.001Hz.

RESULTS AND DISCUSSION

Figure 1 shows typical magnetization (M) versus applied magnetic field (H) curves for the A type single crystal specimens. The FeAl is paramagnetic before rolling and a small hysteresis loop is present after 2.5% thickness reduction, see Figure 1 (a). The hysteresis loop becomes larger as deformation increases, see Figure 1(b). When the thickness reduction is about 30%, the specimens can be picked up using a permanent magnet.

Figure 2 shows the effects of the thickness reduction on the saturation magnetization (M_s), and the initial susceptibility (χ_m). The value of χ_m reported is the maximum slope measured from M versus H curve. For specimen A, M_s increases only slowly until the thickness reduction reached 20%, then increases more rapidly. For the B type specimens, M_s increased more rapidly initially, but appears to be reaching a plateau at the highest strains. Even though the χ_m curves for A and B type crystals show the same trend as for M_s, the highest χ_m for the type B specimen is much lower than that for the type A specimen.

Figure 3(a) shows a typical DSC curve for a rolled Fe-40Al specimen. The large trough represents the enthalpy associated with the removal of APB tubes, see reference 1 for details. This enthalpy is plotted as a function of rolling reduction for the two types of crystals in Figure 3(b). The enthalpy of the A-type crystals is greater than that of the B-type crystals for thickness reductions greater than 25%, and is nearly four times larger at the largest thickness reduction. That the stored energy of cold work due to dislocations depends strongly on crystal orientation is well known [40]. Thus, that the energy stored in APB tubes varies with crystal orientation, as shown in Figure 3(b) is to be expected.

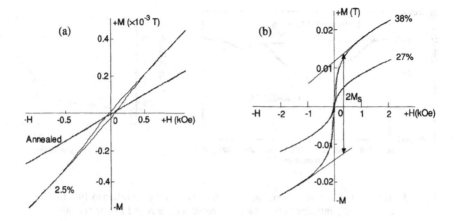

Figure 1. Typical magnetization vs. applied magnetic field curves for A type crystals: (a) as-annealed and after 2.5% thickness reduction; (b) after 27% and 38% thickness reduction. Note that the scales are different in (a) and (b). M_s is defined in (b).

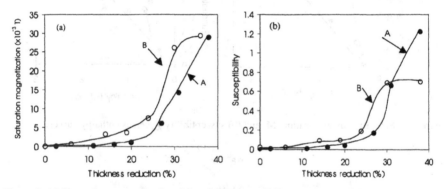

Figure 2. (a) Saturation magnetization, M_s and (b) susceptibility, χ_m versus thickness reduction.

Figure 4 shows M_s and χ_m versus enthalpy graphs. For both M_s and χ_m, B-type crystals generally show much greater values at a given enthalpy than A type crystals. As stated in the APB-tube model [2], M_s is expected to increase with the density of APB tubes. Thus, it is possible that a higher M_s for type A specimens could be obtained if the magnetic field is applied in other than the rolling direction. Such strain-induced magnetic anisotropy is clearly shown in Figure 5 for a Fe-40Al specimen. This specimen has been annealed for 5 days at 673 K before cutting to size, but no further anneal was applied after cutting. M_s is clearly different when the magnetic field is applied along three orthogonal axes.

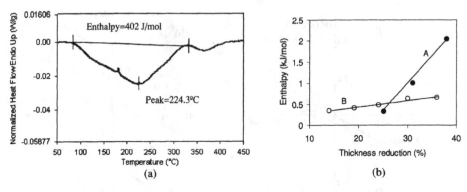

Figure 3. (a) Heat flow versus temperature curve for a type B crystal rolled to 19% thickness reduction, (b) enthalpy versus thickness reduction for A and B type crystals.

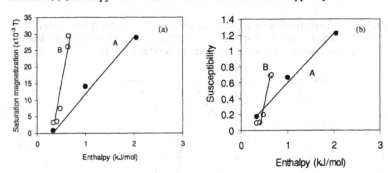

Figure 4 (a) Saturation magnetization, M_s and (b) susceptibility χ_m versus enthalpy curves.

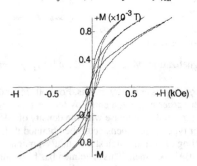

Figure 5. M versus H curves for a specimen when H is applied in three orthogonal directions.

Figure 6(a) shows M versus H curves for type B Fe-40Al single crystal specimens. After cutting to size, the specimen was annealed at 673 K for two days and was paramagnetic. However, when it was sanded on 600 grit paper, ferromagnetic behavior was induced, although the magnitude of M_s was very small. Thus, the induced magnetization is very sensitive to the surface deformation. In contrast, no apparent ferromagnetic behavior was observed for either single slip- or duplex-slip-orientated tensile test specimens even after more than 20% elongation, see Figure 6(b). DSC measurements showed that both the tensile specimen and the sanded specimen have the same magnitude of stored enthalpy due to APB tubes, 47 and 56 J/mol, respectively. By comparison, 402 J/mol was obtained for a type B specimen at a thickness reduction of 19%. The difference in behavior of the tensile specimen and the sanded specimen might be because the VSM measurements are sensitive to the surface behavior of the specimen. Rolled specimens also have more deformation at the surface than in the interior and more slip systems are activated during rolling than tensile testing. As a result, more screw dipoles could cross-slip and more APB tubes are formed. Thus, higher enthalpy is obtained in the DSC and ferromagnetism is observed in the rolled specimens.

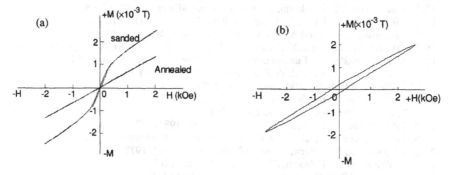

Figure 6. M versus H curves for Fe-40Al crystals: (a) as-annealed and sanded, (b) after 23% elongation.

CONCLUSIONS

Magnetic and calorimetric measurements on two differently orientated B2-structured Fe-40Al single crystals that had been cold-rolled up to 38% thickness reduction were performed. The results show that rolling induces a paramagnetic to ferromagnetic transition. The M_s and χ_m increase slowly until a thickness reduction of 20% is reached, then increase rapidly. Type B crystal seems to be reaching a plateau in M_s and χ_m at the highest strains. DSC measurements showed that the stored energy due to APB tubes varies with crystal orientation and the degree of deformation.

ACKNOWLEDGEMENTS

This research was supported by National Science Foundation though grant DMR 9973977. Dr. E.P. George of the Oak Ridge National Laboratory is gratefully acknowledged for providing ingots for single crystal growth.

REFERENCES

1. Y. Yang and I. Baker, *Proc. Mat. Res. Soc.*, 460 (1997) 367.
2. Y. Yang, I. Baker, and P. Martin, *Phil. Mag. B.*, 79 (1999) 449.
3. K. Yamashita, M. Imai, M. Matsuno and A. Sato, *Philos. Mag. A*, 78 (1998) 285.
4. G. P. Huffman and R.M. Fisher, *J. Appl. Phys.*, 38 (1967) 735.
5. M. J. Besnus, A. Herr and A. J. P Meyer, *J. Phys. F: Metal Phys.*, 5 (1975) 2138.
6. Y. Jirásková, O. Schneeweiss, M. Šobang and I. Novotny, *Acta. Mater.*, 45 (1997) 2147.
7. A. Taylor and R.M. Jones, *J. Phys. Chem Solids*, 6 (1958) 16.
8. M. Vondrácek, O. Schneeweiss and T. Zák, *Sensors and Actuators*, A, 59 (1997) 269.
9. S. Gialanella, M.D. Baro, L. Lutterotti and S. Surinachy, *Proc. Mat. Res. Soc.*, 364 (1995) 213.
10. S. Gialanella, *Intermetallics*, 3 (1995) 73.
11. S. Takahashi, *J. Magm. Mag. Mater.*, 54-57 (1986) 1065.
12. S. Takahashi and Y. Umakoshi, *J. Magm. Mag. Mater.*, 90/91 (1990) 735.
13. S. Takahashi and Y. Umakoshi, *J. Phys.: Condens. Matter.*, 3 (1991) 5805.
14. S. Takahashi, A. Chiba and E. Takahashi, *Phys. Lett.*, A, 197 (1995) 350.
15. S. Takahashi, X. E. Li and A. Chiba, *J. Phys.: Condens. Matter*, 8 (1996) 11243.
16. S. Takahashi, H. Onodera, X. E. Li and S. Miura, *J. Phys.: Condens. Matter*, 9 (1997) 9235.
17. L. M. Di, H. Bakker, Y. Tamminga and F.R. de Boer, *Phys. Rev. B.*, 44 (1991) 2444.
18. L. M. Di, H. Bakker and F.R. de Boer, *Physica B*, 182 (1992) 91.
19. G. F. Zhou and H. Baker, *Phys. Rev. B.*, 49 (1994) 12507.
20. G. E. Bacon and J. Crangle, *Proc. Roy. Soc. A*, 272 (1963) 387.
21. S. Takahashi, *Phys. Stat. Sol. a*, 42 (1977) 201.
22. S. Takahashi and K. Ikeda, *J. Phys. Soc. Japan.*, 52 (1983) 2772.
23. S. Takahashi and Y. Umakoshi, *J. Phys. F: Met. Phys.*, 18 (1988) L257.
24. H. Okamoto and P. A. Beck, *Monashefte für Chemie*, 103 (1972) 907.
25. A. E. Vidoz and L. M. Brown, *Philos. Mag.*, 7 (1962) 1167.
26. C. T. Chou and P. B. Hirsch, *Phil. Mag.*, A, 44 (1981) 1415.
27. C. T. Chou and P. B. Hirsch, *Proc. Royal Soc.*, A, 387 (1983) 91.
28. C. T. Chou and P. M. Hazzledine, P. B. Hirsch, et al, *Phil. Mag. A*, 56 (1987) 799.
29. Z. Y. Song, M. Hida, A. Sakakibara and Y. Takemoto, *Scripta Mater.*, 37 (1997) 1617.
30. Y. Q. Sun, *Philos. Mag. A*, 65 (1992) 287.
31. P. M. Hazzledine and Y.Q. Sun, *Mater. Sci. Eng.*, A152 (1992) 189.
32. A. H. W. Ngan, I. P. Jones and R. E. Smallman, *Proc. Mat. Res. Soc.*, 288 (1993) 269.
33. A. H.W. Ngan, I. P. Jones and R.E. Smallman, *Philos. Mag. B*, 67 (1993) 417.
34. N. Jiang and Y.Q. Sun, *Philos. Mag. Lett.*, 68 (1993) 107.
35. X. Shi, G. Saada and P. Veyssière, *Philos. Mag. A*, 71 (1995) 1.
36. A.H.W. Ngan, *Philos. Mag. A*, 71 (1995) 725.
37. X. Shi, G. Saada and P. Veyssière , *Philos. Mag. A*, 73 (1996) 1159.
38. M. F. Savage, R. Srinivasan, M. S. Daw, T. Lograsso and M. J. Mills, *Mat. Sci. and Eng. A*, 258(1998) 20.
39. P. Nagpal and I. Baker, *Metall. Trans.*, A, 21 (1990) 2281.
40. P. Haessner, G. Hoschek and G. Tölg, *Acta Metal.*, 27(1979) 1539.

Mat. Res. Soc. Symp. Proc. Vol. 646 © 2001 Materials Research Society

Synthesis and Characterization of Mechanically Alloyed and HIP-Consolidated Fe-25Al-10Ti Intermetallic Alloy

Su-Ming Zhu, Makoto Tamura, Kazushi Sakamoto and Kunihiko Iwasaki
Japan Ultra-high Temperature Materials Research Institute, Ube, Yamaguchi 755-0001, Japan

ABSTRACT

The present study is concerned with the processing, microstructural characterization, mechanical and tribological properties of fine-grained Fe-25Al-10Ti intermetallic alloy. The alloy was synthesized from elemental powders by mechanical alloying in an attritor-type ball milling system for 100 h, followed by hot isostatic pressing (HIP). After HIP treatment at 1073 K under an ultra-high pressure of 980 MPa, fully dense compacts with a grain size of about 200 nm were produced. Mechanical properties were evaluated by compression tests from room temperature to 1073 K. At room temperature, the alloy exhibits yield strength as high as 2.4 GPa, together with considerable rupture strain of 0.16. The yield strength decreases monotonically with increasing test temperature with no positive temperature dependence observed. The grain growth after high temperature deformation is not severe, indicating that the alloy has a relatively high thermal stability. Finally, tribological properties of the alloy were evaluated by using a ball-on-disk type wear tester and compared with those for gray cast iron, a currently used material for automotive brake rotors.

INTRODUCTION

Recently, there is some interest in studying the wear properties of iron aluminide based on Fe_3Al [1-3]. The aluminide possesses many attributes necessary for wear resistance - high hardness, high elastic modulus and good environmental resistance, and are thus promising as tribological materials for use in aggressive environments at elevated temperatures. Moreover, in wear related applications, tensile ductility is not so crucial since loads are compressive in nature. It has been shown that Fe_3Al has wear resistances similar to those of a variety of steels, for example 304 SS [1]. To improve the wear resistances of Fe_3Al, Hawk et al. [2] have examined the alloying effects of Ti, Zr, Cr, Ni, Nb and Mo. They demonstrated that additions of Ti to Fe_3Al have a positive influence on their tribological properties. For example, the alloy with 10 at. % Ti substituting for Fe in Fe_3Al shows 40% decrease in volume wear compared with Fe_3Al. However, as pointed out by some researchers [4,5], the addition of Ti into iron aluminides tends to reduce the room temperature ductility, which makes the processing and machining very difficult.

Mechanical alloying has been extensively exploited in the synthesis and processing of intermetallic compounds in recent years [6]. Mechanical alloying is a high-energy ball milling process which involves repeated welding and fracturing of powder particles. This technique can produce fine powders with a nanoscale grain size and a homogeneous distribution of dispersoids, which is expected to improve the ambient temperature ductility and high temperature strength of intermetallic compounds. In this work, bulk Fe-25Al-10Ti intermetallic alloy was synthesized from elemental powders by mechanical alloying and hot isostatic pressing (HIP). An ultra-high pressure (1 GPa) HIP facility was used here in an effort to consolidate the mechanically alloyed powders into fully dense compacts while retaining the nanocrystalline microstructure. Mechanical and tribological properties of the fully densified compacts were then examined.

EXPERIMENTAL PROCEDURE

Commercial metal powders of Fe, Al and Ti (100 mesh, 99.9% purity) were used in the present study. Before mechanical alloying, powders with nominal compositions (at.%) of Fe-25Al-10Ti were blended. Mechanical alloying was carried out in a high-energy attritor-type ball milling system at an agitation speed of 150 rpm for 100 h. More details of mechanical alloying can be found elsewhere [7]. After mechanical alloying, the milled powders were sieved (180 mesh), degassed at 773 K for 1 h, and sealed into steel cans in vacuum. Then the cans were consolidated using an ultra-high pressure HIP equipment. The consolidation was performed at 1073 K, 980 MPa, for a hold time of 2 h. Such HIP conditions have been proven to be capable of producing fully dense binary Fe_3Al compacts without any porosity [8].

Microstructural features of the HIP-consolidated compacts were characterized by using X-ray diffraction (XRD), optical microscopy and transmission electron microscopy (TEM). Mechanical properties of the compacted material were evaluated by compression tests using rectangular-shape specimens with dimensions of $3 \times 3 \times 6$ mm^3. The tests were conducted in air at an initial strain rate of 1.4×10^{-3} s^{-1} from room temperature to 1073 K. Moreover, strain-rate-changing tests were conducted at 873 and 1073 K in order to determine the strain rate sensitivity of the compacted material. Tribological properties of the material were evaluated by using a ball-on-disk type wear tester. Rectangular-shape specimens with dimensions of $20 \times 20 \times 5$ mm^3 were used as rotating disks. Alumina balls were used as the counterface material. The tests were conducted at a sliding speed of 0.1 m s^{-1} to a total distance of 50 m under a load of 10 N. The friction force was measured using a loading cell. The wear rate was calculated by dividing the volume of wear groove by the sliding distance.

RESULTS AND DISCUSSION

Microstructural features

Figure 1 shows the XRD patterns of Fe-25Al-10Ti powders in the as mixed, mechanically alloyed and HIP consolidated conditions. For starting powders, sharp diffraction peaks of Fe, Al and Ti can be seen. Upon milling, these elemental peaks are gradually weakened and broadened. The weakening of Fe, Ti and Al peaks indicates that dissolution due to interdiffusion between Fe, Ti and Al has occurred during milling, whilst the broadening of these peaks indicates the decrease in the effective grain size and/or the increase in the internal strain. After milling for 100 h, only peaks of Fe(Ti,Al) solution can be seen in the XRD profile, indicating that the starting powders have completely transformed into Fe(Ti,Al) solid solution. From the XRD pattern, the lattice parameter and grain size of the 100 h milled powders are determined to be 0.2934 nm and 11.3 nm, respectively. After HIP treatment, the peaks become sharp again and additional superlattice peaks are observed. This reveals that the Fe(Ti,Al) solution is further transformed to Fe_3Al intermetallic compound. The weak superlattice peaks in the XRD pattern of the compacted material indicate the low degree of ordering in the microstructure.

Optical and TEM micrographs of the compacted Fe-25Al-10Ti alloy are shown in Fig. 2. The alloy shows a fully dense and ultra-fine microstructure without any distinct porosity. The diminishing of the prior particle boundaries indicates a good bonding between powder particles. From the TEM micrograph, the grain size is determined to be about 200 nm. By comparing with the initial grain size of powders (11.3 nm), it appears that grain growth has occurred during the HIP process, but the growth rate is not so high. Moreover, fine dispersoids are detected in some areas (Fig. 2(b)). These dispersoids are speculated to be Ti-rich oxide particles.

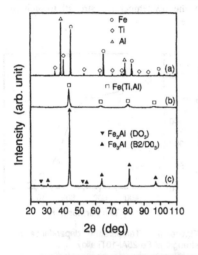

Figure 1. X-ray diffraction patterns of (a) as-mixed powders, (b) 100 h milled powders and (c) HIP consolidated compact.

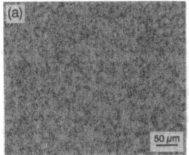
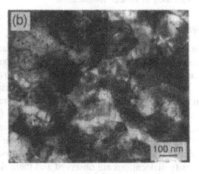

Figure 2. Micrographs of the compacted Fe-25Al-10Ti alloy: (a) optical and (b) TEM.

Mechanical properties

Figure 3 shows the compressive stress-strain curves of the alloy. At room temperature, yield strength as high as 2.4 GPa is attained. Such a high yield strength value is considered to result from the ultra-fine microstructure retained in the compact. It is also noted that the alloy exhibits considerable rupture strain (0.16) at room temperature. Temperature dependence of the yield strength of the alloy is shown in Fig. 4. It can be seen that the yield strength decreases almost monotonically with increasing temperature. The positive temperature dependence of yield strength usually exhibited by cast Fe_3Al-based alloys is not observed. The strength anomaly in Fe_3Al-based alloys is a very complicated phenomenon and is generally thought to be associated with the degree of ordering and the transition of dislocation mechanism with temperature [9,10]. The absence of this behavior in the present alloy is speculated to be due to the relatively low degree of ordering. Another possible reason is the strengthening effect of ultra-fine grain size at low temperatures, which raises the low-temperature strength to such high levels that the strength anomaly is no longer discernible.

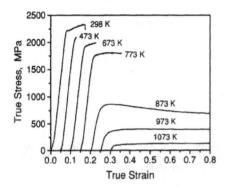

Figure 3. Compressive stress-strain curves of Fe-25Al-10Ti alloy at various temperatures.

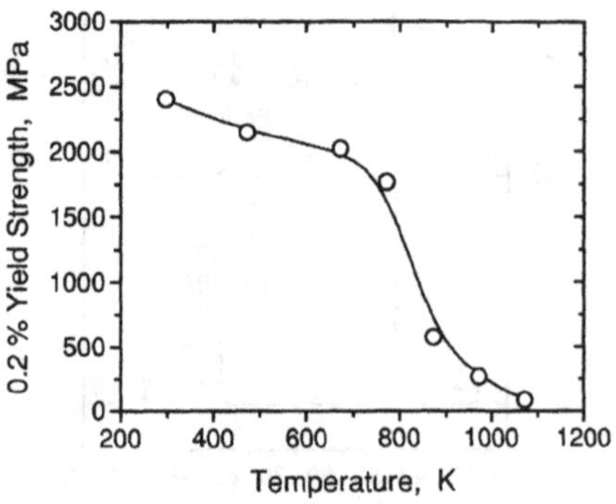

Figure 4. Temperature dependence of yield strength of Fe-25Al-10Ti alloy.

High temperature deformation behavior

The high temperature behavior of the alloy was studied by strain rate change tests at temperatures of 873 K and 1073 K. Examination of the deformed specimens reveals that the alloy can be deformed without surface cracking over the whole temperature and strain rate range used. To analyze the strain rate dependence of flow stress, the flow stress at true strain of 0.2 is used in the present study. Figure 5 shows the strain rate dependence of the compressive flow stress. From the plots, the strain rate sensitivity (m) is determined (0.19 at 873 K and 0.27 at 1073 K). The m value appears to increase with increasing temperature.

The deformed microstructure of the alloy after high temperature compression was examined by TEM and is shown in Fig. 6. The alloy shows a relatively stable grain structure. Low-density dislocations are observed and these dislocations are generally confined within individual grains. These suggest that dislocation movement is not the dominant deformation mechanism. By comparing with the as-consolidated microstructure (Fig. 2(b)), it is found that the grain growth is not obvious. The inhibited grain growth is presumably attributed to the grain boundary pinning by fine dispersoids in a mechanism proposed by Zener and Smith [11].

According to the theories of high temperature deformation [12], when $m = 0.2$, the deformation is controlled by dislocation interactions with the matrix, and when $m = 0.5$, superplasticity is expected since the grain boundary sliding accommodated by slip becomes the controlling deformation mechanism. More recently, Nieh et al have suggested that $m = 0.3$ is a critical value for intermetallics to show superplastic behavior [13]. In view of the m values obtained and the deformation microstructure observed in the present study, grain boundary sliding is believed to play an important role in the deformation of this alloy, especially at temperatures above 1073 K.

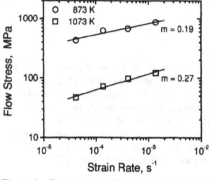

Figure 5. Dependence of flow stress on strain rate in Fe-25Al-10Ti alloy.

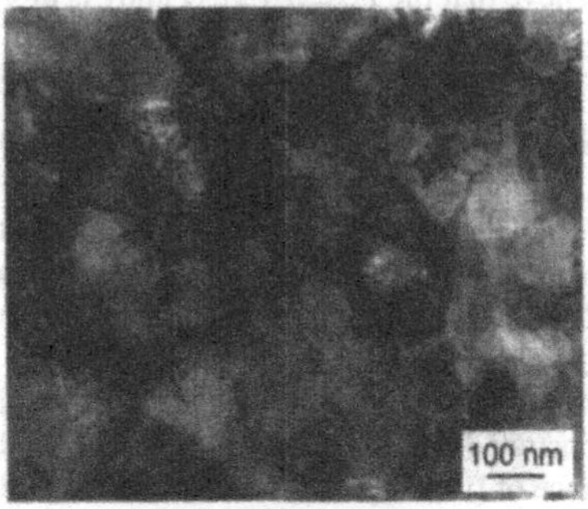

Figure 6. TEM micorgraph of Fe-25Al-10Ti alloy after compression at 1073 K.

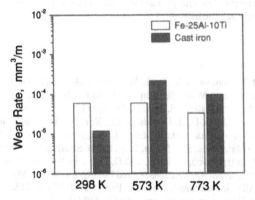

Figure 7. Comparison of the wear resistance of Fe-25Al-10Ti alloy with that of cast iron.

Wear properties

The wear rate of the HIP compacted Fe-25Al-10Ti alloy was measured at ambient and elevated temperatures, and shown in Fig. 7. For comparison purpose, the data of cast iron were also given. The cast iron samples were cut from a commercial automotive brake rotor and had a microstructure of graphite dispersed in α-Fe matrix. It can be seen that the Fe-25Al-10Ti alloy exhibits higher wear rate at room temperature, but lower wear rate at 573 and 773 K than the cast iron. The low wear rate of cast iron at room temperature is attributed to the dispersion of graphite in the α-Fe matrix, while the high wear rates at elevated temperatures are associated with the rapid decrease in strength with increasing temperature. For Fe-25Al-10Ti alloy, there is only a slight decrease in strength from room temperature to 773 K. As a result, the difference in wear rate at ambient and elevated temperatures is very small. Based upon the above results, we believe that the Fe-25Al-10Ti alloy is promising as tribological material for high temperature use.

CONCLUSIONS

Fully dense, bulk Fe-25Al-10Ti alloy with a grain size of about 200 nm was produced from mechanically alloyed powders by HIP consolidation at 1073 K under an ultra-high pressure of 980 MPa. Under compression, the alloy exhibited yield strength as high as 2.4 GPa, together with considerable rupture strain (0.16). The yield strength decreased monotonically with increasing test temperature with no positive temperature dependence observed. It was demonstrated that the alloy had relatively high structural stability at high temperatures due to the presence of fine dispersoids. Compared with commercial gray cast iron, the alloy showed poorer wear resistance at room temperature, but better wear resistance at 573 and 773 K.

ACKNOWLEDGEMENTS

Financial support by the New Energy and Industrial Technology Development Organization (NEDO), Japan, is greatly acknowledged.

REFERENCES

1. H.E. Maupin, R.D. Wilson, and J.A. Hawk, Wear **162-164**, 432 (1993).
2. J.A. Hawk and D.E. Alman, Mater. Sci. Eng. **A239-240**, 899 (1997).
3. Y.-S. Kim and Y.-W. Kim, Mater. Sci. Eng. **A258**, 319 (1998).
4. M.G. Mendiratta, S.K. Ehlers, D.M. Dimiduk, W.R. Kerr, S. Mazdiyasni, and H.A. Lipsitt, in *High Temperature Ordered Intermetallic Alloys II*, edited by N.S. Stoloff, C.C. Koch, C.T. Liu, and O. Izumi, (Mater. Res. Soc. Symp. Proc. **81**, Pittsburgh, PA, 1985), p. 393.
5. C. Testani, A. Di Gianfrancesco, O. Tassa, and D. Pocci, in *Nickel and Iron Aluminides: Processing, Properties, and Applications*, edited by S.C. Deevi, P.J. Maziasz, V.K. Sikka, and R.W. Cahn, (ASM International, Materials Park, OH, 1997), p. 213.
6. C.C. Koch and J.D. Whittenberger, Intermetallics **4**, 339 (1996).
7. S.-M. Zhu and K. Iwasaki, Mater. Sci. Eng. **A270**, 170 (1999).
8. S.-M. Zhu and K. Iwasaki, Mater. Trans. JIM **40**, 1361 (1999).
9. N.S. Stoloff and R.G. Davies, Acta Metall. **12**, 473 (1964).
10. S. Hanada, S. Watanabe, T. Sato, and O. Izumi, Scr. Metall. **15**, 1345 (1981).
11. C. Zener and C.S. Smith, Trans. AIME **175**, 47 (1948).
12. O.D. Sherby and J. Wadsworth, Progr. Mater. Sci. **33**, 169 (1989).
13. T.G. Nieh, J. Wadsworth, and O.D. Sherby, *Superplasticity in Metals and Ceramics*, (Cambridge University Press, Cambridge, 1997), p. 125.

Mat. Res. Soc. Symp. Proc. Vol. 646 © 2001 Materials Research Society

Interaction of boron with crystal defects in B2-ordered FeAl alloys

Anna Fraczkiewicz, Anne-Sophie Gay, Emmanuel Cadel[1], Didier Blavette[1]
ENSMSE, Centre SMS, URA CNRS 1884, 158 Cours Fauriel, F-42023 ST-ETIENNE
[1] Université de Rouen, UMR 6634 CNRS, Faculté des Sciences de Rouen,
F-76 821 MONT ST-AIGNAN

ABSTRACT

Intermetallic alloys are often doped with boron to suppress their intrinsic room-temperature intergranular brittleness. The commonly admitted mechanism of this effect, i.e. an intergranular segregation of boron, seems not to be the only important feature. In this work, boron interactions with numerous kinds of crystal defects (point defects, dislocations, grain boundaries) are studied in B-doped FeAl (B2) alloys containing 40 at. % Al. The intergranular segregation of boron is first characterized. Both an equilibrium and a non-equilibrium (due to a solute atom / thermal vacancy interaction) segregation mechanisms are identified. Strong tendency of boron to segregate to the dislocations lines is shown by direct measurements by atom probe field ion microscopy (AP FIM). This segregation is shown to induce a local depletion in Al in the vicinity of defects.

INTRODUCTION

Doping with small additions of boron (few hundreds at. ppm) is an efficient way to suppress the intergranular weakness of grain boundaries in many intermetallic alloys, like Ni_3Al (L1$_2$), Fe_3Al (DO$_3$), and FeAl (B2). Thus, in B-doped materials, the room temperature fracture mode becomes mainly transgranular, while a brittle intergranular fracture is commonly observed in B-free alloys. Following observations of B-enriched layer at the fracture surfaces of the former alloys, when analyzed by AES (Auger Electron Spectrometry), it is now currently admitted that this reinforcing effect of boron in intermetallic alloys is due to its intergranular segregation [1]. However, the elementary mechanisms of the action of boron, are not really understood. This question was analyzed a few years ago by George et al. [2] for the case of Ni_3Al alloys. Two hypotheses have been proposed, not really contradictory in fact: according to them, the reinforcing effect of boron could consist in (i) an increase of the grain boundary cohesion, or (ii) a modification of conditions of the slip transmission from one grain to another.

In this paper, it will be shown that the effect of boron in FeAl (B2) can not be limited to only intergranular phenomena. Experimental evidence of interactions of boron with different crystal defects (point defects, dislocation lines and grain boundaries) will be given.

EXPERIMENTAL

The B2-ordered FeAl alloys, containing 40 at. % Al ("FeAl" in this text), pure or B-doped (40 - 2000 at ppm), were laboratory prepared by cold-crucible melting. The FeAl alloys are known to retain large concentration of thermal vacancies after quench; therefore, a standard heat treatment was applied to the samples, in accordance to the results of previous work by Rieu and Goux [3].

The samples were first annealed at 950°C/1 hour, and quenched in air; in this way, a maximum concentration of thermal vacancies was retained. Then, a low-temperature annealing (400 °C/24h) was given to the material to allow an efficient elimination of the excess vacancies. In the B-doped alloys, the intergranular segregation of boron is also promoted during this treatment.

Experimental techniques used in this work are described in details elsewhere [5,6]. Isothermal dilatometric experiments were performed for the study of the kinetics of vacancy elimination. The intergranular segregation of boron was measured by Auger Electron Spectrometry (AES) and the nanoscale segregation of boron to dislocation lines was studied by the Atom Probe Field Ion Microscopy (AP FIM).

RESULTS

Boron/Thermal Vacancy Interactions. Origins of Intergranular Segregation of Boron

Table 1 shows the effect of boron on the time of annealing which is necessary to completely eliminate the excess thermal vacancies in isothermal conditions in FeAl alloys. The given values were estimated from shrinkage measurements of quenched samples during isothermal dilatometric experiments performed at 380 °C [4]. It is important to note that the quenched-in concentrations of thermal vacancies are about the same in all the alloys studied, both B-free and B-doped.

Table I. Time to eliminate the excess vacancies during an isothermal annealing at 380 °C. Samples previously quenched from 950 °C. FeAl alloys, 40 at. % Al, B-free or B-doped

Boron content (at. ppm)	Time of elimination of excess vacancies (hours)
0	18
40	8
200	4

The kinetics of vacancy elimination is clearly accelerated by boron. A solute concentration as low as 40 at. ppm is enough to allow to attain the equilibrium vacancy concentration after only 8 hours annealing at 380 °C, while 18 h are necessary to remove the quenched-in vacancies in B-free FeAl alloy, under the same thermal conditions. Thus, it seems likely that thermal vacancies and boron atoms form stable complexes, that migrate together to the vacancy sinks. Their united migration becomes accelerated as compared to the diffusion of separated species. By comparison of kinetics of vacancy elimination in B-free and B-doped alloys, the energy of formation of boron/vacancy complexes has been estimated as being at least equal to − 0.4 eV [4].

Yet, if the vacancy/solute complexes are present in a material, their effects should also be observed as the solute enrichment of internal interfaces, that play the role of vacancy sinks during annihilation of the excess point defects. This is in fact a well-known mechanism of non-equilibrium segregation, first observed (in grain boundaries) and described by Westbrook [6]. The non-equilibrium segregation is generally characterized by fast kinetics and large thickness of solute-enriched interfacial layer. During long annealing which allows to approach the equilibrium state of material, a desegregation process is activated; thus, the thickness and/or the intergranular concentration of segregating solute, decrease in these conditions.

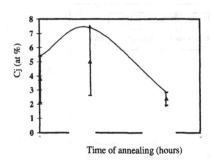

Figure 1. *Effect of time of annealing at 400°C on intergranular concentration of boron. FeAl + 200 at. % B. AES measurements on the in situ open intergranular parts of fracture surface [4].*

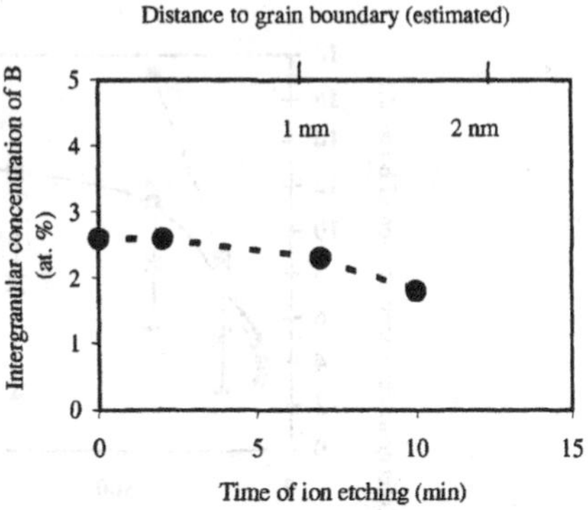

Figure 2. *Intergranular concentration of boron vs. distance from grain boundary; sample annealed during 24 h at 400 °C. Bicrystal FeAl + 50 at. ppm [7].*

The possible effects of non-equilibrium segregation on behavior of FeAl will be checked in the next part of this work. In figure 1, is shown the effect of time of annealing on the intergranular boron concentration in the FeAl + 200 at. ppm B alloy. Figure 2 shows the results of measurements of the thickness of boron-enriched layer in an FeAl alloy annealed at 400°C for 24 h. These measurements were performed by the AES method on the *in situ* open intergranular surfaces; for the thickness measurements, an oriented bicrystal was used to enable the measurements on a flat surface from which atomic layers were successively removed by ion etching.

Already in the as-quenched sample, a relatively high concentration of intergranular boron is measured. A maximum value of intergranular boron is measured after 24 h of annealing. The thickness of B-enriched layer corresponding to these thermal conditions is certainly higher than 1,5 nm (figure 2); this value is rather low for a non-equilibrium mechanism of segregation, but too high for an equilibrium process. Then, a clear decrease of the segregation level after the 3-months long low-temperature annealing is measured. In a previous paper [7], we have analyzed the possible mechanisms of this desegregation process: thanks to a careful analysis of conditions of AES measurements, we could show that the apparent decrease of intergranular concentration of boron after long annealing is in fact due to the decrease of the thickness of the

B-enriched layer only, while the local B concentration in it is stable or even growing with the annealing time.

Fast kinetics of the segregation of boron and, especially, the desegregation process observed after long annealing confirm that a non-equilibrium mechanism of segregation is operating in the FeAl alloys. Still, an equilibrium segregation is also stable in these materials, as was measured after the long annealing.

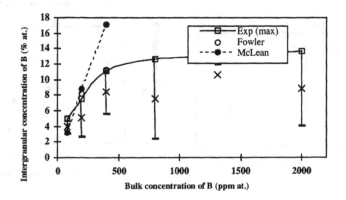

Figure 3. *AES study of boron intergranular segregation in FeAl. Samples annealed 24 h at 400 °C, after quenching from 950 °C. Influence of the bulk boron concentration [4].*

The results of AES measurements of intergranular concentration of boron in FeAl as a function of the bulk content of the solute are shown in figure 3. Even if a strong part of non-equilibrium segregation is certainly present (a 24 h annealing at 400°C was given to the analyzed samples), the relation between intergranular and bulk concentrations of solute is that which is expected for an equilibrium segregation process. The intergranular concentration of boron increases with bulk content of solute until its solubility limit is reached, and stabilizes for higher B contents. Yet, the grain boundary saturation in solute is attained for a very low value, below 14 at. %. This result is not conform to the classic theory of intergranular segregation proposed by McLean [8]. According to this model, which is based on the Gibbs' adsorption at internal interfaces, a complete segregated solute layer should be measured in the saturation conditions. The low-level of saturation of grain boundaries by boron in FeAl may by explained if one takes into consideration that strong repulsive interactions between the segregated atoms exist (Fowler model [9]). With this approach, a satisfactory fit of the experimental curve could be obtained with an average segregation energy E_s = -0.3 eV/at. and the interaction parameter w = 2.3-3.3 eV [8]. This last value is very high (expected value of w is in 1 eV range); further study is necessary to identify the reasons of this effect. Both the intrinsic characteristics of boron in FeAl and the conditions of the AES analysis of fracture surfaces in FeAl, which limit the availability of grain boundaries for analysis to the weakest (less enriched by boron ?) ones, may be blamed.

Boron Segregation to Dislocations

For its exceptional ability of investigation of local chemical composition at a nanometric scale, the atom probe field ion microscopy (AP FIM) technique [10] was used to study the segregation of boron to crystal defects in FeAl. Local enrichment in boron in the vicinity of dislocation lines [5, 11], complex stacking faults [5] and APB [12] was observed. In this work, some of these results dealing with interactions between B and dislocations, will be shown.

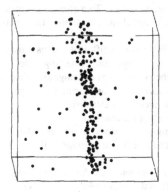

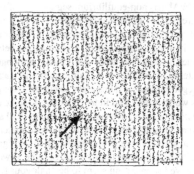

Figure 4. A boron-enriched « finger », in an FeAl + 400 at. ppm B alloy; an AP FIM analysis [13]. Analysed zone : 17x17x17 nm. Only boron atoms are represented on the figure (black dots). The local quantitative analysis of this zone gives : 6 at% of boron (+/- 0.6 %), 33 % of Al (+/- 1.2%)).

Figure 5. Top view of the region analysed in figure 4. Analysed zone : 8x6x10 nm. Only Al atoms are represented. The B-rich "finger" is situated in the zone where Al content is decreased. A <100> edge dislocation is shown by an arrow.

In numerous cases, long, B-rich "fingers" were observed (Figure 4) in the FeAl bulk material. The thickness of B-enriched zone was typically of 3 nm and the local boron concentration in these regions reached 6 at% ; moreover, an Al-depletion and Fe-enrichment was measured there. In the case shown in figure 4, the analyzed zone was turned virtually to put the B-rich axis perpendicular to the figure plane (figure 5). By imaging only Al atoms, a periodic structure of FeAl (100) superplanes was seen. Locally, lower spatial concentration of aluminum is observed in this figure: this zone was identified as that enriched in boron. In fact, the observed B-rich region follows a <100> edge dislocation, which is clearly shown in the analyzed zone with an atomic resolution. The reasons why the dislocation line is slightly shifted from its B-rich Cottrell atmosphere are not clearly identified yet; some plausible hypotheses are discussed in [13].

Thus, an experimental evidence of the segregation of boron to dislocation lines in FeAl could be given. This segregation has probably the same origins as the intergranular one; in particular, its non-equilibrium mechanism seems to be confirmed also in the case of the segregation to dislocation lines, since the boron enrichment was measured on the <100> dislocations, which originate from the thermal vacancies elimination in the studied material at low temperatures [14].

CONCLUSIONS

Boron was shown to interact with all the kinds of crystal defects studied in this work.

In presence of quenched-in vacancies, stable vacancy/boron complexes are formed; their energy of interaction was estimated as being equal to at least −0.4 eV. Two effects of presence of these complexes are observed: (i) a strong acceleration of thermal vacancy annihilation in B-doped FeAl, (ii) non-equilibrium segregation of boron.

Therefore, the intergranular segregation of boron in FeAl has two different origins. First of all, during cooling from high temperature and/or a short low-temperature annealing, a non-equilibrium segregation, due to an interaction between the boron atoms and migrating thermal vacancies, is rapidly established. When a prolonged low temperature heat treatment is applied to the material, in which a non-equilibrium segregation was previously established, the desegregation process takes place to ensure the equilibrium of the material.

It is worth noting that independently of the heat treatment conditions, an intergranular segregation of boron does take place in the FeAl alloys and is efficient to suppress the intergranular brittle fracture of the B-doped materials.

Boron shows also a strong tendency to segregate to dislocation lines. Atomic-resolution, 3-D images of B-rich Cottrell atmospheres in FeAl were obtained thanks to AP FIM. The origins of the observed segregation seem to be analogous to those of the intergranular segregation: especially, the non-equilibrium segregation takes an important place.

REFERENCES:

[1] A. Choudhury, C.L. White, C.R. Brooks, Acta Metall., 33, (1985), p. 213.

[2] E.P. George, C.L. White and J.A. Horton, Scr. Metall. Mater. 25 (1991), p. 1259.

[3] J. Rieu and C. Goux, Mém. Sci. Rev. Métall. LXVI, (1969), p. 869.

[4] A. Fraczkiewicz, A.-S. Gay and M. Biscondi, Mat. Sci. Eng. A 258 (1998), p. 108.

[5] E. Cadel, D. Lemarchand, A.-S. Gay, A. Fraczkiewicz, and D. Blavette, Scripta Metall. Mater. 41 (1999), p. 421.

[6] J.H. Westbrook, K.T. Aust, Acta Met., 11, (1963), p. 1151.

[7] A. Fraczkiewicz, A.-S. Gay and M. Biscondi, J. Phys., EDP 9 (1999), p. 75.

[8] D. McLean, Grain Boundaries in Metals, Oxford, Clarendon Press, (1957), p. 57.

[9] R.G. Faulkner, Int. Mat. Rev. 41, 5, (1996), p. 198.

[10] D. Blavette, A. Bostel, J.M. Sarrau, B. Deconihout, A. Menand, Nature 363, (1993), p. 432.

[11] D. Blavette, E. Cadel, A. Fraczkiewicz and A. Menand, Science, Dec 17th 1999, p. 2317.

[12] E. Cadel, PhD thesis, Universite de Rouen, France, 2000.

[13] E. Cadel, S. Launois, A. Fraczkiewicz, D. Blavette, Phil. Mag. Lett., 80, 11, (2000), p. 725.

[14] A. Fourdeux and P. Lesbats, Phil. Mag. A 45 (1982), p. 81.

Mat. Res. Soc. Symp. Proc. Vol. 646 © 2001 Materials Research Society

The Yield Anomaly in CoTi

M. Wittmann[1]**, I. Baker**[1] **and N.D. Evans**[2]

[1]Thayer School of Engineering, Dartmouth College, Hanover, NH 03755-8000, U.S.A

[2]Oak Ridge National Laboratory, Metals and Ceramics Division, Bldg. 5500, MS 6376, Oak Ridge, TN 37831-6376, U.S.A.

ABSTRACT

Compression tests performed on both stoichiometric and cobalt-rich CoTi over a range of temperatures show a positive temperature dependence of the yield stress with increasing temperature, before a decline occurs at high temperatures. In the region of the peak yield stress, serrated yielding and a negative rate sensitivity of the yield stress were observed. Static strain-aging also occurs. These observations are consistent with strong solute-dislocation interactions. Results from quenching experiments and strain rate change tests are presented, together with transmission electron microscope observations of the dislocation structures below, at, and slightly above the peak temperature. The results suggest that the yield anomaly in CoTi can be accounted for by a classical dynamic strain aging mechanism.

INTRODUCTION

The cobalt–transition metal B2 compounds CoTi, CoHf and CoZr remain ordered up to their melting points, deform by movement of $a<100>$ dislocations [1,2], and exhibit a positive flow stress dependence with temperature [3,4,5]. Since the $a<100>$ dislocations are not dissociated into partials [1,6,7], the yield anomaly cannot be explained by a mechanism that relies on cross slip (locking) to a plane on which the APB energy is lower [8]. The mechanical properties and dislocation behavior of CoTi have been investigated most [1,3,6,7,9]. Based on their experimental observations, Takasugi and Hanada [10] proposed that the yield anomaly arises from increased frictional force due to the spreading of the screw dislocation core, the so-called *core dissociation* mechanism. There is circumstantial evidence to support this mechanism. There is an apparent violation of Schmid's law in CoTi [2], suggesting that the core spreading is stress induced; transmission electron microscope (TEM) observations of specimens strained at intermediate temperatures revealed some bowing of the screw dislocations from apparent pinning points [1]; and high-resolution electron microscopy showed that some screw dislocations had indeed undergone a core-dissociation on a non-slip {100} plane [9]. However, a shortcoming of the core dissociation mechanism is that it is unclear why there should be a yield stress peak, and not merely increasing yield strength with increasing temperature since increased temperatures would presumably cause more core spreading and greater dislocation pinning.

The vacancy-hardening model was developed to describe the yield anomaly in FeAl [11], and it is possible that this other mechanism is responsible for the yield anomaly in CoTi. The vacancy-hardening model is not dependent on a specific dislocation mechanism, but associates the increased yield strength at intermediate temperatures with "solid solution strengthening" by vacancies. A peak in the yield stress versus temperature plot arises because at higher temperatures the vacancies, although still increasing in concentration, become mobile and are no longer effective at pinning dislocations, but allow easier diffusional processes.

Recent compression tests on stoichiometric CoTi polycrystals over wide ranges of temperature and strain rate show a yield strength peak at intermediate temperatures, which shifts to higher temperatures with increasing strain rate [13]. At temperatures close to that of the peak yield stress, serrated yielding, a negative strain-rate sensitivity of the flow stress, and static strain-aging occur. These observations are consistent with strong solute-dislocation interactions. Thus, Wittmann and Baker [13] suggested that the anomalous yield behavior of CoTi is a manifestation of dynamic strain-aging.

In the research presented here, a variety of mechanical tests have been performed on both stoichiometric and cobalt-rich CoTi at different temperatures. TEM observations were performed on polycrystalline specimens deformed, below, at, and slightly above the peak temperature to evaluate dislocation behavior. The observations are consistent with the suggestion that the yield anomaly is caused by dynamic strain aging.

EXPERIMENTAL DETAILS

Two ingots of both stoichiometric CoTi and Co-48Ti were cast into water-cooled copper molds after melting cobalt and titanium in an arc furnace under an argon atmosphere. The resulting grain sizes were ~200 μm (as determined by the linear intercept method). Compression tests showed that there was a significant difference in the yield strengths between castings of the same nominal composition, see Figure 1. Thus, they will be differentiated throughout the text and referred to as castings I and II. Chemical analyses of the two ingots of stoichiometric CoTi, using a Leco GS244 for C and S, and a Leco 436 for O and N, showed that the first contained (in p.p.m.) C - 103, S – 6, O – 761 and N – 61, while the second contained C - 76, S – 6, O – 683 and N – 64.

Compression specimens (2.5 mm x 2.5 mm x 6 mm high) were machined, and polished with 600-grit silicon carbide paper. In order to determine if quenched in vacancies have an effect on mechanical behavior some specimens were annealed at 1473 K for 30 mins, followed by either air-cooling, water-quenching or a 120-hour anneal at 673 K, and compression tests were performed in air on an Instron at an initial strain rate of 5×10^{-4} s^{-1} at room temperature.

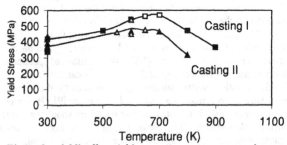

Figure 1: *0.2% offset yield stress versus temperature for two castings of stoichiometric CoTi strained at 4.2×10^{-4} s^{-1} under compression. Open symbols indicate that serrated yielding was observed.*

The effects of temperature and strain rate were also investigated by performing compression tests on Co-48Ti (casting II) and stoichiometric CoTi (casting II) on a MTS 810 at strain rates ranging from 5×10^{-5} to 5×10^{-2} s^{-1} at temperatures from 300-1200 K.

Tensile tests were performed on single crystal specimens of stoichiometric CoTi at room temperature in both air and under vacuum and on polycrystalline specimens of stoichiometric CoTi in air at room temperature and 600K. Fracture occurred before yielding for all tensile specimens, so that no further tensile tests were performed.

Thin foils of polycrystalline stoichiometric CoTi (casting II) were prepared from specimens which had been plastically strained ~ 2% at 4.2×10^{-4} s^{-1} at 298 K, 600 K and 800 K and cooled by forced air after straining, which resulted in a cooling time of ~5 minutes. Three-millimeter diameter discs were electropolished in a Struers Tenupol in 30% nitric acid in methanol at 248 K at 12V, with a resulting current of 140 mA. The thin foils were washed in methanol, and examined using either a Philips CM30 TEM operated at 300 kV or a JEOL 2000FX TEM operated at 200 kV.

RESULTS AND DISCUSSION

It was shown previously that, unlike FeAl [12], quenching stoichiometric CoTi from elevated temperatures does not introduce sufficient vacancies to affect the mechanical behavior [13]. The same result can be seen for Co-48Ti in Figure 2 where water quenching, air cooling, or annealing for 5 days at 673K had no apparent effect on either yield stress or work hardening rate. Based on this result, along with previous results on stoiochiometric CoTi [13], it appears that in B2 CoTi alloys there is not a sufficient concentration of vacancies present at intermediate temperatures to account for the yield anomaly. Thus, it is concluded that the George-Baker vacancy hardening model [11] is not applicable to these alloys.

For Co-48Ti (casting II) it can be seen in Figure 3 that at intermediate temperatures the yield stress increases with increasing temperature reaching a peak at approximately 650 K. Serrated yielding is consistently seen at temperatures around the temperature of the peak yield stress. The serrated yielding observed in these alloys is consistent with the suggestion that dynamic strain ageing is responsible for the yield anomaly.

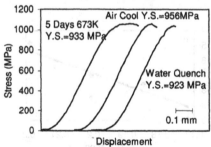

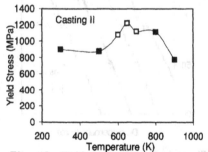

Figure 2: *Stress versus displacement curves for 300 K compression tests of Co-48Ti that underwent three different heat treatments after a 30 minute hold at 1473 K.*

Figure 3: *Yield stress versus temperature for compression tests of Co-48Ti strained at 4.2×10^{-4} s^{-1}. Open symbols indicate that serrated yielding was observed.*

In Figure 4 the serrations seen in stoichiometric CoTi appear as sudden drops in the flow stress, that drop below the general shape of the stress strain curve. There is no abrupt rise in the stress prior to the serration, suggesting that they are associated with dislocation unlocking [14,15]. Considerable effort has been devoted to relating changes in the critical strain, ε_c, (plastic strain at which serrated yielding begins) to changes in temperature and strain rate in order to gain insight to the type of impurity atoms, either interstitial or substitutional, and the migration energy [16,17]. Figure 5a is a plot of ln (ε_c) versus ln (strain rate) at constant temperature, and Figure 5b is a plot of ln (ε_c) versus (temperature)$^{-1}$ at constant strain rate for stoichiometric CoTi (casting II). It is evident that the data is largely scattered and irreproducible providing no meaningful information regarding the type of impurity responsible for the serrated yielding.

Figures 6 and 7 show the dislocation structures of stoichiometric CoTi (casting II) strained at room temperature, 600 K and 800 K. These correspond to temperatures well below the yield anomaly, at the peak yield strength temperature, and at high temperatures where the yield strength drops off rapidly. Analysis showed only <100> dislocations at all temperatures, consistent with previous observations [1,2]. One feature worth noting is that the samples deformed at room temperature and 800 K appeared to contain an increased density of pure edge dislocations. This is most clearly seen in Figure 7c where the residual contrast (i.e. when, $g \cdot b = 0$ but, $g \cdot (b \times u) \neq 0$ where g is the diffraction vector, b is the

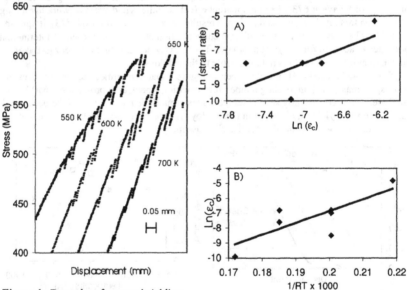

Figure 4: Examples of serrated yielding in stoichiometric CoTi at several temperatures.

Figure 5: Plots of (a) ln (strain rate) versus ln(ε_c), and (b) ln(ε_c) versus 1/RT.

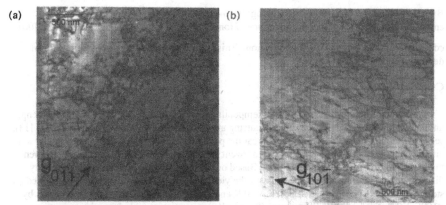

Figure 6: *Transmission electron micrographs of stoichiometric CoTi deformed to ~2% plastic strain at (a) 298K and (b) 600K. Note that only <100> dislocations are present.*

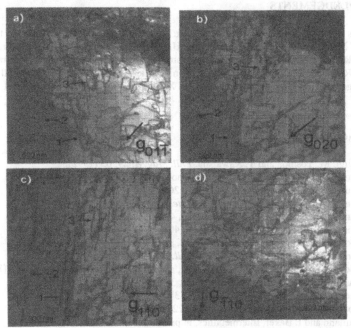

Figure 7: *Transmission electron micrographs of stoichiometric CoTi deformed to ~2% plastic strain at 800 K. Edge dislocations in (a) and the residual contrast in (b) and (c) are arrowed. There is no residual contrast in (d). Only <100> dislocations are present.*

Burger's vector and u is a unit vector along the line direction) from the edge dislocations emphasizes the density of these dislocations. From Figures 7a-d it is evident that these are [001] edge dislocations with a [$\bar{1}$10] line direction. This feature was not observed in the specimen deformed at 600 K.

CONCLUSIONS

Compression tests on Co-48Ti at temperatures from 300 to 900 K exhibit a yield strength anomaly with a peak in yield strength occurring at ~650 K. Similar to stoichiometric CoTi [13], serrated yielding occurs at temperatures near the peak yield strength. Co-48Ti samples that were either water quenched from 1473 K or underwent a 5 day anneal at 673 K showed no difference in yield strength or work hardening rate. Based on these results it is concluded that vacancy-dislocation interactions are not the cause of the yield anomaly. TEM observations performed on stoichiometric CoTi deformed at 300 K, 600 K and 800 K confirm that deformation occurs by <100> dislocations at all temperatures. The results are consistent with dynamic strain-aging being the cause of the yield anomaly in B2 CoTi.

ACKNOWLEDGEMENTS

This work was funded by National Science Foundation grant DMR-9812211, by the Division of Materials Sciences and Engineering, U.S. Department of Energy, under contract DE-AC05-00OR22725 with UT-Battelle, LLC, and through the SHaRE Program under contract DE-AC05-76OR00033 with Oak Ridge Associated Universities. Dr. E.P. George is gratefully acknowledged for providing the castings, and Dr. X. Pierron is thanked for the chemical analyses.

REFERENCES

1. T. Takasugi, M. Yoshida and T. Kawabata, *Phil. Mag. A*, **29**, 65 (1992).
2. T. Takasugi, K. Tsurisake, O. Izumi, and S. Ono, *Phil. Mag. A*, **61**, 785 (1990).
3. T. Takasugi and O. Izumi, *J. Mater. Sci.*, **23**, 1265 (1988).
4. T. Takasugi, O. Izumi, and M. Yoshida, *J. Mater. Sci.*, **26**, 2941 (1991).
5. M. Nakamura and Y. Sakka, *J. Mater. Sci.*, **23**, 4041 (1988).
6. M. Yoshida and T. Taksugi, *Phil. Mag. A*, **68**, 401 (1993).
7. A. François and P. Veyssière, *Intermetallics*, **2**, 9 (1994).
8. V. Paidar, D.P. Pope, and V. Vitek, *Acta Metall.*, **32**, 435 (1984).
9. D. Shindo, M. Yoshida, B.T. Lee, T. Takasugi, and K. Hiraga, *Intermetallics*, **3**, 167 (1995).
10. T. Takasugi and S. Hanada, *Phil. Mag. A*, **71**, 347 (1995).
11. E.P. George and I. Baker, *Phil. Mag. A*, **77**, 737 (1998).
12. P. Nagpal and I. Baker, *Met. Trans. A*, **21A**, 2281 (1990).
13. M. Wittmann and I. Baker, Intermetallics, in press.
14. P. Rodriguez, *Bull. Mater. Sci*, **6**, 653 (1984).
15 B. Russel, *Philos. Mag.*, **8**, 615 (1963).
16. P.G. McCormick, *Acta Metal.*, **20**, 351 (1972).
17. A. Van Den Beukel, *Acta metal.*, **28**, 965 (1980).

Mat. Res. Soc. Symp. Proc. Vol. 646 © 2001 Materials Research Society

High-Temperature Strength of Ir-Based Refractory Superalloys

Yoko Yamabe-Mitarai
High-Temperature Materials 21st Project (HTM21), National Research Institute for Metals,
1-2-1 Sengen, Tsukuba, Ibaraki 305-0047, Japan

ABSTRACT

Ir-based refractory superalloys with an fcc and $L1_2$ two-phase structure similar to that in Ni-based superalloys but with considerably higher melting temperatures are proposed. First, the microstructure and strength are shown for several kinds of binary alloys. Then, the effect that the third element and the combination of Ni-Al with an fcc and $L1_2$ two-phase structure in binary alloys had on the lattice misfit, microstructure, and strength was investigated. The precipitation-hardening effect in Ir-based alloys was investigated and discussed in terms of precipitate morphology and lattice misfit. The creep properties of Ir-based alloys were also investigated.

INTRODUCTION

Interest in pure Ir and Ir-based alloys has increased during the past few decades. In March 2000, the International Symposium of Ir was held [1]. There, the mechanical properties, structures, oxidation behavior, fabrication, processing, refining, and chemistry of Ir and Ir-based alloys were presented. As application fields, medical implants, ignition devices, electrodes, thermometers, and combustion chambers for rocket propulsion were suggested in addition to the classical applications of these alloys as crucibles and catalysts. In Japan, Ir demand has increased since 1995 due to its increasing variety of applications. Ir is used for automobile catalysts, electrodes in the chemical industry, crucibles, thermocouples, and rocket parts [2]. At the beginning of 2000, a symposium concerning platinum group metals and alloys was held in Tokyo. Here, in addition to the above applications, new attempts, structural materials, coating, and ferroelectric memory capability (so-called F-RAM) were presented [3].

Ir is used in many application fields because it has high melting temperature (2447°C), the most stable element for corrosion [4], a modulus with the highest elasticity (570GPa) at room temperature [5], and the second-highest elevated-temperature strength [6]. On the other hand, the disadvantages of Ir are its highest density (22.7 g/cm^3), brittleness (unlike typical fcc metals), and limitation of supply in the world (4 tons/year).

Investigation of the mechanical properties of Ir was started in 1960 and showed that polycrystalline Ir cleaves at grain boundaries similarly to brittle materials after insignificant preliminary deformation but unlike typical fcc metals [6]. On the other hand, the mechanical behavior of a single crystal of Ir, which was shown first in 1961 [7], differs from that of brittle materials. Generally, brittle materials cannot be deformed plastically in either polycrystalline and single forms. However, Ir single crystal showed large elongation (80%) at room temperature; moreover, it showed brittle transcrystalline fractures [8]. This suggests that Ir has both a brittle and a ductile manner. Thus, the fracture behavior and mechanism of Ir have attracted the attention of many researchers and were enthusiastically investigated by a Russian group in the 1990s [9-16]. Their research is summarized in a book concerning Ir [17].

A solution to the brittle fracture of Ir was identified by Oak Ridge National Laboratory [18, 19]. The addition of Th with a ppm level changed the fracture mode from intergranular to transgranular and improved the ductility. Since then, the fracture behavior of Ir alloys, mainly the Ir-0.3%W alloy, and the effect of microalloying elements such as Th, Al, Fe, Ni, Rh, Ce, Lu, and Y have been investigated [for example, 19-26]. The Ir-0.3%W alloy containing Th below 60 ppm (designated as DOP-26) is currently used as a fuel-cladding material in radioisotope thermoelectric generators - the major source of onboard electric power in interplanetary spacecraft. The research conducted at the Oak Ridge National Laboratory was reviewed by E. P. George [27]. The influence of trace impurities such as C, Si, and Fe on creep properties and fracture behavior during creep was also studied by a German group in the 1990s [28-29].

As another trial for Ir, several intermetallics were considered as high-temperature materials. Fleisher et al. suggested a systematic way to find new materials from the point of view of the melting temperature, elastic moduli, and density [30]. They found that some of the platinum-group metal-based intermetallics, such as IrNb with an $L1_0$ structure [31, 32] and RuAl [33-36] and RuTa [37] with a B2 structure, have the potential to be high-temperature materials. Based on their work, some other intermetallics, including IrAl (B2) and Ir_3Nb ($L1_2$), were also investigated. IrAl has been noted because of its good oxidation resistance. A binary phase diagram of Ir-Al was investigated in detail [38]. The oxidation properties [39, 40] and the effect of addition of B [41] and Ni [42] on the oxidation properties of IrAl were investigated. The compression strength [43] and creep behavior [44] of IrAl were also investigated, and it was then shown that their mechanical properties at high temperature were superior than those of other intermetallics such as B2-NiAl, $L1_0$-TiAl, and $L1_2$-Ni_3Al. For Ir-based $L1_2$ intermetallics, the mechanical properties of Ir_3Nb and Ir_3Zr were investigated [45-47]. In addition to the mechanical properties, the thermal expansion coefficient and thermal conductivity in Ir-based $L1_2$ intermetallics were also studied systematically [48].

While these research projects concentrated on single-phase alloys, we noted two-phase alloys consisting of fcc and $L1_2$ phases because it is known that alloys with a single phase are less resistant to creep deformation than alloys with a two-phase even though they show high strength during tensile testing in Ni-based superalloys [49]. To develop a new generation of high-temperature materials, we have proposed a new class of alloys using Ir with an fcc and $L1_2$ coherent two-phase structure similar to that in Ni-based superalloys and named them "refractory superalloys" [50, 51]. The mechanical properties of some Ir-based binary alloys have been investigated. It was shown that the strengthening mechanism is precipitation hardening and that precipitation hardening depends on the lattice misfit, which determines precipitate shape. While these binary alloys showed superior strength at 1200 °C, the strength above 1500°C was not very remarkable considering the high melting temperature [52, 53]. We also found that binary alloys failed mostly as a result of intergranular fracture, in the same way as pure Ir [54]. Furthermore, considering their application, the density of Ir-based alloys should be decreased. To solve these problems, we have performed two trials. First, we have investigated the influence of a third element on the strength at high temperature, fracture mode, and ductility at room temperature [55, 56]. Second, we combined two kinds of alloys, i.e., a Ni-Al alloy (ductile) and an Ir-based binary alloy (high strength at high temperature) [57-59]. Here, both alloys have an fcc and $L1_2$ two-phase structure. Through these trials, we examined whether the fcc and $L1_2$ two-phase structure is formed in each alloy. In this paper, we mainly review the relationship between the microstructure, lattice misfit, and strength behavior.

EXPERIMENTAL PROCEDURE

The alloys tested were prepared as 20g button ingots by arc melting in an argon atmosphere. Cylindrical samples that were 3mm in diameter and 6mm in height were cut from these ingots, heat-treated to produce $L1_2$ precipitates, and then tested between room temperature and 1800 °C in a compression condition. Heating at 1200 °C was carried out with these samples encapsulated in quartz tubes filled with argon gas followed by water quenching. For heat treatments above 1200 °C, the whole sequence of heating and cooling was carried out in vacuum within a furnace, using a tungsten mesh heater to prevent damage of the heater by induction of air above 100 °C. The alloy composition and heat treatment are summarized in Table 1. To observe the precipitate shape and deformation structure, thin discs were cut from the heat-treated samples and the deformed sample, respectively, and then ion-milled for observation by transmission electron microscopy (TEM). Compression tests were carried out in an air atmosphere up to 1200 °C and in an argon atmosphere at 1500 and 1800 °C. At high temperatures, samples were kept at testing temperatures for 15 min before loading. The initial compressive strain rate was 3.0×10^{-4}/s for every test. Compression creep tests were also carried out at 1500 and 1650°C under 137MPa.

Table 1 Alloy composition (at%), component phase, heat treatment, melting temperature (°C), and density (g/cm^3) of tested alloys.

Alloy composition (at%)	Heat treatment (°C - hours)	Melting temperature (°C)	Density (g/cm^3)
Ir-10Nb (fcc)	2000 - 72	-	-
Ir-12, 15Nb (fcc+$L1_2$)	1200 - 72, 168	2400	about 20.0
Ir-25Nb ($L1_2$)	2000 - 72	-	-
Ir-2Zr (fcc)	1800 - 1	-	-
Ir-12, 15Zr (fcc+$L1_2$)	1200 - 10, 168	2120	about 19.0
Ir-25Zr ($L1_2$)	2000 - 72	-	-
Ir-15Nb-5Ni	1300 - 168	>2000	19.4
Ir-15Nb-10Ni	1300 - 168	>2000	18.7
Ir-15Nb-20Ni	1300 - 168	>2000	17.5
Ir-15Nb-30Ni	1300 - 168	>1500	16.3
Ir-15Nb-8Ni-2Al	1600 - 24	about 2100	18.7

RESULTS AND DISCUSSION

Precipitation hardening

The compressive strength of the Ir-Nb alloys is plotted versus the composition of Nb in Figure 1. In the fcc single phase, the strength increased by the addition of Nb, suggesting a solid solution effect. By further addition of Nb, the strength increased in the fcc and $L1_2$ two-phase region. This is attributed to precipitation hardening. Here, we can see that the strength of the two-phase alloys was higher than those of fcc or $L1_2$ single-phase alloys. This also showed the precipitation-hardening effect clearly. The strength decreased with increasing the testing temperature. At 1800 °C, no clear difference between the strengths in single-phase and two-phase alloys was observed, which suggests that precipitation hardening at 1800°C was not very effective. The precipitation hardening up to 1200°C shown by the arrow in Figure 1 is plotted as a function of the volume fraction of the precipitates in Figure 2. The precipitation hardening was higher in the Ir-Zr alloys than in the Ir-Nb alloys at each testing temperature.

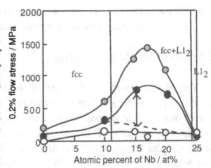

Figure 1. Concentration dependence of compressive strength of Ir-Nb alloys. The shadow, solid, and open symbols show the compressive strength at room temperature, 1200, and 1800°C, respectively.

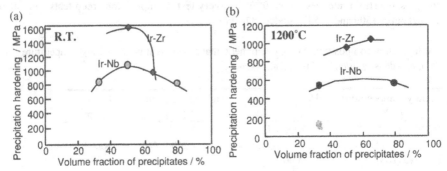

(a)

(b)

Figure 2. Precipitation hardening of Ir-Nb and Ir-Zr alloys as a function of the volume fraction of precipitates at (a) room temperature and (b) 1200°C.

The precipitate shapes of the Ir-Nb and Ir-Zr alloys are shown in Figure 3. Cuboidal $L1_2$ precipitates (bright contrast) were formed and arrayed orderly in the Ir-Nb alloy, similarly to Ni-based superalloys (Figure 3a). In the Ir-Zr alloy, a completely different microstructure was observed. Plate- or rod-like precipitates (bright contrast) arrayed orderly and formed a three-dimensional maze structure (Figure 3b). The habit plane of these cuboidal or plate-like precipitates was the {001} planes. If the precipitates are incoherent with the matrix, the precipitate shape is an oblate spheroid that balances interfacial energy and elastic strain energy. We therefore consider these precipitates to be coherent because they are surrounded by {100} planes. In addition to these coherent microstructures, another microstructure was also observed in the Ir-Zr alloy (Figure 3c). The precipitate shape is not plate or cube anymore, and several misfit dislocations were formed clearly at the interfaces. This structure is clearly a semi-coherent structure. This difference of the microstructure is caused by the lattice misfit. Here, the lattice misfit of the Ir-Nb alloy is 0.33, while that of the Ir-Zr alloy is 2.19% [60]. When the lattice misfit is large, the strain energy at the interface becomes large; as a result, the two-fold shape is more stable at the energetic point [61]. Furthermore, to reduce the strain energy, misfit dislocation is often introduced at the interface and then changes to a semi-

Figure 3. Precipitate shape of (a) Ir-15at%Nb and (b) Ir-15at%Zr alloys. These dark-field images were taken from superlattice reflection from the $L1_2$ phase. (c) Bright-field image of an Ir-15at%Zr alloy.

coherent structure.

The deformation structures of these alloys tested at 1200°C are shown in Figure 4. No clear deformation structure was observed in the Ir-Nb alloy (Figure 4a). On the other hand, a tangle of dislocations at the interface and shearing precipitates by dislocations were clearly observed in the Ir-Zr alloys (Figure 4b). This shows that the interface has high resistance for dislocations moving in the fcc matrix. Dissociation into super-partial dislocations bounding APBs and SISFs were observed in $L1_2$ precipitates [62]. In a previous study, we found that fine fcc precipitates were formed in the cuboidal $L1_2$ precipitates after heat treatment at 1500 °C [52]. The strengths of the alloy with an fcc phase inside $L1_2$ precipitates and the alloy with solid $L1_2$ precipitates (both precipitate size and volume fraction are the same) were almost equivalent. If the alloy with an fcc phase inside the $L1_2$ precipitates is stronger than the alloy with solid $L1_2$ precipitates, the shearing mechanism will be expected because there is more interface in the alloy with an fcc phase inside the $L1_2$ precipitates than in the alloy with a solid $L1_2$ phase. Thus, we have concluded that the Ir-Nb alloy was deformed by a bypass mechanism although there was no evidence in TEM observation.

Effect of Ni and combination with a Ni-Al alloy

Ni was chosen to investigate the effect of a third element on strength at high temperature because Ni has an fcc structure, just like Ir, and is fully miscible with Ir in the Ir-Ni binary system. We expect that it is difficult to form a third phase and keep the fcc and $L1_2$ two-phase structure by addition of Ni. An experimentally determined phase diagram at 1300 °C is shown in Figure 5a. Ni is distributed in the fcc phase more than in the $L1_2$ phase. The fcc and $L1_2$ phase expanded by addition of Ni. However, when too much Ni was added, a third phase (δ phase), that is, δ-$(Ir, Ni)_{11}Nb_9$, was formed, and an fcc $+L1_2+\delta$ three-phase region appeared. The phase diagram shows that plenty of Ni cannot be added in high composition for alloys with a high volume fraction of $L1_2$ precipitates because only a small amount of Ni is distributed in the $L1_2$ phase. If we want to add a third element to the Ir-Nb system as much as

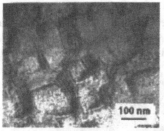

Figure 4. Typical deformation structure of (a) Ir-15at%Nb alloys plastically deformed 3.0% at 1200°C and (b) an Ir-15at%Zr alloy plastically deformed 3.2% at 1200°C.

possible, we must chose an element, which can be distributed in both of the $L1_2$ and fcc phases.

Next, we conducted another trial in which we combined an Ir-Nb alloy with high strength at high temperature and a Ni-Al alloy with good ductility and small density. Here, both alloys have the fcc and $L1_2$ two-phase structure. If the two-phase region connects from the Ir side to the Ni side, we can obtain an alloy with suitable mechanical properties, ductility, and density. An experimentally determined phase diagram is shown in Figure 5b. Contrary to our expectations, the fcc and $L1_2$ two-phase region was not connected from the Ir side to the Ni side. Instead, a three-phase region, fcc+$L1_2$-Ir_3Nb+$L1_2$-Ni_3Al, appeared.

The microstructures of these alloys are shown in Figures 6 and 7. In ternary Ir-Nb-10Ni(at%) alloys, the precipitate size was about 100-200nm, and a semi-coherent structure was formed, that is, a lot of misfit dislocations formed at the interface of the $L1_2$ precipitates and the matrix (Figure 6a, b). However, in the 20Ni alloy, $L1_2$ precipitates were larger than 500nm, and the precipitate shape was irregular (Figure 6c). When the δ phase was formed in

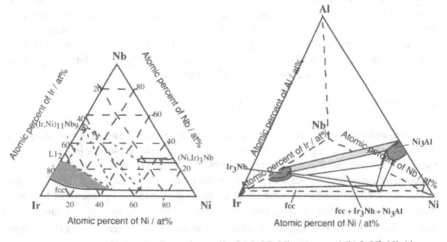

Figure 5 Phase diagram determined experimentally. (a) Ir-Nb-Ni system and (b) Ir-Nb-Ni-Al system at 1300°C.

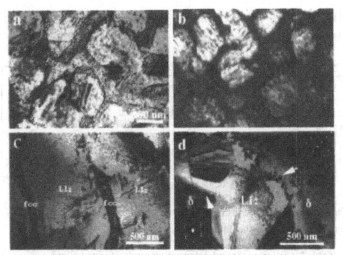

Figure 6. Typical micrographs illustrating the $L1_2$ precipitate shape in $Ir_{85-X}Nb_{15}X$ alloys with X equal to (a) 10, bright-field image (BF), (b) 10, dark-field image, (c) 20, BF, and (d) 30 BF.

the 30Ni alloy, a large phase (more than 500nm) with an irregular shape was formed (Figure 6d). This shows that to form a coherent (or semi-coherent) two-phase structure, Ni can be added up to 10% for Ir-15at%Nb alloys. If too much Ni is added, precipitates change to an incoherent large phase, as shown in 20Ni alloys. In quaternary alloys, a two-phase coherent structure was observed in the two-phase alloy, that is, the fcc and $L1_2$-Ir_3Nb in the Ir side and the fcc and $L1_2$-Ni_3Al in the Ni side. In the three-phase region, rod-like Ir_3Nb with a length of a few μm and fine Ni_3Al precipitates with a size of 200nm were formed in the fcc phase.

The strength behavior of the binary, ternary, and quaternary alloys is plotted as a function of lattice misfit of the fcc and $L1_2$ phases, which is $\delta=(a_p-a_m)/a_m\times100$, in Figure 7a. Here, a_p and a_m are the lattice parameters of the precipitate and matrix determined by X-ray analysis [60]. The lattice misfit decreased to a negative value by the addition of Ni. This shows that the lattice parameter of the precipitates became smaller than that of the fcc matrix by addition of Ni. On the other hand, the lattice misfit increased slightly by addition of Ni-Al. However, the absolute value of the lattice misfit was almost equivalent by addition of Ni and Ni-Al. Below 1200°C, the strength increased by addition of Ni-Al to the binary alloy, and the effect was clearer than by addition of simple Ni. The strength decreased drastically above 1200°C, and the strengths at 1500 and 1800°C were almost equivalent between 100 and 230 MPa. The effect of single Ni was clearer than that by addition of Ni-Al at 1800°C. Precipitation hardening of these alloys up to 1200°C is plotted as a function of the lattice misfit in Figure 8b. Precipitation hardening was highest in the binary alloys. This suggests that the strengthening effect by addition of Ni or Ni-Al is mainly solid-solution hardening. Comparing these three alloys, precipitation hardening was smaller in the alloy with a large absolute lattice misfit. On the other hand, comparing Ir-Nb and Ir-Zr alloys (Figure 2), precipitation hardening was larger in the Ir-Zr alloy with a large lattice misfit. The precipitation-hardening effect appeared in a different manner in Ir-Nb and Ir-Zr binary alloys and between Ir-Nb-based alloys.

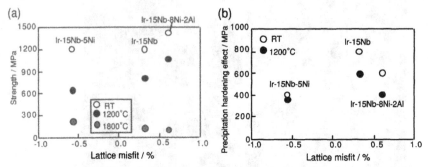

Figure 7. (a) Strength behavior and (b) precipitation hardening of binary, ternary, and quaternary alloys as a function of the lattice misfit.

Table 2 Lattice misfit and deformation mode of tested alloys.

Alloy	Lattice misfit	Deformation mode
Ir-15Nb	0.33	Bypass
Ir-15Nb-5Ni	-0.58	Shearing
Ir-15Nb-8Ni-2Al	0.6	Shearing
Ir-12Zr	2.19	Shearing

To understand the difference, the deformation structure was observed in ternary (Figure 8) and quaternary alloys (Figure 9). In both alloys, shearing of precipitates was clearly observed. The lattice misfit and deformation mode of tested alloys are summarized in Table 2. When the absolute lattice misfit was large, the deformation mode was shearing. Only the Ir-Nb binary alloy did not show a shearing mode. Comparing Ir-Nb and Ir-Zr alloys, precipitate shape changed from cube in the Ir-Nb alloy to plate or rod-like in the Ir-Zr alloy because of a large lattice misfit. The bypass mode seemed to be difficult in the three-dimensional maze structure in the Ir-Zr alloy. No other mode except for the shearing mode seems possible in that case. However, shearing precipitates with a large lattice misfit will be difficult because of the large strain energy in the coherent interface and the many misfit dislocations in the semi-coherent structure. On the other hand, among Ir-Nb-based alloys, the lattice misfit difference by addition of Ni or Ni-Al was smaller than that between the Ir-Nb and Ir-Zr alloys. Thus, the resistance for shearing was not so high, and the precipitation-hardening effect was smaller than that of the Ir-Zr alloy. The results in Figure 8 indicate that the precipitation-hardening effect by the bypass mode in binary alloys is larger than that by the shearing mode in ternary and quaternary alloys. However, the reason is not clear at the moment.

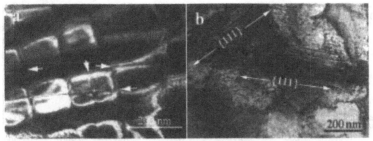

Figure 8. Typical deformation structures of Ir-15Nb-10Ni alloys plastically deformed about 3% at room temperature.

Creep properties and possibilities as high-temperature materials

Finally, the creep properties of these alloys are shown in Figure 10. Above 1500°C, the strengths in all samples were not high enough. However, creep strain at 1500°C was below 2%, and tertiary creep was not observed after 300 hours for coherent structures in Ir-Nb and Ir-Zr binary alloys. This suggests that these binary alloys can be used if the applied stress is under 0.2% flow stress. However, at 1650°C, the binary alloy deformed quickly, and tertiary creep was clearly observed. Creep strain reached 5% only after 24 hours. On the

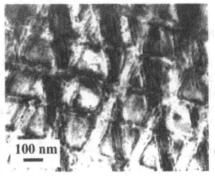

Figure 9. Typical deformation structure of Ir-Nb-Ni-Al alloys plastically deformed about 10% at 1500°C

other hand, ternary alloy with 5at% Ni showed high creep resistance and deformed only 1.8% after 200 hours. This suggests that Ni addition is very effective to improve creep properties although it is not effective for strength above 1500°C. Creep life can be improved drastically by addition of Ni. Another important factor when selecting high-temperature materials is the melting temperature. As shown in Table 2, although the melting temperature decreased slightly by addition of Ni or Ni-Al, it was still above 2000°C. To reduce the density, the added amount of Ni or Ni-Al was not enough.

CONCLUSIONS

We have proposed a new class of superalloys, namely, "refractory superalloys" based on Ir, which are defined as alloys with fcc and $L1_2$ coherent structures similar to Ni-based superalloys and with considerably higher melting temperatures. The strength behavior of binary alloys and the effect of Ni and Ni-Al addition on the microstructure and strength behavior were investigated. The fcc and $L1_2$ two-phase structure remained by addition of Ni or Ni-Al with optimum composition. The addition of Ni and Ni-Al improved the strength at 1200°C, but not above 1500°C. The creep properties drastically improved by addition of Ni.

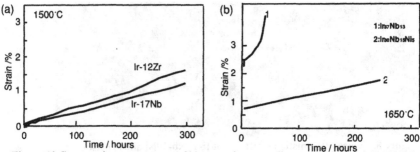

Figure 10 Compressive creep curves of binary and ternary alloys at (a)1500 and (b)1650°C under 137MPa.

ACKNOWLEDGMENTS

I would like to acknowledge the help of Dr. Y. Gu and Dr. X. Yu, who did part of this research. I thank Dr. Y. Ro and Mr. S. Nakazawa for their technical support. I am also grateful to Mr. S. Nishikawa and Mr. T. Maruko of Furuya Metal Co., LTD. for supporting Ir. Finally, this research was conducted under the auspices of the High-Temperature Materials for 21st centry (HTM 21) Project.

REFERENCES

1.   Proc. Inter. Symp. "Iridium," edited by E. V. Ohriner, R. D. Lanam, P. Panfilov, and H. Harada, (TMS, Nashville, USA, 2000).
2.   T. Maruko, Proc. Inter. Symp. "Iridium," edited by E. V. Ohriner, R. D. Lanam, P. Panfilov, and H. Harada, (TMS, Nashville, USA, 2000) 239.
3.   Proc. Symp. "Present Status and Future of Platinum Group Metals and Alloys," (JIM, Tokyo, Japan, 2000, in Japanese).
4.   Metals Handbook, vol. 2, 9th edition (ASM, Metals Park, OH, 1979).
5.   International Tables of Selcted Constants, 16, Metals, Thermal and Mechanical Data.
6.   B. L. Mordike and C. A. Brookes, Platinum Metals Review, **4**, 94 (1960).
7.   R. W. Douglass, A. Krier, and R. I. Jafee, Batelle Memorial Institute, Report NP-10939, August (1961).
8.   C. A. Brookes, J. H. Greenwood, and J. L. Routbout, J. Appl. Phys., **39**, 2391 (1968).
9.   A. Yermakov, P. Panfilov, and R. Adamsdku, J. Matter. Sci. Lett., **9**, 696 (1990).
10. P. Panfilov, A. Yermakov, and G. Baturin, J. Matter. Sci. Lett., **9**, 1162 (1990).
11. P. Panfilov, A. Yermakov, V. Dimitriev, and N. Timofeev, Platinum Metals Rev., **35**, 196 (1991).
12. P. Panfilov, G. Baturin, and A. Yermakov, Int. J. Fracture, **50**, 153 (1991).
13. P. Panfilov, V. Novgorodov, and G. Baturin, Mat. Sci. Lett., **11**, 229 (1992).
14. P. Panfilov, V. Novgorodov, and A. Yermakov, J. Mater. Sci. Lett., **13**, 137 (1994).
15. R. Adamesku, S. Grebenkin, A. Yermakov, and P. Panfilov, J. Mater. Sci. Lett., **13**, 865 (1994).

16. A. V. Ermakov, S. M. Klotsman, V. G. Pushin, A. N. Timofeev, V. N. Kaigorodov, P. Panfilov, and L. I. Yurchenko, Scripta Mater., **42**, 209 (2000).

17. N. I. Timofeev, A. Yermakov, V. Dmitriev, and P. Panfilov, The Metallurgy and Mechanical Behavior of Iridium (Urals Branch of Russian Academy of Science, Ekaterinburg, 1996, in Russian) (1996), ISBN 5-7691-0673-5.

18. C. T. Liu and H. Inouye (Report ORNL-5290, Oak Ridge National Laboratory, October 1977).

19. C. T. Liu, H. Inouye, and A. C. Schaffhauser, Met. Trans. A, **12A**, 993 (1981).

20. S. S. Hecker, D. L. Rohr, and D. F. Sten, Met. Trans. A, **9A**, 481 (1978).

21. D. L. Rohr, L. E. Murr, and S. S. Hecker, Met. Trans. A, **10A**, 399 (1979).

22. C. L. White, R. E. Clausing, and L. Heatherly, Met. Trans. A, **10A**, 683 (1979).

23. D. E. Harasyn, R. L. Heestand, and C. T. Liu, Mat. Sci. and Eng., **A187**, 155 (1994).

24. A. N. Gubbi, E. P. George, E. K. Ohriner, and R. H. Zee, Met. Trans. A, **28A**, 2049 (1997).

25. A. N. Gubbi, E. P. George, E. K. Ohriner, and R. H. Zee, Acta Mater., **46, 3**, 893 (1998).

26. C. G. McKamey, E. P. George, E. H. Lee, E. K. Ohriner, L. Heatherly, and J. W. Cohron, Scripta Mater., **42**, 9 (2000).

27. E. P. George and C. T. Liu, Proc. Inter. Symp. "Iridium," edited by E. V. Ohriner, R. D. Lanam, P. Panfilov, and H. Harada, (TMS, Nashville, USA, 2000) 3.

28. D. Lupton and B. Fischer, (Platinum Group Metals Seminar, IPMI, Philadelphia, 1995) 151.

29. B. Fishcer, A. Behrends, D. Freund, D. Lupton, and J. Merker, Proc. Inter. Symp. "Iridium," edited by E. V. Ohriner, R. D. Lanam, P. Panfilov, and H. Harada, (TMS, Nashville, USA, 2000) 15.

30. R. L. Fleischer, J. of Mat. Sci., **22**, 2281 (1987).

31. R. L. Fleischer, R. D. Field, K. K. Denke, and R. J. Zabala, Met. Trans. A, **21A**, 3063 (1990).

32. R. L. Fleischer and R. J. Zabala, Met. Trans. A, **21A**, 2709 (1990).

33. R. L. Fleischer, R. D. Field, and C. L. Briant, Met. Trans. A, **22A**, 403 (1991).

34. R. L. Fleischer and D. W. McKew, Met. Trans. A, **24A**, 759 (1993).

35. R. L. Fleischer, Acta Metall. Mater., **41, 3**, 863 (1993).

36. R. L. Fleischer, Acta Metall. Mater., **41, 4**, 1197 (1993).

37. R. L. Fleischer, R. D. Field, and C. L. Briant, Met. Trans. A, **21A**, 129 (1991).

38. P. J. Hill, L. A. Cornish, and M. J. Wocomb, J. of Alloys and Compounds, **280**, 240 (1998).

39. H. Hosoda, T. Kingetsu, and S. Hanada, The Third Pacific Rim Inter. Conf. On Advanced Materials and Processing (PRICM-3), edited by M. A. Iman, (Pittsburgh, P: TMS, 1998), 2379.

40. H. Hosoda, S. Watanabe, and S. Hanada, High-Temperature Ordered Intermetallic Alloys VIII, edited by E. P. George et al., 552, (MRS, 1999), KK8.31.1.

41. I. M. Wolff and P. J. Hill, Proc. Inter. Symp. "Iridium," edited by E. V. Ohriner, R. D. Lanam, P. Panfilov, and H. Harada, (TMS, Nashville, USA, 2000) 259.

42. H. Hosoda, S. Miyazaki, S. Waanabe, and S. Hanada, Proc. Inter. Symp. "Iridium," edited by E. V. Ohriner, R. D. Lanam, P. Panfilov, and H. Harada, (TMS, Nashville, USA, 2000) 271.

43. H. Hosoda, T. Takasugi, M. Takehara, T. Kingetsu, and H. Masumoto, Mat. Trans. JIM, 38, 10, 871 (1997).

44. A. Chiba, T. Ono, X. G. Li, and S. Takahashi, Intemetallics, 6, 35 (1998).

45. M. Bruemmer, J. Brimhall, and C. H. Heneger, Jr., Mat. Res. Soc. Symp. Proc. 194, (MRS, 1990), 257.

46. A. M. Gyurko and J. M. Sanches, Mat. Sci. Eng., A170, 169 (1993).

47. Y. Yamabe-Mitarai, M-H. Hong, Y. Ro, and H. Harada, Phil. Mag. Let., 79, 9, 673 (1999).

48. S. Miura, Y. Terada, and T. Mouri, Proc. Symp. "Present status and future of platinum group metals and alloys," (JIM, Tokyo, Japan, 2000, in Japanese) 21.

49. Y. Ro, Y. Koizumi, and H. Harada, Mat. Sci. Eng., A223, 169 (1997).

50. Y. Yamabe, Y. Koizumi, H. Murakami, Y. Ro, T. Maruko, and H. Harada, Scripta Mater., 35, 2, 211 (1996).

51. Y. Yamabe-Mitarai, Y. Ro, T, Maruko, and H. Harada, Metall. Mater. Trans. A, 29A, 537 (1998).

52. Y. Yamabe-Mitarai, Y. Gu, Y. Ro, S. Nakazawa, T. Maruko, and H. Harada, Scripta Mater., 41, 3, 305 (1999).

53. Y. Yamabe-Mitarai, Y. Gu, Y. Ro, S. Nakazawa, T. Maruko, and H. Harada, Proc. Inter. Symp. "Iridium," edited by E. V. Ohriner, R. D. Lanam, P. Panfilov, and H. Harada, (TMS, Nashville, USA, 2000) 41.

54. Y. Gu, Y. Yamabe-Mitarai, Y. Ro, and H. Harada, Scripta Mater., 40, 11, 1313 (1999).

55. Y. Gu, Y. Yamabe-Mitarai, Y. Ro, T. Yokokawa, and H. Harada, Scripta Mater., 39, 4, 6, 723 (1998).

56. Y. Gu, Y. Yamabe-Mitarai, Y. Ro, T. Yokokawa, and H. Harada, 30A, 2629 (1999).

57. X. Yu, Y. Yamabe-Mitarai, Y. Ro, and H. Harada, Scripta Mater., 41, 6, 651 (1999).

58. X. Yu, Y. Yamabe-Mitarai, Y. Ro, and H. Harada, Metall. Mater. Trans. A, 31A, 173 (2000).

59. X. Yu, Y.Yamabe-Mitarai, Y. Ro, and H. Harada, Intermetallics, 8, 619 (2000).

60. Y. Yamabe-Mitarai, Y. Ro, T. Maruko, T. Yokokawa, and H. Harada, Structural Intermetallics 1997, (TMS, edited by M. V. Nathal, R. Darolia, C. T. Liu, P. L. Martin, D. B. Miracle, R. Wagner, and M. Yamaguchi, 1997), 805.

61. M. E. Thompson, C. S. Su, and P. W. Voorhees, Act Metall Mater., 42, 6, 2107 (1994).

62. Y. Yamabe-Mitarai, Y. Ro, T. Maruko, and H. Harada, Scripta Mater., 40, 1, 109 (1999).

Mat. Res. Soc. Symp. Proc. Vol. 646 © 2001 Materials Research Society

High-Temperature Materials Based on Quaternary Ir-Nb-Ni-Al Alloys

X.Yu, Y.Yamabe-Mitarai, Y.Ro, S.Nakaza, and H.Harada
High-Temperature Materials 21 Project, National Research Institute for Metals
Sengen 1-2-1, Tsukuba, Ibaraki 305-0047, Japan

ABSTRACT

A novel method to develop new quaternary alloys with an fcc/L1$_2$ coherent structure is proposed. This paper reviews the development of quaternary Ir-Nb-Ni-Al alloys. The microstructure, lattice misfit, and compressive 0.2% flow stress of 15 kinds of alloys were investigated systematically. Two kinds of coherent structures, fcc/L1$_2$-Ir$_3$Nb and fcc/ L1$_2$-Ni$_3$Al, were observed in most alloys. Two two-phase structures, fcc+L1$_2$-Ir$_3$Nb and fcc+L1$_2$-Ni$_3$Al, were observed in Ir-rich and Ni-rich regions, respectively. The lattice misfits of quaternary Ir-Nb-Ni-Al alloys were higher than those of Ni- or Ir-base binary alloys. The compressive 0.2% flow stresses of quaternary alloys increased dramatically compared with those of Ni-base superalloys. The quaternary alloys located in the Ir-rich region were not only had higher strength but also better ductility than Ir-base binary alloys. The potential use of quaternary alloys is discussed.

INTRODUCTION

Ir-base alloys have attracted the attention of many researchers because of their excellent high-temperature strength and better oxidation resistance. Yamabe-Mitarai et al. [1,2] investigated Ir-base binary refractory superalloys systematically. The L1$_2$ precipitates presented a cuboidal shape, and the fcc/L1$_2$ coherent structure was very similar to that of Ni-base superalloys. Ir-Nb-Ni was used to try to improve the ductility and specific gravity of Ir-base alloys [3]. Ni was found to be beneficial to the properties of Ir-Nb alloys.

However, since improving the ductility and specific gravity is not enough, it was our goal to find a new simple way to improve the high-temperature strength of materials. In the present paper, a novel method to improve high-temperature capability and ductility is proposed [4,5]. Quaternary Ir-Nb-Ni-Al was prepared by combining two kinds of binary alloys, Ir-Nb and Ni-Al. Both of them have an fcc/L1$_2$ coherent structure. The two kinds of two-phase structures were expected to connect the Ni-rich Ir-rich compositions, and the two-phase coherent structure was expected to give a higher creep resistance at high temperature. Thus, the formation of an fcc/L1$_2$ coherent structure was one of our targets, but our final objective was to achieve a material with an excellent balance of high-temperature strength and ductility by combining the advantages of Ir-base alloys (high-temperature strength) and Ni-base alloys (good ductility and lower specific gravity).

In the last two years, we have conducted systematic research work on quaternary Ir-Nb-Ni-Al refractory superalloys. More than 15 quaternary alloys have been investigated systematically [6,7,8]. In this paper, we review the development of quaternary Ir-Nb-Ni-Al alloys based on new results and insights. The strength behavior and microstructural evolution of these alloys as well as their potential use as ultra high-temperature materials are discussed.

EXPERIMENTAL PROCEDURE

The investigated quaternary Ir-Nb-Ni-Al alloys were prepared by combining two kinds of binary alloys, Ir-Nb and Ni-Al, in different proportions. A detailed description can be found in Refs. 4 and 6. Figure 1 shows a schematic diagram of quaternary Ir-Nb-Ni-Al alloys and the locations of the investigated alloys. This research work was divided into four stages. The first stage was to combine Ir-Nb with Ni-Al in different proportions, for example, 25, 50, and 75at.% Ir-based alloys, respectively, and these alloys were identified as S050, S550, and S750. In Fig.1, these alloys are presented from the Ni-rich side to the Ir-rich side in order to determine whether an fcc/ $L1_2$ coherent structure can be formed in the quaternary alloys. The second stage

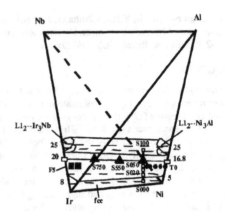

Fig.1 Location of investigated alloys in a schematic Ir-Nb-Ni-Al diagram

focused on the Ni-rich side. Some Ir-Nb and Ni-Al binary alloys were chosen to create quaternary alloys with the same proportion, 25%, of Ir-based alloys to Ni-based alloys. We tried to find an alloy with proper $L1_2$ volume fraction and morphology that would aid in the development of the next generation of Ni-based superalloys. These alloys were named S020, S030, S040, S100, and S000. The third stage of this investigation was conducted on 6 kinds of alloys, T0, T5, T10, T15, T20, and TN, which are located on the Ni-rich side. The fourth stage was carried out along the Ir-side. We wanted to determine whether there was also an fcc/$L1_2$ coherent structure in the quaternary alloys and evataute how the high temperature strength. Therefore, two kinds of alloys, F5 and F10, were investigated. The nominal composition and phase analysis results of all investigated alloys are listed in Table 1. In order to understand the effect of Ni and Al on the properties of the Ir-Nb alloy, 5at.% Ni or 5at.% Al was added to Ir-5Nb, Ir-15Nb, and Ir-25Nb, three kinds of base alloys, respectively corresponding to the fcc, fcc+$L1_2$, and $L1_2$ structures.

All these alloys were prepared as button ingots by arc melting in an argon atmosphere. Cylindrical samples were cut from the quaternary or ternary alloy ingots and heated at different temperatures for different alloys. The microstructures were observed by scanning electron microscopy (SEM) and transmission electron microscopy (TEM).

The lattice misfits of the investigated alloys were determined by the equation $\delta=(a_{L12}-a_{fcc})/a_{fcc}$. Here, a_{L12} and a_{fcc} are the lattice parameters of the $L1_2$ and fcc phases, respectively. Cylindrical samples that were 3 mm in diameter and 6 mm in height were cut from each alloy ingot for compression testing. The samples for the compression tests were heated before the tests. Compression tests were carried out at room temperature and 1200°C in air. The initial compressive strain rate was $3.0 \times 10^{-4} s^{-1}$.

RESULTS

Microstructural evolution

According to the phase analysis, quaternary Ir-Nb-Ni-Al alloys could be divided into three regions: a Ni-rich region (γ + Ni$_3$Al), a three-phase region (γ + Ir$_3$Nb + Ni$_3$Al), and an Ir-rich region (γ + Ir$_3$Nb).

In the Ni-rich region, the microstructure was very similar to that of Ni-base superalloys. It consisted of fcc and L1$_2$-Ni$_3$Al, and the fcc/L1$_2$ formed a coherent structure. The phase size of Ni$_3$Al was similar to that of γ' in Ni-base superalloys. A representative alloy is TN. The effect of cooling and heating rates on the microstructure of the TN alloy was investigated. The CCT curve of the TN alloy and the morphologies of γ' are shown in Fig.2. The L1$_2$ precipitating temperature decreased with the

Table 1 Nominal composition of quaternary Ir-Nb-Ni-Al alloys and phase analysis at 1400°C, at.%

Sample	Composition, at.%	Phase analysis		
		fcc	L1$_2$-Ir$_3$Nb	L1$_2$-Ni$_3$Al
F5	Ir-14.3Nb-4.0Ni-1Al	●	●	O
F10	Ir-13.5Nb-8.1Ni-1.9Al	●	●	O
S750	Ir-15Nb-20.8Ni-4.2Al	●	●	●
S550	Ir-10Nb-41.6Ni-8.4Al	●	●	●
S050	Ir-5Nb-62.4Ni-12.6Al	●	●	●
S100	Ir-7.5Nb-56.3Ni-18.7Al	O	●	●
S040	Ir-4.2Nb-65.3Ni-9.7Al	●	●	●
S030	Ir-3.6Nb-66.7Ni-8.3Al	●	●	●
S020	Ir-3.1Nb-68.3Ni-6.7Al	●	●	●
S000	Ir-2Nb-71.3Ni-3.7Al	●	O	O
T20	Ir-3.4Nb-69.6Ni-10.4Al	●	●	●
T15	Ir-2.5Nb-73.9Ni-11.1Al	●	●	●
T10	Ir-1.7Nb-78.3Ni-11.7Al	●	O	●
T5	Ir-0.84Nb-82.6Ni-12.4Al	●	O	●
T0	Ni-13Al	●	O	●
TN	Ir-2.2Nb-76.9Ni-9.96Al	●	O	●

Note: ●-----Yes, O-----No

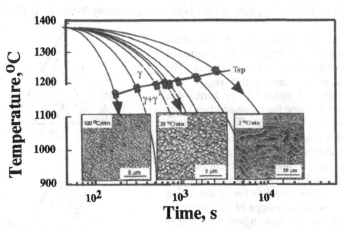

Fig.2 CCT curve of the TN alloy and morphological change of the γ' phase

increase of the cooling rate. The microstructural evolution shows that the morphology of Ni₃Al was fairly coarse, erratic, and heterogeneous in size for a cooling rate lower than 5°C/min. Between the cooling rates of 5 to 20°C/min, the microstructure was a mixture of cubic and coarsened precipitates. For cooling rates of 20-100°C/min, the morphology of Ni₃Al was more cubic in shape with the mean edge length ranging from 0.15 to 0.3 μm.

The TEM observation indicated that the fcc/L1₂ coherent structure formed in Ni- or Ir-rich-side alloys(2,5), respectively, while two kinds of fcc/L1₂ coherent structures were observed at the same time in some three-phase-region alloys. The typical morphologies of the L1₂ phase in the three parts of quaternary Ir-Nb-Ni-Al alloys are shown in Fig.3. In the Ni-rich-region alloy, the microstructure, which consisted of an fcc/Ni₃Al two-phase structure with a cuboidal bulk Ni₃Al shape, is very similar to that of Ni-base superalloys (Fig.3c). The interfacial dislocation shown in this picture came from the rapid cooling at 100°C/min. In the Ir-rich side, the morphology of the precipitates in the dendrite cores showed a special shape (Fig. 3c) in which the fcc phase existed

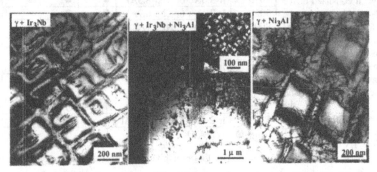

Fig.3 TEM observation of the L1₂ phase in quaternary Ir-Nb-Ni-Al alloys

in the L1₂ phase. The existence of the fcc phase inside L1₂ precipitates increased the number of interfaces between the fcc and L1₂ phases. The microstructure of the three-phase region was different from that of either Ni- or Ir-rich-side alloys. It consisted of one fcc phase and two kinds of L1₂ phases (γ + Ir₃Nb + Ni₃Al). The morphology of the L1₂ phases also exhibited a great difference, i.e., the Ir₃Nb exhibited a plate-like rather than cuboidal shape, and a coherent structure between the fcc and the two kinds of L1₂ phases was also observed (Fig.3b). The investigation of the lattice misfits revealed the lattice misfits of quaternary Ir-Nb-Ni-Al alloys from the Ni-rich (around 1.2%) to the Ir-rich region (0.6%). The lattice misfit gave a maximum value of around 6% in the three-phase region. Ir-rich alloys had smaller lattice misfits than Ni-rich alloys. Figure 4 gives a summary of the phase relationships in a quaternary phase diagram.

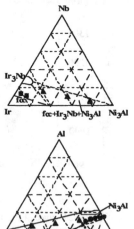

Fig.4 Summary of the phase relationship in a quaternary phase diagram

The lamellar structure appeared in the three-phase region. Both binary regions had an fcc/L1$_2$ coherent structure. An (fcc + L1$_2$) eutectic with a high melting temperature was observed in the Ir-rich side.

COMPRESSION TEST

Compression tests were carried out from room temperature up to 1800°C. The samples with compositions located in the Ni-rich and three-phase regions did not fracture during compression tests. One of these samples, located in the three-phase region (alloy S050), was compressed into a disc shape from a cylindrical shape when we tried to fracture it. Only the samples located in the Ir-rich side could be fractured during the compression tests. The fracture surface presented a sloping shape, indicating that the fracture behavior could be classified as ductile, with compressive fracture ductility around 20%.

Figure 5 gives the dependence of compressive 0.2% flow stress at 1200°C on the mole fraction of Ir-Nb alloy. In the Ni-rich and three-phase regions, the strength increased with the increase of the mole fraction of the Ir-base alloy. The highest strength in

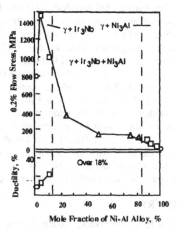

Fig.5 Mole fraction of Ir-Nb alloy dependence of 0.2% flow stress at 1200°C

the quaternary Ir-Nb-Ni-Al alloys appeared at the Ir-rich side with the 5at.% Ni-Al alloy (F5 alloy), in which the 0.2% flow stress at 1200°C was over 1400MPa with a fracture ductility of around 14.5%. All those strength behaviors demonstrated that the combination of an Ir-base alloy with a Ni-base alloy drastically improved the high-temperature strength compared with both Ni-base superalloys and Ir-base binary alloys, and the ductility was improved compared with Ir-base alloys. In particular, the Ir-base quaternary alloy, alloy F10, presented an excellent balance of high-temperature strength (over 1000MPa at 1200°C) and ductility (over 20%).

DISCUSSION

Quaternary Ir-Nb-Ni-Al alloys obtained by combining two kinds of binary alloys with an fcc/L1$_2$ coherent structure serve two purposes. One is that alloys consisting of a two-phase structure with an fcc/L1$_2$ coherent structure could be obtained; the other one is that a balance of high-temperature strength and ductility could be reached.

From the point of view of microstructure, the result exceeded our expectation. The two-phase regions did not connect from the Ni-rich region to the Ir-rich region. A three-phase region existed between the two-phase regions. However, an fcc/L1$_2$ coherent two-phase structure was indeed formed in both the Ni-rich and Ir-rich regions when the mole fraction of the minority binary alloy did not exceed 15at.%. The two minority elements played a solid solution-hardening role in the alloys and did not form a new phase. Moreover, the L1$_2$ precipitates showed a cuboidal shape similar to that of each of the corresponding binary alloys. However, when the mole fraction of the minority binary alloy exceeded 15at.%, two kinds of L1$_2$ phases having different shapes formed in the three-phase region. The coherent structure was also observed in the three regions.

Especially, two kinds of fcc/L1$_2$ coherent structures were observed at the same time in some three-phase alloys. These alloys were of larger lattice misfits. The fcc/L1$_2$ lamellar structures were observed in the three-phase region. It maybe said that lattice misfits of around 3% for γ/Ni$_3$Al and around 6% for γ/Ir$_3$Nb assist in the formation of the lamella structure [8].

Concerning the strength behavior, the quaternary alloys presented an excellent balance of high-temperature strength and ductility both in the Ni-rich and Ir-rich regions. In the Ni-rich region, the strength at 1200°C was over 100MPa, three times of that of the Ni-Al binary alloy and more than two times compared with that of commercial Ni-base superalloys MarM247 [9]. In the Ir-rich region, alloy F10 exhibited over 1000MPa high-temperature strength at 1200°C with 20% compressive ductility, and both the strength and ductility were larger than those of Ir-based binary and ternary alloys. The excellent balance of high-temperature strength and ductility provides more freedom to choose alloy compositions based on the properties required for each application. We can use alloys with a large concentration amount of Ir when ultrahigh-temperature materials are required (i.e., above 1600°C), but low density is not required. In the opposite case, i.e., when we need materials with low density but without ultrahigh-temperature capability, we can use alloys with high proportion of Ni. The next generation of ultrahigh-temperature materials may be developed from these kinds of quaternary superalloys.

SUMMARY

An fcc/L1$_2$ (Ir$_3$Nb, Ni$_3$Al) coherent structure was observed. There were fcc/L1$_2$ two-phase structure and a three-phase volume, fcc, Ni$_3$Al, and Ir$_3$Nb, between the two-phase regions. The alloys with a two-phase structure had small lattice misfits, while alloys with a three-phase structure had larger lattice misfits. A lattice misfit of 3% for fcc/Ni$_3$Al and 6% for fcc/Ir$_3$Nb may assist in the formation of the fcc/L1$_2$ lamella structure. The strength behavior of quaternary alloys exhibited an excellent balance of high-temperature strength and ductility. The Ir-rich quaternary alloys could reach over 1000 MPa at 1200°C with 20% ductility. Therefore, the quaternary Ir-Nb-Ni-Al alloys are promising candidates for development as ultrahigh-temperature materials.

REFERENCES

1. Y. Yamabe-Mitarai, Y. Ro, T. Maruko, and H. Harada: Metall. Trans. A, **29A**, 537 (1998)
2. Y. Yamabe-Mitarai, Y. Ro, T. Maruko, T. Yokokawa, and H. Harada: Structural Intermetallics 1997, edited by M. V. Nathal et al., The Minerals, Metals, & Materials Society, Sept.21-25, 1997, Champion, Pennsylvania, USA, 1997, p.805-813.
3. Y. Yamabe-Mitarai, Y. Gu, Y. Ro, S. Nakazawa, T. Maruko, and H. Harada: Scripta Metall. Mater., **41 (3)**, 305 (1999)
4. Y. F. Gu, Y. Yamabe-Mitarai, Y. Ro, T. Yokokawa, and H. Harada: Metall. Trans. A, **30A**, 2629 (1999).
5. X. H. Yu, Y. Yamabe-Mitarai, Y. Ro, and H. Harada: Metall. Trans. A, , **31A**, 173 (2000)
6. X. H. Yu, Y. Yamabe-Mitarai, Y. Ro, and H. Harada: Key Engineering Materials, **171-174**, 677 (2000)
7. X. H. Yu, Y. Yamabe-Mitarai, T. Yokokawa, M. Osawa, Y. Ro, and H. Harada: Iridium, E. K. Ohriner, R. D. Lanam, P. Panfilov, and H. Harada, eds., TMS, Nashville, TN, 2000, pp. 61-69.
8. X. H. Yu, Y. Yamabe-Mitarai, and H. Harada: Scripta Mater., **43**, 671(2000).
9. W. F. Brown, Jr., H. Mindin, and C. Y. Ho: Aerospace Structural Metals Handbook, CINDAS/Purdue University, 1992, **5**, p. 4218.

Mat. Res. Soc. Symp. Proc. Vol. 646 © 2001 Materials Research Society

Compression Behavior of L1$_2$ Modified Titanium Trialuminides Alloyed with Chromium and Iron

Tohru Takahashi and Tadashi Hasegawa
Department of Mechanical Systems Engineering, Faculty of Engineering,
Tokyo University of Agriculture and Technology,
Koganei, Tokyo 184-8588, Japan

ABSTRACT

L1$_2$ modified titanium trialuminides have been prepared by replacing 9at.% of the aluminum in Al$_3$Ti with chromium and/or iron. The materials were recrystallized into single phase polycrystals after isothermal forging resulting in an average grain diameter of about 40μm. Lattice parameter of the material containing 9 at.% chromium or 9 at.% iron, are 0.3959nm and 0.3939nm, respectively. The lattice parameters varied linearly with composition between these values for additions of both chromium and iron. Uniaxial compression tests were performed at temperatures ranging from 293K to 1300K. The yield strength is not sensitive to chemical composition within the range of compositions tested. Flow stress serrations of a few % were observed at temperatures around 600K, where intermittent drops in flow stress started immediately after yielding and continued to the end. These serrations were observed up to about 800K. At 900K and above the materials became fully deformable. Quasi steady state flow and strain softening were observed at 1200K and 1300K, respectively, due to dynamic recrystallization.

INTRODUCTION

Al$_3$Ti is very brittle material at ambient temperature and the origin of this brittleness is attributed to its of DO$_{22}$ ordered crystallographic structure [1-4]. The deformability can be strikingly improved by modification of the crystal structure into the L1$_2$ structure by replacing some of the aluminum by a third element, such as chromium, manganese, iron, along with several others [5-11]. However, the deformability is not always improved by such alloying, even if the resulting crystal structure is L1$_2$. And the reason for this is unknown. In the present study chromium and/or iron have been used as the third element to produce the L1$_2$ ordered structure, and the amount of aluminum replaced by the third element was fixed at 9 at.%. Two ternary alloys, Al$_{66}$Ti$_{25}$Cr$_9$ and Al$_{66}$Ti$_{25}$Fe$_9$, which will hereafter be referred to as 9Cr and 9Fe materials, respectively, and two quaternary alloys were investigated. The chemical composition of the quaternary alloys is Al$_{66}$Ti$_{25}$Cr$_6$Fe$_3$ and Al$_{66}$Ti$_{25}$Cr$_3$Fe$_6$, and will be referred to as 6Cr3Fe and 3Cr6Fe, respectively. The microstructures of these materials were characterized, and compression tests were performed in air.

EXPERIMENTAL PROCEDURES

The L1$_2$ modified titanium trialuminide phase has been reported to form within a relatively narrow composition range around Al$_{66}$Ti$_{25}$X$_9$, where the third element, X, is chromium, manganese, iron and so on [12]. In the present study, the aluminum and titanium contents have been fixed at 66 and 25 at.%, respectively, and chromium and iron have been added as the third and fourth elements. The total amount of chromium plus iron is thus 9 at.%, and the chromium and iron contents were varied in 3 at.% intervals. The ternary (9Cr and 9Fe) and quaternary (6Cr3Fe and 3Fe6Cr) alloys were prepared by arc melting in an argon atmosphere. All the alloys used in the present study were prepared as button ingots with a mass of around 0.06kg. Thermo-mechanical treatments, based on isothermal-forging and annealing, were performed on 5 mm × 5 mm × 8 mm rectangular pieces cut from the ingot to homogenize the material and to eliminate the dendritic structure. During forging the pieces were hot-compressed by about 50 % at 1300K in air at a strain rate of $10^{-4}s^{-1}$ or $10^{-3}s^{-1}$ and then vacuum annealed at temperature of 1400-1500K for 10ks. This process resulted in a recrystallized, equiaxed polycrystalline microstructure. Small test pieces (typically, 2mm × 2mm × 3mm) having a rectangular parallelepiped shape were cut from these ingots. The microstructures of the materials were observed by an optical microscope under polarized light with Nomarski prism on a lightly etched surface. X-ray diffraction studies were performed on the recrystallized specimens using a RIGAKU RAD II-C diffractometer. Compression tests were carried out in air at various temperatures ranging from 293K to 1300K, and the strain rate was typically $2.0 \times 10^{-4}s^{-1}$. Axial strain was measured with an LVDT sensor connected to the compression jig. The deformed microstructure was subsequently observed optically.

RESULTS AND DISCUSSION

Metallographic Structure

Figure 1 (a)-(d) shows optical micrographs of the recrystallized microstructures in the chromium- and iron-modified titanium trialuminides; (a) 9Cr, (b) 6Cr3Fe, (c) 3Cr6Fe and (d) 9Fe materials, respectively. All materials are single-phase with an equiaxed grain structure. The average grain diameters in the above materials are 40, 38, 42 and 45μm, respectively, as measured by the linear intercept method.

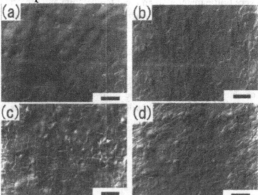

Figure 1. Optical micrographs of the materials that were used in the present paper; (a) 9Cr, (b) 6Cr3Fe, (c) 3Cr6Fe and (d) 9Fe materials, respectively. Markers show 100μm.

Crystallographic Structure

All the prominent diffraction peaks in the X-ray diffraction patterns from all four materials were indexed to the L1$_2$ cubic structure indicating a single phase L1$_2$ structure. The lattice parameters of the quaternary alloys are 0.3952nm (6Cr3Fe) and 0.3945nm (3Cr6Fe). The lattice parameters vary linearly with chemical composition between these two values 0.3959nm (9Cr) and 0.3941nm (9Fe), which suggested that chromium and iron atoms substitute continuously for each other between the ternary end points. Small differences in lattice parameters were determined from the high index peaks, such as (422). If any of the alloys contained a two phase structure, this would be readily discernible in the high order peaks. No peak separation was observed in the diffraction patterns of any alloys, suggesting a continuous substitution between chromium and iron in the chromium- and iron-modified titanium trialuminide intermetallic phases.

Stress vs. Strain Curves at Room Temperature

The compressive true stress-strain curves for the 9Cr, 6Cr3Fe, 3Cr6Fe and 9Fe materials at room temperature (293K) are shown in Figure 2. All samples fractured in a mostly brittle manner at about 0.2 to 0.25 true strain.

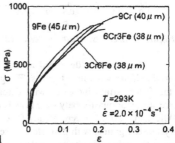

The yield strengths at room temperature of all four materials are 283MPa for 9Cr, 295MPa for 6Cr3Fe, 290MPa for 3Cr6Fe, and 312MPa for 9Fe. Materials containing more iron showed slightly higher values than the 9Cr material, but the differences between them are similar to experimental errors ($\pm$20MPa). Thus the yield strengths of these materials are only weakly dependent on chemical composition in this composition range.

Figure 2. *True stress vs. true strain curves at room temperature.*

The strain hardening rates were calculated to be around 3GPa, which is about 2% of Young's modulus, a rather low value for intermetallics. These strain hardening rates are also varied only slightly with chemical composition. Also, as can be seen from figure 2, the fracture strains of all four materials are similar, with the 9Cr material being slightly more ductile.

Stress vs. Strain Curves at Elevated Temperatures

Figure 3 shows the stress-strain curves obtained on the 6Cr3Fe material at various temperatures. The three other materials showed very similar behavior. The ductility of all four materials is limited below 800K. The work hardening rate observed at 600K is somewhat higher than at the other temperatures, the origin of which is not yet clear. As descibed in the following section, the low amplitude flow stress serrations were observed between 500K and 800K. Above 900K all the materials are fully deformable, and showed nearly steady state deformation between 1100 and 1200K. At 1300K, the highest temperature

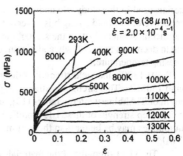

Figure 3. *Stress vs. strain curves of 6Cr3Fe material at various temperatures.*

used in this study, strain softening behavior was seen. Figures 4 and 5 show the 1200K and 1300K stress-strain curves for all four materials used in the present study. Note the similarity in the shapes of the curves in each figure.

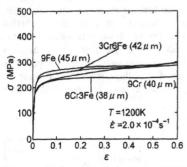

Figure 4. *Stress vs. strain curves at 1200K.*   **Figure 5.** *Stress vs. strain curves at 1300K.*

The materials deformed at and below 1000K (where strain hardening is observed), showed uniformly deformed grains. The grains become increasingly flattened (pancake-shaped) as the failure strain increases with deformation temperature. At temperatures above 1100K, where steady or quasi-steady state flow is obtained, grain boundary serrations were observed. At 1200K and 1300K, where steady-flow and strain softening behavior were observed, fine recrystallized grains, with diameters of several to tens of micrometers, were observed to cover most of the volume in the materials strained up to 0.7 true strain. The average grain diameters were observed to be larger in the materials deformed at 1300K than in the materials deformed at 1200K. These recrystallized grains are equiaxed, and therefore form by dynamic recrystallization during deformation.

Temperature Dependence of the Yield Strength

Figure 6 shows the temperature dependence of the yield strength (0.2% proof stress) of all four materials used in this study. No anomalous temperature dependence of the yield strength is observed (within the experimental scatter). Within the scatter, the yield strength decreases rather slowly and continuously with increasing temperature.

At 1300K the yield strength of the materials is around 150MPa. This strength is slightly higher than the strength of TiAl binary intermetallics [13], and suggests that the creep strength of these $L1_2$ modified titanium trialuminides may be better than those of materials based on TiAl.

The yield strengths of the materials are all contained within a band with a width of about $\pm$50MPa,

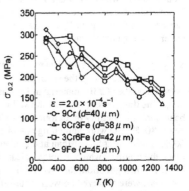

Figure 6. *Temperature dependence of 0.2% proof stresses.*

about the mean. The 9Fe and 3Cr6Fe materials are slightly stronger than the 9Cr and 6Cr3Fe materials over the whole temperature range, but, as a whole, the yield strength is not sensitive to the chemical composition at any temperature.

Flow Stress Fluctuation

Serrations on the stress-strain curve with a magnitude of a few per cent were observed at temperatures around 600K in all the materials. In the 6Cr3Fe material, the amplitude of the flow stress fluctuation is about 2.6% at 600K, and a drop was observed every few seconds at a strain rate of $2.0 \times 10^{-4} s^{-1}$. At 600K flow stress fluctuations were continuously observed from just after yielding to the end of the test. At 500-550K fluctuations were only observed after about 0.1 strain. In contrast to this, at 800K fluctuations were observed only shortly after yielding. At 900K and above the fluctuations totally disappeared. The origin of the flow stress fluctuation and its temperature dependent behavior remains as an open problem at present [14].

SUMMARY

The results of this study show that,
(1) Polycrystals with an equiaxed uniform grain structure can be obtained through thermo-mechanical treatment.
(2) All four materials have a single phase L1$_2$ ordered structure, and the lattice parameter varies linearly with composition.
(3) The materials are somewhat brittle at temperatures below 800K, but are totally deformable at temperatures above 900K.
(4) Steady flow behavior is observed at 1100-1200K, and at 1300K strain softening is seen.
(5) Dynamic recrystallization is observed in the specimens deformed at 1200-1300K.
(6) The yield strengths of all four materials are similar, within experimental error.
(7) Flow stress serrations were observed in all four materials between 500 and 800K. The strain range where the flow stress fluctuations were observed depends on temperature.

ACKNOWLEDGMENTS

The authors are grateful to Nippon Steel Corp. for supplying the samples.

REFERENCES

1. M. Yamaguchi and Y. Umakoshi, Progress in Materials Science, **34** (1990), pp.1-148.
2. M. Yamaguchi, Y. Umakoshi and T. Yamane, in *High-Temperature Ordered Intermetallic Alloys II*, edited by N.S. Stoloff, C.C. Koch, C.T. Liu and O. Izumi, (Mater. Res. Soc. Proc. **81**, Pittsburgh, PA, 1987), pp.275-286.
3. M. Yamaguchi, S.R. Nishitani and Y. Shirai, in *High Temperature Aluminides and Intermetallics*, edited by I. Baker, R. Darolia, J.D. Whittenberger and M.H. Yoo, (TMS, Warrendale, PA, 1990), p.63.
4. Y. Umakoshi, M. Yamaguchi, T. Yamane and T. Hirano, Phil. Mag., **A58**, 651 (1988).

5.   C.D. Turner, W.O. Powers and J.A. Wert, Acta metall., **37**, 2635 (1989).

6.   H. Mabuchi, K. Hirukawa and Y. Nakayama, Scripta metall., **23**, 1761 (1989).

7.   K.S. Kumar, S.A. Brown and J.D. Whittenberger, in *High-Temperature Ordered Intermetallic Alloys IV*, edited by L.A. Johnson, D.P. Pope and J.O. Stiegler, (Mater. Res. Soc. Proc. **213**, Pittsburgh, PA, 1991), pp.481-486.

8.   K.S. Kumar and S.A. Brown, in *High-Temperature Ordered Intermetallic Alloys V*, edited by I. Baker, R. Darolia, J.D. Whittenberger and M.H. Yoo, (Mater. Res. Soc. Proc. **288**, Pittsburgh, PA, 1993), pp.781-786.

9.   S.M. Kim, M. Kogachi, A. Kameyama and D.G. Morris, Acta metal. mater., **43**, 3139 (1995).

10. T. Takahashi, K. Endo, S. Kaizu and T. Hasegawa, in High-Temperature Ordered Intermetallic Alloys V, edited by I. Baker, R. Darolia, J.D. Whittenberger and M.H. Yoo, (Mater. Res. Soc. Proc. **288**, Pittsburgh, PA, 1993), pp.711-716.

11. T. Takahashi and T. Hasegawa, in *High-Temperature Ordered Intermetallic Alloys VI*, edited by J.A. Horton, I. Baker, S. Hanada, R.D. Noebe and D.S. Schwarz, by I. Baker, R. Darolia, J.D. Whittenberger and M.H. Yoo, (Mater. Res. Soc. Proc. **364**, Pittsburgh, PA, 1995), pp.1235-1240.

12. Z.L. Wu and D.P. Pope, in *Structural Intermetallics*, edited by R. Darolia, J.J. Lewandowski, C.T. Liu, P.L. Martin, D.B. Miracle and M.V. Nathal, (TMS, Warrendale, PA, 1993), p.107.

13. T. Takahashi, K. Asano, D. Ashida, T. Murakoshi and T. Hasegawa, presented at the Thermec 2000 International Conference, Las Vegas, NV, 2000 (unpublished).

14. Z.L. Wu, D.P. Pope and V. Vitek, Acta metall. mater., **42**, 3577 (1994).

Mat. Res. Soc. Symp. Proc. Vol. 646 © 2001 Materials Research Society

Development of Tough and Strong Cubic Titanium Trialuminides

Robert A. Varin[1], Les Zbroniec[3] and Zhi Gang Wang[3]
[1]Department of Mechanical Engineering, University of Waterloo, Waterloo, Ontario, Canada N2L 3G1
[2]National Institute of Materials and Chemical Research, 1-1 Higashi, Tsukuba, Ibaraki 305-8565, Japan
[3]Institute of Aeronautical Materials, Bejing 100095, P.R. China

ABSTRACT

In this work, the recent breakthroughs in the understanding of the fracture behavior and fracture toughness of $L1_2$-ordered titanium trialuminides are described and discussed. First, it is shown that, as opposed to many other intermetallics and specifically those with an $L1_2$ crystal structure, the fracture toughness of $L1_2$ titanium trialuminides is insensitive to testing in various environments such as air, water, argon, oxygen and vacuum ($\sim 1.3 \times 10^{-5}$ Pa). Second, it is reported here that by increasing the concentration of Ti combined with boron (B) doping, the room temperature fracture toughness of a Mn-stabilized titanium trialuminide can be improved by 100% from ~ 4 MPam$^{1/2}$ to ~ 8 MPam$^{1/2}$ and by 150-250% at 1000^{0}C to $\sim$(10-12) MPam$^{1/2}$ with a simultaneous suppression of intergranular fracture (IGF) to $\sim$(40-50%). Almost three fold increase in yield strength to ~ 550 MPa is attained at room temperature for high Ti, boron-doped trialuminides. Both Vickers microhardness and strength increase linearly with increasing concentration of (Ti+B) indicating a classical solid solution strengthening response.

INTRODUCTION

The $L1_2$-ordered titanium trialuminides are derived from $D0_{22}$-ordered Al_3Ti by alloying with fourth-period transition elements such as Cr, Mn, Fe, Co, Ni, Cu, and Zn [1,2]. They have attracted much attention as high temperature structural materials because of their low density, high melting point, oxidation resistance better than that for TiAl-based alloys, and expected improvement of room temperature tensile ductility and fracture toughness due to their cubic lattice structure. Regrettably, despite numerous efforts in the last ten years their tensile ductility and fracture toughness have not been improved. Most surprisingly, such brittle materials also exhibit quite a low strength (hardness) [3]. The first breakthrough that has been achieved in our laboratory demonstrates that, as opposed to many other intermetallics and specifically those with $L1_2$ crystal structure, the $L1_2$ titanium trialuminides are immune to the environmental embrittlement [4,5]. This important finding prompted us to focus further research efforts on other factors. Specifically, some years ago we found that the microhardness of cubic titanium trialuminides stabilized with various transition elements Fe, Mn, Cr and Cu depends nearly linearly on the Ti concentration [6]. Regarding boron effects Winnicka and Varin [7,8] found that the boron–doped, Cu-stabilized cubic titanium trialuminide exhibited entirely different fracture surface than its boron-free counterpart. In addition, fine cracks were formed around a 20 kg indentation in a boron-doped trialuminide as opposed to very long, edge cracks observed in a boron–free alloy [7]. In view of the above findings it has been decided to investigate the combined influence of Ti and boron (B) on the fracture toughness of cubic titanium trialuminide stabilized with Mn.

EXPERIMENTAL DETAILS

Three batches of cubic titanium trialuminide alloys stabilized with Mn whose overall composition (from a fully quantitative energy dispersive spectroscopy-EDS, and neutron activation analysis [9]) is listed in Table I, referred to as 9Mn, 14Mn and 18Mn, respectively, were induction melted in an alumina crucible from pure elements (Al-99.99999%, Ti-99.7% and Mn-99.9%) under high purity argon and cast into a pre-heated (~500^{0}C) austenitic stainless steel mold [9]. Some alloys were doped with boron as shown in Table I. The cast ingots were wrapped with stainless steel pouches (Sen/Pak$^{™}$ heat treatment containers) and homogenized for 100 h at 1000^{0}C under a high purity argon atmosphere [9]. Subsequently, the homogenized ingots were HIP-ed at 1250^{0}C for 2 h under 170–180 MPa pressure in argon. Fracture toughness was measured by four-point bending of chevron-notched specimens (CNB) (more details see [9]).

Table I. Composition of cubic titanium trialuminide alloys [9].

Alloy	Composition						
	Overall (at. %)				L1$_2$ Matrix (at. %)		
	Al	Ti	Mn	B	Al	Ti	Mn
9Mn	66.2±0.2	24.7±0.2	9.2±0.2	-	66.2±0.2	24.7±0.2	9.2±0.2 (H)
9Mn-0.004B	65.8±0.1	25.5±0.1	8.8±0.1	0.004±0.001	65.8±0.1	25.5±0.1	8.8±0.1 (H)
9Mn-0.25B	66.7±2.1	24.2±1.0	8.9±0.3	0.25±0.007	65.7±0.3	25.0±0.2	9.3±0.2 (H)
9Mn-0.66B	65.1±2.1	25.2±1.0	9.0±0.3	0.66±0.02	65.4±0.3	25.1±0.5	9.5±0.1 (H)
14Mn	56.4±0.1	29.4±0.1	14.2±0.1	-	56.1±0.2	29.0±0.1	14.9±0.2 (H)
14Mn-0.24B	57.6±1.9	28.5±1.1	13.7±0.4	0.24±0.007	56.3±0.4	29.2±0.1	14.4±0.4 (H)
14Mn-0.65B	53.6±1.8	31.9±1.2	13.8±0.4	0.65±0.02	55.2±0.2	30.3±0.3	14.5±0.4 (H)
18Mn-0.005B	48.8±1.2	32.5±0.5	18.7±0.7	0.005±0.0006	51.6±0.4	31.6±0.4	16.8±0.2 (H)
18Mn-0.21B	49.5±0.6	32.2±0.5	18.3±0.5	0.21±0.02	51.4±0.3	31.9±0.3	16.6±0.5 (H)
18Mn-0.34B	49.4±1.1	32.9±0.6	17.7±0.6	0.34±0.03	51.3±0.3	32.1±0.3	16.6±0.2 (H)

Note: (H)-HIP-ed (1250 ^{0}C/2 h/170-180MPa). The overall composition of boron–doped 9Mn-0.25 and 0.66B and 14Mn alloys was obtained from a neutron activation analysis (Becquerel Labs Inc., Mississauga, Ontario).

Compression testing at room temperature was conducted for 9Mn and 14Mn. Approximate dimensions of the test specimen were 4 mm x 4 mm x 7 mm. Specimens were cut out from the end sections of the broken CNB bars. Initial strain rate was ~1.3 x 10^{-4} s^{-1}. Four specimens were tested for each composition.

RESULTS AND DISCUSSION

Both boron–free and boron–doped 9Mn and 14Mn alloys after HIP-ing exhibited a typical as-cast polycrystalline near-single phase L1$_2$ structure with either equiaxed or columnar grains [9]. A small amount of minor constituent phases such as γ_1 (most probably the hexagonal Laves phase Ti(Mn,Al)$_2$) and borides were also observed [9]. Alloys 18Mn after HIP-ing contained about 20 vol.% of equilibrium γ_1 phase [9], i.e. they were dual-phase alloys.

Figure 1 is a summarizing plot of the results [4,5] of fracture toughness testing of the Mn-stabilized, L1$_2$ 9Mn-25Ti trialuminides in various environments. It is quite clear that the environment does not affect the fracture toughness of L1$_2$ titanium trialuminides. In other words, environment is not a primary cause of brittle behavior of L1$_2$ Al$_3$Ti(Mn) intermetallic alloys. This

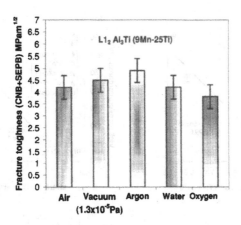

Figure 1. *Fracture toughness dependence at room temperature on testing environment (SEPB-single-edge pre-cracked beam; CNB-chevron-notched beam) (results extracted from [4,5]).*

behavior is quite beneficial in contrast to many other cubic $L1_2$ and B2 intermetallics such as Ni_3Al and FeAl, respectively, whose fracture toughness is detrimentally affected by the environment, i.e. moisture in air [10].

Figure 2 shows dependence of Vickers microhardness (Fig.2a,b) and compressive yield strength (Fig.2c,d) on the square root of total concentration of either (Ti+Mn+B) or (Ti+B) in the $L1_2$ matrix (based on a simplified assumption that all boron is dissolved in the matrix; small amount of borides was indeed observed in the alloy [9]) of nearly single-phase 9Mn and 14Mn (data for a dual-phase18Mn alloy are not included since it contains ~20% γ_1 phase that would affect mechanical properties). There are two important features clearly seen in Fig.2. First, there is excellent linear regression correlation between both microhardness and yield strength and the square root of total matrix concentration of solute atoms Ti, Mn and B. The observed linear increase of strength parameters with the root of concentration of solutes strongly suggests a classical solid-solution strengthening effect [11]. Second, a better linear regression coefficient R^2 for both microhardness and yield strength is obtained if the microhardness and yield strength are fitted to the concentration of (Ti+B) rather than (Ti+Mn+B). That strongly suggests that the effect of Mn on the solid solution strengthening in the investigated trialuminides is smaller than the combined effect of (Ti+B). It must also be pointed out that the yield strength increment by boron doping for high Ti 14Mn is much higher (~1.74 MPa/0.01at%B) than that for low Ti 9Mn (~0.4 MPa/0.01at%B). Similar behavior is observed for microhardness. Apparently, the strengthening by boron seems to be synergistically enhanced by a high concentration of Ti [9].

Fracture toughness dependence on the Ti concentration and boron-doping at room temperature is shown in Fig.3a and a relationship between fracture toughness and yield strength is shown in Fig.3b. Comparing boron-free 9Mn and 14Mn alloys it is clear that their respective fracture toughness is not affected by quite a substantial increase in Ti concentration (Table I). However,

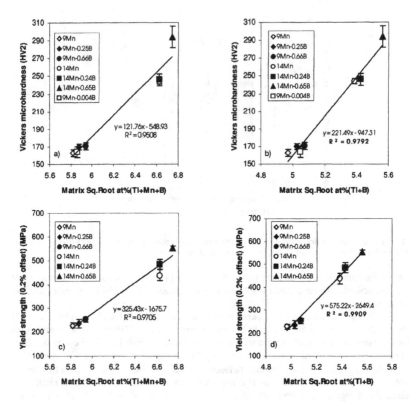

Figure 2. *Linear regression fitting of Vickers microhardness (a,b) and compressive yield strength (c,d) to the square root of total concentration of either(Ti+Mn+B) or (Ti+B) in the Ll_2 matrix.*

much higher fracture toughness is observed for boron–doped high Ti 14Mn and 18 Mn than for boron–doped low Ti 9Mn alloy, although the boron doping to the latter also slightly increases its toughness. That strongly suggests that the presence of *both* high Ti and boron is needed for the effective increase of room temperature fracture toughness in these trialuminides. Interestingly, the fracture mode in boron–doped high Ti 14Mn alloys exhibiting increase in fracture toughness remains transgranular cleavage at room temperature [9]. A very high matrix Ti concentration in boron–doped 18Mn does not lead to enhanced toughening over that of boron–doped 14Mn (Fig.3a). This behavior could be related to the presence in this particular microstructure of high volume fraction (~20%) of more brittle γ_l phase degrading toughness. Alternatively, the overall boron level in 18Mn might be insufficient at such a high volume fraction of intragranular γ_l to provide a required boron saturation of the Ll_2 matrix. Boron might have segregated to either γ_l or the Ll_2/γ_l interfaces, depleting the Ll_2 matrix [9]. It seems that a small addition of boron on

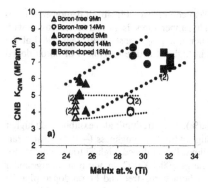
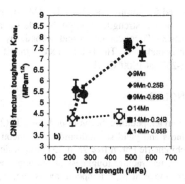

Figure 3. *Fracture toughness (CNB) dependence at room temperature on (a) matrix concentration of Ti and (b) on yield strength. Number of overlapping data points in parentheses.*

the order of 0.2at% is sufficient to improve the fracture toughness of high Ti 14Mn alloy. Higher concentration of boron (~0.65at%) does not increase further the average room temperature fracture toughness of both low Ti 9Mn and high Ti 14Mn alloys. Fig.3b summarizes clearly that a simultaneous increase of both yield strength and fracture toughness, at least at room temperature, occurs only in boron-doped trialuminides. Such a behavior is quite peculiar because in general, it is observed for conventional metallic materials that plane–strain fracture toughness (K_{Ic}) decreases as the yield strength of a material increases [12].

At elevated temperatures, the intergranular failure mode (very characteristic at high temperatures for 9Mn-25Ti trialuminide alloys [4,5]) is strongly suppressed by high Ti concentration and the transgranular fracture-intergranular failure transition temperature is shifted from ~200°C to ~600°C (Fig.4a). This might be one of the causes of higher fracture toughness

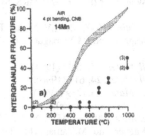
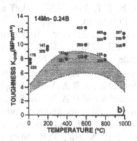

Figure 4. *Surface fraction of intergranular failure mode of boron-free (a) and CNB fracture toughness of boron-doped (number of overlapping data in parentheses) (b) high Ti 14Mn alloys vs. testing temperature. Shaded bands:data ranges for low Ti 9Mn-25Ti alloy;number beside each data point indicates the average grain size in μm. After[9].*

observed for boron-doped high Ti 14Mn over its 9Mn counterpart (Fig.4b). It demonstrates that the cohesive strength of grain boundaries at high temperatures is primarily improved with increasing Ti concentration and secondarily with boron doping (most likely, segregated at the grain boundaries). The latter effect is manifested in higher fracture toughness of boron-doped 9Mn-25Ti at 1000°C [9].

CONCLUSIONS

At room temperature, boron-doped, high Ti (29-33at%), cubic trialuminides exhibit much higher hardness and yield strength and simultaneously, higher fracture toughness (~100%) than near-stoichiometric boron-free 9Mn-25Ti trialuminide alloys. Boron-doped, high Ti trialuminides also have better fracture toughness at the temperature range 200-1000°C. In addition, intergranular failure mode at high temperatures is suppressed in both boron-free and boron-doped high Ti trialuminides. It is possible that by further optimization of chemical composition and microstructure, tough and strong cubic titanium trialuminides could be developed.

ACKNOWLEDGMENTS

This work was supported by grant from the Natural Sciences and Engineering Research Council of Canada, which is gratefully acknowledged.

REFERENCES

1. K.S. Kumar, *Structural Intermetallics 1997*, ed. M.V. Nathal, R. Darolia, C.T. Liu, P.L. Martin, D.B. Miracle, R. Wagner and M.Yamaguchi (TMS, 1997), Warrendale, USA, pp.87-96.
2. R.A. Varin, M.B. Winnicka and I.S. Virk, *ibid*, pp.117-124.
3. M.B. Winnicka and R.A. Varin, *Metall. Trans. A*, **24A**, 935 (1993).
4. R.A. Varin and L. Zbroniec, *High-Temperature Ordered Intermetallic Alloys VII*, ed. C.C. Koch, C.T. Liu, N.S. Stoloff and A. Wanner (MRS, 1997), MRS Symp. Proc. Vol.460, Pittsburgh, USA, pp. 121-126.
5. R.A. Varin and L. Zbroniec, *Structural Intermetallics 1997*, ed. M.V. Nathal, R. Darolia, C.T. Liu, P.L. Martin, D.B. Miracle, R. Wagner and M.Yamaguchi (TMS, 1997), Warrendale, USA, pp. 787-793.
6. M.B. Winnicka and R.A. Varin, *Scripta Metall. Mater.* **25**, 2297 (1991).
7. M.B. Winnicka and R.A. Varin, *Scripta Metall. Mater.* **24**, 611 (1990).
8. M.B. Winnicka and R.A. Varin, *Scripta Metall. Mater.* **25**,1289 (1991).
9. R.A. Varin, L. Zbroniec and Z.G. Whang, *Intermetallics* (in press).
10. E.P. George and C.T. Liu, *Structural Intermetallics 1997*, ed. M.V. Nathal, R. Darolia, C.T. Liu, P.L. Martin, D.B. Miracle, R. Wagner and M.Yamaguchi (TMS, 1997), Warrendale, USA, pp. 693-702.
11. T.H. Courtney, *Mechanical Behavior of Materials*, (McGraw-Hill, 2000), pp.186.
12. M.A. Meyers and K.K. Chawla, *Mechanical Behavior of Materials*, (Prentice Hall, 1999), pp. 365.

Metal Silicides II

Mat. Res. Soc. Symp. Proc. Vol. 646 © 2001 Materials Research Society

Elastic and Plastic Properties of Mo₃Si Measured by Nanoindentation

J. G. Swadener, Isaí Rosales and Joachim H. Schneibel

Oak Ridge National Laboratory, Metal and Ceramics Division, P. O. Box 2008, Oak Ridge, TN 37831

ABSTRACT

Single crystal Mo_3Si specimens were grown and tested at room temperature using established nanoindentation techniques at various crystallographic orientations. The indentation modulus and hardness were obtained for loads that were large enough to determine bulk properties, yet small enough to avoid cracking in the specimens. From the indentation modulus results, anisotropic elastic constants were determined. As load was initially increased to approximately 1.5 mN, the hardness exhibited a sudden drop that corresponded to a jump in displacement. The resolved shear stress that was determined from initial yielding was 10-15% of the shear modulus, but 3 to 4 times the value obtained from the bulk hardness. Non-contact atomic force microscopy images in the vicinity of indents revealed features consistent with $\{100\}(010)$ slip.

INTRODUCTION

The A15 phase Mo_3Si is present in several boron-containing molybdenum silicide systems (i.e. $Mo-Mo_3Si-Mo_5SiB_2$ [1-3] and $Mo_3Si-Mo_5Si_3-Mo_5SiB_2$ [4, 5]) that have recently been investigated for their high temperature strength and oxidation properties. However, the room temperature mechanical properties of the Mo_3Si phase have not been previously determined. The objective of this study is to determine the elastic and plastic properties of Mo_3Si. Additional goals are to investigate the nucleation of dislocations and the slip systems involved in plastic deformation of Mo_3Si at room temperature.

Other examples of A15 phases include Cr_3Si [6] and Nb_3Al [7]. A15 phases can be quite strong at high temperatures. For example, Cr_3Si single crystals at 1300° C have a yield strength on the order of 450 MPa [6] and the compressive strength of $Cr_{40}Mo_{30}Si_{30}$ is approximately 200 MPa at 1323° C [8]. In general, A15 phases are brittle at room temperature. For example, the room temperature fracture toughness of Nb_3Al is 1.1 MPa m$^{1/2}$ [7], whereas that of Mo_3Si is on the order of 3 MPa m$^{1/2}$ [9]. Single crystal specimens of Cr_3Si usually fracture without any detectable deformation at temperatures below 1200° C [6]. Polycrystalline specimens of $Cr_{40}Mo_{30}Si_{30}$ are also brittle below 1200° C [10]. Therefore, room temperature slip systems can not be examined by conventional testing methods (i.e. uniaxial compression testing of single crystals and slip trace analysis). However, room temperature indentations at very low loads (less than 20 mN), which induce triaxial compressive stress, may result in limited room temperature slip in A15 compounds.

EXPERIMENTAL

Mo_3Si bars were fabricated by arc-melting elemental materials in a copper mold in an argon atmosphere. Using 25.5 at. % Si in the initial mixture compensated for evaporative losses and resulted in the correct stoichiometry. The bars were electro-discharge machined (EDM) into 6.5 mm diameter by 100 mm long rods. Single crystals were grown from the rods using an optical float zone furnace in an argon atmosphere (ASGAL). The growth rate was 10 mm/hr and the upper and lower ends of the specimen were rotated at 10 rpm in opposite directions. Laue back reflection revealed that the growth direction was near <102>. Specimens were cut by EDM in <100>, <110> and <111> orientations and polished using 0.5 μm diamond paste followed by colloidal silica.

Ultra-low load indentation (nanoindentation) experiments were conducted at 23° C. Displacements and loads were measured with a resolution of 0.04 nm and 75 μN, respectively. The experiments were run in load control at a constant loading rate to a prescribed maximum load that ranged from 0.6 to 20 mN. For some of the experiments, the continuous stiffness measurement mode was used, which applied an oscillatory displacement of 2 nm peak-to peak at a 45 Hertz frequency.

A Berkovich diamond indenter (3 sided pyramidal tip) was used. The tip shape and the compliance of the nanoindentation system were calibrated by conducting experiments on fused quartz and sapphire [11, 12]. The tip was determined to be rounded with a 200±10 nm radius of curvature. The rounded portion of the indenter extended to a depth of approximately 10 nm.

The specimen surfaces were imaged after indentation using non-contact atomic force microscopy (AFM). Contact AFM was found to wear the silicon tip and leave deposits on the specimen. Using the non-contact mode eliminated this problem. After 36 hours in air, some specimen surfaces began to grow islands of what appeared to be oxides. To alleviate this problem, the specimens were kept in an inert atmosphere after polishing and imaged within 24 hours after indentation in air.

THEORY

Indentation of anisotropic materials by spherical and conical indenters in the elastic regime has been analyzed in depth [13-15]. The modulus measured by indentation (M) is a combination of all the material elastic constants weighted more heavily by the elastic modulus in the indentation direction. For cubic materials, measurement of the indentation modulus in three independent directions can be used to determine the three independent elastic constants.

Vlassak and Nix [14] have shown that the contact stiffness (S) for anisotropic materials can be written as:

$$S = \frac{dP}{dh} = \frac{2}{\pi} M \sqrt{A} \qquad (1)$$

where P is the applied load, h is the total depth (rigid body displacement), M is the indentation modulus, and A is the area in contact. For isotropic materials, Oliver and Pharr [11] previously determined that $M = E/(1-v^2)$. For cubic anisotropic materials where the modulus in different directions varies by a factor of three or less, Swadener and Pharr [15] have shown that the value of M measured by a conical indenter is within 1% of the value of M measured by a spherical indenter. King [16] has shown that a Berkovich indenter gives a value of M that is approximately 3% greater than the value measured by a conical indenter. Therefore, when contact is elastic, the indenter is modeled as a sphere, and when inelastic deformation occurs, the indenter is modeled as a pyramid and the King factor of 1.03 is applied. Considering the moderate anisotropy that will be shown for Mo_3Si, these models should be accurate to within 3%. For spherical indentation, the load displacement relation can be described by a generalization of the Hertzian relation as:

$$P = \frac{4}{3} M \sqrt{R} \left(h^{\frac{3}{2}} \right), \qquad (2)$$

where R is the radius of the sphere.

RESULTS AND DISCUSSION

Indentation with the rounded Berkovich tip was generally in the elastic regime up to loads of 1.2 mN. The load-displacement results for five indentations in the specimen oriented in the <110> direction are shown in figure 1. The initial approximately 10 nm of displacement shows some hysteresis, while the remaining portion shows complete recovery. This indicates that there is some initial inelastic deformation of asperities on the specimen surface at low loads. At greater loads (up to 1.2 mN) no additional hysteresis is observed despite the greater stresses, which indicates that the deformation in this regime is elastic. The unloading curve is in agreement with eq. (2), but with a slightly greater radius (R = 213±10 nm) due to the slight indention of the surface that accompanies the inelastic compression of the asperities.

For a slightly greater load, the elastic regime ends abruptly with a jump in displacement of approximately 8-10 nm. Figure 2 shows ten load-displacement curves to a maximum load of 1.5 mN in a specimen oriented in the <110> direction. The displacement jump occurred at loads between 1.2 and 1.5 mN for all orientations in 85% of the experiments. For the remaining 15% of the experiments to a load of 1.5 mN, the displacement jump occurred at load below 1 mN (see fig. 2), which was probably caused by a defect in the region near those indentations.

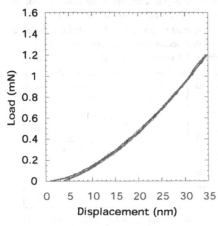

Figure 1. Load-displacement results for 1.2 mN indents in <110> Mo₃Si

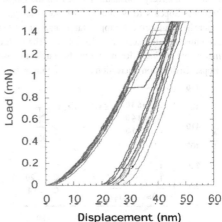

Figure 2. Load-displacement results for 1.5 mN indents in <110> Mo₃Si

The indentation modulus and hardness of Mo_3Si single crystals was determined for three different surface orientations by using the Oliver-Pharr method [11]. The results from continuous stiffness measurements at 20 mN load, which are listed in Table 1, were found to give the least scatter. Both the modulus and hardness are greatest in the <100> direction and least in the <111> direction. Results for other maximum loads were similar. A load of 20 mN resulted in a maximum depth of approximately 200 nm for all three directions. Cracking was rarely observed at 20 mN loads, however cracking usually occurred for loads between 50 and 100 mN.

If the elastic constants of an isotropic material are known, the indentation modulus in any direction can be determined, however the inverse relation is not known. Since the elastic constants cannot be directly determined from the indentation results, a trial and error procedure was adapted in which the indentation moduli (C_{ij}) were calculated for trial sets of elastic constants using the method given by Swadener and Pharr [15]. The values of C_{ij} were then compared to the experimental results. The results for the two most extreme sets of C_{ij} that agreed within one standard deviation of the experimental results are compared to the experimental results in figure 3. The elastic constants obtained by this approach are $C_{11} = 505 \pm 35$ GPa, $C_{12} = 80 \pm 60$ GPa and $C_{44} = 130 \pm 15$ GPa (68% confidence). There is a wide variation possible in the obtained value of C_{12}, because it only weakly effects the indentation modulus, and its effect can be mitigated by small changes in the other two elastic constants. The hardness values listed in Table 1 represent conventional measures of material hardness. However at small loads, the hardness can be significantly greater. Typical hardness-displacement results for a maximum load of 4 mN are plotted in figure 4. During the elastic portion of the loading cycle, the measured hardness increases, because the hardness during elastic loading does not represent a material property, but merely reflects the increase in stress applied to the specimen. At approximately 40 nm displacement, the displacement jump occurs accompanied by a sudden increase in the hardness for all three specimen orientations. After the drop in hardness, the hardness value remains relatively constant. The reason the hardness at small depths can exceed the plateau value of hardness appears to be the lack of sites that promote dislocation nucleation in the virgin material. After the displacement jump, a large number of dislocations must be present in the material in order to accommodate the surface deformation. The dislocations that accompany the displacement jump can then serve as Frank-Read sources for nucleation of additional dislocations, and therefore the hardness drops. Dislocation nucleation accompanying a displacement jump has been observed in sapphire [17].

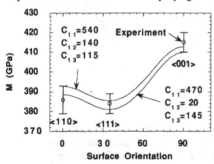

Figure 3. Comparison of experimental and
calculated indentation moduli

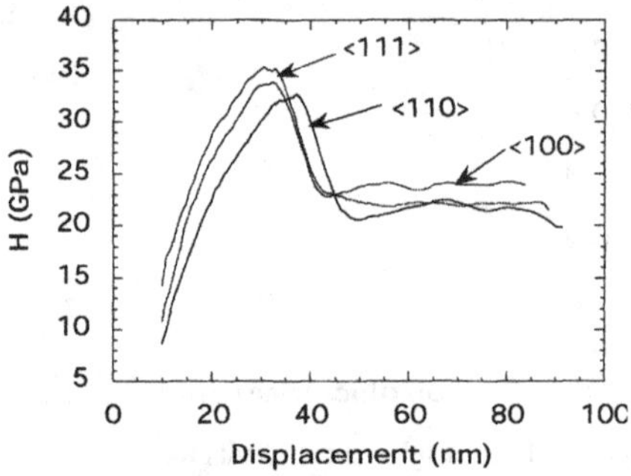

Figure 4. Variation of measured hardness
with displacement

Table 1. Indentation Modulus and Hardness of Mo_3Si Measured by Nanoindentation
Mean Values Shown (Standard Deviation in Parenthesis)

Surface	<100>	<110>	<111>
Indentation Modulus (GPa)	415 (4)	386 (7)	384 (5)
Hardness (GPa)	22.5 (0.5)	22.0 (0.7)	20.8 (0.4)

The indented <111> surface in the region surrounding a 1.5 mN indent was examined by using non-contact AFM. The surface showed three approximately 80 nm deep troughs corresponding to the locations of the indenter edges. In the areas between the troughs, the surface was raised approximately 80 nm. This indicates that the displacement jump was not due to the sudden nucleation of a large number of dislocations underneath the indenter, but rather that stress concentrations from the indenter edges (that are sharper than the rounded tip) triggered the displacement jump. The fact that the displacement jump occurred at approximately the same depth for the three surface orientations is further evidence that the sharp edges caused the jump, because the resolved shear stress differed by approximately 50% for the three directions. The mean values of the maximum resolved shear stress at the initiation of yielding for the {100}(010) family of slip systems were: 11GPa for indentation in the <100> direction, 17GPa for the <110> direction and 18 GPa for the <111> direction. The resolved shear stress for indentation in the <100> direction appears low, indicating that a localized resolved shear stress greater than 11 GPa was probably present near the edges of the indenter. The resolved shear stresses were calculated based on the simplified assumption of a Hertzian stress field, which may underestimate the stress near the edges of the indenter. On the other hand, the hardness-displacement curves show a knee prior to maximum hardness (see fig, 4), which indicates that plastic deformation may begin prior to the maximum load. Calculating the resolved shear stress from the knee of the curve would reduce the resolved shear stress by approximately 30%. Overall, the results indicate that a maximum resolved shear stress of 11-18 GPa can be obtained for the initiation of yielding in small volumes of Mo_3Si. This value is 3 to 4 times higher than the value determined from the plateau value of hardness using the Tabor method [17], which is 4.0-4.5 GPa and represents the bulk yield strength.

Non-contact AFM images showed that slip occurred in the (100) directions, although slip was not necessarily limited to {100} planes. The raised material around a 4 mN indent on a [110] surface shown in figure 5 appears to indicate {100}(010) slip. Some other indents show similar features, while others have features that indicate that slip could occur in either {110}(100) or {100}(010) slip systems.

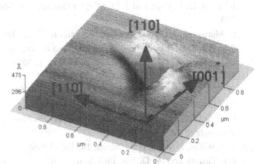

Figure 5. Non-contact AFM image in the vicinity of a 4 mN indent on a [110] surface.

SUMMARY

Established nanoindentation techniques were used to determine the mechanical properties of single crystal Mo_3Si specimens at various crystallographic orientations. For specimens oriented in the <100>, <110> and <111> directions, the indentation modulus and hardness were obtained for 20 mN loads, which was large enough to determine bulk properties, yet small enough to avoid cracking in the specimens. From the indentation modulus results, anisotropic elastic constants were determined. As load was initially increased to approximately 1.5 mN, the hardness exhibited a sudden drop that corresponded to a jump in displacement. After the drop, the hardness remained approximately constant to a load of 20 mN. The peak hardness at loads below 1.5 mN can be attributed to a lack of sources for dislocation nucleation in the small volume tested. The resolved shear stress that was determined from the peak hardness was 10-15% of the shear modulus, but 3 to 4 times the value obtained from the bulk hardness. Non-contact AFM images in the vicinity of indents revealed localized raised areas (pile-up). These features were generally aligned with the crystallographic directions consistent with {100}(010) slip.

ACKNOWLEDGEMENT

Research at the Oak Ridge National Laboratory SHaRE User Facility was sponsored by the Division of Materials Sciences and Engineering, U.S. Department of Energy, under contract DE-AC05-00OR22725 with UT-Battelle, LLC. I. Rosales acknowledges partial support by the Universidad Nacional Autonoma de México (UNAM—DGEP)

REFERENCES

1. D. M. Berczik, United States Patent 5,595,616 (1997).
2. D. M. Berczik, United States Patent 5,693,156 (1997).
3. J. H. Schneibel, M. J. Kramer, Ö. Ünal and R. N. Wright, Intermatallics, in press.
4. M. K. Meyer, M. J. Kramer and M. Akinca [sic], Intermetallics, **4**, 273 (1996).
5. M. K. Meyer, A. J. Thom and M. Akinc, Intermetallics, **7**, 153 (1999).
6. C. S. Chang and D. P. Pope, in *High-temperature Ordered Intermetallic Alloys IV*, edited by L. A. Johnson, D. P. Pope and J. D. Stiegler (Mater. Res. Soc. Symp. Proc. **213**, Pittsburgh, PA 1991), p. 745.
7. L. Murugesh, K. T. Venkteswara Rao and R. O. Ritchie, Scripta Metall. Mater. **29**, 1107 (1993).
8. S.V. Raj, J. D. Whittenberg, B. Zeumer and G. Sauthoff, Intermetallics, **7**, 743 (1999).
9. I. Rosales and J. H. Schneibel, Intermetallics, **8**, 885 (2000).
10. S.V. Raj, Mater. Sci. Eng. **A201**, 229 (1995).
11. W. C. Oliver and G. M. Pharr, J. Mater. Res. **7**, 1564 (1992).
12. J. G. Swadener and G. M. Pharr, in *Thin Films: Stresses and Mechanical Properties VII*, edited by R. Vinci, O. Kraft, N. Moody, and E. Shaffer VIII (Mater. Res. Soc. Symp. Proc. **594**, Warrendale, PA 2000), p. 525.
13. J. R. Willis, J. Mech. Phys. Solids **14**, 163 (1966).
14. J. J. Vlassak and W. D. Nix, Phil. Mag. A, **67**, 1045 (1993).
15. J. G. Swadener and G. M. Pharr, Phil. Mag. A, in press.
16. R. B. King, Int. J. Solids Structures **3**, 1657 (1988).
17. T. F. Page, W. C. Oliver and C. J. McHargue, J. Mater. Res. **7**, 450 (1992).
18. D. Tabor, *The Hardness of Metals* (Clarendon Press, Oxford 1951).

Mat. Res. Soc. Symp. Proc. Vol. 646 © 2001 Materials Research Society

Refinement of Crystallographic Parameters in Refractory Metal Disilicides

K. Tanaka, K. Nawata, H. Inui, M. Yamaguchi and M. Koiwa
Department of Materials Science and Engineering,
Kyoto University, Kyoto 606-8501, Japan.

ABSTRACT

The crystallographic structures of seven refractory metal (Ti, V, Cr, Nb, Mo, Ta and W) disilicides with the $C11_b$, C40 and C54 structures have been refined through analysis of single-crystal X-ray diffraction data. Crystallographic parameters refined are lattice constants, atomic parameters and the space group of the C40 disilicides. In most of previous studies, silicon atoms have been considered to locate at the ideal positions so that the refractory metal atoms are perfectly six-fold coordinated in RSi_2 layers prevailing in all the three structures. The present analysis shows that the silicon atoms are displaced from the ideal positions. The magnitude of such displacement is found to be closely related to the interatomic distance in these pseudo-hexagonally arranged RSi_2 layers. The space group of three of the four C40 disilicides, VSi_2, $CrSi_2$ and $TaSi_2$, is determined to be $P6_422$, which is of chirality with respect to that ($P6_222$) assigned in the previous studies.

INTRODUCTION

There is a growing interest in refractory metal disilicides in recent years not only as candidate materials for very high-temperature structural applications [1, 2] but also as materials used in microelectronic applications [3, 4]. These disilicides include $MoSi_2$ and WSi_2 with the tetragonal $C11_b$ structure, VSi_2, $CrSi_2$, $NbSi_2$ and $TaSi_2$ with the hexagonal C40 structure, and $TiSi_2$ with the orthorhombic C54 structure. The space group of $C11_b$ structure is I4/mmm and the basis has refractory metal atoms in $2a$ positions and Si atom in $4e$ positions with the atom position parameter z. The space group of the C40 structure can be either $P6_222$ (prototype: $CrSi_2$) or $P6_422$ (prototype: $NbSi_2$), depending on the chirality related to the six-fold axis parallel to the c-axis. For the former case, the basis consists of refractory metal atom in $3d$ positions and Si atoms in $6j$ positions, while for the latter case, it consists of refractory metal atoms in $3c$ positions and Si atoms in $6i$ positions. For both cases, the position of Si atoms is expressed with the parameter x. The basis of the C54 structure with the space group Fddd consists of refractory metal atoms in $8a$ positions and Si atoms in $16e$ positions with the atom position parameter x.

According to Pearson's Handbook [5], the Si atoms in most of these disilicides are located the ideal positions. Recent first principle calculations, however, have indicated that the total energy of both $MoSi_2$ [6-8] and WSi_2 [6, 8] with the $C11_b$ structure is minimized when the position is slightly shifted from the ideal one, which has very recently been confirmed experimentally for $MoSi_2$ by the Rietveld analysis of X-ray powder diffraction data [9]. Similar deviation of the Si atom position from the ideal one has been reported also for $NbSi_2$ with the C40 structure [10] and for $TiSi_2$ with the C54 structure [11]. In structure refinements so far reported, the positions of Si atoms for most of these silicides seem to be assumed *a priori* as ideal since the perfect six-fold coordination is achieved for the atomic arrangement in RSi_2 layers. However, there is no justification for such an assumption, as the position parameters for Si atoms are variable in each of the assigned space group. In addition, the space group of all binary C40 disilicides seems to be assumed *a priori* as $P6_222$ although $P6_422$, which is of chirality with respect to the six-fold axis parallel to the c-axis, is also possible. The exception for this is $NbSi_2$, for which two controversial reports are listed in Pearson's Handbook [5]. One report has assigned the space group of $P6_222$ with the *x* value significantly deviated from the ideal one while the other has assigned the space group of $P6_422$ with the ideal *x* value (1/6).

In the present study, we refine the crystallographic parameters in seven binary transition-metal disilicides with the $C11_b$, C40 and C54 structures by a single-crystal X-ray diffraction method, paying special attention to the position parameters for Si atoms and the chirality of C40 disilicides.

EXPERIMENTAL PROCEDURE

Single crystals of seven binary transition-metal disilicides formed with W, Mo, Cr, V, Nb, Ta and Ti were grown in an optical floating-zone furnace under a purified argon flow. Spherical specimens, 0.15 to 0.32 mm in diameter, were prepared from the single crystals.

X-ray diffraction data were collected on a Rigaku four-circle automatic diffractometer. The angular data for the lattice parameter determination were collected with germanium-monochromatized Mo-$K\alpha_1$ radiation. The intensity data for refinement of the Si atom positions were collected with graphite-monochromatized Mo-$K\alpha$ or Ag-$K\alpha$ radiation. Since the structure factors are sensitive to the displacement of Si atoms for higher-indexed reflections, it is desirable to measure intensities with diffraction angles as high as possible. However, the intensity decreases with the increase in diffraction angle, deteriorating the reliability in the intensity measurement. We thus collected intensities within the range of $1.1 < \lambda/\sin\theta < 1.2$ in the ω-2θ scan mode. In general, only the structure factors of some limited indices, termed the asymmetry units, are required for structural analysis. However, we have measured the structure factors for all the indices including equivalent ones, in order to reduce any possible experimental errors, particularly due to the deviation of the specimen shape from the ideal sphere.

The usual Lorentz, polarization and absorption corrections were applied in analyzing the measured structure factors. Then, the structure factors of equivalent indices were averaged. The refinement of the position of Si atoms were made by the full-matrix least squares method using the XTAL3.6.1 program [12] with seven or ten independent parameters; scaling factor, anisotropic temperature factor of each constituent atom, extinction parameter (Zachariasen type) and position of silicon atoms, incorporating neutral atomic scattering factors (Cromer-Mann coefficients) and anomalous-scattering factors taken from International Tables for X-ray Crystallography. Table 1 shows space groups and atomic coordinates used in the refinement. For the ease of comparison with the space group $P6_422$, the origin of the space group $P6_222$ is displaced by (0, 0, 1/2) so that the 3*d* and 6*j* positions are converted to the 3*c* and 6*i* positions.

Table 1. Space groups and atomic coordinates for the $C11_b$, C40 and C54 structures.

Crystal structure	$C11_b$	C40		C54
Space group	I4/mmm (139)	$P6_222$ (180)	$P6_422$ (181)	Fddd (70)
Refractory metal	2*a*	3*c*	3*c*	8*a*
	0, 0, 0	1/2, 0, 0	1/2, 0, 0	0, 0, 0
Silicon	4*e*	6*i*	6*i*	16*e*
	0, 0, *z*	*x*, 2*x*, 0	*x*, 2*x*, 0	*x*, 0, 0

When Si atoms are in the ideal positions, these positions are denoted as $z = 1/3$, $x = 1/6$ and $x = 1/3$ for the C11$_b$, C40 and C54 structures, respectively. We now introduce a parameter, Δ, which represents the displacement of Si atoms from the ideal position. Then the positions of Si atoms are written as $(0, 0, 1/3 + \Delta)$, $(1/6 - \Delta, 1/3 - 2\Delta, 0)$ and $(1/3 + \Delta, 0, 0)$ for the C11$_b$, C40 and C54 structures, respectively.

In order to determine the absolute configuration of the C40 disilicides, the intensities of selected reflections were measured separately. For the space group of P622, the signs of the indices for the Bijvoet pair are divided into two groups, *viz* $F(+)$ for ($hkil$, $\overline{hk}il$, $hki\overline{l}$, $hik\overline{l}$), and $F(-)$ for ($\overline{hkil}$, $hki\overline{l}$, $\overline{hik}l$, $hikl$). We chose four types of indices that exhibit significant anomalous-dispersion effects; $51\overline{6}1$, $51\overline{6}2$, $51\overline{6}4$ and $51\overline{6}5$. For each type of indices, the diffraction intensities were measured for all twelve combinations of plus and minus signs.

The lattice constants are determined by the least-squares method (full matrix) by utilizing the following empirical relationship.

$$a_{ij} = a_0 + C_i \cos^2(\omega_{ij}), \tag{1}$$

where a_{ij} is the lattice constant obtained from ω_{ij} which is the j-th reflection of the i-th systematic reflections; a_0 and C_i are the most reliable lattice constant and coefficients which are common for the systematic reflections. The weighting scheme of $1/\cos^2(\omega_{ij})$ was adopted in the calculation.

RESULTS

Figure 1 shows a part of observed and calculated structure factors of MoSi$_2$. If silicon atom is located at the ideal position, the calculated structure factors of the indices indicated in this figure have almost the same value. However, observed structure factors are different from each other. When a certain amount of displacement is considered, all the calculated structure factors well coincide to the observed ones. Then, it is clear that silicon atom is displaced from the ideal position. Such a displacement was observed for all the materials.

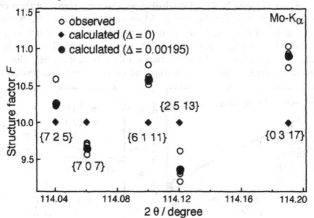

Figure 1. A part of observed and calculated structure factors of MoSi$_2$.

Table 2. Comparison of the observed, Fo, and calculated, Fc, structure factors of four C40 disilicides for reflections with significant anomalous-dispersion effects.

(a) CrSi$_2$

hkil	$|Fo(+)|$	$|Fo(-)|$	P6$_2$22: $|Fc(+)|$ P6$_4$22: $|Fc(-)|$	$|Fc(-)|$ $|Fc(+)|$
5$\bar{1}$61	7.13(4)	6.97(4)	6.97	7.11
5$\bar{1}$62	6.95(4)	7.11(4)	7.09	6.94
5$\bar{1}$64	7.08(4)	6.96(3)	6.97	7.09
5$\bar{1}$65	6.91(4)	7.03(4)	7.05	6.93

(b) VSi$_2$

hkil	$|Fo(+)|$	$|Fo(-)|$	P6$_2$22: $|Fc(+)|$ P6$_4$22: $|Fc(-)|$	$|Fc(-)|$ $|Fc(+)|$
5$\bar{1}$61	6.87(4)	6.31(4)	6.28	6.85
5$\bar{1}$62	6.23(2)	6.79(2)	6.80	6.25
5$\bar{1}$64	6.72(2)	6.21(2)	6.22	6.70
5$\bar{1}$65	6.14(3)	6.59(2)	6.60	6.16

(c) NbSi$_2$

hkil	$|Fo(+)|$	$|Fo(-)|$	P6$_2$22: $|Fc(+)|$ P6$_4$22: $|Fc(-)|$	$|Fc(-)|$ $|Fc(+)|$
5$\bar{1}$61	20.06(8)	20.37(4)	19.93	20.25
5$\bar{1}$62	19.98(4)	19.64(4)	19.94	19.62
5$\bar{1}$64	18.55(5)	18.84(5)	18.56	18.85
5$\bar{1}$65	17.91(5)	17.61(3)	17.99	17.70

(d) TaSi$_2$

hkil	$|Fo(+)|$	$|Fo(-)|$	P6$_2$22: $|Fc(+)|$ P6$_4$22: $|Fc(-)|$	$|Fc(-)|$ $|Fc(+)|$
5$\bar{1}$61	53.6(2)	52.4(1)	52.24	53.47
5$\bar{1}$62	51.9(2)	53.1(1)	52.73	51.50
5$\bar{1}$64	49.9(2)	48.7(2)	48.85	50.01
5$\bar{1}$65	46.4(2)	47.7(2)	48.03	46.88

Table 2 summarizes determined and calculated structure factors of the Bijvoet pair for the four C40 disilicides. For three of the four C40 disilicides, VSi$_2$, CrSi$_2$ and TaSi$_2$, the space group is determined to be P6$_4$22, in contrast to what is listed in Pearson's Handbook. Only NbSi$_2$ exhibits the space group of P6$_2$22. Since the observed values of $|F(+)|$ and $|F(-)|$ agree well with the calculated ones, it is concluded that all the specimens used contain only a crystal belonging to either of the two space groups (P6$_2$22 and P6$_4$22) and that crystals belonging to these two space groups do not coexist in any of individual specimens. However, we do not know whether or not the space group determined in the present study is the unique one. Since we have examined only one crystal for each of the C40 disilicides, there is a possibility that crystals with the other space group (with the other chiral) exist. More work is needed to clarify this.

Figure 2 shows an example of the result by which the lattice constants of MoSi$_2$ are determined. All the experimental values are well fitted by the empirical relationship of eq. (1). This fact indicates that deviations of the experimental values from the ideal value mainly come from systematic errors caused by machinery set-up, and then the most of experimental errors are

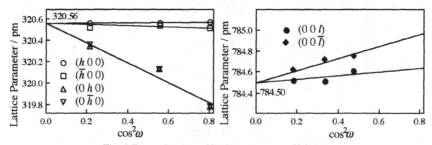

Fig. 2 Determination of lattice constants of MoSi$_2$.

Table. 3. Space group, lattice constants and atomic parameters determined for the seven refractory metal disilicides.

Material	Space group	a / pm	b / pm	c / pm	Δ	R (%)
MoSi$_2$	I4/mmm	320.56(3)	a	784.50(4)	0.00195(5)	0.659
WSi$_2$	I4/mmm	321.38(2)	a	782.99(3)	0.00137(7)	0.333
VSi$_2$	P6$_4$22	457.26(6)	a	637.44(7)	0.00407(4)	1.14
CrSi$_2$	P6$_4$22	442.83(1)	a	636.80(9)	0.00089(3)	1.08
NbSi$_2$	P6$_2$22	479.74(1)	a	659.23(9)	0.00741(4)	1.16
TaSi$_2$	P6$_4$22	478.39(3)	a	657.00(3)	0.0077(2)	1.42
TiSi$_2$	Fddd	826.80(3)	480.02(1)	855.21(6)	0.00339(2)	0.687

correctly compensated by the fitting.

The lattice constants and atomic parameters of transition-metal disilicides with the C11$_b$, C40 and C54 structures refined in the present study are listed in Table 3. The lattice constants determined in the present study for all the disilicides are not so much different from those listed in Pearson's Handbook. It is remarkable to note that all the disilicides exhibit displacement of Si atoms from the ideal position, again in contrast to what is listed in Pearson's Handbook. The value of Δ determined for MoSi$_2$, NbSi$_2$ and TiSi$_2$ are in good agreement with those reported previously; Δ = 0.0020 [9], 0.0068 [10] and 0.0033 [11], respectively.

DISCUSSION

The values of Δ, the displacement of the Si atom, increase in the order of CrSi$_2$ < WSi$_2$ < MoSi$_2$ < TiSi$_2$ < VSi$_2$ < NbSi$_2$ < TaSi$_2$. This order can be related neither to the difference in crystal structure nor atomic number and electro negativity of the constituent refractory metal atom. However, a rather clear relationship is found between the absolute values of the displacement (the value of Δ multiplied by the corresponding lattice parameter) and the interatomic distance in hexagonally-arranged RSi$_2$ layer for each compound as seen in Fig. 3, which layer is the most characteristic feature commonly observed in the C11$_b$, C40 and C54

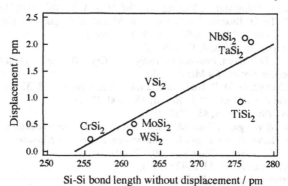

Fig. 3. The absolute values of displacement of the Si atom (Δ) plotted as a function of the interatomic distance of Si-Si bond without displacement.

structures. The interatomic distance in Fig. 3 corresponds to the R-Si (Si-Si) distances in RSi_2 layers, which is calculated by assuming that Si atoms locate in the ideal positions; the absolute values of Δ increase with the increase in the interatomic distance in RSi_2 layers. The Si-Si distance in RSi_2 layers ranges from 256 to 277 pm, which is by far larger than the Si-Si distance (202 pm) in pure Si [5]. Thus, the Si-Si distances in these refractory metal disilicides are uncomfortably elongated and the absolute values of such displacement tend to be higher for disilicides with larger interatomic distance in RSi_2 layers.

CONCLUSIONS

The crystal structures of seven binary refractory metal disilicides with the $C11_b$ ($MoSi_2$ and WSi_2), C40 (VSi_2, $CrSi_2$, $NbSi_2$ and $TaSi_2$) and C54 ($TiSi_2$) structures have been refined through analysis of single-crystal X-ray diffraction data. The present investigation has revealed definite displacements of Si atoms from the ideal positions in such a way that the Si-Si distance in RSi_2 layers is shortened. The absolute values of the displacement increase with the increase in the Si-Si interatomic distance in RSi_2 layers.

The space group of three of the four C40 disilicides, VSi_2, $CrSi_2$ and $TaSi_2$, is determined to be $P6_422$, which is of chirality with respect to $P6_222$ that was assigned in the previous studies. $NbSi_2$ is only the C40 disilicide with the space group of $P6_222$.

ACKNOWLEDGEMENTS

The authors would like to thank Dr. K. Yamamoto, Nara Women's University for helpful advice in sample preparation. This work was supported in part by Grant-in-Aid for Scientific Research (C) from the Ministry of Education, Science, Sports and Culture of Japan and in part by the Japan Society for Promotion of Science Grant on Advantage High-temperature Intermetallics. The authors would like to thank Dr. Y. Yukawa and Mr. T. Shiraki, Nikko Superior Metals, Co. Ltd. for supplying high-purity Mo, Nb and Ta, and Dr. H. Shiraishi, Sumitomo Sitix Co. Ltd., for supplying high-purity Si.

REFERENCES

1. A. K. Vasudevan and J. J. Petrovic, *Mater. Sci. Eng. A*, **155**, 1 (1992).
2. K. Ito, H. Inui, Y. Shirai and M. Yamaguchi, *Phil. Mag. A*, **72**, 1075 (1995).
3. F. Nava, K. N. Tu, O. Thomas, J. P. Senateur, R. Madar, A. Borghesi, G. Guizzetti, U. Gottlieb, O. Laborde and O. Bisi, *Sci. Rep.*, **9**, 141 (1993).
4. S. P. Murarka, *Intermetallics*, **3**, 173 (1995).
5. P. Villars and L. D. Calvert, *Pearson's Handbook of Crystallographic Data for Intermetallic Phases*, (American Society for Metal, 1985).
6. B. K. Bhattacharyya, D. M. Bylander and L. Kleinman, *Phys. Rev. B*, **32**, 7973 (1985).
7. M. Alouani and R. C. Albers, *Phys. Rev. B*, **43**, 6500 (1991).
8. L. F. Mattheiss, *Phys. Rev. B*, **45**, 3252 (1992).
9. Y. Harada, M. Morinaga, D. Saso, M. Takata and M. Sakata, *Intermetallics*, **6**, 523 (1998).
10. R. Kubiak, R. Horyn, H. Broda and K. Lukaszewicz, *Bulletin de L'academie Polonaise des Sciences Serie des Sciences Chimiques*, **20**, 429 (1972).
11. W. Jeitschko, *Acta Cryst. B*, **33**, 2347 (1977).
12. http://www.crystal.uwa.edu.au/xtal

Mat. Res. Soc. Symp. Proc. Vol. 646 © 2001 Materials Research Society

Phase Stability in Processing and Microstructure Control in High Temperature Mo-Si-B Alloys

J.H. Perepezko, R. Sakidja and S. Kim
Department of Materials Science and Engineering, University of Wisconsin-Madison,
1509 University Ave, Madison, WI 53706, USA

ABSTRACT

For applications at ultrahigh temperatures the multiphase microstructural options that can be developed in the Mo-Si-B system have demonstrated an effective and attractive balance of essential characteristics. The coexistence of the high melting point (>2100°C) ternary intermetallic Mo_5SiB_2 (T_2) phase with Mo provides a useful option for in-situ toughening. A further enhancement is available from a precipitation reaction of Mo within the T_2 phase that develops due to the temperature dependence of the solubility behavior of the T_2 phase. However, direct access to $Mo+T_2$ microstructures is not possible in ingot castings due to solidification segregation reactions that yield nonequilibrium boride and silicide phases with sluggish dissolution. Alternate routes involving rapid solidification of powders are effective in suppressing the segregation induced phases. The processing and microstructure options can also be augmented by selected refractory metal substitutional alloying, such as the incorporation of Nb, that alters the solubility of the T_2 phase and the relative phase stability to yield solidification of two phase refractory solid solution + T_2 structures directly. The observed alloying trends highlight the role of atomic size in influencing the relative stability of the T_2 phase. A key component of the overall microstructural control and long term microstructural stability is determined by the kinetics of diffusional processes. The analysis of selected diffusion couples involving binary boride and silicide phases has been used to assess the relative diffusivities in the T_2 phase and coexisting phases over the range of solubility and to provide a basis for the examination of the kinetics of reactions involved in coatings and oxidation.

INTRODUCTION

The performance requirements for structural materials in an elevated temperature environment represent some of the most demanding challenges facing contemporary materials design strategies. In this respect the evolutionary development of superalloys over the past 4-5 decades represents a remarkable achievement and provides important lessons to guide future materials design efforts. One clear message is the importance of multiphase microstructures and the capability to control phase fractions and morphologies within the overall structure [1-3]. The flexibility in microstructure control has been shown to be critical in tailoring alloy performance in order to satisfy a number of mechanical property characteristics that sometimes present conflicting demands [4,6]. Besides the essential structural requirements, elevated temperatures also involve aggressive environments that require a material to display an inherent oxidation protection that can be enhanced further by coatings [7].

The importance of the guidance from the superalloy experience has been demonstrated repeatedly over the past 15 years in several efforts to develop new classes of elevated temperature materials. In particular, the considerable efforts that have been directed to examine the Ni and Ti aluminides clearly reflect this guidance [5,6]. From repeated experiences the most attractive performance in comparison to a monolithic single phase is always demonstrated by a multiphase design where carefully controlled phase fractions and morphologies are required for optimum performance [5,6,8-10]. However, even with the most successful materials designs, the Ni and Ti aluminides do not provide any significant increase in operating temperature that would justify their substitution for superalloys. The case for substitution is stronger for NiAl and TiAl in some applications where a reduced density without significant increase in operating temperature can be used to advantage to benefit overall performance.

At the same time the most challenging demand of a significant increase in operating temperature has not been resolved to date in a successful manner. One important constraint is simply that as the operating temperature increases significantly the number of possible candidate materials decreases [11]. In terms of metallic system candidates (similar considerations apply for ceramics) there are several high melting temperature intermetallics, but there is a much smaller number of intermetallic phases that offer a level of inherent environmental resistance. At elevated temperature alloy phases that contain Al or Si are most attractive for developing stable Al_2O_3 and SiO_2 coatings. Moreover, above about 1400°C, SiO_2 films are preferred since the parabolic rate constant for oxidation is lower for SiO_2 than for Al_2O_3 [12]. In fact this selection is supported by the superior oxidation resistance available with monolithic $MoSi_2$ where an SiO_2 surface provides for useful operation up to about 1700°C (i.e. 0.8 T_m). At high temperatures creep strength is insufficient and at low temperature $MoSi_2$ suffers from brittle behavior [13-15]. Nonetheless the attractive oxidation resistance has motivated efforts to develop multiphase structures based on $MoSi_2$ [16]. However, an effective combination of structures has not been identified for $MoSi_2$ that addresses the mechanical property deficiencies without compromising the oxidation resistance.

At the same time, the multiphase microstructures that can be developed in the Mo-Si-B system offer useful options for high temperature applications [17-18]. Two phase alloys based upon the coexistence of the high melting temperature (>2100°C) ternary intermetallic Mo_5SiB_2 (T_2) phase with Mo allows for in-situ toughening and a further possibility for strengthening through a precipitation of Mo within the T_2 phase [19-21]. Three phase alloys comprised of Mo, T_2 and Mo_3Si offer favorable oxidation resistance [21-23]. A focal point of the microstructural designs is the T_2 phase which develops upon solidification through a peritectic reaction and exhibits a range of solubility. One part of the current effort is directed to an examination of the diffusion behavior in the T_2 phase by determining the kinetics of reactive diffusion between binary silicides and borides. In ternary alloys direct formation of Mo+T_2 structures is not possible due to severe segregation under usual solidification processing conditions [24]. In addressing this issue it has been established that rapid solidification processing such as that available in powders is effective in suppressing the solidification segregation [24-25]. However, alternate approaches are also of interest for bulk ingots. For example, in the current work selected refractory metal substitutional alloying, such as the incorporation of Nb has been examined to alter the solubility of the T_2 phase and the relative phase stability as a method to control the solidification of two phase refractory metal solid solution + T_2 structures. The observed alloying trends also highlight the fundamental factors that influence the relative stability of the T_2 phase.

PHASE STABILITY OF THE Mo-RICH Mo-Si-B SYSTEM AT 1600°C

Based upon the EPMA examination of phase compositions of the long-term annealed as-cast samples and rapidly solidified samples and x-ray diffraction determination of phase identity, the Mo-Si-B system at 1600°C has been constructed as shown in Figure 1. The phase boundaries are plotted on the basis of composition data obtained using EPMA as a broken line. Moreover, the present study has concentrated on the Mo-MoB-Mo$_5$Si$_3$ region. Accordingly the ternary isothermal section does not contain a complete boundary of MoB-Mo$_5$Si$_3$ two-phase equilibrium region as yet. The compositional homogeneity region of Mo$_2$B ranges from ~32 at%B to ~33 at%B and has Si solubility range of <1 at%Si. The MoB phase has a compositional homogeneity region higher than ~48.5 at%B and Si solubility range of <1 at%Si. The Mo-ss phase does have negligible B solubility, but appreciable Si solubility to ~3 at% Si. The Mo$_3$Si phase has a compositional homogeneity region of ~24 – 25 at% Si and a negligible B solubility of <1 at% B. The compositional boundary of Mo$_5$Si$_3$ (T$_1$ phase) on the Mo-rich side extends to ~37 at% Si. The T$_2$ phase has a Si solubility range of ~9 - 13.8 at% Si as well as a B solubility range of ~24 – 27.5 at% B. The T$_2$ phase has a relatively appreciable compositional homogeneity range around the stoichiometric composition (shown as a solid triangle in figure 1). The baseline of phase stability that has been established at 1600°C appears to be maintained over a range of temperatures which are being examined in ongoing studies.

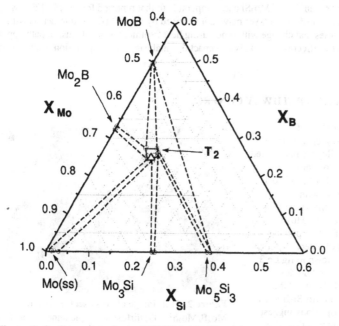

Figure 1. Isothermal section at 1600°C for the Mo-rich portion of Mo-Si-B system [26]

DIFFUSION BEHAVIOR DURING T_2 PHASE FORMATION

A diffusion study has been initiated to examine the formation of the T_2 phase as a diffusion reaction product between the Mo_2B and Mo_5Si_3 phases. The annealing of Mo_2B/Mo_5Si_3 diffusion couples yields the T_2 phase, but two diffusion pathways are observed that involve either the $Mo_2B/T_2/Mo_3Si/Mo_5Si_3$ or $Mo_2B/T_2/Mo_5Si_3$ phase sequences. The T_2 phase initiates and grows from the Mo_2B phase. The T_2 phase grains grow with a columnar structure (figure 2) with a growth direction that has been identified from TEM analysis as normal to <001>. The Mo_3Si phase initiates and grows from the Mo_5Si_3 phase. The BSE image of $Mo_2B/T_2/Mo_5Si_3$ in figure 2 also indicates that the T_2 phase has a relatively planar interface with the Mo_2B phase and a somewhat irregular interface with the Mo_5Si_3 phase. The second diffusion pathway of $Mo_2B/T_2/Mo_3Si/Mo_5Si_3$ was observed in diffusion couples annealed at 1600 °C for 100 and 200 hrs [26]. A key consequence of changing diffusion pathway with annealing time is that the initial diffusion pathway is not the steady state path, but reflects transient condition. The diffusion pathway that appears to represent steady state is given in figure 2, but this assignment requires a further confirmation.

The overall growth kinetics for the T_2 phase demonstrates diffusion control and the estimated growth rate constant for T_2 is 9×10^{-16} m^2/sec at 1600°C [27]. According to Bartlett *et al* [28], the growth rate constant of Mo_3Si produced in the $MoSi_2$-coated Mo system is approximately 3.75×10^{-14} m^2/sec at 1600°C. The growth rate of T_2 is about two orders of magnitude lower than that of Mo_3Si and comparable to that reported for Mo_5Si_3 [29]. Moreover, the growth behavior of the T_2 layer may be influenced by the diffusional interaction with neighboring phases that change with time during heat treatment as the diffusion path approaches steady state. This effect is also likely to impact the analysis and interpretation of oxidation behavior.

SOLIDIFICATION PATHWAY CONTROL

The large extent of solidification segregation in Mo-rich Mo-Si-B alloys is manifested by the presence of boride primary solidification (Mo_2B and MoB) products in alloys with compositions in the $Mo(ss)+T_2$ two-phase field. The Mo_2B primary solidification precludes the attainment of eutectic $Mo(ss) + T_2$ alloys as shown in figure 3 which presents the liquidus projection [24]. To avoid the Mo_2B primary in Mo-Si-B alloys, the liquidus projections suggest that at least the formation of the four-phase equilibria of Class II

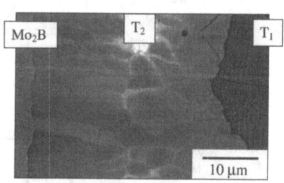

Figure 2. BSE image of cross-section of the Mo_2B/Mo_5Si_3 (T_1) diffusion couple annealed at 1600°C for 400 hours

reaction ($Mo_2B + L \Rightarrow Mo(ss) + T_2$) must be avoided. Since the equilibrium point is outside the two-phase field, the compositions used will always be within the three-phase field of $Mo(ss) + T_2$ + Mo_3Si. This may present some difficulties particularly since the preceding reaction of Class I ($L \Rightarrow Mo(ss) + T_2 + Mo_3Si$) may be bypassed due to the relatively shallow liquidus surfaces for Mo_3Si and T_2 . Therefore, only a slight undercooling is needed to suppress the invariant ternary eutectic and as a consequence the eutectic structure of $Mo_3Si + T_2$ forms instead. The presence of a $Mo_3Si + T_2$ eutectic structure with relatively brittle constituents engulfing the $Mo(ss) + T_2$ eutectic may not be desirable. Furthermore, even if the Class I reaction can be preserved, the volume percentage of the $Mo(ss)$ in the ternary eutectic is quite low since the composition of the four-phase equilibrium point (Class I) is near to that of the Mo_3Si phase. It is therefore necessary to alter the primary solidification event in the two-phase field so that the Mo_2B primary region no longer "interferes" with two-phase field region.

One strategy to directly produce two-phase microstructure from the melt is to suppress the primary solidification of the boride phases by rapid solidification [24, 25]. The Rapid Solidification Processing (RSP) such as splat-quenching can yield a fine two-phase

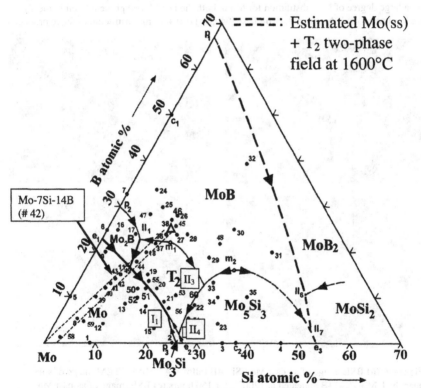

Figure 3. Liquidus projection on the Mo-rich portion of Mo-Si-B system

microstructure by bypassing the formation of the boride phase and by facilitating rapid solidification velocity due to the high cooling rate (10^6K/s). The solidification path of the undercooled melt is simplified into either Mo(ss) or T_2 primary solidification which is then followed by the Mo(ss) + T_2 eutectic formation. In alloys with compositions close to that of the Mo(ss) + T_2 eutectic composition, an amorphous phase may be formed initially which contains a dispersion of a high density dendritic Mo(ss) particles and fine spherical eutectic structures. It is important to point out here that due to the very short diffusion distances in this fine microstructure, subsequent annealing at a relatively low temperature such 1200°C for 150 hours is sufficient to yield a uniform sub-micron two-phase Mo(ss) + T_2 microstructure as shown in Figure 4b.

Another effective design strategy for altering the solidification pathway is to apply selected quarternary additions to alter the extent of the primary boride reactions. In fact, the systematic substitution of Mo by Nb has been shown to reduce the extent of Mo_2B primary on the two-phase field as exemplified in Figure 4c-d. The reduction in the Mo_2B liquidus extension may be due to the fact that the amount of Nb substitution for Mo in Mo_2B is limited [31]. On the other hand, a large degree of Nb substitution for Mo in both the bcc Mo(ss) phase as well in the T_2 phase has been found. As a result, two-phase [Mo,Nb] (ss) + T_2 microstructures can be produced

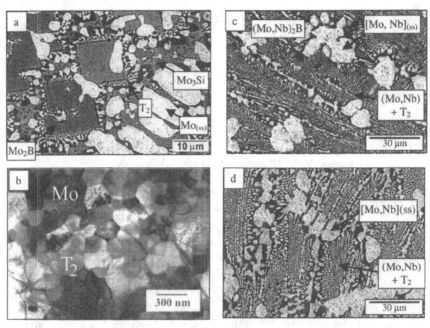

Figure 4. (a) BSE image of as-cast Mo-7Si-14B (alloy # 42) **(b)** BF TEM image of splat-quenched Mo-7Si-14B annealed at 1200°C for 150 hours **(c)** BSE image of as-cast Mo-7Si-14B-5Nb **(d)** as-cast Mo-7Si-14B –20Nb.

directly from the melt.

In alloys where the nominal composition is in the three-phase field of Mo(ss) + T_2 + Mo_3Si, Nb substitution also alters the solidification reactions. The substitution modifies the final solidification reactions that involve the formation of Mo_3Si phase. Similar to the Mo_2B phase, it appears that the liquidus surface extension of Mo_3Si phase is quite limited and therefore with the Nb substitution, the monovariant $(Mo,Nb)_3Si$ + T_2 + [Mo,Nb] (ss) eutectic reaction terminates and the solidification continues with the five-phase equilibria of : L + $(Mo,Nb)_3Si$ => [Mo,Nb](ss) + T_2 + T_1. This reaction allows for the (Mo,Nb) + T_2 eutectic structure in the as-cast alloys to be surrounded with the ternary eutectic that contains a ductile [Mo,Nb] (ss) phase which has a larger volume fraction than that of Mo(ss) phase in the Mo(ss) + T_2 + Mo_3Si ternary eutectic as shown in Figure 5 a-b. Furthermore since the T_1 phase has a much higher oxidation and creep resistance than the Mo_3Si phase, the multi-phase eutectic may serve as a more effective matrix structure. A notable example is shown in figure 6 a-b where the two alloys exhibit solidification pathways that are initiated by the co-precipitation of T_2 and T_1 phases. In the Mo-Si-B alloy, the subsequent solidification reactions lead to the formation of Mo_3Si-rich eutectic serving as the multi-phase matrix. On the other hand, in the Nb-substituted Mo-Si-B alloy, the solidification pathway proceeds with the formation of the [Mo,Nb] (ss) + T_2 + T_1 three-phase eutectic engulfing the two primary phases. In the Mo-Si-B system, the formation of T_2 phase is always initiated by the MoB primary solidification. In contrast, in the Nb-Si-B system, over a certain range of compositions, single phase T_2 can be directly produced from the melt. The Nb substitutional alloying may therefore open up the potential for direct solidification of the Mo-rich T_2 single phase by reducing the extension of the MoB primary region. Furthermore, it is clear that other refractory metals which are known to form continuous solid solution with Mo in the bcc phase and may stabilize the T_2 phase can also have an influence on the solidification pathways.

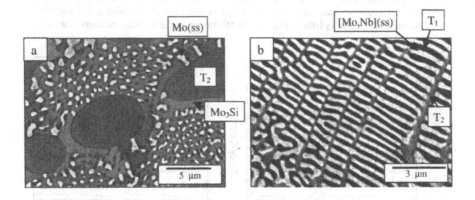

Figure 5. BSE image of (a) the invariant three-phase Mo(ss) + T_2 + Mo_3Si ternary eutectic (b) the three-phase [Mo,Nb](ss) + T_2 + T_1 eutectic

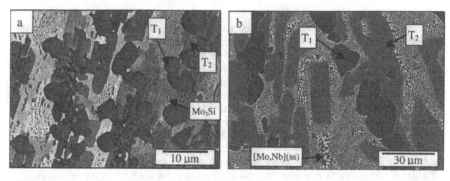

Figure 6. BSE images of (**a**) Mo-Si-B and (**b**) Nb-substituted Mo-Si-B alloys with co-primary solidification of $T_2 + T_1$. In the Mo-Si-B alloy, the primary phases are surrounded by Mo_3Si-rich eutectic whereas in the Nb-substituted Mo-Si-B alloy, the three phase [Mo,Nb](ss) + T_2 + T_1 eutectic serves as the multi-phase matrix.

INFLUENCE OF SUBSTITUTIONAL ALLOYING ON PHASE STABILITY

Two important aspects concerning the extension of the Mo(ss) + T_2 two-phase field in the quaternary systems are the stabilization of the T_2 phase by Nb substitution for Mo and the solubility behavior of metalloid constituents in the T_2 phase. The substitution of Mo by Nb in alloys (e.g. Mo-10Si-20B) with compositions in the two-phase field shows the continuous solid solution in both bcc phase (Mo,Nb) and the T_2 phase as indicated by the continuous shift in X-ray peak positions of the two phases after annealing at 1600°C for 200 hours (Figure 7a). The solubility trends also indicate an increase in both the c and a parameters with increasing Nb substitution level for the T_2 lattice. The total metal content in the T_2 phase remains relatively

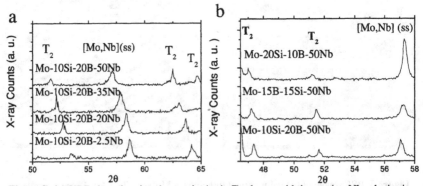

Figure 7. (**a**) XRD data showing the continuity in T_2 phases with increasing Nb substitution (**b**) XRD data showing the expansion of the metalloid solubility in the T_2 phase with Nb substitution

constant at 61-63 at. % indicating that Nb fully substitutes Mo in the T_2 phase. In addition, with an increase in Nb substitution, there is an apparent enlargement of the compositional range of the T_2 phase field involving a shift to the Si-rich region (Figure 7b). The resulting expansion of the two-phase field to the Si-rich regions is also an indication of the increasing ease of metalloid exchange with increasing Nb content.

This finding confirms a continuous solubility between both the Mo and Nb solid solutions and the respective T_2 phases in the Mo-Si-B and Nb-Si-B ternary systems as illustrated in Figure 8. The increased accommodation of Si in the metalloid sites in the T_2 phase reflects the importance of the atomic size on the stability of the T_2 phase. While other considerations clearly should also be taken into account such as the electronegativity and the tendency for the ordered arrangements of metalloid atoms in the T_2 structure, there appear to be geometrically necessary conditions for the stabilization of the T_2 structure. The geometrical constraint can be viewed as a means to allow for optimum packing between the transition metal constituents with the metalloids such as B,Si and Ge. It has been shown previously [30] that a large portion of the

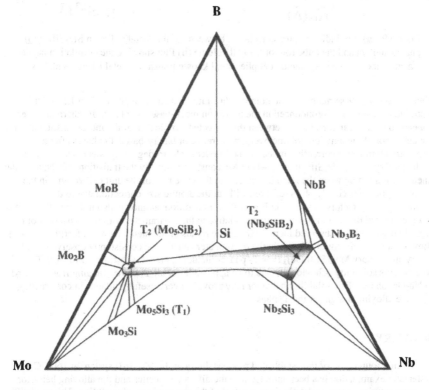

Figure 8. Schematic illustrating the continuity in solubility of T_2 phases from the Mo-Si-B and Nb-Si-B ternary systems.

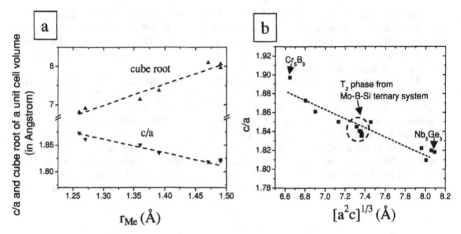

Figure 9. (a) Plot indicating the correlation between r_{Me} the metal radius in Me_5SiB_2 (T_2) phases and c/a and the cube root of the cell volume (b) Plot showing the correlation between c/a and cube root cell volume that applies to all known transition-metal based T_2 phases

binary-based T_2 phase maintains an atomic radius ratio in a strict range of 1.2 to 1.5 and the restriction is even more pronounced in the transition metal-based T_2 phase. Furthermore, there appears to be a linear correlation between the corrected atomic size of the metal constituents and the cube root of the unit cell volume over a wide range of binary-based T_2 phases after a necessary atomic size correction based on the Miedema electronegativity was made. In the case of the ternary-based Me_5SiB_2 phases where Me constitutes strictly the transition metals, a similar linear correlation between the metallic radius and the cube root of the unit cell volume in the ordered T_2 phase is clearly observed even without the atomic size correction due to the electronegativity factors as shown in Figure 9(a). In addition, as shown also in Figure 9(a), there is a preferential dilation of the *a* parameter relative to the *c* parameter with the expansion of the crystal because of the observed increase in atomic radius of the metal atom. In fact, Figure 9(b) shows that the preferential dilation in the *a* parameter can also be consistently observed in all binary and ternary Me_5 (Si,B,Ge)$_3$ type T_2 phases including the Nb-substituted Mo-Si-B T_2 phases. This solubility behavior appears to be applicable for all T_2 phases ranging from the Cr_5B_3 to Nb_5Ge_3 phase. This solubility behavior may provide a very useful guidance in constructing an effective alloying design for the T_2 phase.

SUMMARY

From a foundation of the established phase stability in the Mo-Si-B system at 1600°C a systematic examination has been initiated into the diffusion kinetics and the alloying behavior that are essential factors in developing a control of the microstructural evolution. The ternary T_2 (Mo_5SiB_2) phase is clearly a key member of any alloy design since it develops equilibria with each of the binary boride and silicide phases for Mo-rich alloys in the Mo-Si-B system. The T_2 phase is not a stoichiometric compound, but shows a compositional existence range that extends mainly to B-rich and Si-rich departures from stoichiometry. During reactive diffusion between Mo_2B and Mo_5Si_3, the T_2 phase is synthesized with a columnar growth structure that exhibits a preferred growth approximately normal to the c axis. Analysis of the growth kinetics yields diffusivities of about 10^{-16} m^2/s at 1600°C in the T_2 phase. Following conventional solidification processing Mo-Si-B alloys exhibit extensive segregation in the form of non-equilibrium phases. For example, the stoichiometric T_2 alloy composition follows a solidification path involving: $L \rightarrow MoB + L_1 \rightarrow MoB + T_2 + L_2 \rightarrow MoB + T_2 + Mo_5Si_3 + L_3 \rightarrow MoB + T_2 + Mo_5Si_3 + Mo_3Si + L_4 \rightarrow MoB + T_2 + Mo_5Si_3 + Mo(ss) + Mo_3Si$. The homogenization of the as-cast structure to establish the equilibrium phase constitution is difficult during solid state annealing due to sluggish diffusion kinetics. Alternate processing by rapid solidification or selected alloying such as partial substitution of Mo by Nb has been demonstrated to be effective in reducing segregation to yield two-phase (Mo(ss) +T_2) microstructures directly during solidification. At the same time the observed modification of the relative phase stability due to alloying has revealed the central role that the component atomic sizes play in the structural stability of the T_2 phase and has provided guidance for the further exploration of alloying effects.

ACKNOWLEDGEMENT

This work is sponsored by the Air Force Office of Scientific Research, USAF under grant number F49620-00-1-0077 which is most gratefully acknowledged. We thank Dr. J. Fournelle for expert guidance with EPMA measurements.

REFERENCES

[1] N. A. Stoloff, in *Superalloy II*, edited by C. T. Sims (John Wiley, New York, 1987), p. 61.
[2] E. R. Ross and C. T. Sims, ibid, p. 97.
[3] B. F. Dyson and M. McLean, JISI Int., **30**, 802 (1990).
[4] D. M. Diminuk, D. B. Miracle and C. H. Ward, Mater. Sci. and Tech., **8**, 367 (1992).
[5] E. P. George, M. Yamaguchi, K. S. Kumar and C. T. Liu, Annu. Rev. Mater. Sci., **24**, 409 (1994).
[6] Y. W. Kim and D. M. Diminuk, J. Metals, **43**, 40 (1991).
[7] G. H. Meier and F. S. Pettit, Mater. Sci and Tech., **8**, 331 (1992).
[8] S. Naka, M. Thomas and T. Khan, Mater. Sci. and Tech., **8**, 291 (1992).
[9] R. Yang, N. Saulders, J. A. Leake and R. W. Cahn, Acta Metall. Mater., **40**, 1553 (1992).
[10] R. Yang, J. A. Leake, and R. W. Cahn, Mater. Sci. Engr. A, **A152**, 227 (1992).
[11] R. L. Fleischer, J. Mater. Sci., **22**, 2281 (1987).
[12] N. Birks and G. H. Meier, *Introduction to High Temperature Oxidation of Metals*, (E. Arnolds, London, 1983) p. 54.
[13] A. K. Vasudevan and J. J. Petrovic, Mater. Sci. and Eng. A, **A155**, 1 (1992).
[14] D. M. Shah, D. Berczik, D. Anton and R. Hect, Mater. Sci. Eng. A, **A155**, 45 (1992).
[15] D. E. Alman and N. S. Stoloff, Mat. Res. Soc. Symp. Proc., **322**, 255 (1994).
[16] W. J. Boettinger, J. H. Perepezko and P. S. Frankwicz, Mater. Sci. Eng. A, **A155**, 33 (1992).

[17] J. H. Perepezko, C. A. Nunes, S. H. Yi, and D. J. Thoma, in *High-Temperature Ordered Intermetallic Alloys VII*, edited by C.C. Koch, C. T. Liu, N. S. Stoloff and A. Wanner, (Mater. Res. Soc. Proc. **460**, Pittsburgh, PA, 1997) pp. 1-14.

[18] C. A. Nunes, R. Sakidja and J. H. Perepezko, in *Structural Intermetallics 1997*, edited by M. V. Nathal, R. Darolia, C. T. Liu, P. L. Martin, D. B. Miracle, R. Wagner and M. Yamaguchi (TMS, Warrendale, PA, 1997) p. 831.

[19] R. Sakidja, H. Sieber, and J. H. Perepezko, in *Molybdenum and Molybdenum Alloys*, edited by A. Crowson, E. S. Chen, J.A Shield and P. R. Subramanian (TMS, Warrendale, PA, 1998) pp. 99-110.

[20] R. Sakidja, H. Sieber, J. H. Perepezko, Philosophical Magazine Letters, **79** (6), 351-357 (1999).

[21] J. H. Schneibel, C. T. Liu, D. S. Easton, and C. A. Carmichael, Mat. Sci. & Eng. A, **A1-2**, 78-83 (1999).

[22] J. H. Schneibel, C. T. Liu, L. Heatherly, and M. J. Kramer, Scripta Materialia, **38** (7), 1169-76 (1998).

[23] A. J. Thom, M. K. Meyer, M. Akinc and Y. Kim, in *Processing and Fabrication of Advanced Materials for High Temperature Applications III*, edited by T. S. Srivitsan and V. A. Ravi, (TMS, Warrendale, PA, 1993) pp. 413.

[24] C. A. Nunes, R. Sakidja, Z. Dong and J. H. Perepezko, Intermetallics, **8** (4), 327-337 (2000).

[25] R. Sakidja, G. Wilde, H. Sieber and J. H. Perepezko, in *High-Temperature Ordered Intermetallic Alloys VIII*, edited by E. P. George, M. Yamaguchi and M.J. Mills, (Mater. Res. Soc. Proc. **522**, Pittsburgh, PA, 1999) pp. 1–6.

[26] S. Kim, R. Sakidja, Z. Dong, J. H. Perepezko and Y. W. Kim, this symposium.

[27] S. Kim and J. H. Perepezko, to be published

[28] R. W. Barnett and P. A. Larssen, Trans AIME, **230**, 1528 (1964).

[29] P. C. Tortorici and M A. Dayananda, Mater. Sci. & Eng. A, **A261**, 64-77 (1999).

[30] E. A. Franceschi and F. Ricaldone, Revue de Chimie minerale, **21**, 202-220 (1984).

[31] Y. B. Kuz'ma, Poroshkovaya Metallurgiya [Soviet Powder Metallurgy and Metal Ceramics, **10** (4), 298 (1971)].

Mat. Res. Soc. Symp. Proc. Vol. 646 © 2001 Materials Research Society

Plastic deformation of single crystals with the C11$_b$ structure : Effect of the c/a axial ratio

Kazuhiro Ito, Hironori Yoshioka and Masaharu Yamaguchi
Department of Materials Science and Engineering, Kyoto University, Sakyo-ku, Kyoto 606-8501, JAPAN.

ABSTRACT

MoSi$_2$ has a great potential for very high temperature structural applications. Plastic deformation of MoSi$_2$ single crystals with the C11$_b$ structure is extremely anisotropic. It is caused by non-Schmid behavior of slip on {013}<331> with the higher CRSS values for orientations closer to [001]. In order to provide better understanding of key factors on such non-Schmid behavior in MoSi$_2$ (c/a=2.45), we chose PdZr$_2$ with a c/a axial ratio higher than 3 (c/a=3.30) and characterized the plastic deformation. Compression tests were conducted at various temperatures along [001], [010] and [110] axes. Slip on {013}<100> has the shortest Burgers vector and the largest interplanar spacing in PdZr$_2$ and was observed to be activated for [110] with the lowest CRSS. While slip on {013}<331> can be activated even at –196°C for [001]. Although {013}<331> slip has the same Schmid factors for [001] and [010], the yield stress of the [010]-oriented crystals is about twice higher than that of the [001]-oriented crystals. Thus non-Schmid behavior of slip on {013}<331> is also observed in PdZr$_2$, and the manner is opposite to that in MoSi$_2$. Plastic anisotropy in the C11$_b$ structure will be discussed in terms of the c/a axial ratio.

INTRODUCTION

MoSi$_2$ has a great potential for very high temperature structural applications. This stems from its excellent oxidation resistance, high melting point and relatively low density [1]. The plastic deformation of MoSi$_2$ single crystals with the C11$_b$ structure is extremely anisotropic; the [001]-oriented crystals can be plastically deformed only at 900°C, while plastic flow is possible at ambient temperature for single crystals with orientations other than [001] [2,3]. It is caused by non-Schmid behavior of slip on {013}<331> with the higher CRSS values for orientations closer to [001]. This gives rise to not only high creep strength at high temperatures in a [001]-oriented crystal, but also poor polycrystalline ductility below 1000°C. The best hope for improving the polycrystalline ductility may be reducing the slip asymmetry. In order to provide better understanding of key factors on such non-Schmid behavior, currently atomistic studies based on calculations using ab inito [4-6] and a modified embedded atom method [6] are in progress.

The C11$_b$ structure is derived by stacking up three bcc unit cells (c/a=3) and then compressing them along the c-axis yields MoSi$_2$ (c/a=2.45). In this paper, we introduce another parameter: - *the effect of the c/a axial ratio* -, since slip asymmetry is observed in bcc transition metals, but the plastic deformation tends to be less anisotropic in bcc transition metals [7-10] than in MoSi$_2$ [3]. Figure 1 shows the variation of c/a axial ratio with a-lattice constant for binary intermetallic compounds with the C11$_b$ structure [11,12]. The range of c/a axial ratio is from 2.2 to 4.0, while the ratio can be divided into two groups. One group has a constant c/a axial ratio of about 2.5, while compounds of the other group have variable c/a axial ratios of more than 3. Of

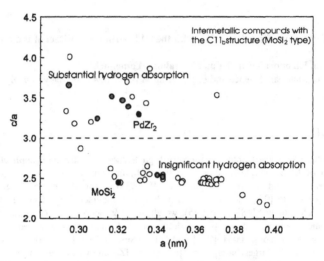

Figure 1. Variation of c/a axial ratio with a-lattice constant for intermetallic compounds with the C11ₗ structure.

interest is that hydrogen absorption behavior also depends on axial ratio such that insignificant hydrogen absorption was observed in the former group, while substantial hydrogen absorption in the latter group, including $PdZr_2$, as shown in Fig. 1 (gray circles) [13]. We chose $PdZr_2$ with a c/a axial ratio higher than 3 (c/a=3.3) and almost the same a-lattice constant as $MoSi_2$. The interplanar spacing of {013} increases with increasing c/a axial ratio, while that of {110} does not vary with it. Thus {013} has a larger interplanar spacing than {110} when c/a>3 and vice versa. We characterized the plastic deformation of single crystalline $PdZr_2$, and slip systems and plastic anisotropy in the C11ₗ structure will be discussed in terms of the c/a axial ratio.

EXPERIMENTAL PROCEDURE

Rods, 10 mm in diameter and 80 mm long, were prepared by Ar arc-melting of high-purity Pd and Zr. Single crystals of $PdZr_2$ were grown from the rods, using our ASGAL FZ-SS35W optical floating-zone furnace at a growth rate of 4 mm h^{-1} under an Ar gas flow. Compression specimens with dimensions 1.5 x 1.5 x 5 mm^3 were cut from as-grown single crystals after determining their crystal orientations by the back-Laue x-ray diffraction method, and then mechanically polished with diamond paste. The compression axes investigated were [001], [110] and [010] in the [001]-[010]-[110] standard triangle. These axes have the same Schmid factor of 0.413 for {013}<331> slip. Compression tests were conducted on an Instron-type testing machine in the temperature range from −196°C to 800°C at a strain rate of 1x10^{-4} s^{-1} in vacuum. Operative slip planes were determined by optical microscope observations of slip traces on two orthogonal surfaces of specimens. Dislocation structures and dislocation

Burgers vectors were examined by transmission electron microscopy, JEM-2000FX. Thin foils for TEM observations were prepared by Ar ion-milling with 5 keV.

RESULTS

Yield stresses obtained at a strain rate of 10^{-4} s^{-1} for [001]-, [110]- and [010]-oriented specimens, corresponding to the flow stress at 0.2 % plastic strain in the stress-strain curves, are plotted in Fig. 2 as a function of temperature. For the specimens with orientations other than [001], the plastic deformation is observed above room temperature (RT) and the yield stress decreases with increasing temperature. While, the [001]-oriented specimens can be plastically deformed even at −196°C. Their yield stress slightly increases with increasing temperature in the temperature range between RT and 600°C. The yield stress is the highest for [010], followed by that for [001] and for [110] in the temperature range between RT and 600°C. The tendency of orientation dependence of yield stress in PdZr$_2$ is different from that in MoSi$_2$, which has a much higher yield stress for orientations closer to [001].

Slip trace observation on the two orthogonal faces indicates that slip on {013} was activated at RT for the specimens with three orientations. Fine slip lines are observed and are homogeneously distributed for [001] and [110], while coarse slip lines are observed and are heterogeneously distributed for [010]. The activated slip plane and morphology of slip lines do not change in the temperature range investigated for each orientation, although activated slip plane could not be identified at 800°C for both [110] and [010]. Of interest is that cleavage fracture occurs on (001) plane in compression along [001] with considerable plastic deformation and along [010] and [110] at −196°C.

Figure 3 shows a typical dislocation structure for slip on (013) observed in a

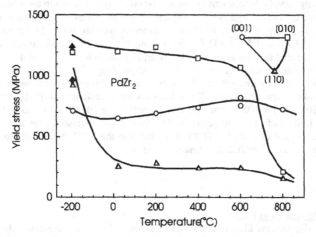

Figure 2. *Temperature dependence of yield stress for PdZr$_2$ single crystals with orientations [001], [110] and [010]. Stresses for orientations [110] and [010] at −196° are those at fracture.*

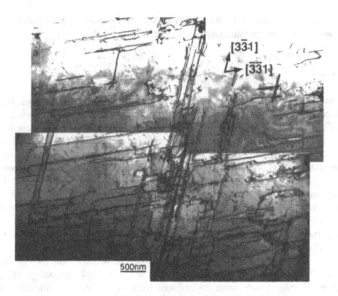

Figure 3. Dislocation structure observed in a [001]-oriented crystal deformed at 200°C. Thin foil was cut parallel to (013).

[001]-oriented crystal deformed at 200°C. The thin foil was cut parallel to (013) macroscopic slip plane. Long straight dislocations are seen to lie on (013) slip planes in the whole area imaged. These dislocations tend to align along [3 $\overline{3}$ 1] and [$\overline{3}$ $\overline{3}$ 1] directions. Contrast analyses carried out to determine the Burgers vector of these dislocations indicate b=[3 $\overline{3}$ 1] for the dislocations aligned along [3$\overline{3}$ 1] and b=[$\overline{3}$ $\overline{3}$ 1] for those along [$\overline{3}$ $\overline{3}$ 1]. Thus, dislocations with b=<331> gliding on (013) tend to align along their screw orientation. In a [110]-oriented crystal deformed at 200°C, long dislocations lying on (013) tend to align along [100] directions. Contrast analysis indicates b=[100] for the dislocations. Thus, dislocations with b=[100] gliding on (013) tend to align along their screw orientation. Short dislocation segments with b=[010] are also seen to align along [010] in the [001] zone. We thus identified the {013}<331> and {013}<100> slip systems for [001] and [110], respectively. Different dislocation structures are observed in a [010]-oriented crystal deformed at 200°C from those in the [001]- and [110]-oriented crystals. However, we have not identified deformation mode yet.

DISCUSSION

Operative slip systems in PdZr$_2$

The slip systems identified at a strain rate of 10^{-4} s^{-1} in the present study for the [001] and [110] orientations are {013}<331> and {013}<100>, respectively. Slip on {013}<100> has a lower CRSS than {013}<331>, since it has a lower Schmid factor for [110] than {013}<331>.

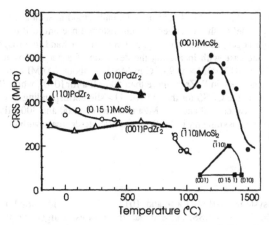

Figure 4. *Temperature dependence of CRSS obtained at a strain rate of 1×10^{-4} s^{-1} for slip on $\{013\}<331>$ in MoSi$_2$ (c/a=2.45) and PdZr$_2$ (c/a=3.30) single crystals with various orientations.*

Also it has the lowest CRSS of the slip systems with the shortest Burgers vector of <100>. It is in sharp contrast to the operation of slip on $\{011\}<100>$ and $\{010\}<100>$ for [110] in MoSi$_2$ [2,3]. With increasing c/a axial ratio in the C11$_b$ structure, the interplanar spacing of $\{013\}$ increases more than those of $\{011\}$ and $\{010\}$. Thus it is attributed to the effect of the c/a axial ratio that $\{013\}<100>$ other than $\{011\}<100>$ and $\{010\}<100>$ is operative in PdZr$_2$ (c/a=3.30). For [001] and [010], on the other hand, slip systems with Burgers vector of <100> could not be activated. The slip on $\{013\}<331>$ was observed for [001] and supposed to be activated for [010]. The CRSS for the slip for both orientations was calculated using the corresponding Schmid factor of 0.413 as a function of temperature and the results obtained are shown in Fig. 4. The CRSS for $\{013\}<331>$ strongly depends on crystal orientation such that the CRSS for [010] is higher than that for [001]. Also the CRSS for $\{013\}<331>$ at $-196°C$ for [110] would be higher than that for [001] ($\{013\}<331>$ has the same Schmid factor in the three orientation).

Effect of the c/a axial ratio on slip asymmetry

The manner of the non-Schmid behavior observed in PdZr$_2$ (c/a=3.30) is opposite to that

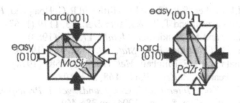

Figure 5. *Distorted bcc unit cells for MoSi$_2$ (c/a=2.45) and PdZr$_2$ (c/a=3.30) and their easy and hard orientations for compression.*

N4.6.5

in $MoSi_2$ (c/a=2.45), as shown in Fig.4. This may be understood in such a way that compression along the direction parallel to shorter edges of their distorted bcc unit cell is more difficult than that parallel to longer edges, as shown in Fig. 5. On the other hand, the magnitude of the slip asymmetry tends to increase with increasing the deviation of a c/a axial ratio from c/a=3. For example, the CRSS for {013}<331> at 900°C in $MoSi_2$ is about 700 MPa for [001], while 250 MPa for [110] [3]. That at RT in $PdZr_2$ is more than 500 MPa for [010], while 300 MPa for [001]. In high purity Fe and Ta, the CRSS for the {110}<111> slip corresponding to {013}<331> in the $C11_b$ structure varies between 130 and 180 MPa and between 200 and 300 MPa, respectively, at 77K [8,10]. Thus, the slip asymmetry in $MoSi_2$ is expected to decrease with increasing a c/a axial ratio from 2.45 to 3.

CONCLUSIONS

Two slip systems, {013}<100> and {013}<331>, are identified in $PdZr_2$. Slip on {013}<100> has the lowest CRSS, since it has the shortest Burgers vector and the largest interplanar spacing. The CRSS for {013}<331> depends on crystal orientation such that the CRSS for [010] and [110] is higher than that for [001]. The manner of the non- Schmid behavior observed in $PdZr_2$ (c/a=3.30) is opposite to that in $MoSi_2$ (c/a=2.45). This may be understand in such a way that compression along the direction parallel to shorter edges of their distorted bcc unit cell is more difficult than that parallel to longer edges.

ACKNOWLEDGMENTS

This work was supported by Grant-in-Aid for Scientific Research from the Ministry of Education, Science and Culture (No. 11750612) and JSPS-RFTF96R12301.

REFERENCES

1. A.K. Vasudevan and J.J. Petrovic, *Mater. Sci. Eng. A*, **155**, 1 (1992).
2. K.Ito, H. Inui, Y. Shirai and M. Yamaguchi, *Phil. Mag. A*, **72**, 4, 1075 (1995).
3. K.Ito, T. Yano, T. Nakamoto, H. Inui and M. Yamaguchi, *Intermetallics*, **4**, S119 (1996).
4. U.V. Waghmare, V. Bulatov, E. Kaxiras and M.S. Duesbery, *Phil. Mag. A*, **79**, 3, 655 (1999).
5. U.V. Waghmare, E. Kaxiras, V.Bulatov and M.S. Duesbery, *Modelling Simul. Mater. Sci. Eng.*, **6**, 493 (1998).
6. T.E. Mithchell, M.I. Baskes, S.P. Chen, J.P. Hirth and R.G. Hoagland, *Phil. Mag. A*, in press.
7. S. Takeuchi, E. Furubayashi and T. Taoka, *Acta Met.*, **15**, 1179 (1967).
8. M. Feller-Kniepmeier and M. Hundt, *Scr. Metall.*, **17**, 905 (1983).
9. M.H.A. Nawaz and B.L. Mordike, *Phys. Stat. Sol. a*, **32**, 449 (1975).
10. G.L. Webb, R. Gibala and T.E. Mitchell, *Met. Trans.*, **5**, 1581 (1974).
11. H. Heller and W.B. Pearson, *Z. Kristallogr.*, **168**, 273 (1984).
12. M.V. Nevitt and C.C. Koch, *Intermetallic Compounds: vol. 1, Principles*, ed. J.H. Westbrook and R.L. Fleischer (John Wiley & Sons, 1994) pp.385-401.
13. A.J. Maeland and G.G. Libowitz, *J. Less-Common Met.*, **74**, 295 (1980).

Mat. Res. Soc. Symp. Proc. Vol. 646 © 2001 Materials Research Society

Mechanical Behavior of Molybdenum Disilicide-Based Alloys

A. Misra, A.A. Sharif[1], J. J. Petrovic, and T. E. Mitchell,
MST Division, Los Alamos National Laboratory, Los Alamos, NM 87545
[1] University of Michigan, Department of Engineering Science, Flint, MI 48502

ABSTRACT

We have investigated the mechanical behavior of the following single-phase polycrystalline alloys with the $MoSi_2$ body-center tetragonal structure: $MoSi_2$ alloyed with ~2.5 at.% Re, $MoSi_2$ alloyed with 2 at.% Al, $MoSi_2$ alloyed with 1 at.% Nb, and $MoSi_2$ alloyed with 1 at.% Re and 2 at.% Al. Several anomalies in the mechanical behavior of alloyed materials were observed. For example, (i) addition of only ~2.5 at. % Re results in an order of magnitude increase in compressive strength at 1600 °C, (ii) additions of Nb and Al cause solution softening at near-ambient temperatures, and (iii) quaternary $MoSi_2$-Re-Al alloys show strengthening at elevated temperatures and reduction in flow stress with enhanced plasticity at near-ambient temperatures in compression. The mechanisms of anomalous solution hardening and softening are discussed.

INTRODUCTION

Significant increases in the operating temperatures of high-temperature structural components will only be possible if superalloys are replaced with higher melting temperature intermetallic or ceramic materials [1-2]. Refractory metal silicides such as $MoSi_2$, Mo_5SiB_2, Nb_5Si_3, etc. typically have melting temperatures in excess of 2000 °C and are potential materials for use at temperatures exceeding the melting point of Ni-base superalloys [1-3]. In addition to the low room temperature fracture toughness, the strengths of monolithic silicides at temperatures above ~1400 °C are typically very low. Silicides in the composite form, either with ceramic reinforcements (e.g., $MoSi_2$-SiC) or eutectics ($MoSi_2$-Mo_5Si_3), often show enhanced high temperature strengths but the room temperature fracture toughness is still very low [4]. Ductile metallic reinforcements such as Nb react with the disilicide matrix to form other silicides and have limited practical use for long-term high temperature applications, although improvements in toughness have been demonstrated in these composites [5]. This approach, however, may be promising for Si-poor silicides, as shown in systems such as Nb-Nb_5Si_3, Mo-Mo_5SiB_2, etc. [6].

Due to the complex crystal structures and presumably limited deformation capability of these compounds, little work has been done on studying how alloying in the single-phase region influences the deformation behavior. Recent studies on single crystal $MoSi_2$ have shown compressive plasticity even at temperatures below ambient [7,8]. Furthermore, first principles calculations have predicted that solutes such as Al, Nb, Mg and V may enhance the ductility of $MoSi_2$ [9]. In the present investigation, we have chosen $MoSi_2$ as a model material (Fig. 1) to study selected substitutional alloying effects that are anomalous as compared to the classical theories for metals.

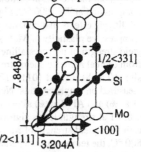

Fig. 1 Body-centered tetragonal unit cell of $MoSi_2$; slip vectors are shown by arrows.

EXPERIMENTAL PROCEDURES

Pure $MoSi_2$, ternary $(Mo,Re)Si_2$, $(Mo,Nb)Si_2$, $Mo(Si,Al)_2$ and quaternary $(Mo,Re)(Al,Si)_2$ alloys were prepared by arc-melting high-purity elements in an argon atmosphere. The starting melt compositions were $MoSi_x$, with x=2.01 to account for the Si loss during melting. The buttons were turned over and remelted 4 times for homogeneity. Alloying levels were in the 1-2 at.% range. Compression testing was performed on 2 x 2 x 4 mm^3 samples, polished to 0.05 μm finish, in air using an Instron 1125 machine at an initial strain rate of ~1 x 10^{-4} /s. Transmission electron microscopy (TEM) was performed on a Philips CM30 microscope operating at 300 kV.

RESULTS

All materials tested were polycrystalline with large grain sizes on the order of 100 μm. Compression testing revealed three unusual phenomena in these alloys that are described below:

Anomalous Hardening

Alloying with Re in the 1-2% level resulted in large increases in the high temperature (T) yield strength of $MoSi_2$ (Fig. 2). No tests were conducted at T > 1600 °C. At T < 900 °C, fracture before yield was typically observed and data for $MoSi_2$ are estimated from tests on single crystals [7,8]. Fig. 2 is a log-linear plot, where linear fits to the data are consistent

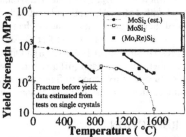

Fig. 2 Effect of 2.5 at.% Re on the high T strength of polycrystalline $MoSi_2$. Data at T < 900 °C are estimated (est.) from single crystal tests [7,8].

with a Peierls-stress controlled behavior [10]. Note that the strength of $MoSi_2$ drops significantly at T >1400 °C, while the Re-alloys show better strength retention. Even at 1600 °C, the addition of only 2.5 at.% Re resulted in an order of magnitude increase in high temperature strength.

Solution Softening

Fig. 2 shows that Re is a potent hardener in $MoSi_2$, but the Re-alloys exhibit no plasticity in compression at T <1000 °C. Fig. 3 shows examples of alloying effects that enhance the room temperature compressive plasticity of $MoSi_2$. At T ≤ 900 °C, the yield strength of polycrystalline $MoSi_2$ is estimated from critical resolved shear stresses for {110}<111> slip at T>500 °C, and {011}<100] at T < 500 °C, measured on single crystals. Note that both Nb (1 at.%) and Al (2 at.%) alloys are deformable at room T at strength levels significantly lower than the estimated yield strength of $MoSi_2$ at room T. At T < 800 °C, the temperature dependence of flow stress of $MoSi_2$ is also lowered by Nb and Al additions.

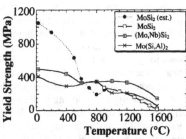

Fig. 3 Effects of 1 at.% Nb and 2 at.% Al alloying, respectively, on the yield strength of polycrystalline $MoSi_2$. The alloys exhibit compressive plasticity at room temperature.

High T Hardening and Low T Softening

Fig. 4 shows two examples where alloying additions have both increased the high T strength and reduced the low T yield strength enough to cause compressive plasticity at room T. This behavior is best exhibited by the quaternary $(Mo,Re)(Si,Al)_2$ alloys that contained only 1 at.% Re and 2 at.% Al. In other words, these alloys combine the beneficial effects of Re (high T hardening) and Al (low T softening) alloying shown in Fig. 2 and 3. Similar effect is seen with Nb additions (Fig. 3 and 4), where only 1 at.% Nb lowers the minimum T at which compressive plasticity is observed from ~900 °C to ambient, and resulted in high T hardening. The Al-containing ternary alloys (Fig. 3),

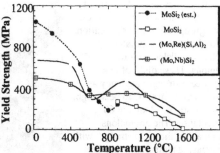

Fig. 4 Quaternary $(Mo,Re)(Si,Al)_2$ alloys with ~ 1 at.%Re and 2 at.% Al are stronger than $MoSi_2$ at high T and softer at near-ambient T. Ternary Nb-containing alloys exhibit similar behavior.

however, have insignificant effect on high T strength.

Dislocation Substructures

In unalloyed $MoSi_2$, dislocation substructures were similar to those observed in pure metals deformed at elevated temperatures [11]. Regular cell structures and sub-grains were observed at deformation T > 1300 °C consistent with climb controlled deformation, i.e., glide is easy and climb of dislocations over substructure obstacles is the rate controlling step. The Re-containing alloys show "viscous glide" controlled substructures, i.e., lots of free dislocations and tangles of the <100] and $1/2$<111> type, with little tendency to form cell structures even at 1600 °C. Furthermore, weak beam imaging of the $1/2$<111> dislocations indicated that Re alloying increases the stacking fault energy [11].

In the ternary $(Mo,Nb)Si_2$, $Mo(Si,Al)_2$ and the quaternary $(Mo,Re)(Al,Si)_2$ alloys, the dislocation substructures at temperatures where solution softening is observed (i.e., T < ~600 °C) consisted primarily of $1/2$<111> dislocations, with some <100> dislocations as shown in Fig. 5. The preferred line direction of the $1/2$<111> dislocations appeared to be 60° from screw, consistent with similar observations on unalloyed $MoSi_2$ [7,8]. While polycrystalline $MoSi_2$ could not be deformed at low temperatures in compression, studies of dislocations under hardness indents in polycrystalline $MoSi_2$ have revealed primarily <100> type dislocations at these temperatures [12]. This suggests that alloying elements such as Nb and Al that cause solution softening at low temperatures may promote $1/2$<111> slip in $MoSi_2$. A similar inference was made through an analysis of slip traces around room temperature hardness indents on single crystalline $Mo(Si,Al)_2$ [13], and more recently, through compression testing on single crystals [14]. Another important effect revealed in weak beam imaging of $1/2$<111> dislocations in Nb-alloyed materials, Fig. 5, (and also, Al-alloyed materials not shown here) is that partial spacing is increased, i.e., stacking fault energy is lowered, consistent with other studies [14,15].

At intermediate temperatures (~800-1000 °C), when the flow stress increases anomalously with increasing temperatures, still predominantly $1/2$<111> dislocations are observed in the Nb and Al-alloyed $MoSi_2$. Note that at these temperatures, solution softening is not observed. Most of the 1/2<111> dislocations on {110} planes are in the form of "bundles" at these temperatures, i.e., groups of several dislocations closely spaced, separated by relatively dislocation free regions. These dislocations showed a stronger tendency as compared to 400 °C to be straight and mostly be along the 60° orientation. The anomalous increase in the yield strength with increasing temperature has been interpreted as a Portevin-Le Chatelier effect in unalloyed MoSi$_2$, presumably caused by interstitial impurities, even though impurity levels as low as 40 wtppm were reported in float-zone crystals studied by Ito *et al.*

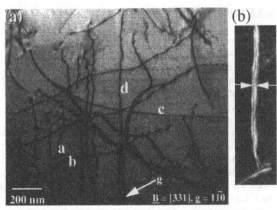

Fig. 5 (a) Bright field and (b) weak beam TEM micrographs showing the dislocation substructures in a MoSi$_2$-1 at.% Nb alloy compressed ~0.5% at 400 °C. The labels a and d correspond to <100], and b and c correspond to 1/2<111> type dislocations respectively. The weak beam image shows that the 1/2<111> dislocations are dissociated into 1/4<111>partials, with lower fault energy than unalloyed MoSi$_2$.

[7]. An alternate view is that the 60° non-screw dislocations lie along the <110] directions which are common to several low-index intersecting planes such as {110), (001) and {331). Hence, the possibility of a non-planar dislocation core at these temperatures with segments on planes where the critical resolved shear stress for glide is very high cannot be ruled out. Our present work shows that anomalous yielding is also observed in alloyed MoSi$_2$. Needless to add, the mechanisms of anomalous yielding in MoSi$_2$ need to be studied in more detail including first principles and atomistic simulations of dislocations, as shown in a recent study by Mitchell *et al.* [16].

At elevated temperatures (>1200 °C), the Nb-alloyed materials are harder than MoSi$_2$ and dislocation substructures (Fig. 6) show predominantly <100] type dislocations and dipole loops that may be pinched-off gliding <100> dislocations. Several short <110> dislocation segments are also seen formed as a reaction product between [100] and [010] dislocations. Since this work is on polycrystals, the critical resolved shear stresses for <111> and <100] dislocations could not be determined as a function of temperature. Also, more statistics is needed to quantitatively assess the relative amounts of <111> and <100] slip at elevated temperatures. From Fig. 6, it does, however, appear that <111> slip is not as favored as it is at lower temperatures compared to <100] slip.

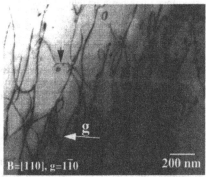

Fig. 6 Bright field TEM micrograph from a MoSi$_2$-1at.% Nb alloy deformed ~ 1% at 1200 °C. All dislocations in this micrograph have a <100] type Burgers vector. Reactions between [100] and [010] produce [110] dislocations (e.g., marked with an arrow), and dipole loops, that may have been "pinched-off" from glide dislocations are also seen.

DISCUSSION

The rapid solution hardening of $MoSi_2$ by Re at elevated temperatures is anomalous and may not be explained within the framework of the classical solution hardening theories based on atomic size and modulus misfits. Since rhenium "disilicide" has a Si-deficient stoichiometry corresponding to $ReSi_{1.75}$ [17], a Si vacancy may form with every four Re atoms added. The pairing of Re substitutionals with Si constitutional vacancies will result in point defect complexes with elliptical strain fields and hence, strong interaction with both edge and screw dislocations. The fact that hardening is observed even at 1600 °C indicates that these defect complexes are very stable. A dislocation model for this rapid hardening has been presented elsewhere [11], and we only discuss the solution softening here.

With regard to solution softening in $MoSi_2$, we note the following features: (i) softening is caused by elements such as Nb, Al, Ta, etc. that change the structure of $MoSi_2$ from body-centered tetragonal ($C11_b$) to hexagonal, (ii) softening occurs only at low concentrations (typically < 2-3 at.%) in the single-phase $C11_b$ structure, i.e., at higher concentrations of Nb or Al when the structure changes to hexagonal, the normal hardening behavior is observed, (iii) softening occurs only at low temperatures (< ~600 °C), and (iv) all solutes that soften $MoSi_2$ lower the stacking fault energy, as inferred from the 1/4<111> partial spacing. Furthermore, Harada et al [18] have shown that elements such as Zr with a strong affinity for interstitial impurities did not result in any significant softening. This led Harada et al to conclude that the scavenging of interstitial solutes by Al (or, Nb) may not be the dominant mechanism for solution softening in $MoSi_2$. The more likely mechanism is the lowering of the Peierls stress and/or easier double kink nucleation in the presence of solutes. Waghmare et al. [9] have shown, through first-principles calculations, that elements such as Nb, Al, Mg, and V change the generalized stacking fault energy (γ-surface) of dislocations in $MoSi_2$. Since Peierls stress is given directly by the maximum gradient of the γ-surface, the net effect of solutes such as Nb and Al was predicted to be a lowering of the Peierls barrier at low homologous temperatures. Experimentally, we observed that Nb and Al lowered the stacking fault energy, and the increased 1/4<111> partial spacing may result in increased mobility of these partials, consistent with the lower Peierls barrier predicted by Waghmare et al [9]. The reason why these solutes segregate to the 1/2<111> dislocation may be related to the higher solubility of these solutes in the hexagonal structure of $MoSi_2$. The difference in the tetragonal and hexagonal structures of $MoSi_2$ is that former has ABAB stacking while the latter has ABCABC stacking of {110} planes. The stacking sequence in the fault that separates the two 1/4<111> partials in the otherwise tetragonal $MoSi_2$ is also ABCABC. Thus, solutes that stabilize the hexagonal structure are likely to segregate to the fault. We further speculate that at elevated temperatures, changes in the core structure (e.g., climb-assisted) and/or solute mobility may reduce the effect of solute on the γ-surface and the Peierls stress.

SUMMARY

The anomalous effects of substitutional alloying, at < 2 at.% level, on the mechanical behavior of polycrystalline $MoSi_2$ are summarized as follows:

(i) <u>Rapid hardening by Re at elevated temperatures</u>: an order of magnitude increase in compressive strength was observed at 1600 °C, and is interpreted as due to strong interactions of both screw and edge dislocations with Re substitutional-Si vacancy point defect complexes.

(ii) <u>Solution Softening at temperatures < ~ 600 °C</u>: alloying with either Nb or Al caused a decrease in yield strength at these temperatures. The Nb or Al-alloyed $MoSi_2$ exhibited compressive plasticity at room temperature, while the unalloyed $MoSi_2$ fractured before yield at $T < ~ 900$ °C. We hypothesize that these solutes, through a lowering of the stacking fault energy, are able to lower the Peierls barrier at low temperatures.

(iii) <u>Low Temperature Softening and High Temperature Hardening</u>: observed in ternary $(Mo,Nb)Si_2$ and quaternary $(Mo,Re)(Si,Al)_2$ alloys. These results, particularly the quaternary alloys, indicate a non-linear addition of the effects of different solutes, e.g., Al effect dominates at low temperatures and Re effect at high temperatures.

ACKNOWLEDGEMENT

This research was funded by Department of Energy, Office of Basic Energy Sciences. We acknowledge discussions with M. Baskes, S.P. Chen and R.G. Hoagland.

REFERENCES

1. J.J. Petrovic and A.K. Vasudevan, *Mat.Sci.Eng.A*, **261**, 1-5 (1999).
2. T.E. Mitchell, R.G. Castro, J.J. Petrovic, S.A. Maloy, O. Unal and M.M. Chadwick, *Mat.Sci.Eng.A*, **155**, 241 (1992).
3. R. Gibala, A. K. Ghosh, D. C. Van Aken, D. J. Srolovitz, A. Basu, H. Chang, D. P. Mason and W. Yang, *Mater. Sci. Eng.A*, **155**, 147 (1992).
4. K. Ito, T. Yano, T. Nakamoto, M. Moriwaki, H. Inui and M. Yamaguchi, *Prog. Mat. Sci.*, **42**, 193 (1997).
5. K.T.V. Rao, W.O. Soboyejo and R.O. Ritchie, *Met.Trans.A*, **23**, 2249-2257 (1992).
6. J.H. Schneibel, C.T. Liu, D.S. Easton and C.A. Carmichael, *Mat.Sci.Eng.A*, **261**, 78 (1999).
7. K. Ito, H. Inui, Y. Shirai and M. Yamaguchi, *Phil. Mag. A*, **72**, p 1075 (1995).
8. S.A. Maloy, T.E. Mitchell and A.H. Heuer, *Acta Metall.Mater.*, **43**, 657 (1995).
9. U.V. Waghmare, V. Bulatov, E.Kaxiras and M.S. Duesbery, *Mat.Sci.Eng.A*, **261**, 147 (1999).
10. T.E. Mitchell, P. Peralta and J.P. Hirth, *Acta Mat.*, **47**, 3687 (1999).
11. A. Misra, A.A. Sharif, J.J. Petrovic and T.E. Mitchell, *Acta Mat.*,**48**, 925-932 (2000).
12. S.A. Maloy, A.H. Heuer, J.J. Lewandowski and T.E. Mitchell, *Acta Metall. Mater.*, **40**, 3159 (1992).
13. P. Peralta, S.A. Maloy, F. Chu, J.J. Petrovic and T.E. Mitchell, *Scripta Mat.*, **37**,1599 (1997).
14. H. Inui, *et al.*, this symposium proceedings.
15. D.J. Evans, F.J. Scheltens, J.B. Woodhouse and H.L. Fraser, *Phil. Mag. A*, **75**, 17 (1997).
16. T. E. Mitchell, M. I. Baskes, S. P. Chen, J. P. Hirth and R. G. Hoagland, Phil. Mag.A, in press.
17. A. Misra, F. Chu and T.E. Mitchell, Phil. Mag.A, **79**, 1411 (1999).
18. Y. Harada, Y. Murata and M. Morinaga, Intermetallics, 6, 529-535 (1998).

Mat. Res. Soc. Symp. Proc. Vol. 646 © 2001 Materials Research Society

Ab initio simulation of a tensile test in MoSi$_2$ and WSi$_2$

M. Friák[*†], M. Šob[*], and V. Vitek[‡]
[*]Institute of Physics of Materials, Academy of Sciences of the Czech Republic, Žižkova 22, CZ-616 62 Brno, Czech Republic, mafri@ipm.cz
[†]Department of Solid State Physics, Faculty of Science, Masaryk University, Kotlářská 2, CZ-611 37 Brno, Czech Republic
[‡]Department of Materials Science and Engineering, University of Pennsylvania, 3231 Walnut St., Philadelphia, PA 19104-6272, U. S. A.

ABSTRACT

The tensile test in transition metal disilicides with C11$_b$ structure is simulated by *ab initio* electronic structure calculations using full potential linearized augmented plane wave method (FLAPW). Full relaxation of both external and internal parameters is performed. The theoretical tensile strength of MoSi$_2$ and WSi$_2$ for [001] loading is determined and compared with those of other materials.

INTRODUCTION

Transition metal (TM) silicides are considered as a very promising basis for a new generation of high-temperature structural materials that can significantly improve the thermal efficiency of energy conversion systems and advanced engines. The reason is that at high temperature they combine the ductility and thermal conductivity of metals with high strength and corrosion resistance of ceramics. Intrinsic oxidation resistance is due to the formation of silicon oxide films at surfaces and high creep strength is related to low diffusion coefficients. The melting temperature is much higher than that of Ni-based superalloys or Ni and Ti-based aluminides and is comparable to that of silicon-based ceramics. The largest impediment is low ductility and/or toughness at ambient temperatures.

The purpose of this paper is to investigate, from first principles, the electronic structure and ground state of TM-disilicides, MoSi$_2$ and WSi$_2$, with C11$_b$ structure, to simulate a tensile test for ideal crystal without defects, including full relaxation of both external and internal parameters, and to determine thus the theoretical tensile strength for the [001] loading.

TENSILE TEST SIMULATION

The tensile strength of materials is usually limited by presence of internal defects, mostly dislocations. In a defect-free crystal, the tensile strength is several order of magnitude higher and is comparable with elastic moduli. Most of the calculations of theoretical (ideal) strength is based on empirical potentials with the parameters adjusted to experimental data. However, most of these experimental data correspond to the equilibrium ground state. Therefore, the semiempirical approaches adapted to the equilibrium state may not be valid for materials loaded close to their theoretical strength limits.

In the first-principles (*ab initio*) electronic structure calculations, we start from the fundamental quantum theory. The only input is atomic numbers of the constituent atoms and, usually, some structural information. This approach is reliable even for highly non-equilibrium states.

To simulate the tensile test, we first calculate the total energy of the material in the ground state. Then, in the second step, we apply some elongation of the crystal along the loading axis (in the [001] direction in the present case; the loading axis is denoted as 3) by a fixed amount ε_3 that is equivalent to the application of a tensile stress σ_3. Subsequently, we fully relax both the stresses σ_1 and σ_2 in the directions perpendicular to the axis 3 as well as the internal degrees of freedom. In this way, we find the contractions ε_1 and ε_2 which correspond to zero tensile stresses σ_1 and σ_2 and the new values of internal parameters. The $C11_b$ structure is tetragonal and keeps its tetragonal symmetry during the tensile test along the [001] axis. Therefore, $\sigma_1 = \sigma_2$ and we minimize the total energy as a function of lattice parameter a and internal parameter Δ (defined e.g. in [1, 2], see also Fig. 2).

The tensile stress σ_3 is then given by

$$\sigma_3 = \frac{2}{c_0} \frac{\partial E}{\partial \varepsilon_3} \frac{1}{a^2},$$

where E is the total energy per basis (i.e. one TM atom and two Si atoms) and c_0 is the lattice parameter in the direction of loading in the ground state. The inflexion point in the total energy dependence yields the maximum of the tensile stress; if some other instability does not occur before reaching the inflexion point, it also corresponds to the theoretical tensile strength.

DETAILS OF THE CALCULATIONS

In order to obtain reliable *ab initio* total energies of materials during tensile test simulations, the methods using a shape approximation of the crystal potential (for example spheroidization, as in the LMTO-ASA method or in standard KKR and APW approaches) are not adequate [3, 4]. Instead, full-potential treatments must be employed. In this study we utilized the full-potential linearized augmented plane waves (FLAPW) code described in detail in [5]. The electronic structure calculations were performed self-consistently within the local density approximation (LDA).

When simulating the tensile test, crystal lattices are severely distorted and some atoms may move very close together. Therefore, the muffin-tin radii must be sufficiently small to guarantee non-overlapping of the muffin-tin spheres at every stage of the test. We use the muffin-tin radii equal to 2.3 a.u. for transition metal atoms and 2.1 a.u. for silicon. These are kept constant in all calculations presented here. The product of muffin-tin radius and the maximum reciprocal space vector, $R_{MT}k_{max}$, is equal to 10, the maximum l value for the waves inside the atomic sphere, l_{max}, and the largest reciprocal vector $\mathbf{G}$ in the Fourier expansion of the charge, G_{max}, are set to 12 and 15, respectively, and the number of k-points in the first Brillouin zone is equal to 2000.

RESULTS AND DISCUSSION

First we determined the ground state properties for $MoSi_2$ and WSi_2. The values of lattice parameters, a and c, internal parameter Δ, and ratio of calculated ground-state volume over the experimental one, V/V_{exp} (Ref. [2]), are summarized and compared with other calculations and experiments in Table 1. Our results are in very good agreement with both experimental and previous theoretical data.

	a	c	Δ	V/V_{exp} [2]
	parameters of $MoSi_2$			
FLAPW-this work	6.004	14.689	0.0021	0.974
exp. Ref. [2]	6.051	14.836	0.0019	1.000
exp. Ref. [6]	6.059	14.830	0.0020	1.002
exp. Ref. [7]	6.047	14.855	–	0.999
FLAPW Ref. [1]	6.089	14.897	0.0022	1.017
LMTO-FP Ref. [8]	6.021	14.740		0.984
LMTO-ASA Ref. [9]	6.026	14.780	–	0.988
	parameters of WSi_2			
FLAPW-this work	6.020	14.721	0.0013	0.975
exp. Ref. [2]	6.068	14.870	0.0014	1.000
exp. Ref. [7]	6.070	14.891	–	1.010
FLAPW Ref. [1]	6.104	14.866	0.0018	1.020

Table 1: Ground-state lattice parameters of $MoSi_2$ and WSi_2.

The theoretical tensile strengths are summarized in Table 2. They are comparable with the tensile strengths predicted for, e.g., NiAl [10] or W [11]. Unfortunately, to the best of our knowledge there are no measurements of ideal tensile strength of $MoSi_2$ and WSi_2 (whiskers, nanoindentation) and, therefore, we were not able to compare our results with the experimental ones.

material	$MoSi_2$	WSi_2	NiAl [10]	W [11]
structure	$C11_b$	$C11_b$	B2	A2
σ_{th} (GPa)	37	38	46	29
ε_3	0.18	0.18	0.21	0.12

Table 2: Theoretical tensile strengths σ_{th} of $MoSi_2$ and WSi_2 and the corresponding ε_3 compared with those of other materials.

In Figure 1, we display the dependences of total energy E, internal parameter Δ, tensile stress σ_3 and the lattice parameter a on ε_3 in $MoSi_2$. The total energy has a parabolic shape around the minimum; it becomes almost flat in the neighbourhood of the inflexion point corresponding to the maximum of tensile stress. During the deformation the value of lattice constant perpendicular to the loading axis decreases nearly monotonously. On the other hand, the internal parameter increases with increasing ε_3 except for small neighbourhood of the ground state where it is almost constant. Let us note that the tendency of Δ to increase with ε_3 is in agreement with the interpretation of recent experimental data [12].

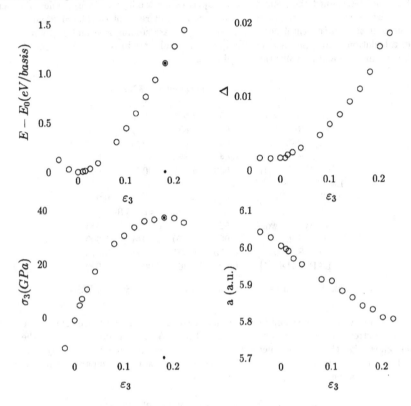

Figure 1: Variations of total energy E per basis (one Mo and two Si atoms), internal parameter Δ, tensile stress σ_3 and lattice parameter a during the simulation of the tensile test in MoSi$_2$. Here E_0 is the ground-state energy and ε_3 is the strain in the [001] direction. The position of the inflexion point in the energy dependence and the maximum of the tensile stress are denoted by a dashed line.

In Figure 2 we remind the definition of the internal parameter Δ [1, 2] and show four different types of bonds between constituent atoms of the repeat cell in MoSi$_2$. It was proposed by Tanaka et al. [2] that the Mo-Si bonds along the [001] direction (thin dash-dotted lines in Fig. 2) are weaker than the Mo-Si bonds in the other directions (thick dashed lines in Fig. 2) and, on the contrary, Si-Si bonds along the [001] direction (thick full lines in Fig. 2) are stronger than those in the other directions (thin dashed lines in Fig. 2). Our calculations confirm this suggestion regarding the Mo-Si bonds. In the C11$_b$ structure, four Mo atoms in the (001) plane form a square of side a which constitutes the basis of a bipyramid completed by two Si atoms above and below the center of this square. It seems that the atoms forming these bipyramids try to keep together during the tensile test in the [001] direction. Namely, the edges of the bipyramid are the Mo-Si bonds the length of which is nearly constant during the test even when exceeding the theoretical tensile strength (full circles in the left-hand part of Fig. 2). In agreement with Ref. [2], they can be denoted as "strong" bonds. The Si-Si bonds in the [001] direction, constituting the height of the bipyramid, are elongated under tension (open triangles in Fig. 2); this is related to the decrease of the lattice parameter a (Fig. 1). However, the Si-Si bonds in the other directions are extended approximately in the same way (open circles in Fig. 2).

The behavior of the Si-Si bonds is somewhat different under compression. The length of the non-[001] Si-Si bonds is nearly constant whereas the length of the [001] Si-Si bonds changes significantly, similarly as the length of the [001] Mo-Si bonds. Therefore, we can distinguish the "strong" and "weak" Mo-Si bonds, but it is not possible to introduce the "strong" and "weak" Si-Si bonds. The situation in WSi_2 is similar.

Let us note here that the relaxation of the internal parameter Δ during the tensile test is crucial. If Δ were kept constant the [001] Si-Si and Mo-Si bonds would behave very similarly. The same would be true for non-[001] Si-Si and Mo-Si bonds.

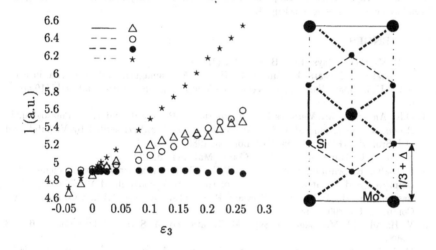

Figure 2: Variations of the length of atomic bonds during the tensile test simulation in $MoSi_2$. The right-hand side of the figure displays the (110) plane in $MoSi_2$ with the $C11_b$ structure; the Si atoms are represented by small circles and Mo atoms by large circles. Bonds between Mo-Si and Si-Si atoms are shown as thick dashed and thin dash-dotted lines (the "strong" and "weak" Mo-Si bonds) and thick full and thin dashed lines (the [001] and the other Si-Si bonds). The internal parameter Δ is defined as deviation from the ideal value of 1/3 [1, 2].

CONCLUSIONS

We simulated the tensile test in ideal $MoSi_2$ and WSi_2 loaded along the $[001]$ axis using the first-principles full-potential electronic structure calculations and determined theoretical tensile strength of those materials. The analysis of bond lengths variation under an uniaxial stress shows that, in accordance with Ref. [2], it is possible to distinguish "strong" and "weak" TM-Si bonds but the behavior of Si-Si bonds is more complex and exhibits a tension-compression asymmetry.

ACKNOWLEDGEMENTS

This research was supported by the Ministry of Education of the Czech Republic (Project No. ME-264), by the National Science Foundation–International Programs (Grant No. INT-96-05232), by the Grant Agency of the Academy of Sciences of the Czech Republic (Project No. A1010817), by the Grant Agency of the Czech Republic (Project No. 106/99/1178), and by the U.S. Department of Energy, Basic Energy Sciences (Grant No. DE-FG02-98ER45702). A part of this study has been performed in the framework of the COST Action P3 (Project No. OC P3.10). The use of the computer facility at the MetaCenter of the Masaryk University, Brno, and at the Boston University Scientific Computing and Visualization Center is acknowledged.

REFERENCES

1. L. F. Mattheiss, Phys. Rev. B, **45**, 3252 (1992).
2. K. Tanaka, K. Nawata, K. Yamamoto, H. Inui, M. Yamaguchi and M. Koiwa, in Proc. of the U.S.-Japan Workshop on Very High Temperature Structural Materials (1999), p. 67.
3. O.K. Andersen, M. Methfessel, C.O. Rodriguez, P. Blöchl, and H.M. Polatoglou, in *Atomistic Simulations of Materials: Beyond Pair Potentials*, edited by V. Vitek and D.J. Srolovitz (Plenum, New York-London, 1989), p. 1.
4. M. Šob, L.G. Wang, and V. Vitek, Comp. Mat. Sci., **8**, 100 (1997).
5. P. Blaha, K. Schwarz, and J. Luitz, WIEN97, Technical University of Vienna 1997 (improved and updated Unix version of the original copyrighted WIEN-code, which was published by P. Blaha, K. Schwarz, P. Sorantin, and S.B. Trickey, Comput. Phys. Commun., **59**, 399 (1990).
6. Y. Harada, M. Morinaga, D. Saso, M. Takata and M. Sakata, Intermetallics, **6**, 523 (1998).
7. M.-A. Nicolet and S.S. Lau, in VLSI Electronics: Microstructure Science, ed. N.G. Einspruch and G.B. Larrabee (Academic, New York, 1983), Vol. 6, p. 329.
8. M. Alouani, R. C. Albers, and M. Methfessel, Phys. Rev. B, **43**, 6500 (1991).
9. S. Tang, K. Zhang, and X. Xie, J. Phys. C, **21**, L777 (1988).
10. M. Šob, L.G. Wang, and V. Vitek, Phil. Mag. B, **78**, 653 (1998).
11. M. Šob, L.G. Wang, and V. Vitek, Mat. Sci. Eng. A, **234-236**, 1075 (1997).
12. K. Tanaka et al., this Proceedings.

Mat. Res. Soc. Symp. Proc. Vol. 646 © 2001 Materials Research Society

Effects of ternary additions on the deformation behavior of single crystals of MoSi$_2$

Haruyuki Inui, Koji Ishikawa and Masaharu Yamaguchi
Department of Materials Science and Engineering, Kyoto University, Sakyo-ku, Kyoto 606-8501, Japan

ABSTRACT

Effects of ternary additions on the deformation behavior of single crystals of MoSi$_2$ with the hard [001] and soft [0 15 1] orientations have been investigated in compression and compression creep. The alloying elements studied include V, Cr, Nb and Al that form a C40 disilicide with Si and W and Re that form a C11$_b$ disilicide with Si. The addition of Al is found to decrease the yield strength of MoSi$_2$ at all temperatures while the additions of V, Cr and Nb are found to decrease the yield strength at low temperatures and to increase the yield strength at high temperatures. In contrast, the additions of W and Re are found to increase the yield strength at all temperatures. The creep strain rate for the [001] orientation is significantly lower than that for the [0 15 1] orientation. The creep strain rate for both orientations is significantly improved by alloying with ternary elements such as Re and Nb.

INTRODUCTION

MoSi$_2$ with the C11$_b$ structure has received a great deal of attention as a candidate for structural materials to be used in oxidizing environments at temperatures higher than the upper limit for Ni-base superalloys because of its high melting temperature, relatively low density, good oxidation resistance and high thermal conductivity [1-3]. However, monolithic MoSi$_2$ exhibits only a modest value of fracture toughness at low temperatures and inadequate strength at high temperatures. Thus, many of recent studies on the development of MoSi$_2$-based alloys have focused on improving these poor mechanical properties through forming composites with ceramics [1,4]. However, the volume fraction of Si$_3$N$_4$ and SiC ceramic reinforcements in these MoSi$_2$-composites generally exceeds 50 % [5,6]. Further improvements in mechanical properties of these composites will be achieved if those of the MoSi$_2$ matrix phase are improved. The present study was undertaken to achieve this by alloying additions to MoSi$_2$.

Transition-metal atoms that form disilicides with tetragonal C11$_b$ and hexagonal C40 structures are considered as alloying elements to MoSi$_2$. The two structures commonly possess (pseudo-) hexagonally arranged TMSi$_2$ layers and differ from each other only in the stacking sequence of these TMSi$_2$ layers; the C11$_b$ and C40 structures are based on the AB and ABC stacking of these layers, respectively. W and Re have been known to form a C11$_b$ disilicide with Si and they are believed to form a complete C11$_b$ solid-solution with MoSi$_2$. Large amounts of alloying additions are possible for these alloying elements, and hence high-temperature strength is expected to be improved through a solid-solution hardening mechanism. V, Cr, Nb and Ta have been known to form a C40 disilicide with Si. Al is also known to transform MoSi$_2$ from the C11$_b$ to the C40 structures by substituting it for Si. 1/2<111> dislocations that carry slip on {110}<111> dissociate into two identical 1/4<111> partials separated by a stacking fault [3]. The stacking across the fault is ABC and resembles the stacking of (0001) in the C40 structure. Hence, the addition of elements that form a C40 disilicide may cause the energy difference between C11$_b$ and C40 structures to decrease so that the energy of the stacking fault would also be decreased. From this point of view, we may expect that the deformability of MoSi$_2$ at low temperatures increases upon alloying with elements that form a C40 disilicide.

In the present study, we have chosen V, Cr, Nb and Al that form a disilicide with the C40 structure and W and Re that form a disilicide with the C11$_b$ structure as alloying elements to MoSi$_2$ [3], and investigated the deformation behavior of single crystals of MoSi$_2$ containing these elements in a wide temperature range from room temperature to 1500°C. The crystal orientations investigated were the [0 15 1] orientation, in which slip on {110}<111> is operative, and the [001] orientation, in which the highest strength is obtained at high temperatures for binary MoSi$_2$ [3].

EXPERIMENTAL PROCEDURES

Single crystals of binary and ternary MoSi$_2$ were grown with our ASGAL FZ-SS35W optical floating-zone furnace at a growth rate of 10 mmh^{-1} under an Ar gas flow. The nominal compositions of single crystals were (Mo$_{0.97}$TM$_{0.03}$)Si$_2$ where TM stands for W, Re, V, Cr and Nb, and Mo(Si$_{0.97}$Al$_{0.03}$)$_2$. Specimens with [0 15 1] and [001] orientations were cut from as-grown crystals measuring 1.7×1.7×5 mm^3 and 3×3×6 mm^3 for compression and creep tests, respectively. Compression tests were conducted on an Instron-type testing machine at a strain rate of 1×10^{-4} s^{-1} in vacuum. The temperature ranges employed for compression tests were from room temperature to 1500°C for [0 15 1]-oriented crystals and from 1300 to 1500°C for [001]-oriented crystals. Creep tests were conducted in compression in air at 1200, 1300 and 1400°C under applied stresses of 50-100 and 300-500 MPa for the [0 15 1] and [001] orientations, respectively. Deformation microstructures were examined by optical microscopy and transmission electron microscopy (TEM).

RESULTS AND DISCUSSION

Deformation behavior of the [0 15 1] orientation
Critical resolved shear stresses (CRSSs) for slip on {110}<111> calculated with the corresponding Schmid factor are plotted in Fig. 1 as a function of temperature for binary, Al-, V-, Cr-, Nb-, W- and Re-bearing MoSi$_2$. Although all the crystals, except for Re-bearing MoSi$_2$, exhibit deformability even at room temperature by slip on {013}<331>, the data points plotted in Fig. 1 are only those for slip on {110}<111>. At low temperatures below 800°C, all ternary MoSi$_2$ with C40 formers (Al, V, Cr and Nb) exhibit CRSS values lower than those for binary MoSi$_2$. The extent of the decrease in CRSS is significant for Al- and

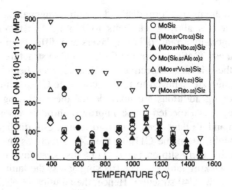

Fig. 1. Temperature dependence of CRSS for slip on {110}<111> in binary and some ternary MoSi$_2$ single crystals with the [0 15 1] orientation.

Table 1. The CRSS values for slip on $\{110\}<111>$ at 500 and 1500 °C and energies of stacking faults on $\{110\}$ for binary and some ternary $MoSi_2$, and the atomic radius, shear modulus and melting temperatures for the corresponding alloying elements.

$C11_b$ disilicide	$MoSi_2$	$(Mo,Re)Si_2$	$(Mo,W)Si_2$	$(Mo,V)Si_2$	$(Mo,Nb)Si_2$	$(Mo,Cr)Si_2$	$Mo(Si,Al)_2$
CRSS for $\{110\}<111>$ slip at 500°C (MPa)	232	405	248	150	95	99	75
CRSS for $\{110\}<111>$ slip at 1500°C (MPa)	18	74	25	10	45	27	16
Energy for stacking faults on $\{110\}$ (mJ/m²)	365	382	357	321	315	297	281
Alloying element	Mo	Re	W	V	Nb	Cr	Al
Atomic radius (nm)	0.139	0.137	0.139	0.134	0.146	0.127	0.143
Shear modulus (GPa)	122.9	178.6	160.2	47.1	37.7	115.4	26.2
Melting temperature (°C)	2617	3180	3410	1890	2468	1857	660

Nb-bearing $MoSi_2$. In contrast, ternary $MoSi_2$ with $C11_b$ formers (W and Re) exhibit CRSS values higher than those for binary $MoSi_2$ at low temperatures with the extent of the increase in CRSS significantly higher for Re-bearing $MoSi_2$. At high temperatures above 1300°C, while Al-bearing $MoSi_2$ exhibits CRSS values lower than those for binary $MoSi_2$, all the other ternary $MoSi_2$ exhibit CRSS values higher than those for binary $MoSi_2$. The extent of the increase in CRSS at high temperatures is significant for Re- and Nb-bearing $MoSi_2$.

Dislocations observed in all the crystals after deformation at low temperatures are determined exclusively to be those with b (Burgers vector)$=1/2[\bar{1}11]$. $1/2[\bar{1}11]$ dislocations with their 60° and screw orientations are frequently observed in all the crystals, except for Re-bearing $MoSi_2$ in which $1/2[\bar{1}11]$ dislocations tend to align parallel to the $[\bar{3}\bar{3}1]$ direction corresponding to the edge orientation. Weak-beam microscopy indicates that $1/2[\bar{1}11]$ dislocations in all the crystals are dissociated into two identical partial dislocations separated by a stacking fault (SF) on (110), as described below,

$$1/2[\bar{1}11] \rightarrow 1/4[\bar{1}11] + SF + 1/4[\bar{1}11]. \qquad (1)$$

All ternary $MoSi_2$ with C40 formers exhibit the separation distance between two-coupled partials wider than the binary counterpart, as we expected. While W-bearing $MoSi_2$ exhibits the separation distance a little wider than the binary counterpart, Re-bearing $MoSi_2$ exhibits the separation distance narrower than the binary counterpart.

The CRSS values for slip on $\{110\}<111>$ at 500°C (the lowest temperature at which this slip is observed for binary and all ternary $MoSi_2$) and energies of stacking faults on $\{110\}$ obtained for binary and ternary $MoSi_2$ are listed in Table 1 together with the atomic radius and shear modulus for each alloying element. As seen in Table 1, the observed softening and hardening behaviors upon alloying can be simply correlated with neither the atomic radius nor the shear modulus of alloying elements. However, a good correlation can be found between the CRSS for slip on $\{110\}<111>$ at 500°C and the energy of stacking faults on $\{110\}$. As the stacking fault energy is lowered, the CRSS value tends to decrease. This indicates that the softening and hardening behaviors are associated with changes in dislocation core structure, which may result from electronic structure changes occurring upon alloying. This view is consistent with the theoretical prediction by Waghmare et al. [7] that alloying elements with the number of valence electrons less than Mo improve the deformability of $MoSi_2$ while the those with the number of valence electrons more than Mo give rise to the worsening of deformability.

The dislocation arrangements after deformation at high temperatures for binary and all ternary $MoSi_2$, except for Nb- and Re-bearing $MoSi_2$, are very similar to each other and involves dislocations with $b=<111>$, $<100>$ and, to a lesser extent, $<110>$, which frequently form nodes. These dislocations are considered to originate from the following reactions,

$$1/2<11\bar{1}> + 1/2<111> \rightarrow <110> \qquad (2)$$

and

$$<110> \rightarrow \langle 100 \rangle + <010>. \qquad (3)$$

Most of these dislocations are observed to climb out from their slip planes, indicating a contribution of atomic diffusion to deformation. The dislocation arrangement in Nb- and Re-bearing $MoSi_2$, which exhibit the significantly higher high-temperature strength than the binary counterpart, is characterized by rather straight dislocations parallel to [001]. Contrast analysis indicates that their Burgers vectors are <110> and that the incidence of dislocations with b=<111> and <100> is less marked. Dislocations with b=<110> aligned parallel to [001] are considered to be formed via a reaction between two different $1/2<111>$ dislocations, as in (2). However, further decomposition of <110> dislocations into two <100> dislocations via (3) seems to be retarded in Nb- and Re-bearing $MoSi_2$, probably because of the slower atomic diffusion. However, diffusion data are not available for the relevant alloy systems. We thus list in Table 1 the melting temperature of each alloying element investigated as a guide to infer the diffusivity, together with the CRSS values for slip on {110}<111> at 1500°C for binary and ternary $MoSi_2$. But, the observed behaviors upon alloying can not be simply correlated with the melting temperatures of the alloying elements. Alternative explanation for this is that the core structures of <110> dislocations in Nb- and Re-bearing $MoSi_2$ are different from those in binary and the other ternary $MoSi_2$ and are not favorable for the decomposition to occur. In fact, some of <110> dislocations in Re-bearing $MoSi_2$ are observed to decompose into two $1/6<331>$ dislocations instead of decomposing into <100> dislocations [8].

Deformation behavior of the [001] orientation

The [001] orientation is very brittle as the onset temperature for plastic flow is as high as 900°C even for binary $MoSi_2$. [001]-oriented crystals usually exhibit a very high work-hardening rate and fail within a few % plastic strain after exhibiting ultimate strength. The ultimate strengths observed at 1400 and 1500°C are plotted in Fig. 2 for binary and ternary $MoSi_2$. All ternary elements tend to increase the ultimate strength of $MoSi_2$. The extent of the increase in the ultimate strength is significantly large for Nb- and Re-bearing $MoSi_2$. Although the ultimate strength defined for the [001] orientation is physically different from the yield stress defined for the [0 15 1] orientation, Nb- and Re-bearing $MoSi_2$ exhibit significantly high strength at high temperatures both for the [0 15 1] and [001] orientations.

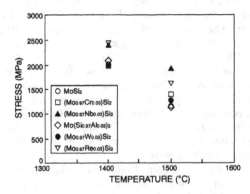

Fig. 2. The ultimate strengths obtained for binary and some ternary $MoSi_2$ single crystals with the [001] orientation.

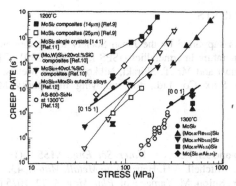

Fi g. 3. Creep strain rates at 1300°C for binary and some ternary MoSi₂ single crystals with [0 15 1] and [001] orientations plotted as a function of applied stress.

Creep behavior of the [0 15 1] and [001] orientations

Creep strain rates at 1300°C for binary and some ternary MoSi₂ single crystals with [0 15 1] and [001] orientations are plotted in Fig. 3 as a function of applied stress. Creep strain rates obtained at 1200°C for some relevant MoSi₂-based materials (monolithic MoSi₂ compacts [9], MoSi₂-SiC composites [10], single crystals with the soft [141] orientation [11] and MoSi₂-Mo₅Si₃ eutectic alloys [12]) as well as those obtained at 1300°C in tension for the most advanced Si₃N₄-based ceramic composites (AS-800) [13] are shown in the figure for comparison. The creep rates for [0 15 1]-oriented binary MoSi₂ single crystals locate in the range of strain rates attained by many relevant MoSi₂-based materials. This indicates that single crystals with soft orientations do not have any significant advantages over the relevant MoSi₂-based materials produced by powder metallurgy methods in terms of creep resistance. However, when Re and Nb are added to MoSi₂, the creep rates of [0 15 1]-oriented crystals are more than an order of magnitude reduced and thus these materials exhibit modest advantage over the relevant MoSi₂-based materials. In contrast to [0 15 1]-oriented crystals, the creep strain rates for [001]-oriented binary MoSi₂ single crystals are by far smaller than those attained by any other MoSi₂-based materials and are comparable to or a little larger than those attained by the most advanced Si₃N₄-based ceramic composites (AS-800) [11]. When ternary elements such as Re are added to MoSi₂, the creep strain rates is further reduced to a level where ternary MoSi₂ single crystals may have an advantage in creep resistance over the Si₃Ni₄-based composites.

CONCLUSIONS

(1) Additions of Al decreases the yield strength of MoSi₂ at all temperatures. Thus, Al seems to be effective in improving the low-temperature deformability of MoSi₂.

(2) Alloying elements (V, Cr and Nb) that form a C40 disilicide are effective both in decreasing the yield strength at low temperatures and in increasing the yield stress at high temperatures. This effect occurs most significantly for Nb.

(3) A small amount of W addition does not significantly affect the deformation behavior of MoSi₂. In contrast, a small amount of Re addition significantly increases the yield strength of MoSi₂. However, at the same time, the low-temperature deformability drastically declines.

(4) The creep strain rate is significantly lower for the hard [001] orientation than for the soft [0 15 1] orientation. The creep rate for both orientations is significantly reduced upon alloying MoSi₂ with ternary elements such as Re and Nb.

ACKNOWLEDGEMENTS

This work was supported by the Japan Society for Promotion of Science grant on Advantage High-temperature Intermetallics (JSPS-RFTF96R12301). The authors would like to thank Dr. Y. Yukawa and Mr. T. Shiraki, Nikko Superior Metals, Co. Ltd., and Dr. H. Shiraishi, Sumitomo Sitix Co. Ltd., for supplying high-purity Mo, Nb, Ta and Si, respectively.

REFERENCES

[1]   A.K. Vasudevan and J.J. Petrovic, *Mater. Sci. Eng.*, **A155**, 1 (1992).

[2]   S.A. Maloy, T.E. Mitchell, A.H. Heuer, *Acta Metall. Mater.*, **43**, 657 (1995).

[3]   K. Ito, H. Inui, Y. Shirai, M. Yamaguchi, *Phil. Mag. A*, **72**, 1075 (1995).

[4]   J.J. Petrovic and A.K. Vasudevan, *Mater. Sci. Eng.*, **A261**, 1 (1999).

[5]   M.G. Hebsur, M.V. Nathal, in *Structural Intermetallics 1997*, edited by M.V. Nathal et al. (TMS, Warrendale, PA, 1997) p. 949.

[6]   J.J. Petrovic, M.I. Pena, I.E. Reimanis, M.S. Sandlin, S.D. Conzone, H.H. Kung and D.P. Butt, *J. Am. Ceram. Soc.*, **80**, 3070 (1997).

[7]   U.V. Waghmare, V. Bulatov, E. Kaxiras and M.S. Duesbery, *Mater. Sci. Eng.*, **A261**, 147 (1999).

[8]   H. Inui, K. Ishikawa and M. Yamaguchi, *Intermetallics*, **8**, 1131 (2000).

[9]   K. Sadananda and C.R. Feng, in *High Temperature Silicides and Refractory Alloys*, edited by C.L. Briant, J.J. Petrovic, B.P. Bewlay, A.K. Vasudevan and H.A. Lipsitt (Mater. Res. Soc. Proc. **322** , Pittsburgh, PA, 1994) pp.157-162.

[10] S.M. Wiederhorn, R.J. Gettings, D.E. Roberts, C. Ostertag and J.J. Petrovic, *Mater. Sci. Eng.*, **A155**, 209 (1992).

[11] Y. Umakoshi, T. Nakashima, T. Nakano and E. Yanagisawa, E., in *High Temperature Silicides and Refractory Alloys*, edited by C.L. Briant, J.J. Petrovic, B.P. Bewlay, A.K. Vasudevan and H.A. Lipsitt (Mater. Res. Soc. Proc. **322** , Pittsburgh, PA, 1994) pp. 9-16.

[12] D.P. Mason and D.C. Aken, *Acta Metall. Mater.*, **43**, 1201 (1995).

[13] K. Sadananda and C.R. Feng, R. Mitra and S.C. Deevi, *Mater. Sci. Eng.*, **A261**, 223 (1999).

Mat. Res. Soc. Symp. Proc. Vol. 646 © 2001 Materials Research Society

Dislocation Interaction with Point Defects in Transition-Metal Disilicides

Man H. Yoo
Metals and Ceramics Division, Oak Ridge National Laboratory
Oak Ridge, TN 37831-6115, U.S.A.

ABSTRACT

Energetics of the formation of jog-pairs and kink-pairs on a straight dislocation are analyzed using anisotropic elasticity theory for the equivalent slip systems of seven transition-metal disilicides. While glide loops of the active slip systems are stable in all cases, having positive line tension, the interaction energies of two opposite segments in kink-pairs and jog-pairs are found to be very anisotropic with respect to dislocation orientation. The anisotropic interaction plays an important role in the glide resistance due to dislocation-point defect interactions. A dislocation model is proposed for the glide resistance on edge and near edge dislocations based on jog-pairs resulting from the contact interaction between dislocations and intrinsic point defects. The available data of yield strength anomaly and dislocation structures of disilicide crystals are discussed in view of the proposed model for jog-pair pinning and dynamic breakaway.

INTRODUCTION

Transition-metal disilicides (MSi_2, M: transition metals in group IV-VII) have non-cubic crystal structures, viz., hexagonal C40 for the four elements in groups V-VI (M = V, Nb, Ta, and Cr), tetragonal $C11_b$ for the three elements in groups VI-VII (M = Mo, W, and Re), and orthorhombic C54 (M = Ti) and C49 (M = Zr and Hf) for the three elements in group IV. The results of *ab initio* self-consistent band-structure calculations by Carlsson and Meschter [1] show that the structural energies of these disilicides are determined primarily by electronic-band effects rather than by atomic-size effects. Single-crystal elastic constants have been determined for most of these disilicides [2,3], and the directional nature of atomic bonds in MSi_2 compounds with the $C11_b$, C40 and C54 structures were analyzed in terms of the bond-stretching and bond-bending components of interatomic forces [4].

Geometrically, the atomic arrangements of a (110) plane in the $C11_b$ structure are identical to those of the (0001) plane in the C40 structure if the axial ratio of the tetragonal structure is set to c/a = $\sqrt{6}$ = 2.449 instead of the actual value (e.g., 2.450 for WSi_2 or 2.452 for $MoSi_2$). Moreover, the atomic arrangements of the (001) plane, the pseudo-hexagonal plane, in the C54 structure correspond also to those of the C40 basal plane. So, $(1\bar{1}0)[111]$, $(0001)[11\bar{2}0]$, and (001)[110] are called the equivalent slip systems in the $C11_b$, C40, and C54 structures, respectively [5]. Plastic deformation behavior of MSi_2 single crystals has been reviewed by Ito et al. [6], including the role of the equivalent slip systems. In many cases (M = V, Nb, Ta, Mo, and Ti), an anomalous increase in the critical resolved shear stress (CRSS) with increasing temperature was observed over the temperature ranges in which serrated stress-strain curves were recorded. Therefore, the yield strength anomaly may be related to the Portevin-Le Chatelier (PLC) effect, i.e., dislocation interaction with point defects, either intrinsic defects or solute atoms [7,8].

The purpose of this paper is to analyze energetics of the formation of kink-pairs and jog-pairs on straight dislocations of the equivalent slip systems using the anisotropic elasticity theory and to develop a strengthening model based on jog-pair formation.

DISLOCATION GEOMETRY

The numerical method used for calculating the energy factor (K) of straight dislocations and the line tension factor (K + K") for curved dislocations was described earlier [9]. It is

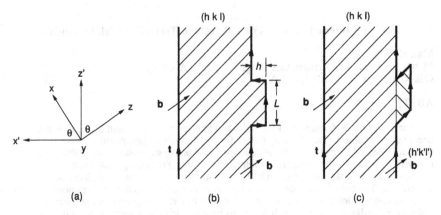

Figure. 1 *SISF-type colinear dissociation (B --> b + b) in the equivalent slip system. (a) dislocation coordinate axes, (b) a kink-pair and (c) a jog-pair on the leading superpartial.*

shown in Fig. 1(a) that the reference right-handed Cartesian coordinate system is used for a straight dislocation lying in the (hkl) slip plane (y = 0) and having its Burgers vector, **b**, parallel to the z-axis. The shaded areas in Figs. 1(a) and 1(b) designate superlattice intrinsic stacking faults (SISFs). A rotation of the coordinate system about the y-axis is made by the angle θ measured from the Burgers vector (the z-axis) in the counterclockwise direction. The energy and line tension factors, $K(\theta)$ and $K(\theta) + \partial^2 K/\partial \theta^2$, become K_s and $K_s + K''_s$ for the screw orientation (θ = 0°) and K_e and $K_e + K''_e$ for the edge orientation (θ = 90°).

The elastic interaction energy between the two opposite segments of a kink-pair is given by $W = (b^2/8\pi)[K + K'']h^2/L$, where b is the magnitude of Burgers vector, and h and L are the height and length, respectively, of the kink pair on the leading superpartial dislocation as shown schematically in Fig. 1(b). For a given angle θ, the transformed Cartesian coordinate system is rotated about the x' axis by the angle φ in the y'z' plane, sketched as the (h'k'l') plane in Fig. 1(c). At φ = 0°, $K(\theta, \phi) + \partial^2 K/\partial \phi^2$ gives the line tension factor for the elastic interaction between the two opposite segments of a jog-pair as shown in Fig. 1(c).

LINE TENSION FACTORS

Table 1 lists the calculated results of K and K + K" for those disilicides for which elastic constants are available in the literature [2,3,10]. In all cases, the line tension is positive (K + K" > 0), and no concavity in the inverse Wulff plot (1/K) is observed. The ratio K_e/K_s is in the range of 1.15 - 1.25, and $[K_s + K_s''(\theta)]/[K_e + K_e''(\theta)]$ is in the range of 1.52 -1.95. The last column in Table 1 shows the line tension factors for a jog-pair on the edge orientation, and these are larger than those for a kink-pair by the factor in the range between $[K_e + K_e''(\phi)]/[K_e + K_e''(\theta)]$ =1.39 for $CrSi_2$ and1.90 for $TiSi_2$.

Figure 2 shows the orientation dependence of the in-plane line tension factors, K + K"(θ), for the (1̄10)[111] slip system in $MoSi_2$ and WSi_2. The minimum for $MoSi_2$ at 300 K is at θ = 107°, and that at 1200 K is at θ = 112°. In the latter case, the second minimum occurs at θ = 60°. The out-of-plane line tension factors, $K(\theta) + K''(\phi)$, are shown in Fig. 3, which show the maxima at near-screw orientations and the minima at near-edge orientations. In $MoSi_2$, the anisotropy of $K(\theta) + K''(\phi)$ at 300 K with the maximum at θ = 2° and the minimum of 87° increase slightly to that at 1200 K with θ = 3° and 95°, respectively.

Table 1. Line energy factors (K) and tension factors (K + K") of MSi_2 at room temperature for two special cases of pure screw and edge orientations (in units of 10^2 GPa)

Structure	Slip System	M	Screw ($\theta = 0°$)		Edge ($\theta = 90°$)		
			K_s	$K_s + K_s''(\theta)$	K_e	$K_e + K_e''(\theta)$	$K_e + K_e''(\phi)$
C11$_b$	$(1\bar{1}0)[111]$	Mo	1.78	2.49	2.10	1.60	2.52
		W	1.87	2.83	2.25	1.65	2.67
C40	$(0001)[11\bar{2}0]$	V	1.44	1.99	1.72	1.17	1.69
		Nb	1.49	2.18	1.83	1.14	1.74
		Ta	1.46	2.17	1.82	1.11	1.69
		Cr	1.56	2.02	1.79	1.33	1.85
C54	$(001)[110]$	Ti	1.07	1.50	1.31	0.83	1.58

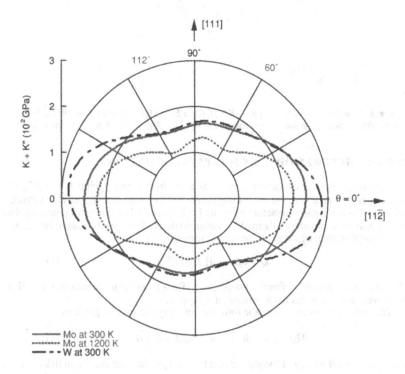

Figure 2 Line tension factors, $K(\theta) + K''(\theta)$, for a $b = [111]/2$ glide loop on the $(1\bar{1}0)$ slip plane, with respect to θ in the plane, in MSi_2 (M = Mo or Si) of the C11$_b$ structure

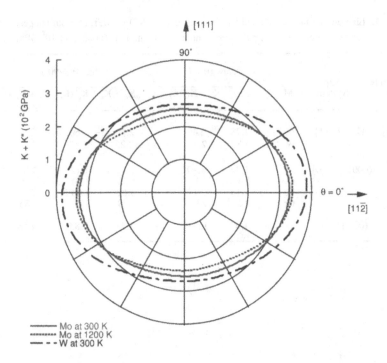

Figure 3 *Line tension factors, $K(\theta) + K''(\phi)$, for a $\boldsymbol{b}=[111]/2$ glide loop on the $(1\bar{1}0)$ slip plane, with respect to ϕ out of the plane, in MSi_2 (M = Mo or Si) of the $C11_b$ structure*

STRENGTHENING MODEL BY JOG-PAIR FORMATION

The critical shear stress necessary to break away from pinning points is $\tau_c = \alpha T_L / bl$, where T_L is the line tension on the leading partial, l is the mean spacing between pinning points, and α is a numerical factor near unity [11]. The number density of pinning points, $N = 1/l$, can be obtained by a thermal activation analysis of jog-pair formation based on Boltzmann statistics, which gives the critical stress as [11]

$$\tau_c = \tau_0 \exp(-H_c/3kT) \quad , \tag{1}$$

where τ_0 is the stress at a reference temperature (T_0), kT has the usual meaning, and H_c is the activation enthalpy for the formation of a jog-pair.

The enthalpy associated with the formation of a jog pair may be given as

$$H(h,L) = 2ch - (\Gamma_c + \sigma_x b)hL - Mh^2/L , \tag{2}$$

where c is the self energy of the jog per unit length, Γ_c is the fault energy on (h'k'l') plane, σ_x is the compressive internal stress perpendicular to the (h'k'l') plane, and $M = (b^2/8\pi)[K_e + K_e''(\phi)]$. Assuming that h is a constant of the atomic dimension (i.e., $h \approx b$), one finds the critical length, L_c, and the activation enthalpy to be

$$L_c = [Mb/(\Gamma_c + \sigma_x b)]^{1/2} , \qquad (3)$$

$$H_c = Mh\{c' - 2[(\Gamma_c + \sigma_x b)h/M]^{1/2}\} , \qquad (4)$$

where c' is the normalized self energy of the two jogs (c' = 2c/M).

The internal stress has two possible origins, namely, the climb stress of mechanical origin due to the resolved normal stress component, σ_x^e, of an externally applied stress and that of chemical origin due to the vacancy supersaturation, c/c_0, at temperature, T, as discussed earlier [9],

$$\sigma_x = n\sigma_x^e - (kT/\Omega)\ln(c/c_0) , \qquad (5)$$

where n is the stress concentration factor, Ω is the atomic volume, and c_0 is the equilibrium vacancy concentration. In certain cases, the out-of-plane component, F_y, of the elastic interaction force between the two superpartials contributes an additional term in Eq. (4), e.g., in the $(1\bar{1}0)[111]$ case of the $C11_b$ structure, but not in the $(0001)[11\bar{2}0]$ case of the C40 structure.

In the so-called uncorrelated jog-pair formation [9], as depicted in Fig. 1(c), a flux of $n_v = L_c b^2/\Omega$ vacancies from the surrounding matrix to the core of the leading superpartial must be involved. After the jog-pair formation, the mobile segments of the superpartial on either sides of the jog-pair bow out and reach a critical curvature due to kink pile-ups. It is at this critical condition that dynamic break away of the superpartial from its pinning center occurs, leaving the same n_v vacancies behind which are subsequently dissolved into the matrix.

DISCUSSION AND SUMMARY

Interaction of a dislocation with point defects may be classified into two parts: (a) long-range elastic interaction by size and modulus effects, and (b) contact interaction by either solute segregation to stacking fault or jog-pair formation. Two jog-pairs of opposite sign may be formed by two different correlated processes of short-circuit self-diffusion. First, when the diffusion path is within SISF ribbon, a climb dissociation may result in pair at the superpartials. Second, on the other hand, when the diffusion path is along dislocation core, the so-called Z-shaped climb instability may result along one of the superpartials [9]. These jog-pairs may act as effective barriers to dislocation motion, particularly at relatively low temperatures (i.e., T < $0.6T_m$, where T_m is the melting temperature) at which lattice diffusion is sluggish.

Transmission electron microscopy (TEM) data available in the literature (e.g., M = Mo [6], W [12], and Nb [8]) show that *post mortem* dislocation configurations of the equivalent (hkl)**B** slip systems are of the SISF-type colinear glide dissociation (**B** --> **b** + **b**, where **b** = **B**/2), as schematically depicted in Fig. 1. For basal slip, with **B** = $[11\bar{2}0]/3$ in NbSi$_2$, at low temperatures, the preferred orientation of dislocations was either $\theta = 60°$ or $\theta = 0°$ [8]. For $(1\bar{1}0)[111]$ slip with **B** = [111]/2 in MoSi$_2$ at 1100 K (where CRSS increase with temperature occurs), most of dislocations were very straight and parallel to [110] with $\theta = 60°$ [6]. *In situ* straining experiments using MoSi$_2$ single crystals of a soft orientation at T = 1100 - 1300 K in a high-voltage electron microscope operated at 1 MV [13,14], $(1\bar{1}0)[111]/2$ dislocations were observed to align along 60° and edge orientations and moved at high velocities or in a viscous way. These direct observations are consistent with the calculated line tension factors in the fact that superdislocations at the edge and 60° mixed orientations are more mobile than those at the screw orientation (Fig. 2), and jog-pair formation is slightly easier at near edge orientations than at near screw orientations (Fig. 3).

As an alternative explanation for the preferred orientation of active slip dislocations, climb dissociation of the $(1\bar{1}0)[111]/2$ dislocation in $MoSi_2$ is a possibility whether it is the two-fold type [15] or the three-fold type [14]. In either case, an antiphase boundary (APB) is involved. Therefore, the specific values of the SISF and APB energies on the $(1\bar{1}0)$ plane are important factors that determine the probability of glide or climb dissociation.

Weak-beam TEM analyses of the two-fold glide dissociation (Fig. 1) were made using $MoSi_2$ specimens deformed at high temperatures [8,16], which yield the (110) SISF energy estimate of 365 ± 6 mJ/m^2. The separation for the screw orientation was always widest, and the scatter was very small. In contrast, the separation for non-screw orientations showed a large scatter, strongly suggesting the presence of jog-pairs on superpartial dislocations.

About the recently proposed mechanism of the formation of dragging atmosphere for anomalous strengthening in $NbSi_2$ and ternary disilicide single crystals of the C40 structure, readers are referred to Nakano et al. [17].

In summary, the elastic interaction energies between the two opposite segments of jog-pair and kink-pair are found to be very anisotropic with respect to the dislocation orientation in transition-metal disilicides. The strengthening mechanism proposed here on the basis of the jog-pair formation along the edge and near-edge orientations is consistent with the experimental data of dislocation structures available in the literature.

ACKNOWLEDGMENT

The author thanks H. Inui, U. Messerschmidt, S. R. Agnew, and E. P. George for helpful comments on the manuscript. This research was sponsored by the Division of Materials Sciences and Engineering, Office of Basic Energy Sciences, U. S. Department of Energy under Contract DE-AC05-00OR22725 with UT-Battelle, LLC.

REFERENCES

1. A. E. Carlsson and P. J. Meschter, *J. Mater. Res.* **6**, 1512 (1991).
2. M. Nakamura, S. Matsumoto, and T. Hirano, *J. Mater. Sci.* **25**, 3309 (1990).
3. F. Chu, M. Lei, S. A. Maloy, J. J. Petrovic, and T. E. Mitchell, *Acta Mater.* **44**, 3035 (1996).
4. K. Tanaka, H. Inui, M. Yamaguchi, and M. Koiwa *Mater. Sci. Eng.* **A261**, 158 (1999).
5. H. Inui, M. Moriwaki, K. Ishikawa, and M. Yamaguchi, *Electron Microscopy 1998*, Vol. II, ed. H. A. Calderon-Benavides and M. J. Yacaman (Bristol, 1998), pp.49-50.
6. K. Ito, M. Moriwaki, T. Nakamoto, H. Inui, and M. Yamaguchi, *Mater. Sci. Eng.* **A233**, 33 (1997).
7. K. Ito, H. Inui, Y. Shirai, M. Yamaguchi, *Phil. Mag.* **A72**, 1075 (1995).
8. M. Moriwaki, K. Ito, H. Inui, and M. Yamaguchi, *Mater. Sci. Eng.* **A239-240**, 69 (1997).
9. M. H. Yoo, K. Yoshimi, and S. Hanada, *Acta Mater.* **47**, 3579 (1999).
10. K. Tanaka, H. Onome, H. Inui, M. Yamaguchi, and M. Koiwa, *Mater. Sci. Eng.* **A239-240**, 188 (1997).
11. M. H. Yoo, J. A. Horton, and C. T. Liu, *Acta Metall.* **36**, 2945 (1988).
12. K. Ito, T. Yano, T. Nakamoto, H. Inui, and M. Yamaguchi, *Acta Mater.* **47**, 937 (1999).
13. S. Guder, M. Bartsch, M. Yamaguchi, and U. Messerschmidt, *Mater. Sci. Eng.* **A261**, 139 (1999).
14. U. Messerschmidt, S. Guder, L. Junker, M. Bartsch, and M. Yamaguchi, *Mater. Sci. Eng.* A (2001) (in press)
15. S. Maloy, T. E. Mitchell, and A. H. Heuer, *Acta Metall. Mater.* **43**, 657 (1995).
16. D. J. Evans, S. A. Court, P. M. Hazzledine, and H. L. Frazer, *Phil. Mag. Lett.* **67**, 331 (1993).
17. T. Nakano, M. Kishimoto, D. Furuta, and Y. Umakoshi, *Acta Mater.* **48**, 3465 (2000).

High-Temperature Ordered
Intermetallic Alloys

Mat. Res. Soc. Symp. Proc. Vol. 646 © 2001 Materials Research Society

THE EFFECTS OF ENVIRONMENT ON THE ROOM TEMPERATURE DEFORMATION OF B2-STRUCTURED Fe-43Al SINGLE-CRYSTALS

I. Baker[1], D. Wu[1] and E. P. George[2]

[1]Thayer School of Engineering, Dartmouth College, Hanover, NH 03755, U.S.A.
[2]Metals and Ceramics Division, Oak Ridge National Laboratory, Oak Ridge, TN 37831-6093, U.S.A.

ABSTRACT

The effects of the environment on the room temperature mechanical behavior of Fe-43Al single crystals have been studied. In both single slip and duplex slip crystals, fracture strains greater than 40% were obtained in specimens tested in oxygen, whereas elongations of ~10% and ~20% were obtained in air and vacuum, respectively. By comparison, similar elongations were obtained in boron-doped single-slip-oriented single crystals in both air and vacuum, but more ductility was obtained in air at slow strain rate. Fractography showed that testing in different environments produced marked differences in the fracture surfaces. Alternate loading of tensile specimens in air and under vacuum was performed at slow strain rates and showed changes in the flow stress between the two environments. The results are discussed in terms of the effects of moisture-produced hydrogen on the flow and fracture of FeAl.

INTRODUCTION

The deleterious effect of water vapor on the ductility of iron-rich B2-structured FeAl alloys is now well documented [1-11]. The aluminum in FeAl reacts with water vapor to produce atomic hydrogen, which diffuses into the metal at crack tips and results in embrittlement [1, 12, 13]. The water vapor not only reduces the strain to failure but also changes the crack initiation site in single crystals [4, 5, 7, 14]. For example, Fe-40Al crystals tested in air were reported to exhibit failure strains of less than 1%, with crack initiation occurring at the surface [2], whereas, in oxygen, failure strains 10% occurred in similar crystals with crack nucleation occurring internally [4]. More recently, Baker et al. [11] obtained elongations of greater than 40% in single-slip-oriented single crystals of iron-rich FeAl tested in oxygen but as little as 0.5% elongation in similar crystals tested in air.

Interestingly, Pike and Liu [15] reported that the yield strength of polycrystalline Fe-40Al was independent of strain rate in vacuum, but in air was both lower and increased with increasing strain rate. Similarly, Wu and Baker [16] found that when tensile testing single crystals of Fe-40Al, Fe-40Al-1Y and Fe-43Al (compositions given in atomic percent throughout) the yield strength increased with increasing strain rate up to 1×10^{-2} s^{-1} in air, but was independent of strain rate in vacuum. More recently, Baker et al. [11] showed that the yield strengths of single-slip oriented single crystals of Fe-40Al and Fe-43Al were lower when tensile tested at slow strain rates in air compared with tests in oxygen or vacuum. In contrast, several earlier studies [1, 9, 17-22] did not show any effect of a water vapor-containing atmosphere on the yield strength of FeAl

How hydrogen is transported into the lattice and how far ahead of the crack tip it needs to travel in order to produce embrittlement is unclear. For single crystals the diffusion paths are limited to either pipe diffusion along dislocations or bulk diffusion. It is possible that hydrogen transport into the lattice is by gliding dislocations that originate at the surface. However, results from tensile tests of single-slip-oriented Fe-40Al single crystals oriented so that the majority of the strain was produced either by screw dislocations or by edge dislocations showed no difference in strength or ductility [10].

In this paper we explore further the effects of the environment on both the room temperature ductility and strength of single crystals of Fe-43Al both in single-slip and duplex-slip orientations, and with and without boron.

EXPERIMENTAL

Single crystals of Fe-43Al with and without 300 p.p.m. boron were grown in alumina crucibles under argon using a modified Bridgeman technique. They were annealed for 10 hrs at 1473 K, followed by slow furnace cooling, and further annealed at 673 K for 5 days to minimize the thermal vacancies [23].

Tensile specimens, 0.64 mm thick by 2 mm wide with a gauge length of 9.5 mm, were produced from crystals whose orientations were determined by X-ray Laue back reflection. For single-slip-oriented specimens, two orientations of tensile axes were used $\overline{8}95$ and $\overline{1}\,\overline{1}\,32$. For these orientations slip occurs on $(011)[1\overline{1}\,1]$ and $(\overline{1}\,01)[1\overline{1}\,1]$, with Schmid factors of 0.40 and 0.48, respectively. For the boron-doped crystal, the orientation was $\overline{8}\overline{2}3$, which produces slip on $(\overline{1}\,10)[111]$only, with a Schmid factor of 0.48. For the duplex-slip-oriented crystals, the orientation was $\overline{1}\,\overline{1}4$, which produces slip on both $(0\overline{1}\,1)(\overline{1}\,11)and(\overline{1}\,01)[1\overline{1}\,1]$, with Schmid factors of 0.45.

Before mechanical testing, the surfaces of the specimens were prepared by mechanical polishing with SiC paper to 600 grit, followed by electropolishing at 16 V in a 10% perchloric acid-methanol solution at about 248 K, then washed successively in methanol and ethanol.

Tests were performed under monotonic loading in air, vacuum (~1 x 10^{-4} Pa) and oxygen (~7 x 10^4 Pa) at initial strain rates of either 2.5 x 10^{-3} s^{-1} or 4.4 x 10^{-6} s^{-1}. A single-slip oriented unalloyed specimen was also strained at 5 x 10^{-6} s^{-1} alternately in air and vacuum, for 30-60 minutes each, until failure. Elongations were measured directly from fiducial marks on the shoulders of the fractured specimens using an optical microscope equipped with a micrometer. Slip lines were observed using an optical microscope. The fracture surfaces were examined using a JEOL JSM-5310LV scanning electron microscope (SEM) operated at 20 kV.

RESULTS AND DISCUSSION

Table 1 shows the results of tensile tests on the single-slip-oriented single crystals with and without boron and duplex-slip-oriented single crystals strained at 2.5 x 10^{-3} s^{-1}. The environment had little, if any, effect on the yield stress (also shown as the critical resolved shear stress, CRSS, for the single-slip-oriented crystals). In air, vacuum and oxygen, the average yield stress was 90 MPa, 90 MPa and 94 MPa for undoped crystals, respectively, and 98 MPa, 98 MPa and 99 MPa for boron-doped crystals. The boron increased the CRSS only slightly. In contrast, the elongation to failure was markedly greater under vacuum than in air, and was greater still in oxygen. The elongation of 46.3% obtained for the Fe-43Al single-slip-oriented single crystals tested in oxygen is, by far, the largest reported for FeAl. There was a small improving effect of boron on the ductility of the single slip-oriented crystals in air, but boron-doped crystals actually showed a little less ductility in oxygen, probably because of their slightly greater strength. However, duplex slip oriented crystals showed similar ductility. Hydrogen charging had little effect on the ductility of the specimens tested in air, a result consistent with environmental embrittlement. It is clear that sufficient hydrogen is generated in air to embrittle FeAl to a similar extent as hydrogen charging. That more ductility was obtained in oxygen than in vacuum may be because of residual water vapor in the vacuum or because oxygen has a protective effect, reacting with the aluminum in preference to any water vapor.

Straight prominent slip lines, consistent with the operation of single slip on {110}, were observed in the undoped, single-slip-oriented single crystals, see Figure 1(a). In the duplex-slip single crystals, slip lines were finer, somewhat wavy and more closely spaced, see Figure 1(b). For the boron-doped specimens, no slip lines could be observed even though substantial elongations occurred, suggesting that one role of the boron was to refine or homogenize the slip.

Fracture appeared to initiate from the surface when tested in either air or vacuum. Fracture occurred along the {110} slip planes in all cases. In general, all the fracture surfaces consist of two regions. One was characterized by smooth planar fracture with river marks, the other consisted of non-planar fracture with a rougher topography. Usually larger extents of rougher regions were observed on the specimens tested in vacuum than in air. Suggesting that hydrogen can lower the cohesive force of the cleavage plane. For tests in air or vacuum, numerous parallel secondary cracks were observed in the undoped Fe-43Al specimens see Figure 2(a). A few secondary cracks sometimes run through the river pattern instead of parallel to it. In contrast, fracture originated from the surface as well as inside the specimens tested in oxygen. No secondary cracks were present and ductile dimples, about 1 um in diameter, were observed in some regions on the surfaces, which is consistent with the greater ductility for test in oxygen. In contrast, boron-doped specimens contained very few secondary cracks on the fracture surfaces, and much rougher fracture surfaces were seen when tested in either vacuum or oxygen, see Figure 2(b).

Table 1. *Room-temperature strength and ductility of Fe-43Al single crystals with and without 300 p.p.m. boron strained at room temperature in different environments at 2.5 x 10^{-3} s^{-1}*

Specimen	Environment	Elongation (%)	Yield Stress (MPa)	CRSS (MPa)
Single slip	Air	11.5	242	97
		9.7	224	90
		8.7	179	86
		7.7	179	86
	H-charged	6.9	178	85
	Vacuum	31.4	185	89
		17.3	221	88
		20.4	221	88
		12.6	239	96
	Oxygen	43.6	234	94
		46.3	232	93
Duplex slip	Air	14.1	184	

	Vacuum	31.8	169	
	Oxygen	33.7	193	
Single slip + B	*Air*	13.8	203	97
		13.2	205	98
	Vacuum	28.5	213	102
		19.9	203	97
		22.8	202	96
	Oxygen	25.7	211	101
		39.5	194	93
		33.4	213	102

Figure 1. *Optical micrograph of slip lines on the surface of (a) single-slip-oriented, and (b) duplex-slip-oriented Fe-43Al single crystals tensile tested to failure.*

Faceted holes were observed on the fracture surfaces of a specimen of Fe-43Al tested in oxygen. The holes appear to reflect the cubic symmetry of the crystal structure, see Figure 2(c). The holes look somewhat like particles but were clearly observed to be holes when tilting the fracture surface to $60°$ in the SEM. Although these holes may have served as crack initiation sites, it is noteworthy that the specimen in which a hole was observed exhibited over 40% elongation.

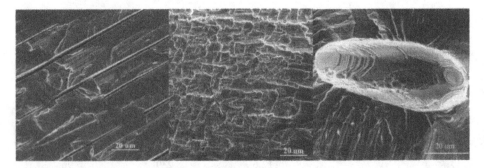

Figure 2. *Fractographs of tensile tested Fe-43Al crystals: (a) duplex-slip crystals in air; (b) boron-doped in air; and (c) undoped single slip-oriented in oxygen.*

Table 2 shows the results of tensile tests on single-slip-oriented, single crystals with and without boron tested at 4.4×10^{-6} s^{-1}. As noted previously for undoped Fe-40Al and Fe-43Al single crystals [11,16], testing in air at slow strain rates leads to a reduction in yield strength. In this case, the average yield stress in air for the undoped single-slip-oriented single crystals strained at 4.4×10^{-6} s^{-1} is 82 MPa, compared with 90 MPa for tests at 2.5×10^{-3} s^{-1} (and 94 MPa in oxygen). In contrast, the yield stress of the boron-doped crystals was found to be the same in air at 4.4×10^{-6} s^{-1} (98 MPa) as in air, vacuum or oxygen at 2.5×10^{-3} s^{-1}, indicating that boron must ameliorate the environmental effect. At the slower strain rate, the average ductility of both the boron-doped and the undoped single crystals was reduced compared to tests at 2.5×10^{-3} s^{-1}, with again the boron-doped crystals showing more ductility. It is worth noting that although the average ductility obtained in the latter crystals is less (7.9%) than at the higher strain rate (13.5%), the largest elongation obtained at the slower strain rate (12.6%) is little different from the values obtained at the higher strain rate. Again, it appears that boron can produce some very slight amelioration of the environmental effect in single crystals, as has been observed in FeAl polycrystals [24,25].

Table 2. *Room-temperature strength and ductility of single-slip-oriented Fe-43Al single crystals with and without 300 p.p.m. boron strained at room temperature in air at 4.4 $\times 10^{-6}$ s^{-1}.*

Specimen	Elongation, %	Yield Stress (MPa)	CRSS (MPa)
Fe-43Al	2.9	210	84
	6.5	208	83
	8.8	200	80
	4.1	205	82
Fe-43Al+B	5.5	210	100
	12.6	196	93
	10.2	198	94
	3.3	220	105

As at the higher strain rates, {110} slip lines could be observed on the surface of the undoped crystals, whereas slip lines could not be observed on the doped crystals, indicating a refinement of slip due to boron. Again, as at the higher strain rate, the fracture surfaces of undoped crystals showed transgranular cleavage with numerous secondary cracks, whereas the boron-doped crystals did not exhibit the secondary cracking.

Figure 3 shows the stress-strain curves for single-slip-oriented undoped single crystals strained at 5×10^{-6} s^{-1} alternately in air and vacuum, for 30-60 minutes in each, with about 40 minutes before testing in vacuum (in order to achieve the desired low pressure). Note that initial straining was in air but yield occurred under vacuum. The flow stress after reloading was not the same as that prior to the previous unloading: in air the subsequent flow stress upon reloading was always lower than the flow stress under vacuum before unloading; whereas the flow stress under vacuum upon reloading was always higher than the prior flow stress in air. These results corroborate previous suggestions that hydrogen introduced into FeAl during testing in air lowers the flow stress [11,15,16] and that testing in vacuum reduces the hydrogen concentration, allowing the flow stress to increase. For comparison, similar alternate loading in air and vacuum was performed on 6061 aluminum specimens. The changes in flow stress observed in Fe-40Al in different environments were not present in this alloy [11].

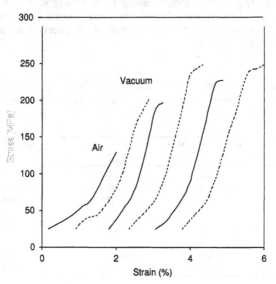

Figure 3. *Tests of undoped single-slip-oriented Fe-43Al single crystals under tension at 5×10^{-6} s^{-1} alternately in air and vacuum.*

CONCLUSIONS

Tensile tests have been performed on single crystals of Fe-43Al in various environments in order to further elucidate the effects of moisture-produced hydrogen in FeAl. Moisture-produced hydrogen in FeAl dramatically lowers the ductility and caused a change in fracture mode. Numerous secondary cracks on fracture surfaces and the greater ductility observed in tests performed alternately in air and vacuum or intermittently in air suggest that the role of hydrogen is not merely to enhance crack initiation but also to ease crack propagation. The presence of water vapor also lowers the yield stress and subsequent flow stress of undoped FeAl single crystals at slow strain rates. Doping with boron both produces a slight improvement in elongation in air (only) and prevents water vapor reducing the strength of FeAl in air at slow strain rates.

ACKNOWLEDGMENTS

Research supported at Dartmouth College by NSF grant DMR-9973977 and by the Oak Ridge Associated Universities SHaRE program contract DE-AC05-76OR00033; and by the Division of Materials Science and Engineering at the Oak Ridge National Laboratory, managed by UT-Battelle, LLC, for the U.S. Department of Energy under contract DE-AC05-00OR22725.

REFERENCES

1. C.T. Liu, E.H. Lee, and C.G. McKamey, *Scripta Metall.*, **23** (1989) 875.
2. D.J. Gaydosh, and M.V. Nathal, *Scripta Metall.*, **24** (1990) 1281.
3. N.S. Stoloff, C.T. Liu, *Intermetall,.* **2** (1994) 75.
4. M.V. Nathal, and C.T. Liu, *Intermetall,.* **3** (1995) 77.
5. R.J. Lynch, K.A. Gee, L.A. Heldt, *Scripta Metall. Mater.*, **30** (1994) 945.
6. R.J. Lynch, M. Harburn, L. Maucione, L.A. Heldt, *Scripta Metall. Mater.*, **30** (1994) 1157.
7. H. Saka, and T. Nishizaki, *Philos. Mag. A*, **73** (1996) 1173.
8. J.W. Cohron, Y. Lin, R.H. Zee, and E.P. George, *Acta mater.*, **46** (1998) 6245.
9. C.T. Liu, and E.P. George, *Scripta Metall. Mater.*, **24** (1990) 1285.
10. M. Wittmann, D. Wu, I. Baker, E.P. George, and L. Heatherly, Proceedings of the 12th International Conference on the Strength of Metals and Alloys, in press.
11. I. Baker, D. Wu, S.O. Kruijver and E.P. George, *Intermetall.*, (2001), in press.
12. M. Nakamura and T. Kumagai, *Metall. Mater. Trans A.*, **30A** (1999) 3089.
13. Y.F. Zhu, C.T. Liu, and C.H. Chen, *Scripta Mater.*, **35** (1996) 1435.
14. P. Specht, M. Brede and P. Neumann, *MRS Symp. Proc.* **364** (1995) 207.
15. L.M. Pike and C.T. Liu, *Scripta Mater.*, **38** (1998) 1475.
16. D. Wu and I. Baker, *Intermetall.*, (2000), in press.
17. Y. Yang, I. Baker and E.P. George, *Mater. Char.* **42** (1999) 161.
18. O. Klein and I. Baker, *Scripta Metall Mater.*, **27** (1992) 1823.
19. D. Li, D. Lin, A. Shan, and Y. Liu, *Scripta Metall. Mater.*, **30** (1994) 655.
20. O. Klein, P. Nagpal and I. Baker, Proc MRS, **288** (1993) 935.
21. P. Nagpal and I. Baker, *Scripta Metall. Mater.*, **25** (1991) 2577.
22. I. Baker, O. Klein, C. Nelson, and E.P. George, *Scripta Metall. Mater.*, **30** (1994) 863.
23. P. Nagpal and I. Baker, *Metall. Trans.* **A21** (1990) 2281.
24. C.T. Liu and E.P. George, Proc. MRS, **213** (1991) 527.
25. E.P. George, M. Yamaguchi, K.S. Kumar and C.T. Liu, *Ann. Rev. Mater.*, **24** (1994) 409.

Mat. Res. Soc. Symp. Proc. Vol. 646 © 2001 Materials Research Society

The Influence of some Microstructural and Test Parameters on the Tensile Stress and Ductility Behaviour of a MA FeAl Intermetallic

J. Chao, M.A. Muñoz-Morris, J.L. Gonzalez-Carrasco and D.G. Morris,
Department of Physical Metallurgy, CENIM, CSIC, Avenida Gregorio del Amo 8, 28040 Madrid, SPAIN

ABSTRACT

This study examines the influence of microstructural parameters (grain and dispersoid size, vacancy content) and some test parameters (strain rate, protective oxide coatings, air and water vapour excluding films, and surface geometrical quality) on the tensile behaviour (yield stress, work hardening rate, tensile stress, ductility) of a mechanically-alloyed, fine-grained Fe-40Al intermetallic. Major changes of strength and ductility are obtained by changing grain size (1% and 10% for grain sizes of 100μm and 1μm) and by avoiding premature stress/strain concentrators (ductility increased from 5% to 10% for imperfectly machined to prepolished samples). Ductility variations are interpreted using a slow-crack-propagation-to-instability model, where the roles of environment, surface state, deformation processes, and fracture mechanisms can be distinguished.

INTRODUCTION

The intermetallic FeAl has many attractive properties for use as a structural material but suffers from a relatively low ductility [1-3]. Some of the factors involved in producing this low ductility are environmental embrittlement [2,3], brought about by the reaction of atmospheric water vapour with fresh Al near a crack tip and leading to hydrogen injection below the surface, weak grain boundaries tending to lead to intergranular failure, particularly for materials of high Al content, and hardening-embrittlement by quenched-in vacancies [4-6]. Many studies have examined various aspects of failure in Fe_3Al and FeAl alloys and show how, for example, higher strain rates lead to increased flow stress and ductility [7-9] and Boron additions improve grain boundary cohesion leading to a suppression of intergranular failure and increased ductility [3,4,10,11]. These aspects have been described in a recent review of iron aluminides [12].
The present study re-examines the influence of many such test and material variables on the behaviour of a mechanically alloyed (MA) Fe-40Al material which is known to show high strength, because of the fine grain size and the presence of fine Y_2O_3 particles, as well as reasonably good, for FeAl, ductility, presumably because of its fine grain size [13-15]. In addition, special attention is given to a careful examination of the fracture surfaces obtained by failure and the evolution of fracture surface morphology throughout failure. Little attention has in fact been given to details of fracture surfaces apart from the common distinction as transgranular cleavage or as intergranular failure. Attention is also given to the influence of pre-oxidation treatments on the mechanical properties [16], since such oxide coatings may be expected to diminish the environmental sensitivity of the material and hence improve ductility.

EXPERIMENTAL DETAILS

The mechanically alloyed Fe-40Al material was supplied, in extruded bar form of diameter 19mm, by CEREM, CEA (Centre de Recherche en Materiaux, Centre d'Energie Atomique), Grenoble, France. Material was tested in the as-supplied state (extruded at 1100ºC and air cooled) and after a recrystallization treatment of 1h at 1300ºC. The two material states are characterised by a grain size of 1μm and 200μm and by Y_2O_3 particles of size about 18nm and 150nm [15]. Following machining to tensile samples of gauge length 20mm and diameter 3mm, samples were mechanically polished by turning on a lathe with progressively finer abrasive papers, finishing with 1600 grade, and then with 1μm diamond paste, giving a good polished surface. Samples were then given an oxidising treatment of 3h at 1100°C in air followed by a furnace cool to 900°C, then air cool. This treatment produces a protective oxide film of 1μm thickness. Some oxidised samples were then "de-oxidised" by repolishing mechanically, using the same sequence of abrasive papers and diamond paste to remove the oxide. This procedure ensured identical heat treatments for all samples. A few samples were alternatively "de-oxidised" using coarse (600 grade) abrasive paper. Others were given additional vacancy-removing anneals after oxidising, involving cooling from the 1100°C oxidation treatment to 475°C in the furnace, and holding for 1 week at 475°C. Tensile testing was carried out in air and fracture surfaces examined by scanning electron microscopy. 2-3 samples were tested for each condition, confirming the good data reproducibility. Ductility data showed scatter of ± 10-20% relative variation about the average values, and stress values of ± 1-3% about average values.

RESULTS

Experimental results of tensile testing are summarised in Tables 1 and 2, where various data deduced from fracture surfaces are also shown. For the most part the results confirm variations reported earlier during fracture testing of FeAl. Comparing samples tested at different strain rates (Table 2): there is no significant effect on yield stress and only a slightly smaller ductility for the slower strain rate. Surface protection by oil or by pre-oxidation (Table 1) does not alter the yield stress but allows an improvement of ductility, especially for the more brittle, recrystallised state. (The heat treatment associated with oxidation - 3h at 1100°C - does, however, lead to a slight reduction of yield stress.) Surface geometry or quality plays an important role (Table 1) with the ductility of sanded and especially machined samples being much lower than for the polished samples. Vacancy removal has a minor effect for as-extruded materials, but allows a significant fall of strength and also improvement of ductility for the very brittle recrystallised, de-oxidised samples (Table 2). Finally, as-extruded material is seen to be very much stronger and more ductile than recrystallised material (Tables 1 and 2).

Micrographs illustrating important features of fracture surfaces are given in Figs. 1-4. Fig. 1 shows a low magnification view of an as-extruded, de-oxidised sample where an evolution of fracture morphology from smooth failure on an outer ring about 10μm deep, followed by crystallographic cleavage over 300-500μm, and followed by final transgranular, ductile failure has been distinguished. Fig. 2 emphasizes the overall smooth nature - with river lines - of the initial ring zone (the grain size of the bulk material is 1μm) and shows the transition to fine grain-size dependent cleavage. Samples tested after machining and polishing showed similar mechanical behaviour, but did not show this outer smooth ring. The origin of the outer, brittle ring was shown to be the surface layer of recrystallized material produced by the sub-surface deformation left after machining and the 1100°C oxidation heat treatment. The only difference between samples pre-oxidised and de-oxidised is that the pre-oxidised samples show several thumbnail initial surface failure regions, presumably where the oxide first failed during straining, instead of a continuous surface ring of initial failure. Recrystallized samples showed similar changes of fracture surface morphology - initial fairly smooth surface ring, later coarser crystallographic facetted cleavage (where {100} cleavage planes are deduced from the facet distributions) and final ductile failure, as shown in Figs. 3 and 4. Table 1 records the depth of the initial brittle surface ring or thumbnail (10-15μm for as-extruded and recrystallised samples) as well as the crack depth and estimated stress intensity factor, and the true fracture stress, when the brittle to ductile transition occurs.

DISCUSSION

A detailed discussion of the fracture process occurring during tensile testing the present FeAl samples is not possible here, in view of the limited space available, and will be reported elsewhere [17]. A brief description of the processes occurring is given below, emphasizing the importance of the various material and test parameters.

Table 1

Influence of surface state on mechanical properties (average data of 2-3 tests) and fracture morphology

Material State	Test Conditions	Yield Stress (MPa)	Nominal Fracture Stress (MPa)	True Fracture Stress (MPa)	Ductility (%)	Work Hardening Rate (GPa)	Small Facet Size (μm)	Crack Length at B-D transition (mm)	Stress Intensity Factor at B-D transition (MPa$\sqrt{m}$)
Extruded	DeOx	795	1315	1580	10	5.2	10	0.33	41
	Sanded	760	1170		8.2				
	Ox	800	1340	1600	10.7	5.5	10	0.31	42
	Oil	790	1375	1700	12.3	5		0.35	45
	Machined	910	1190	1360	6.3	6		0.3	40
	Fine-Repol	925	1360	1600	10.2	5	0	0.35	45
Recryst	DeOx	585	755	1400	1.3	13	15	1.0	43
	Ox	575	990	1250	5	8.3	13	0.8	50
	Oil	600	900	1350	3	10	15	1.0	50

Table 2

Influence of test conditions and material state on mechanical properties (average data of 2-3 tests) and fracture morphology

Material State	Test Conditions	Yield Stress (MPa)	Nominal Fracture Stress (MPa)	True Fracture Stress (MPa)	Ductility (%)	Work Hardening Rate (GPa)	Small Facet Size (μm)	Crack Length at B-D transition (mm)	Stress Intensity Factor at B-D transition (MPa$\sqrt{m}$)
Extruded	DeOx	795	1315	1580	10	5.2	10	0.33	41
	DeOx/Slow	790	1260	1530	8.5	6	10	0.5	48
	Ox	800	1340	1600	10.7	5.5	10	0.31	42
	Ox/Slow	820	1340	1690	10.5	5.5		0.42	46
	Ox/NoVacs	755	1215		10				
Recryst	DeOx	585	755	1400	1.3	13	15	1.0	43
	DeOx/NoVacs	430	700		2.3				
	Ox	575	990	1250	5	8.3	13	0.8	50
	Ox/NoVacs	440	840		4.5				

DeOx:: means specimen polished, then oxidised 3h at 1100°C, then repolished
Ox:: means specimen polished, then oxidised 3h at 1100°C
Sanded/Oil means, respectively: oxidised, de-oxidised, then abraded with 600 grade paper/coated in silicone oil
Machined/Fine-Repol means after final machining/ or machined and then repolished
Normal/Slow Strain rate: 5 x 10^{-4}/s / 5 x 10^{-6}/s
No Vacs: means vacancies removed by 475°C anneal

50 μm

Evolution of fracture surface from
brittle cleavage ring to ductile
interior on tensile failed Extruded
MA Fe-40Al

Figure 1: *Fracture surface of De-oxidised as-extruded material.*

Brittle surface ring on Extruded sample   10 μm

Figure 2: *Brittle surface ring of De-oxidised as-extruded material.*

100 μm

Sequence of Fracture Evolution in Recrystallized
MA Fe-40Al:

Initial Smooth Surface Failure;
Later Crystallographic Cleavage;
Final Fine-Dimple Ductile Failure

Figure 3: *Stages of fracture of De-oxidised recrystallised material.*

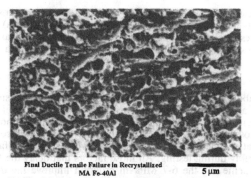

Final Ductile Tensile Failure in Recrystallized
MA Fe-40Al   5 μm

Figure 4: *Ductile final failure of recrystallised material.*

An important feature of the fracture surfaces is the evolution of morphology seen, for most as-extruded samples from the flat initial fracture ring or thumbnail, to cleavage and later transgranular failure. It can be deduced that cleavage growth, from 10μm to 300-500μm for these materials does not occur until the final seconds of testing, as rapid failure occurs. Equally, for the machined-and-repolished as-extruded samples, and for recrystallized samples, cleavage crack growth from a crack nucleation stage to a length of about 0.35mm, respectively 1mm, does not occur until the same final seconds of testing. This deduction is based on the very good linearity of work hardening from yield to near failure (≈ 2% or ≈ 10% strain for recrystallized and as-extruded samples, respectively). Therefore we can see that deformation up to near-failure occurs whilst the initial cracking stage is occurring. We cannot distinguish whether such cracking begins at yield and slowly propagates to a critical length near the failure stage, or whether such crack initiation occurs only much closer to final failure. Such crack initiation will occur due to stress and strain concentrations during the initial stage of plastic deformation and hence this stage is sensitive to both the grain size, determining the scale of the stress and strain concentrations, and the yield stress, with softer initial materials able to deform more for a given grain size [17]. Crack initiation occurs at the surface, perhaps where the internal stress-strain concentration meets a grain boundary or other stress-resisting obstacle, and grows in a manner strongly affected by environmental attack. The difference between most of the as-extruded samples tested here (typically machined, heat-treated, then polished) and the machined-and-polished samples was the presence of the ring of recrystallized material at the surface of the first type of samples, where the sub-surface strains due to machining allowed recrystallization during high temperature oxidation, and since the normal polishing procedures had not removed all the recrystallized material. The depth of this layer was about 10 μm after normal polishing and since the grain size was large (~100 μm), fracture initiated there early during the test, but clearly only through the recrystallized zone, 10 μm deep. The similar ductility for samples both with and without this recrystallized surface ring, but with similar surface roughness (compare in Table 1 the DeOx and the Fine-Repol samples) confirms that the 10 μm deep crack did not adversely affect fracture.

Using the concept of local stress intensity factor at the crack tip, the transition from crack nucleation to cleavage growth occurs for the as-extruded samples here when the stress intensity factor K_{IC} is about 6-7 MPa√m. At this state failure is deduced to occur by crack growth at a rate accelerating from its initial value to above 100μm/s (i.e. producing the large cleavage crack in the final few seconds) as the stress intensity factor increases from 6-7 MPa√m to about 45-50 MPa√m (see Table 1, the point where ductile, dimple failure takes over). Such values of crack growth rate and stress intensity factors are of similar magnitude to those determined by Kasul and Heldt [18], Schneibel and Jenkins [19] and Tonneau [20] for crack growth initiation near a threshold level of stress intensity factor, and accelerating crack growth at high levels of stress intensity factor. Rapid crack growth has been explained by hydrogen embrittlement at the crack tip, with hydrogen ingress mostly determined by transport by dislocations [18]. Final ductile failure occurs as the crack growth rate becomes too fast ($>10^{-3}$ m/s) or the stress intensity factor becomes too large ($>$45-50 MPa√m - Table 1) or simply as the average stress on the remaining sample cross section becomes so high ($>$1500 MPa - Table 1).

CONCLUSIONS

From the discussion above it can be understood that the creation of a surface crack, more than 10μm deep, and leading to stress intensity factors of the order of 6-7 MPa√m as the applied load increases, is the critical stage leading to failure of most of the as-extruded samples. For other samples, it is crack initiation by stress/strain concentrations during plastic deformation that is the critical stage. Beyond this point failure occurs rapidly and uncontrollably during testing. All those factors that delay the formation of the initial defect and achieving the 6-7 MPa√m criterion improve the ductility: e.g. reducing initial geometrical stress raisers, allowing less or slower environmental attack by removing active gaseous species and testing more quickly, softening the material by vacancy elimination, ensuring fine grain size in the bulk and at the sample surface. Microstructural factors (determining yield stress, e.g. grain size) are important during the stage of crack initiation, since it is stress and strain localisation, dependent on these microstructural factors, that leads to crack formation. Hence grain size refinement has an important effect on improving ductility, since it allows only fine levels of deformation inhomogeneity. Similarly, softer materials for a given grain size improve ductility, possibly because of more homogeneous deformation and the need for work hardening to increase stress levels before the onset of stress and strain concentrations.

REFERENCES

1: C.G. McKamey, J.H.CeVan, P.F. Tortorelli and V.K. Sikka, J. Mater. Res. 6, 1779 (1991).
2: C.T. Liu, E.H. Lee and C.G. McKamey, Scripta Metall. 23, 875 (1989).
3: C.T. Liu and E.P. George, Scripta Metall. 24, 1285 (1990).
4: M.A. Crimp, K.M. Vedula and D.J. Gaydosh, in High Temperature Ordered Intermetallic Alloys II, Vol. 81, eds. N.S. Stoloff, C.C. Koch, C.T. Liu and O. Izumi, p. 499, MRS Symposium Proceedings, Pittsburgh, PA, 1987.
5: P. Nagpal and I. Baker, Mater. Char. 27, 167 (1991).
6: D.J. Gaydosh, S.L. Draper, R.D. Noebe and M.V. Nathal, Mater. Sci. Eng. A150, 7 (1992).
7: P. Nagpal and I. Baker, Scripta Metall. Mater. 25, 2577 (1991).
8: I. Baker, O. Klein, C. Nelson and E.P. George, Scripta Metall. Mater. 30, 863 (1994).
9: L.M. Pike and C.T. Liu, Scripta Mater. 38, 1475 (1998).
10: M.A. Crimp and K.M. Vedula, Mater. Sci. Eng. 78, 193 (1986).
11: D.V. Gaydosh, S.L: Draper and M.V. Nathal, Metall. Trans. 20A, 1701 (1989).
12: I. Baker and P. Munroe, Int. Mater. Rev. 42, 181 (1997).
13: R. Baccino, D. San Filippo, F. Moret, A. Lefort and G. Webb, Proc. PM'94, Powder Metallurgy World Congress, Paris, June 1994, Vol. II, Editions de Physique, Les Ulis, 1994, p. 1239.
14: D.G. Morris and S. Gunther, Mater. Sci. and Eng. A208, 7 (1996).
15: D.G. Morris, S. Gunther and C. Briguet, Scripta Mater. 37, 71 (1997).
16: M.A. Montealegre, J.L. Gonzalez-Carrasco, M.A. Morris-Muñoz, J. Chao and D.G. Morris, Intermetallics 8, 439 (2000).
17: J. Chao, M.A. Muñoz-Morris, J.L. Gonzalez-Carrasco and D.G. Morris, submitted to Intermetallics.
18: D.B. Kasul and L.A. Heldt, Metall. and Mater. Trans. 25A, 1285 (1994).
19: J.H. Schneibel and M.G. Jenkins, Scripta Metall. Mater. 28, 389 (1993).
20: A. Tonneau, Ph. D. Thesis, University of Poitiers, 20th Oct. 1999.

Mat. Res. Soc. Symp. Proc. Vol. 646 © 2001 Materials Research Society

Low Temperature Positron Lifetime and Doppler Broadening Measurements in B2-type FeAl Alloys

Tomohide Haraguchi, Fuminobu Hori[1], Ryuichiro Oshima[1] and Mineo Kogachi[2]
Graduate School of Science, Osaka Prefecture University,
Sakai 599-8531, Japan.
[1]Research Institute for Advanced Science and Technology, Osaka Prefecture University,
Sakai 599-8570, Japan.
[2]College of Integrated Arts and Sciences, Osaka Prefecture University,
Sakai 599-8531, Japan.

ABSTRACT

Positron lifetime and Doppler broadening measurements were carried out at a low temperature, 100K, for B2-type intermetallic compounds FeAl with composition ranges from 41.2 to 50.7 at%Al systematically in order to clarify the feature of lattice defects of the alloys. Quenching temperature dependences of positron lifetimes ranging from 573 to 1173 K was examined for alloys with 41.2, 49.0 and 50.7 at%Al. Two lifetime component analyses could not be made except for a few spectra, indicating a saturation trapping of positrons at atomic vacancies. The mean positron lifetimes are in a range of 175-195 psec, which are not corresponding to vacancy clusters but to mono and/or di-vacancies. The tendency for S-parameter to increase with increase in Al content was found. This suggests that such an increase of S-parameters is attributed to change in the fraction of atoms around the vacancies, not to increase in vacancy concentration.

INTRODUCTION

Many intermetallic compounds have attracted a number of investigators because of their possible application capabilities as high temperature structural materials. Considerable efforts have been directed toward practical applications of ordered titanium aluminide, metal silicides, nickel aluminide and iron aluminide. A B2-type intermetallic compound FeAl is one of the practical application candidates because of its excellent features, e.g., good oxidation resistance, relative high melting point and low cost. One of difficulties for practical use is in its poor ductility at ambient temperature.

As is well known, the physical and mechanical properties are strongly affected by lattice defects, such as point defects and dislocations. Recently, vacancy hardening [1, 2] and yield stress anomaly [3] were reported as the topics of FeAl system, and participation of lattice defects is suggested in these phenomena. However, accumulation of information about the lattice defects for intermetallic compounds may be deficient compared with pure metals and disordered alloys. One of the reasons is that investigation of lattice defects for intermetallic compounds is more complex than pure metals and disordered alloys because of a variety of defects in the ordered alloys, such as constitutional defects, super dislocations and anti-phase boundaries. In addition, it is well known that some B2-type alloys of NiAl, CoAl and FeAl have curious point defect structures on off-stoichiometric composition region. The constitutional defects consist of changes from anti-site atoms of transition metal atom in the transition metal rich region to vacancies on transition metal site in the Al-rich side.

So far our group has studied the point defect structures and the behaviors in B2-type alloys

[4, 5]. For the B2-type FeAl alloys, density measurements [6] and neutron diffraction method [7] and further, thermodynamic calculations [8] have been performed. Under the circumstances, however, it seems to be requested to confirm the defect structures and the behaviors in more detail by employing other experimental techniques. Positron is an excellent probe to detect the atomic vacancies sensitively. Recently, there were reported the results of positron lifetimes and S-parameters for iron aluminide system [9-13]. In this paper, the positron lifetimes and S-parameters measured at 100 K in several B2-type FeAl alloys with different compositions from Fe-rich region to Al-rich side are reported.

EXPERIMENTAL DETAILS

Eight FeAl samples with compositions ranging from 41.2 to 50.7 at%Al were prepared from aluminum of 99.999 % purity and iron of 99.99 % purity by an arc melting method. Each ingot was homogenized at 1273 K for 3-4 days under an argon gas flow. They were cut into slice with the thickness of 1.0-1.3 mm. Each slice was sealed in a silica tube filled with argon gas, and heat-treated at 1173 K for 1 hour followed by slow-cooling to room temperature at a rate of 1 K/min in order to remove mechanical stresses introduced by sample cutting. Compositions were determined by using electron probe micro analyzer (EPMA). Since formation of iron silicide has been suggested when using a silica tube [10-12, 14], we checked the Si concentration. However, it was an order of 0.001 at% or less in each sample. Further, these samples were confirmed to be a B2 single phase by means of x-ray diffraction. Accordingly, such an order of contaminated Si was accounted to be negligible. In order to improve the sensitivity of the measurements by inhibiting the atomic vibration effects as well as possible, a series of positron annihilation measurements were performed at a low temperature, 100 K, for the specimens quenched from various temperatures.

In FeAl system, it is well known that a large amount of vacancies are retained easily even by very slow cooling (e.g., [6]). Therefore, careful long time annealing at 573 K is given to all specimens. In addition, we also attempted to hold the samples at 610 K for special long annealing times because Schaefer and Sprengel suggested that such a heat treatment is the best way to remove the retained vacancies [15]. Quenching temperatures and holding times are listed in table I.

A ^{22}NaCl positron source with an activity of 680 kBq, sealed by 6 μm thick kapton films was used in the present experiments. Positron lifetime measurements were made with the conventional fast-fast coincidence circuit with a BaF$_2$ scintillator with the time resolution of 190-200 psec. For the Doppler broadening measurements, a Ge semiconductor detector with energy resolution of 581 eV for 122 keV line of ^{57}Co was used to monitor the 511 keV γ-ray annihilation. Total counts for each measurement were 1×10^6 for the positron lifetime measurements, and 1×10^7 for the Doppler broadening measurements, respectively. Each data point was taken the average value of at least two measurements.

Table I. List of quenching temperatures and holding times. Heat treatments at 973, 1073 and 1173 K are applied for three samples with 41.2 (Fe-rich), 49.0 (near stoichiometry) and 50.7 (Al-rich) at%Al composition. Special long time annealing at 610 K in FeAl is the best way to remove the retained vacancies suggested by Schaefer and Sprengel [13].

Quenching Temperature (K)	573	673	773	873	973	1073	1173	610
Holding Time (hours)	168	96	72	48	24	12	1	444

RESULTS AND DISCUSSION

Mean positron lifetime

Composition dependence of mean positron lifetimes is shown in Figure 1. No marked composition changes in the mean positron lifetimes are seen although slight increase with Al concentration for the temperature range concerned seems to exist in some temperature regions. The numerical values of the mean positron lifetimes are in a range of 175-195 psec, being similar to the published data [9-11, 13]. Broska et al. [10, 11] examined the *in-situ* positron lifetime measurements for $Fe_{58}Al_{42}$ and $Fe_{52}Al_{48}$ alloys, and found that the mean positron lifetimes increased with temperature higher than 1000 K, suggesting that the defect type was changed from mono-vacancy to di-vacancy. Taking into account of previous results [16], our lifetimes must correspond to single or double vacancies. We carried out the two-lifetime component analyses but could not obtain two components except for a few spectra, indicating a saturation of trapping by vacancies, which considerably exist in these alloys in spite of the careful long time heat treatments at 573 K and 610 K. This result well agrees with previous reports regarding the vacancy concentration obtained by means of density measurements [6].

Figure 2 shows the mean positron lifetime change with quenching temperature, but no remarkable temperature dependence is found. The values of the positron lifetimes at temperatures higher than 973 K are also in a range of 175-195 psec, indicating annihilation at mono-vacancies or di-vacancies as mentioned above. Accordingly it is exhibited that the vacancies do not grow into the vacancy cluster, despite that the vacancy concentration amounts more than 2 % in Al-rich region at 1173 K [6].

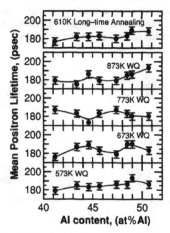

Figure 1. *Composition dependence of mean positron lifetimes for specimens quenched from various temperatures. "WQ" stands for water-quenching.*

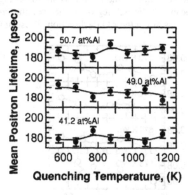

Figure 2. *Quenching temperature dependence of mean positron lifetime for three samples with 41.2, 49.0 and 50.7 at%Al. Each composition corresponds to Fe-rich, near stoichiometry and Al-rich region, respectively.*

Composition dependence of S-parameter

We can determine the shape parameter S (S-parameter) as the ratio of the counts in a central part of the Doppler broadening spectrum to the total counts. In Figure 3 is shown the composition dependence of S-parameter, ΔS, which is defined as:

$$\Delta S = S_{FeAl} - S_{pure\,Fe} \qquad (1)$$

where S_{FeAl} and $S_{pure\,Fe}$ represent the S-parameter obtained from FeAl alloys and high purity Fe, respectively. This procedure can remove the effect of positron annihilation in the kapton films sealing ^{22}NaCl crystal. In this time, we use the S-parameter for pure Fe instead of that for no-defect FeAl alloys, because of existing the remnant vacancies in this system mentioned previously. A tendency for S-parameter to increase with increase in Al composition is noticed for all quenching temperatures. The increase in S-parameter is generally considered to reflect the increase in defect concentrations [17, 18]. However, it was confirmed from our density measurements [6] that large amount of vacancies are residual in this FeAl system, and the concentration amounts to about 1% in the alloys with higher Al content even after a long time annealing at 573 K as mentioned above. Accordingly, it is thought that such an increase in S-parameter is not associated with increase in vacancy concentration. One of possibilities of increase of the S-parameter will be attributed to a change in the fraction of Fe and Al atoms occupied around the vacancies. It has been recognized that high momentum core electrons affect the S-parameter to decrease [18]. According to the first principle calculation for B2-type FeAl• alloy by Fu and Yoo [19], there is an on-site charge transfer from the non-bonding d^2_z state to the bonding d-orbital pointed along the nearest-neighbor Fe-Al direction. Assuming the case of vacancies sitting on the Fe-site as shown in Figure 4, there are many anti-site Fe atoms around the vacancy in Fe-rich region as the constitutional defects, which is having high momentum core d-electrons along the <111> direction. These anti-site Fe atoms make S-parameter small decrease.

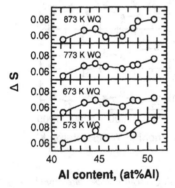

Al content, (at%Al)

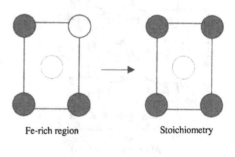

Fe-rich region   Stoichiometry

Figure 3. Composition dependence of S-parameters, ΔS, which is defined as equation (1). "WQ" stands for water-quenching.

Figure 4. Illustration of possibility of S-parameter increase with increase in Al composition from Fe-rich region to stoichiometry. Open circle and closed circle represent the Fe-atom and Al-atom, respectively. Dashed circle represents the vacancy.

As the Al composition approaches to the stoichiometry from an Fe-rich region, the concentration of anti-site Fe atom decreases, resulting in the increase of S-parameter. On the other hand, in the case of vacancies on the Al-site, we may not find the change of S-parameter since the fraction of atoms around the vacancies is not changed. This will give an explanation of S-parameter increase with Al-composition. However, this does not mean that all of vacancies occupy only the Fe-sites. It is difficult to conclude clearly from these results that either the vacancies coexist on both sites or not.

Furthermore, the Al concentration dependence of S-parameter seems to be grouped three; a rapid increase in the first stage (41.2-45.0 at%Al), flat or slight decrease in the second stage (45.0-47.5 at%Al) and then a rapid increase in the third stage (47.5-50.7 at%Al). Detailed experiments are now under way to confirm if this reflects the change in defect configuration with composition, that is, a change from mono-vacancies to di-vacancies with increase Al content.

CONCLUSIONS

We have presented results of mean positron lifetimes and S-parameters in B2-type intermetallic compounds FeAl as functions of compositions, and quenching temperatures. They are summarized in the following.
(1) Two lifetime components could not be resolved with the two-lifetime component analysis except for a few spectra. A large amount of vacancies already exist in the alloys, and most of positrons are annihilated with electrons at the vacancy sites.
(2) The observed mean positron lifetimes are in a range of 175-195 psec for the compositions and the quenching temperature regions concerned. These values correspond to the lifetimes for mono or di-vacancies, not but for vacancy clusters.
(3) A tendency for the S-parameter to increase with increase in Al composition is observed for all quenching temperatures, suggesting the change in the fraction of Fe and Al atoms occupied around the vacancies.

ACKNOWLEDGEMENTS

Authors would like to thank Professor M. Yamaguchi, Drs. H. Inui and K. Ito (Kyoto University) and Drs. H. Mabuchi and H. Tsuda (Osaka Prefecture University) for their kind helps in sample preparation. This work was partly supported by the Grant-in-Aid for the General Scientific Research from the Ministry of Education, Science and Culture of Japan.

REFERENCES

1. P. Nagpal and I. Baker, Metall. Trans., A21, 2281 (1990).
2. Y. A. Chang, L. M. Pike, C. T. Liu, A. R. Bilbrey and D. S. Stone, Intermetallics 1, 107 (1993).
3. K. Yoshimi, S. Hanada and M.H. Yoo, Intermetallics, 4, S159 (1996).
4. M. Kogachi and T. Haraguchi, Mat. Scie. and Eng. A, (2001) (in press)
5. T. Haraguchi and M. Kogachi, Mat. Scie. and Eng. A, (2001) (accepted)
6. M. Kogachi and T. Haraguchi, Mat. Scie. and Eng. A230, 124 (1997).
7. T. Haraguchi and M. Kogachi, Proc. of Inter. Conf. on Solid-Solid Phase Transformation '99, 521 (1999).
8. M. Kogachi and T. Haraguchi, Intermetallics 7, 981 (1999).

9. R. Würschum, C. Grupp and H. -E. Schaefer, Phys. Rev. Lett. **75**, 97 (1995).
10. A. Broska, J. Wolff, M. Franz and Th. Hehenlamp, Intermetallics **7**, 259 (1999).
11. J. Wolff, M. Franz, A. Broska, R. Kerl, M. Weinhagen, B. Köhler, M. Brauer, F. Faupel and Th. Hehenkamp, Intermetallics **7**, 289 (1999).
12. B. Köhler, J. Wolff, M. Franz, A. Broska and Th. Hehenkamp, Intermetallics **7**, 269 (1999).
13. B. Somieski, J. H. Schneibel and L. D. Hulett, Phil. Mag. Lett., 79, 3, 115 (1999)
14. M. Fähnle, B. Meyer and G. Bester, Intermetallics **7**, 1307 (1999).
15. H. -E. Schaefer and W. Sprengel, (private communication).
16. M. J. Puska and R. M. Nieminen, J. Phys. F **14**, 333 (1983).
17. I. K. MacKenzie, J. A. Eady and R. R. Gingerich, Phys. Lett., Vol. 33A, **5**, 279 (1970).
18. R. N. West, in *Positron in Solid*, edited by P. Hautojrävi (Springer-Verlag Berlin Heidelberg New York 1979) p. 90.
19. C. L. Fu and M. H. Yoo, Acta Metall. Mater., Vol.40, **4**, 703 (1992).

Mat. Res. Soc. Symp. Proc. Vol. 646 © 2001 Materials Research Society

Void Morphology In NiAl

M. Zakaria and P.R. Munroe
Electron Microscope Unit
University of New South Wales
Sydney NSW 2052, Australia

ABSTRACT

Void formation in stoichiometric NiAl was studied through controlled heat treatments and transmission electron microscopy. Voids formed at temperatures as low as 400°C, but dissolved during annealing at 900°C. Both cuboidal and rhombic dodecahedral voids were observed, often at the same annealing temperature. At higher annealing temperatures (≥800°C) extensive dislocation climb was noted. The relative incidence of void formation and dislocation climb can be related to the mobility of vacancies at each annealing temperature. Further, differences in void shape can be described in terms of their relative surface energy and mode of nucleation.

INTRODUCTION

Large supersaturations of vacancies can be quenched into NiAl following high temperature (>1000°C) heat treatment. During subsequent annealing at lower temperatures the excess vacancies are removed from the lattice through either dislocation climb or void formation. A number of workers have studied voids in NiAl, but the behaviour of these defects has not been unambiguously defined [1-5]. That is, the conditions under which voids form, their shape, size and range of stability have not been clearly defined. Two distinct void shapes, cuboidal and rhombic dodecahedral, are observed, but their incidence cannot be unambiguously related to particular heat treatment conditions or alloy composition. In some cases, dislocation loops, rather than voids, are formed [6-8]. We have examined the structure of voids in stoichiometric NiAl as a function of annealing temperature over a range from 400°C to 900°C. Detailed descriptions of the observed microstructures can be found elsewhere [9]. In this paper experimental observations will be summarised more briefly, the principal aim of this paper is to discuss the relative incidence of void formation and dislocation climb, and the variations in void shape.

EXPERIMENTAL METHODS

Nominally stoichiometric NiAl was prepared by arc melting under an argon atmosphere. The material was remelted several times to improve homogeneity. Chemical analysis indicated that composition was close to stoichiometry. Samples were annealed, in air, at 1300°C for 2 hours and cooled to room temperature by air-cooling. Subsequent annealing was performed at temperatures ranging from 400°C to 900°C for 1, 5 or 24 hours at each temperature. Thin foils for transmission electron microscopy (TEM) study were prepared and examined in a JEOL 2000FX TEM operating at 200kV.

RESULTS AND DISCUSSION

The microstructure of NiAl prior to annealing was examined. As expected, the microstructure was single phase with equaixed grains about 200μm in diameter. Following homogenization at 1300°C the microstructure contained fine (~ 20nm diameter) dislocation loops, with <001> Burgers vectors. A very small number of fine (~20nm diameter) voids were also noted. These voids were presumably formed during cooling, where there is presumably sufficient time for some vacancy agglomeration to occur.

Following annealing at 400°C for times up to 5 hours, the defect structures were broadly similar to the specimen air-cooled from 1300°C. However, following annealing for 24 hours at 400°C a number of small (~10nm) cuboidal voids was observed (figure 1). In contrast, other workers have observed dislocation loops in single crystal Ni-53Al annealed at 425°C for 1 hour after prior air-cooling from 1175°C. However, Eibner et al. [5] also observed cuboidal voids (~10nm in diameter) following annealing at 400°C for 32 hours following water quenching from 1600°C. On the other hand, Epperson et al. [3] observed rhombic dodecahedral voids in single crystal Ni-50.4Al annealed at 400°C for 22 hours following water-quenching from 1600°C.

Figure 1. Bright field TEM image of NiAl following annealing at 1300°C and subsequent annealing at 400°C for 24 hours. Very fine (10nm diameter) voids can be observed.

Following annealing at 500°C for 1 hour, cuboidal voids, ~40nm in diameter, were noted (figure 2a). A few larger voids (~100 nm) were also observed (marked 'v'), possibly associated with growth of pre-existing voids formed during initial heat treatment. In contrast, after annealing for either 5 or 24 hours rhombic dodecahedral voids were seen, typically ~50-100 nm in diameter (figure 2b). Yang and Dodd did not observe voids in NiAl following annealing at 500°C [2], but others have noted rhombic dodecahedral voids after quenching from high temperature and annealing at 500°C [1,4]. In contrast, Ball and Smallman [6] observed dislocation loops in near-stoichiometric NiAl containing 0.06%C annealed at 500°C for 15 minutes after quenching from 1300°C.

After exposure for 1 hour at 600°C rod-shaped voids with lengths parallel to {001} were observed (figure 3a). These voids were between 600 and 1200 nm in length and ~40 nm in width. However, following annealing for 5 hours only rhombic dodecahedral voids were observed. The void diameter was about 50-100nm, and some void coalescence was noted (figure 3b). The elongated voids were not observed. These elongated voids were very similar to those observed by Yang and Dodd [2] in near-stoichiometric NiAl containing 0.05%C following the same heat treatment conditions. However, it is not clear why these voids are not present after annealing for longer times. The void size, shape and density following annealing at 600°C for 5 hours is also similar to that noted by Yang and Dodd [2] and Epperson *et al.* [3].

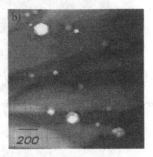

Figure 2. *Bright field TEM images of NiAl following annealing at 1300°C and subsequent annealing at 500°C for a) 1 hour and b) 5 hours.*

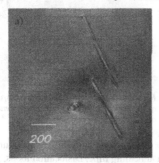
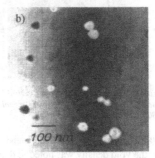

Figure 3. *Bright field TEM images of NiAl following annealing at 1300°C and subsequent annealing at 600°C for a) 1 hour and b) 5 hours*

Following annealing at 700°C for 1 hour, cuboidal voids, with diameters between 50 and 100nm, were observed (figure 4a) A similar defect structure was noted after 5 hours. This is consistent with the observations of Eiber *et al.* [5]. However, following annealing for 24 hours rhombic dodecahedral voids were noted, although in some regions of this specimen a higher dislocation density was noted (figure 4b). The dislocations exhibited a <001> Burgers vector and were edge in character. Numerous jogs were observed along their line length. Often these dislocations were noted to interact with any voids present (see region marked V).

Heat treatment at 800°C lead to the formation of very large (100-300nm) cuboidal voids following annealing for 24 hours (figure 5). In contrast, Yang and Dodd [2] observed larger (~160nm diameter) rhombic dodecahedral voids in near-stoichiometric alloys containing 0.05%C following annealing at 800°C. Whilst other workers observed only dislocation loops in near-stoichiometric NiAl annealed at 800°C for 30 minutes [1].

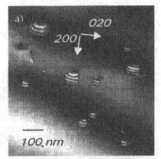 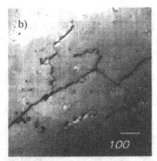

Figure 4. *Bright field TEM images of NiAl following annealing at 1300°C and subsequent annealing at 700°C for a) 1 hour, and b) 24 hours*

 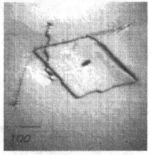

Figure 5. *Bright field TEM images of NiAl following annealing at 1300°C and subsequent annealing at 800°C for 24 hours.*

Figure 6. *Bright field TEM images of NiAl following annealing at 1300°C and subsequent annealing at 900°C for 24 hours.*

At 900°C the void density was noted to be lower than at lower annealing temperatures. Further, the void density was noted to decrease with annealing time. Dislocation loops, which exhibited jogs along their length, were more commonly noted in specimens annealed at this temperature (figure 6).

It is clear that variable void formation (size, shape and density) took place at different annealing temperatures, or even at the same temperature for different annealing times. Both rhombic dodecahedral and cuboidal voids were observed. It was also observed that void shrinkage occurred at 900°C, while at lower to intermediate temperatures (400-700°C), void nucleation and growth occurred. In spite of the large variations in void shape and size noted, these observations are broadly consistent with those of other workers [1-5].

The equilibrium thermal vacancy concentrations at the annealing temperatures used can be calculated (assuming an energy of formation, E_f, of 1.45eV [4]), these are presented in Table 1. Clearly, a high concentration of vacancies form during heat treatment at 1300°C. The large density of voids subsequently formed indicates that a significant fraction of these were retained on air-cooling. The excess vacancies were removed during subsequent annealing by either void nucleation and growth, or dislocation climb. Vacancies can also be removed by migration to vacancy sinks such as grain boundaries, although the large grain size in these alloys would mitigate against this.

Table 1. *Vacancy concentration (%) and number of vacancy jumps, J_v, at different annealing temperatures, and for different annealing times*

Temperature (°C)	Vacancy concentration n/N (%)	Vacancy jumps, J_v		
		1 hr	5 hr	24 hr
400	7.92×10^{-4}	3.94×10^4	1.97×10^5	9.47×10^5
500	3.96×10^{-3}	6.64×10^6	3.32×10^7	1.59×10^8
600	6.84×10^{-3}	3.45×10^8	1.73×10^9	8.29×10^9
700	0.0183	7.97×10^9	3.99×10^{10}	1.91×10^{11}
800	0.0408	1.03×10^{11}	5.13×10^{11}	2.46×10^{12}
900	0.0795	8.53×10^{11}	4.27×10^{12}	2.05×10^{13}
1300	0.49	-	-	-

The vacancy migration energy, E_m has been estimated for NiAl to be about 2.3eV [4]. On this basis the number of jumps, J_v for each annealing condition was calculated and the results are also summarised in Table 1. On the basis of a random walk process, and assuming that <100> jumps take place, the vacancy migration distance can be calculated. For 1 hour at 400°C, a vacancy will move ~50 nm, but at 900°C for 24 hours a vacancy can migrate about 1mm. The high density of voids observed at lower annealing temperatures presumably corresponds to the short distances over which the vacancies were able to move; that is, they cluster to form voids. However, at higher temperatures, vacancies migrate much further, thus they would be able to diffuse to dislocations, where they can be annihilated by climb, or they migrate to grain boundaries.

The highest void densities were noted at low or intermediate temperatures ($\leq 700°C$), where the equilibrium thermal vacancy concentration is relatively low. Thus, the driving force for vacancy removal is high, but vacancy mobility is low. A large number of vacancies may be driven out of supersaturation and with their limited mobility may cluster locally to form a high density of voids. At higher annealing temperatures ($\geq 800°C$) lower void densities were noted. Here, the equilibrium thermal vacancy concentration is much higher, so the driving force for vacancy removal decreases. These vacancies are much more mobile and may diffuse to grain boundaries or to dislocations.

At 900°C voids dissolved as annealing time increased. This is perhaps related to both mechanisms of vacancy removal being in operation together. That is, voids form initially at this temperature, but as annealing time increases more vacancies migrate to dislocations where they are annihilated. This lowers the retained vacancy concentration so voids may dissolve and go back into solution to maintain the equilibrium vacancy concentration at this temperature.

Both rhombic dodecahedral and cuboidal voids were observed in this study. Often, different void shapes were observed at different times at the same annealing temperature. The origins of these different shapes are unclear. It is possible that void shape is affected by both ease of nucleation and surface energy effects. Turning firstly to surface energy effects, cuboidal voids, with faces parallel to {001}, will have a different surface energy to that of rhombic dodecahedral voids with most faces parallel to {011}. Clapp et al. [10] estimated that for NiAl the surface energy of {001} was ~1 J/m^2, while the surface energy of {011} was ~1.5 J/m^2. If the void diameter is taken nominally as 50 nm then for cuboidal voids, there will be six faces, each of which has an area of $(50 \times 50)10^{-9}$ m^2. Using the model for shape of the rhombic dodecahedral voids suggested by Epperson et al. [3], there will be six {011} "prism" faces with an area of $(50 \times 50)10^{-9}$ m^2 plus two {001} "basal" faces with an area of 7.5×10^{-15} m^2. Thus, a cuboidal void has an energy of 1.5×10^{-14} J and a rhombic dodecahedral void an energy of 3×10^{-14} J. On this basis, cuboidal voids should be energetically more stable. It is possible that rhombic dodecahedral voids exist due to the differing nucleation mechanisms that may operate.

The B2 structure of NiAl consists of two interpenetrating simple cubic cells, where Al atoms occupy one sublattice and Ni atoms the other sublattice. If two vacancies substitute on to nearest-neighbour positions, this may lead to a plane of vacancies on {011}. Vacancy agglomeration on these planes may then ultimately result in the nucleation and growth of rhombic dodecahedral voids. Alternatively, if vacancies substitute on to next nearest-neighbour positions, that is two vacancies on the same sublattice, a plane of vacancies on {001} is more likely to result and this may then lead to the nucleation and growth of cuboidal voids. It would appear, therefore, that both mechanisms occur here. This is consistent with the work of Fu et al. who suggested that thermal vacancies in NiAl do not exhibit a preference for any specific lattice site [11].

CONCLUSIONS

Vacancy formation has been studied in stoichiometric NiAl, heat treated to produce a supersaturation of thermal vacancies, over a temperature range from 400°C to 900°C. Both cuboidal and rhombic dodecahedral shaped voids were noted, often both void types were noted at a single annealing temperature. At lower annealing temperatures void formation was the preferred method of removal of thermal vacancies, but at higher temperatures vacancies were more likely to be removed by dislocation climb. The shapes of the two vacancy types observed were rationalized in terms of their relative surface energy and possible methods of nucleation.

REFERENCES

1. W. Yang, R.A. Dodd and P.R. Strutt, Metall. Trans. 3A, 2049 (1972).
2. W. Yang and R.A. Dodd, Scripta. Metall. 8, 237 (1974).
3. J.E. Epperson, K.W. Gerstenberg, D. Berner, G. Kostroz and C. Ortiz, Phil. Mag. A 38, 529 (1978).
4. A. Parthasarathi and H.L. Fraser, Phil. Mag. A 50, 89 (1984).
5. J.E. Eibner, H.J. Engell, H. Schultz, H. Jacobi and G. Schlatte, G, Phil. Mag. 31, 739 (1975).
6. A. Ball and R.E. Smallman, Acta. Metall. 14, 1517 (1966).
7. G.W. Marshall and J.O Brittain, Metall. Trans. 7A, 1013 (1976).
8. T.C. Tisone, G.W. Marshall and J.O Brittain, J. Applied Phys. 39, 3714 (1968).
9. M. Zakaria and P.R. Munroe, J. Mater. Sci., in submission (2000).
10. P.C. Clapp, M.J. Rubins, S. Charpenay, J.A. Rifkin and Z.Z. Yu, in High Temperature Ordered Intermetallic Alloys III, edited by C.T. Liu, A.I. Taub, N.S. Stoloff and C.C. Koch, (Mater. Res. Soc. Proc. 133, Pittsburgh PA, 1989) pp. 29-35.
11. C.L. Fu, Y.Y.Ye and M.H. Yoo, in High Temperature Ordered Intermetallic Alloys V edited by I. Baker, R. Darolia, J.D. Whittenberger and M.H. Yoo, (Mater. Res. Soc. Proc. 288, Pittsburgh PA, 1993) pp. 21-32.

Mat. Res. Soc. Symp. Proc. Vol. 646 © 2001 Materials Research Society

Advancement of the Directional Solidification process of a NiAl-Alloy

F. Scheppe, I. Wagner and P.R. Sahm
RWTH Aachen, Giesserei-Institut, Intzestr. 5, D-52056 Aachen Germany

ABSTRACT

The directional solidification process of high temperature intermetallic alloys was investigated and discussed for a hypo eutectic NiAl-Cr-alloy. An unexpected solidification behavior was found which may be peculiar to the selected alloy group. The attempts show that due to the required high furnace temperatures only a narrow range of tolerance exists, in which the variation of the process parameters led to directional solidification. Aligned primary NiAl dendrites were then observed, embedded in randomly oriented interdendritic lamellas as known from conventional equiaxed castings.
The directional solidification process was basically evaluated for high temperature intermetallics.

INTRODUCTION

In order to improve the efficiency of modern gas turbines for energy transformation with simultaneously decreasing ecological damage, higher material demands are inevitable [1]. Intermetallic compounds such as NiAl offer new opportunities for developing low density, high-strength structural alloys with higher temperature capability when compared to conventional Ti- and Ni-base alloys.

The advantages of NiAl-base intermetallics are the high melting points of up to 1650 °C which are about 100 K to 250 K higher than these of the conventional Ni-base superalloys, a thermal conductivity of about four times than that of Ni-base alloys as well as an excellent oxidation and hot corrosion resistance. Furthermore, these materials exhibit a density of 6.20 g g/cm^3 to 6.35 g/cm^3 which is approximately 75% the density of state of the art superalloys and high temperature strength is provided above the ductile-to-brittle transition temperature of 900°C to 1000°C. In contrast to Ni-base alloys NiAl-base intermetallics exhibit an excellent microstructural stability without coarsening or dissolution of second phases like Ni$_3$Al at temperatures up to 1350°C. The high strength, however, is usually associated with low ductility at room temperature [3]. This requires a special adjustment of the mechanical machining due to the high hardness.

Intermetallic NiAl-base compounds like NiAl-Ta-Cr [5] are subject of an ongoing development for low density, high strength structural alloys with additional second phase strengthening for applications in gas turbines. Strong bonding between aluminum and nickel, which persists at elevated temperatures yields excellent high temperature properties with specific strength competitive to superalloys and ceramics. Thus, these alloys offer new opportunities for the application in gas turbines at temperatures higher than currently possible with conventional Ni-base superalloys [2].

At the foundry institute of RWTH Aachen the casting technology of NiAl-base intermetallics has been developed and optimized in the last years for different applications. High quality components for land based gas turbines can be produced reliably by an adapted precision casting technology [6]. The current paper focusses on the extension of the casting technology to directional solidification (DS) of high temperature intermetallics. Basic process parameters have to be correlated to the resulting microstructure with the currently available facilities which are optimized for conventional Ni-base-superalloys. The directional solidification is typically controlled by the thermal gradient G (dT/dx) at and the growth velocity v (dx/dt) of the solidification front. The vertical gradient is given by the temperature differences between the heating and the cooling zone. Assuming stationary solidification the growth velocity is estimated to be the withdrawal rate of the specimen relatively to the furnace.

In the context of this test series for directional solidification the NiAl base alloy FG 27 with a nominal melting temperature of 1550 C has been chosen [7]. The microstructure of the hypoeutectic FG 27 alloy (36.5 at.% Ni, 36.5 at.% Al, 27 at.% Cr) in the as precision cast state consists of NiAl primary crystals with an average size of $\overline{d}_{NiAl}$ = 50μm. The NiAl crystals are embedded in eutectic cells of an average size of $\overline{d}_{eut}$ =300μm. The average diameter of the Cr-fibers in the eutectic is $\overline{d}_{eut}(Cr)$ = 1μm. Inside the primary NiAl crystals precipitate fine dispersed Cr-particles, of which the biggest can be resolved in optical micrographs, Figure 1. NiAlCr alloys with Cr-concentration up to the eutectic concentration show a high oxidation resistance up to 1300°C and Cr promotes the forming of the stable α-Al_2O_3-modification [4].

EXPERIMENTAL

Numerous experiments were executed in a laboratory size Bridgman furnace using LMC (liquid metal cooling), Figure 2. The furnace consists of a heating zone with two separately controllable graphite heaters and a cooling zone. The cooling zone consists of a water-cooled copper die with liquid GaIn-alloy (75% Ga, 25% In) inside. The alloy is characterized by a particularly low melting (15.6°C), but high boiling point (above 2000°C).

The practical work covered three steps:

a) Production of cylindrical preforms;
b) Directional solidification of the remelted preforms in the LMC furnace;
c) Metallography.

For the production of the preform material at first ceramic shell molds were manufactured by investment casting method. The castings were done in a centrifugal casting system. The mold temperature was 1280°C and the casting temperature 1650°C for all attempts. The final, optimized casting parameters are the result of extensive investigations for the NiAl base alloy FG 27 on the centrifugal casting unit.

For the DS experiments in the LMC furnace the samples were placed in Al_2O_3-tubes and preheated in Ar-atmosphere to 1650°C (100K above the nominal T_{liq}). Then the samples were withdrawn at constant rates between 1 and 5 mm/min from the heating zone into the cooling bath through an insulating zone (Baffle). Furthermore during some selected attempts temperatures were measured. PtRh30-PtRh6 themocouples of 0.4 mm in diameter were exactly positioned within the specimens.

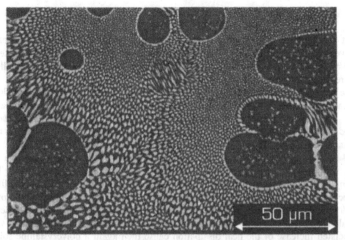

Figure 1: Optical micrograph of the microstructure of the hypoeutectic FG 27-alloy in the as precision cast state, NiAl: dark, Cr light phase [2].

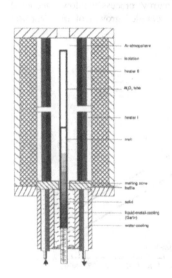 

Figure 2: The experiments were carried out in a lobatory size bridgman type furnace for directional solidification. The furnace consists of a heating and a cooling zone divided by an insulating baffle. The heaters can be controlled separately up to 1750 °C. A characterisitc withdrawal velocity is between 1 and 5 mm/min.

RESULTS AND DISCUSSION

The test series in the laboratory size LMC furnace demonstrated the influence of a varying growth velocity on the formation of aligned microstructure for constant preheating temperatures of 1650 °C. In the experiments with the FG 27-alloy the withdrawal rate was gradually increased and the resulting microstructure was examined.

With the help of the temperature measurement it can be stated that the growth velocity of the solidification front is equivalent to the withdrawal rate. Furthermore a determination of the liquidus and solidus temperature could take place by the cooling curves. These could be indicated as values of 1496°C for the liquidus temperature and 1427°C for the solidus temperature which both are below the nominal temperatures of the alloy, Figure 3.

Figure 4 shows the correlation of local process parameters for defined locations of a sample. By using a withdrawal rate of 1mm/min, polycrystalline microstructure developed. The further increasing of the lowering rate shows however the tendency towards directional solidification. At values between 3 mm/min and 4mm/min, directional solidification finally can be proven, Figure 5. This lowering speed is to be classified as ideal for the unimpaired growth of the dendrites and the avoidance of spurious grain formation. The further increase of the solidification velocity and the associated increase of the heat dissipation cause then again a polycrystalline solidification. This uncertain behavior may be explained by the peculiar solidification of intermetallic alloys.

The experiments show a sensitive correlation between the process parameters and the directional solidification of a hypo eutectic NiAl-Cr alloy. Within a narrow process window directional solidification of primary NiAl dendrites was realized. The coupled growth of interdendritic eutectic lamellas was not observed.

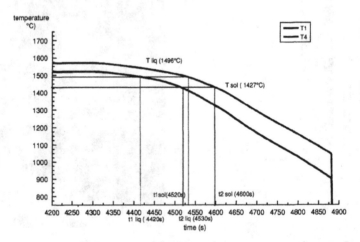

Figure 3: The cooling curves were calculated for two thermal couples, which were in a distance from 5mm. With the help of these curves a solidification length could be calculated of 4,95 mm/s.

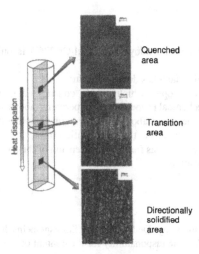

Quenched area

Transition area

Heat dissipation

Directionally solidified area

Figure 4: At defined places samples were taken to characterize the structure. All samples were withdrawn at constant rates through the Baffle into the liquid metal bath and then the upper half of the samples were quenched in the liquid metal cooling bath. The transient area of a structure to the next is clearly recognizable.

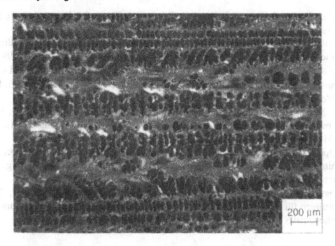

Figure 5: The directionally solidified microstructure is clearly recognizable (solidification velocity = 3mm/min). Dendritic arrays of primary NiAl (dark) are visible.

CONCLUSION

- The microstructure of hypo eutectic alloys like NiAl-Cr(27%) is only possible within a narrow process window;
- Aligned growth of interdendritic lamellas does not occur;
- Special alloys have to be developed which enable constarined growth of intermetallic compounds with superior mechanical properties at temperatures above 1000°C;
- The increased processing temperatures above 1700°C require improved facilities enabling a greater variety of temperature gradients at the solidification front;
- The investigations are an excellent basis for further extension of the solidification technology with regard to potential applications.

ACKNOWLEDGMENT

The work was promoted by funds by Deutsche Forschungsgemeinschaft through Sonder-forschungsbereich 561 (SFB 561). The responsibility for the content of this publication lies with the authors.

REFERENCES

[1] Grünling, H.W., Bremer, F.J.: Intermetallische Phasen als Strukturwerkstoffe für hohe Temperaturen, Forschungszentrum Jülich, S.41-45, 1991

[2] Hermann, W., et al.: Intermetallische NiAl-Komponenten zur umweltfreundlich Energiewandlung, Tagungsband Werkstoffwoche 98, Band III, München, 1998

[3] Darolia, R.: NiAl Alloys for High Temperature Structural Application, JOM 43(3), 1991

[4] Brumm, M.: Oxidationsverhalten von β-NiAl und von NiAl-Cr-Legierungen, Düsseldorf, VDI Verlag, 1992

[5] Palm, M., Sauthoff, G.: Werkstoffcharakterisierung und –optmierung von NiAl-Ta-Cr-Legierungen für Anwendungen im Gasturbinenbau, Werkstoffwoche 98, Band 6, S.503-508, München, 1998

[6] Scheppe, F. et. al.: Comparison of the numerical simulation and the cast process of Nickel Aluminides, eds. P.R. Sahm, P.N. Hansen, J.G. Coinley, Shaker Verlag, S.207-214, Aachen, 2000

[7] Rablbauer, R. et. al.: Strukturen und Eigenschaften von NiAl-α (Cr, Mo, Re) – Legierungen für den Hochtemperatureinsatz, Werkstoffwoche 98, Symposium 3, S.55-60, München, 1998

Mat. Res. Soc. Symp. Proc. Vol. 646 © 2001 Materials Research Society

The Effect of Cu-Macroalloying Additions to Rapidly Solidified NiAl Intermetallic Compound

J. Colin, B. Campillo, C. Gonzalez, O. Alvarez-Fregoso and J. A. Juarez-Islas
Instituto de Investigaciones en Materiales-UNAM
Circuito Exterior S/N, Cd. Universitaria
Mexico, D. F., 04510, MÉXICO.

ABSTRACT

The effects of two variables on the NiAl intermetallic compound were studied: 1) copper macroalloying additions and 2) rapid solidification processing. For that purpose, several NiCuAl alloys were vacuum induction melted and rapidly solidified by using a copper wheel, rotating at 15 m/s, under an argon atmosphere. Chemical analysis of as-rapidly solidified ribbons indicated, that four alloy compositions lie in the β-(Ni, Cu)Al field, one alloy composition lie in the boundary of the β-(Ni, Cu)Al/(Ni, Cu)$_2$Al$_3$ fields, one alloy composition lies in the boundary of the β-(Ni, Cu)Al/β-(Ni, Cu)Al + (Ni, Cu)$_3$Al fields and two alloy compositions lie in the β-(Ni, Cu)Al + (Ni, Cu)$_3$Al field. Transmission electron microscopic observations carried out in as-rapidly solidified ribbons, revealed the presence of at least three main structures: i) β-(Ni, Cu)Al, ii) β-(Ni, Cu)Al + martensite (Ni, Cu)Al and iii) (Ni, Cu)$_3$Al + martensite (Ni, Cu)Al. Microhardness Vickers and tensile test data indicated that alloys with a β-(Ni, Cu)Al + martensite (Ni, Cu)Al microstructure have improved room temperature ductility, reaching values of elongation up to 3.28 %.

INTRODUCTION

It is well known, that NiAl intermetallic compound possesses high melting temperature, low density, good oxidation resistance, metal-like properties, attractive modulus, high thermal conductivity and low raw material cost. These characteristics place it, as an interesting and potential structural material, for a wide range of applications, including those for high temperature in aerospace structures. The NiAl intermetallic compound shows a simple ordered B2 (cP2) CsCl crystal structure and a wide range of composition stability [1-3].

This B2 structure is stable for large deviations from stoichiometry with a significant long-range order [4]. In intermetallic compounds, different atomic species occupy different regular lattice sites and the strong bonds between their atoms result in attractive properties. In the NiAl intermetallic compound, the strong bonds and the lack of five independent slip systems, on the basis of ordered structure, give rise to low temperature embrittlement [5] and inadequate strength and inadequate creep resistance at elevated temperature [6].

Most ordered B2 intermetallic compounds exhibit at room temperature, a minima in both their yield strength and hardness at, or close to their stoichiometric composition [7]. In the NiAl intermetallic compound, the strength increases on either side of the stoichiometry composition [8]. Polycrystalline NiAl is generally referred as an intermetallic compound which is brittle at room temperature [9], and limited elongation of about 2% has been observed in stoichiometric composition. Since intermetallic compounds possesses a variety of properties, which would otherwise be beneficial for high temperature structural applications, a major objective of this work is to study, the effect of rapid solidification processing and copper macroalloying additions on room temperature ductility of polycrystalline NiAl.

EXPERIMENTAL DETAILS

NiCuAl alloys were prepared by vacuum induction melting, and then, rapidly solidified by using the melt spinning technique. Melt spun ribbons of 50 μm in thickness were obtained after impelling the liquid alloy melt onto a copper rotating wheel (15 m/s), under an argon atmosphere. Microstructural characterization of the as-rapidly solidified ribbons was carried out by using a Jeol 2100 scanning transmission electron microscope (STEM) and a Siemens D-5000 X-ray diffractometer. Evaluation of the mechanical properties of as-cast ribbons were carried out by using a Matzusawa microhardness tester (50 g load), and an Instron 1125 testing machine (10 kg load). During the experiments a crosshead speed of 0.10 mm/min was employed.

RESULTS AND DISCUSSION

The various ribbons obtained were analyzed in terms of their chemical composition, and the results of those analyses are summarized in Table I. The first five alloys showed a composition close to $(Ni_{50-y}Cu_y)Al$, where y varied from 5 to 25 at. %. The other three alloys showed a composition close to $Ni(Al_{50-y}Cu_y)$, where y varied from 10 to 25 at. %.

Table I Chemical composition of rapidly solidified ribbons (in at. %).

Alloy	Symbol	Ni	Cu	Al
A1	●	45 ± 0.5	5 ± 0.4	bal.
A2	●	40 ± 0.2	10 ± 0.2	bal.
A3	●	35 ± 0.1	15 ± 0.1	bal
A4	●	30 ± 0.1	20 ± 0.1	bal.
B1	■	25 ± 0.3	25 ± 0.4	bal.
C1	□	bal.	10 ± 0.3	40 ± 0.3
D1	○	bal.	20 ± 0.2	30 ± 0.2
D2	○	bal.	25 ± 0.4	25 ± 0.2

Figure 1 shows part of the ternary NiCuAl phase diagram [10], where the numbers indicate the different fields of interest, and the figures show the position of the alloys under study. For instance, the full circle indicates alloys in the β-(Ni, Cu)Al field (β-field from now on), the full square locates an alloy just outside the β-field. The open square shows the position of the alloy in the boundary between the β-field and the β + (Ni, Cu)₃Al field. The open circle indicates the location of the alloys in the β + (Ni, Cu)₃Al field.

Table II shows the phases identified by X-ray diffractometry on as-rapidly solidified ribbons. For instance, alloys A1 to A4 showed mainly the presence of the β-solid solution, with a lattice parameter increasing as the Cu-content increases. With respect to these lattice parameter behavior, it can be said, that the lattice value at the stoichiometry composition of the β-NiAl intermetallic compound at room temperature is 2.887 Å [10] and as reported in [9], when Ni is substituted by Cu, the lattice parameter as a function of alloy concentration is shifted to higher values. Alloy C1 also showed the presence of the β-solid solution, but the lattice parameter decreased from 2.887 A (for the stoichiometry NiAl) to 2.852 A, indicating that probably another phase is present. Alloy B1 showed the presence of β-solid solution + Cu₃Al. In alloys D1 and D2, it was detected the presence of the cubic (Ni, Cu)₃Al + martensite (Ni, Cu)Al; [M-(Ni, Cu)Al from now on].

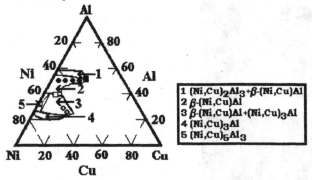

1 (Ni,Cu)₂Al₃+β-(Ni,Cu)Al
2 β-(Ni,Cu)Al
3 β-(Ni,Cu)Al+(Ni,Cu)₃Al
4 (Ni,Cu)₃Al
5 (Ni,Cu)₅Al₃

Figure 1 Part of the ternary NiCuAl phase diagram [10], where the fields of interest are indicated by numbers. The geometrical figures indicated the location of the alloys under study.

Table II Lattice parameter of phases detected in the alloys under study.

Alloy	Symbol	β-(Ni, Cu)Al (a, Å)	Cu₃Al (a, Å)	Ni₃Al (a, Å)	Martensite (Ni, Cu)Al (d, Å)
A1	●	2.888	---	---	---
A2	●	2.893	---	---	---
A3	●	2.901	---	---	---
A4	●	2.907	---	---	---
B1	■	2.911	2.96	---	---
C1	□	2.852	---	---	---
D1	○	---	---	3.56	2.10, 1.79,
D2	○	---	---	3.55	1.79, 1.23

a = lattice parameter, d = d spacing

TEM observations carried out on melt spun ribbons, of alloys A1 to A4, showed the presence of β-solid solution grains with long grain boundaries (figure 2a). These grains showed a poor grain boundary cohesion (figure 2b) and sometimes in matrix was observed the presence of vacancy cluaters (figure 2c). Alloy B1 showed the presence of β-solid solution grains with the presence of spherical Cu₃Al precipitates in matrix, as shown in figure 2d (Cu₃Al was identified by EDAX microanalysis giving 74.5 wt. % Cu and 25.5 wt. % Al). Dark field images (figure 2e) resembling antiphase boundary domains were obtained when the beam was diffracted in the (110) plane (figure 2f).

Alloy C1 showed the presence of β-solid solution grains and M-(Ni, Cu)Al grains, as can be observed in figure 3a. β and M grains were mainly identified by its diffraction pattern, as that shown in figure 3b. Alloys D1 and D2 showed the presence of cubic Ni₃Al and M-(Ni, Cu)Al grains (see figure 3c). A characteristic diffraction pattern for alloy D is shown in figure 3d.

With respect to the microstructure observed in alloy C1, it can be mentioned, that T. Cheng [11] carried out studies on melt spun ribbons of an Ni-34.6 at. % Al alloy. The resulting microstructure, depending on cooling rate of annealed melt spun ribbons (at 1250 °C) was of the β-solid solution, β + γ´ (Ni₃Al) or β + γ´ + M-NiAl type. He also reported values up to 7.6 % of elongation in melt spun ribbons after an annealing treatment of 2 hours at 1250°C.

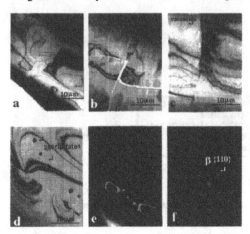

Figure 2 TEM micrographs of alloy A and B1, (a) β-solid solution with long grain boundaries, (b) poor grain boundary cohesion, (c) vacancy clusters, (d) spherical particles in β-matrix, (e) dark field image, (f) diffraction pattern.

Table III shows microhardness Vickers values for melt spun ribbons. For instance, alloys A1 to A4 and B1, presented a continuous increase in their microhardness Vickers values. That increase was from 320 Kg/mm² (alloy A1) to 517 kg/mm² (alloy B1). With respect to alloys C1, D1 and D2, it can be said, that microhardness values decreased as the Al content increased, from 381 kg/mm² (25 at. % Al) to 337 kg/mm² (40 at. % Al).

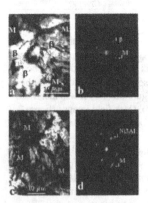

Figure 3 (a) Alloy C1 showing β + M grains, (b) diffraction pattern of alloy C1, (c) Alloy D1 showing Ni₃Al + M grains, (d) diffraction pattern of alloy D1.

In order to evaluate the room temperature ductility of as-rapidly solidified ribbons, tensile test were performed on them. Ribbons of alloys A1 to A4 and alloy B1 showed a very brittle behavior (these ribbons broken just as the test started). Alloy D1 and D2 showed a partial ductility as high as 1.47 %. Alloy C1 showed ductilities between 2.68 % to 3.28 %, depending of the amount of M-(NiAl). In the case of elongations of 3.28 %, the amount of martensite was close to 15% vol. %.

Table III Microhardness Vickers of as-rapidly solidified ribbons for the alloys under study (50 g load).

Alloy	A1	A2	A3	A4	B1	C1	D1	D2
Kg/mm^2	320 ± 15	370 ± 33	430 ± 30	465 ± 15	517 ± 25	337 ± 18	365 ± 12	381 ± 10

As mentioned before, several studies have been carried out in order to improve ductility of NiAl intermetallic compounds, including those on the NiAl-Ni$_3$Al two-phase alloy [12] and refining of grain size [13]. In our case, the increase in room temperature ductility can be attributed to the presence of small grain size (< 10 μm), the effect of Cu-macroalloying additions and solidification processing which gave rise to a microstructure of β-(Ni, Cu)Al plus M-(Ni, Cu)Al grains.

CONCLUSIONS

From the microstructural and mechanical characterization of melt spun ribbons the effect of Cu-macroalloying additions and rapidly solidified processing was noticed, for instance, alloy C1, with a microstructure consisting of β-(Ni, Cu)Al grains plus M-(Ni, Cu)Al grains, showed an important improvement in room temperature ductility, reaching values of elongation up to 3.28 %.

ACKNOWLEDGMENTS

The authors would like to thanks Mr. E. Caballero, Eng. L. Baños, A. Maciel and C. Vazquez for the experimental work. This research was supported by a Conacyt grant U-31346..

REFERENCES

1. E. M. Schulson, The Int. J. of Powder Met., 23, 25 (1987).
2. D. B. Miracle, Acta Metall., 41, 649 (1993).
3. R. D. Neobe, R. R. Bowman and M. V. Nathal, Inter. Mater. Rev., 38, 193 (1993).
4. T. Hughes, E. P. Lautenschlager, J. B. Cohen and J. O. Brittain, J. of Appl. Phys., 42, 3705 (1971).
5. Y. Umakoshi, in Materials Science and Technology, eds. R. W. Cahn, P. Haasen and E. J. Kramer, VHC Publishers Inc., 1, 254 (1991).
6. D. B. Miracle and R. Darolia, Intermetallic Compounds, eds. J. H. Westbrook and R. L. Fleisher, John Wiley and Sons Ltd., 2, 53 (1994).

7 P. Nagoal and I. Baker, Metall. Trans., **21A**, 2281 (1990).

8 I. Baker, P. Nagpal, F. Liu and P. R. Munroe, Acta Metall. Matter., **39**, 1637 (1991).

9 H. Jacobi and H. J. Engell, Acta Metallurgica, **19**, 701 (1971).

10 A. J. Bradley and H. Lipson, Proc. Roy. Soc., **177A**, 421 (1938).

11. T. Cheng, J. of Mat. Sci., **30**, 2877 (1995).

12. K. S. Kumar, S. K. Mannar and R. K. Viswanadhan, Acta Metall, **40**, 1201 (1992).

13. T. Cheng, Scripta Metall., **27**, 771 (1992)

Mat. Res. Soc. Symp. Proc. Vol. 646 © 2001 Materials Research Society

Multi-Modal "Order-Order" Kinetics in Ni_3Al Studied by Monte Carlo Computer Simulation

P. Oramus, R. Kozubski, V.Pierron-Bohnes[1], M.C.Cadeville[1], W.Pfeiler[2]
Institute of Physics, Jagellonian University, Reymonta 4, 30-059 Kraków, Poland.
[1]Institut de Physique et Chimie des Matériaux de Strasbourg, 23, rue du Loess, 67037 Strasbourg, France.
[2]Institut für Materialphysik, University of Vienna, Strudlhofgasse 4, A-1090 Vienna, Austria.

ABSTRACT

"Order-order" relaxations in γ-Ni_3Al previously extensively studied by means of resistometry, are simulated within a model of vacancy mechanism of atomic migration in a superstructure implemented with Monte Carlo technique and the Glauber algorithm. The observed operation of two simultaneous relaxation processes showing different rates, as well as the theoretically predicted effect of vacancy ordering have been definitely reproduced and analysed in detail in terms of the dynamics of particular kinds of atomic jumps. The proposed model scenario for the creation and elimination of antisite atoms in the relaxing $L1_2$-type superstructure shows that the experimentally observed features of the "order-order" processes in Ni_3Al follow from an interplay between two dominating and coupled modes of long- and short-range ordering: the creation/elimination of nn pairs of antisites (SRO) and the change of the "overall" number of antisites (LRO). High profile of the first process results in a high contribution of the fast component of LRO kinetics.

INTRODUCTION

If a long-range ordered system annealed at temperature T_i is abruptly cooled-down or heated-up to temperature T_f, the degree η of its long-range order (LRO) - i.e. the number of antisite atoms, evolves from the initial equilibrium value to the final one. If both T_i and T_f temperatures are lower than the "order-disorder" transition point, the process is called an "order-order" relaxation.

The present Monte Carlo simulation study concerns "order-order" relaxations in a homogeneous A_3B binary system with a superstructure of $L1_2$-type and refers to the previous experimental works on Ni_3Al [1]. The interest is focused on the origin of the complex character of the $\eta(t)$ relaxation isotherms measured by means of residual resistometry, which fitted weighted sums of exactly two single exponentials with substantially different relaxation times. Preliminary results of the study have been presented on MRS Spring Meeting in 1998 [2]; an extended paper containing the complete material will appear soon [3].

MONTE CARLO SIMULATION PROCEDURE

The sample was simulated by arranging two kinds of atom (Ni and Al) taken in a stoichiometric proportion $N_{Ni} / N_{Al} = 3/1$ over an $L1_2$-type superlattice containing 256000 sites. Subsequently, a fixed number of 10 vacancies were introduced by emptying at random the corresponding number of lattice sites.

Following the tradition of the pioneering works of the sixties [4,5], the equilibrium configuration of the system corresponding to a given temperature was then generated by imposing periodic boundary conditions upon the system and letting it relax according to the Glauber-dynamics algorithm applied to the vacancy mechanism of atomic jumps:

$$\Pi_{i \to j} = \frac{\exp\left[-\dfrac{\Delta E}{kT}\right]}{1 + \exp\left[-\dfrac{\Delta E}{kT}\right]} \tag{1}$$

where: $\Pi_{i \to j}$ is the probability for an atomic jump from a site "i" to a vacancy residing in the nearest-neighbouring (nn) site "j", $\Delta E = E_j - E_i$, E_i and E_j denote the system energies before and after the jump, respectively, k and T are Boltzmann constant and absolute temperature, respectively.

The regular "order-order" relaxations were simulated by pursuing the procedure at temperature T_f starting from a system previously equilibrated (be means of the same procedure) at another temperature T_i. All the results were averaged over 20 independent realisations of the procedure.

The energy changes ΔE were calculated within the Ising approximation with atomic pair interactions $V_{A-B}^{(v)}$ in two co-ordination zones (v=1,2) evaluated in the way that the reality of Ni_3Al was reproduced (for details see ref.[2]). Interaction with vacancies was neglected. Consequently, the simulations were run with $V_{Al-Al}^{(1)} = -0.15$ eV and $V_{Al-Al}^{(2)}$ varied between -0.06 eV and $+0.08$ eV.

The current configuration of the system and its MC-time evolution were analysed by monitoring a set of parameters:

- a Bragg-Williams-type LRO parameter η:

$$\eta = 1 - \frac{N_{Ni}^{(Al)}}{0.75 \times N^{(Al)}} \tag{2}$$

where $N^{(Al)}$ and $N_{Ni}^{(Al)}$ denote the number of Al-type sublattice sites and the number of Ni-antisites (Ni-atoms on the Al-sublattice), respectively,

- a specific short-range order (SRO) –type parameter APC (Antisite-Pair-Correlation):

$$APC = \frac{N_{NiAl}^{(Al)(Ni)}}{N_{Ni}^{(Al)}}, \tag{3}$$

where $N_{NiAl}^{(Al)(Ni)}$ denotes the number of nearest-neighbour pairs of Ni- and Al-antisites.

- the "jump-frequency" parameters $P_{Ni(Al);i \to j}$:

$$P_{Ni(Al);i \to j} = \frac{N_{Ni(Al);i \to j}^{exec}}{N_{att}} \tag{4}$$

where: $N_{Ni(Al);i \to j}^{exec}$ denotes a number of Ni(Al)-atom jumps from an "i"-type sublattice site to a nn vacancy residing on "j"-type sublattice *executed* within a fixed number of MC steps,

N_{att} denotes the total number of *jump attempts* (executed and not executed) during the same MC-time period.

- the „jump-efficiency" parameters $E_{Ni(Al)}^{ord(dis)}$ of particular ordering/disordering jumps:

$$E_{Ni}^{ord} = \frac{P_{Ni:Al \to Ni} - P_{Ni:Ni \to Al}}{P_{Ni:Al \to Ni}} \qquad E_{Ni}^{dis} = \frac{P_{Ni:Ni \to Al} - P_{Ni:Al \to Ni}}{P_{Ni:Ni \to Al}} \qquad (5)$$

$$E_{Al}^{ord} = \frac{P_{Al:Ni \to Al} - P_{Al:Al \to Ni}}{P_{Al:Ni \to Al}} \qquad E_{Al}^{dis} = \frac{P_{Al:Al \to Ni} - P_{Al:Ni \to Al}}{P_{Al:Al \to Ni}}$$

RESULTS

Laplace analysis of the simulated $\eta(t)$ relaxation isotherms indicated a definite domination of two relaxation times (Fig.1).

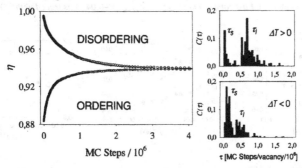

Figure 1. *$\eta(t)$ curves simulated at common T_f for $T_f/T_i = 0.78$ and corresponding spectra of relaxation times.*

A contribution $C(\tau_s)$ of the fast relaxation process to the overall kinetics decreased with an increase of $V_{Al-Al}^{(2)}$ parameters and this effect was the main tool for finding the origin of the complexity of LRO kinetics in Ni$_3$Al.

In parallel, $\eta(t)$ relaxations were accompanied by specific evolutions of APC also affected by the value of $V_{Al-Al}^{(2)}$ (Fig.2).

Detailed analysis of the results consisted of an inspection of the relationships between the atomic-jump-frequencies $P_{Ni(Al)i \to j}$ (Eq.4), efficiencies $E_{Ni(Al)}^{ord(dis)}$ (Eq.5) and the "energetics" of the system represented by the pair-interaction energies $V_{Al-Al}^{(2)}$ correlated to the weight-factor $C(\tau_s)$. Considered were $V_{Al-Al}^{(2)}$-dependence of $P_{Ni(Al)i \to j}$ and $E_{Ni(Al)}^{ord(dis)}$ calculated within the first 5000 MC steps per vacancy - i.e. within the MC-time period shorter by an order of magnitude than the shortest relaxation time observed in any performed simulation, but covering the highest activity of the fast process. The results are displayed in Figs.3.

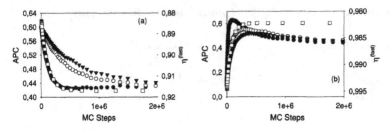

Figure 2. APC *against MC-time during isothermal "order-order" relaxations: (a) ordering at* $T_i = 1500$ K *and* $T_f = 1350$ K; *(b) disordering at* $T_i = 900$ K, $T_f = 1350$ K *with* $V_{Al-Al}^{(2)} = -0.04$ eV ($\bullet$), $+0.06$ eV (O), $+0.08$ eV ($\blacktriangledown$). *Fast component of* $\eta(t)$ *at* $V_{Al-Al}^{(2)} = -0.04$ eV *is traced with ( ).*

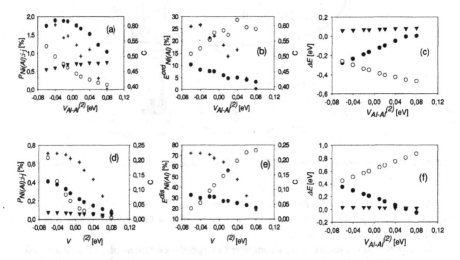

Figure 3. *Initial values of atomic-jump frequencies* $P_{Ni(Al):i\rightarrow j}$, *efficiencies* $E_{Ni(Al)}^{ord(dis)}$ *and related average system-energy changes* ΔE *against* $V_{Al-Al}^{(2)}$: *(a), (b), (c) ordering,* $P_{Ni:Al\rightarrow Ni}$, E_{Ni}^{ord} ($\bullet$), $P_{Al:Ni\rightarrow Al}$, E_{Al}^{ord} (O), $P_{Al:Ni\rightarrow Ni}$ ($\blacktriangledown$); *(d), (e), (f) disordering,* $P_{Ni:Ni\rightarrow Al}$, E_{Ni}^{dis} ($\bullet$), $P_{Al:Al\rightarrow Ni}$, E_{Al}^{dis} (O), $P_{Al:Ni\rightarrow Ni}$ ($\blacktriangledown$). $V_{Al-Al}^{(2)}$ - *dependence of C is traced in diagrams (a), (b), (d) and (e) with "+" symbols.*

The atomic-jump frequencies $P_{Ni(Al):i\rightarrow j}$ and efficiencies $E_{Ni(Al)}^{ord(dis)}$ were also monitored as evolving with MC time in simulations run with extreme values of $V_{Al-Al}^{(2)}$ (the diagrams are not shown in the present paper). All through the ordering process simulated at the lowest $V_{Al-Al}^{(2)}$ it held: $P_{Ni:Al\rightarrow Ni} > P_{Al:Ni\rightarrow Al} > P_{Al:Ni\rightarrow Ni}$ and $E_{Al}^{ord} > E_{Ni}^{ord}$. While the process progressed, the differences between the jump frequencies and efficiencies definitely reduced. In the case of the highest $V_{Al-Al}^{(2)}$ $P_{Al:Ni\rightarrow Al}$ was considerably lower: $P_{Ni:Al\rightarrow Ni} > P_{Al:Ni\rightarrow Ni} >> P_{Al:Ni\rightarrow Al}$. It was, however, remarkable that the corresponding efficiencies E_{Ni}^{ord} and E_{Al}^{ord} were substantially reduced and enhanced, respectively.

The disordering relaxation simulated at the lowest $V_{Al-Al}^{(2)}$ started again with highly efficient disordering jumps with $P_{Al:Al\rightarrow Ni} > P_{Ni:Ni\rightarrow Al} > P_{Al:Ni\rightarrow Ni}$. While the process progressed, both disordering-jump frequencies increased, however, $P_{Ni:Ni\rightarrow Al}$ increases faster and quite soon became higher than $P_{Al:Al\rightarrow Ni}$. Both disordering-jump efficiencies $E_{Ni(Al)}^{dis}$ gradually decreased, being, however, almost equal all through the relaxation. In the case of the highest $V_{Al-Al}^{(2)}$ the frequencies of Ni- and Al-atom disordering jumps again increased and their efficiencies decreased with decreasing η, but: $P_{Ni:Ni\rightarrow Al} > P_{Al:Ni\rightarrow Ni} >> P_{Al:Al\rightarrow Ni}$ and $E_{Al}^{dis} > E_{Ni}^{dis}$. Similarly as in the case of ordering, the considerable reduction of $P_{Al:Al\rightarrow Ni}$ was followed by a substantial increase of E_{Al}^{dis}.

DISCUSSION

Monte Carlo computer simulations of "order-order" relaxations in the $L1_2$ -long-range ordered intermetallic compound Ni_3Al based on the vacancy mechanism of atomic migration reveal the parallel operation of two coupled "order-order" relaxation modes observed in experiment: short-range-ordering (time evolution of the antisite-pair-correlation APC) and long-range-ordering (time evolution of the single-site correlation η).

The predominating mechanism of any change of the degree of long-range order in the system observed e.g. by means of resistometry - is the creation (disordering) or elimination (ordering) of nearest-neighbour pairs of Ni- and Al-antisites by means of correlated jumps of Ni- and Al-atoms to nn vacancies [2]. This process is a mechanism for the simultaneous and correlated evolution of APC and η (see Fig.2) and shows up as the *fast* component of "order-order" relaxation in Ni_3Al.

It is easy to show that if nn antisite pairs are created/eliminated *exclusively* due to creation/elimination of Ni- and Al-antisites, the parameters η and APC fulfil the following relationship:

$$(APC - 1) \times (1 - \eta) = const. \tag{6}$$

with the value of *const* determined by initial values of η and APC.

The picture below shows APC of Fig.2 against η with the solid curves given by Eq.6, which almost perfectly coincide with the simulated ones in early stages of relaxations.

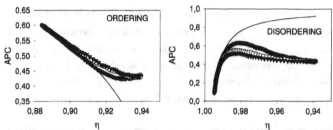

Although the concentration of the nn antisite-pairs (APC) is correlated with the concentration of antisites (η) APC and η have independent equilibrium levels at any temperature. The mechanism enabling the establishment of their independent equilibria is an easy migration of Al-antisites within the Ni-sublattice (Fig.3). Due to the Al-antisite migration the nn antisite pairs are permanently coupled or uncoupled, which leads to and then maintains a dynamical equilibrium of APC and enables the possible further evolution of η towards its equilibrium level. The evolution of η enabled by the latter mechanism is observed as the *slow* component of "order-order" relaxation in Ni_3Al.

The contribution of the fast process to the simulated "order-order" relaxation in a Ll_2-long-range-ordered A_3B system may be controlled by the values of atomic pair-interaction-energies (Fig.3). These parameters determine the balance between the frequencies of ordering/disordering and "A-sublattice migration" jumps of the minority B-atoms within the initial stage of the process. The contribution of the fast process is reduced if $P_{B:A \to A}$ increases in relation to $P_{B:A \leftrightarrow B}$ (see Fig.3). Complete elimination of the fast process is, however, possible only in the case of disordering, where the nn antisite pairs are currently "produced". If $P_{B:A \to A}$ is sufficiently high, this "production" may be at once compensated by the uncoupling mechanism. In the case of ordering, certain concentration APC of the nn antisite pairs to be quickly eliminated always exist in the system and, therefore, an increase of $P_{B:A \to A}$ causes only a limited reduction of the contribution of the fast process.

Acknowledgments

The work was partially supported by State Committee for Scientific Research, (grant no. 2P03B 088 19) and the governments of France (grant no. 76411) and Austria (grant no. 96022).

References

1. R.Kozubski, W.Pfeiler, Acta Mater. **44**, 1573, (1996).
2. P.Oramus, R. Kozubski, M.C.Cadeville, V. Pierron-Bohnes, W.Pfeiler, Mat.Res.Soc.Symp.Proc., **527**, 185, (1998).
3. P. Oramus, R. Kozubski, V.Pierron-Bohnes, M.C.Cadeville, W.Pfeiler, Phys.Rev.B, **63** (2001) – in press
4. A.Flinn, P.G.M.McManus, Phys.Rev. **124**, 54, (1961).
5. J.R.Beeler, Phys.Rev. **A138**, 1259, (1965).

Mat. Res. Soc. Symp. Proc. Vol. 646 © 2001 Materials Research Society

First Principles Calculation of Cooperative Atom Migration in L1$_2$ Ni$_3$Al

H. Schweiger[1,2], R. Podloucky[1], W. Wolf[1], W. Püschl[2] and W. Pfeiler[2]
[1]Institut für Physikalische Chemie, University of Vienna,
Liechtensteinstrasse 22a/I/3, A-1090 Vienna, Austria
[2]Institut für Materialphysik, University of Vienna,
Strudlhofgasse 4, A-1090 Vienna, Austria

ABSTRACT

Recent Monte-Carlo simulations of order relaxations in L1$_2$-ordered Ni$_3$Al reproduced the simultaneous action of two processes as experimentally observed by residual resistometry. It was shown that the fast process is related to the fast annihilation/creation of nearest neighbour antisite pairs. These findings are now strongly corroborated by a new supercell approach of *ab initio* quantum mechanical calculations describing the simultaneous displacement of Ni and Al atoms on their way to their respective antisite positions. Studies of single jumps suggest that such a cooperative migration of Ni and Al is necessary in order to prevent Al antisites from jumping back into their regular position. Relaxation of neighbouring atoms was taken into account. Thus, a minimum migration barrier of about 3 eV was derived which together with the calculated formation enthalpy of a Ni vacancy of 1.5 eV amounts to 4.5 eV, in remarkable agreement with the high activation enthalpy of 4.6 eV as observed experimentally.

INTRODUCTION

The extraordinary high-temperature mechanical and corrosion properties of Ni$_3$Al make this intermetallic compound a leading candidate from a technological and a scientific standpoint [1]. These properties mainly are a consequence of chemical long-range ordering of the alloy atoms in the L1$_2$ superstructure. For thermodynamic reasons the state of order depends on temperature. It is changed by jumps of atoms between the two different sublattices, this way creating or annihilating antisite defects. Therefore the investigation of so-called 'order-order' relaxations yield results complementary to usual tracer diffusion experiments using a Ni* tracer. Ni* can easily diffuse via its own sublattice; the same holds true for Al antisite atoms. Presuming a sufficient number of Al-antisites to be present self diffusion in Ni$_3$Al was recently explained in this way [2]. Changes in the degree of order, however, need atom jumps between different sublattices that change the concentration of antisite atoms correspondingly. In order-order relaxation experiments the system is kicked out of its current equilibrium state of order by a small and sudden temperature change resulting in a subsequent relaxation to a new equilibrium state of order. For such experiments in a so-called directly ordering alloy like Ni$_3$Al, where the order/disorder temperature equals the melting temperature or even virtually lies above it [3], extremely small changes in the degree of order with temperature are to be expected.

It turned out in recent years that measuring residual electrical resistivity is a very sensitive indicator for these fine variations of order, indeed at present the only suitable experimental method for studying order-order relaxations in Ni$_3$Al with sufficient accuracy [4]. It was found by careful analysis of order-order relaxation between true equilibrium niveaus of long-range order in Ni$_{76}$Al$_{24}$+0.19at.% B that two first order relaxation processes are involved showing an equal activation enthalpy of 4.6 eV [5]. As expected, this is in contrast to standard diffusion experiments with Ni* tracer but corresponds well with diffusion experiments where the tracer elements substitute Al [6,7]. In disagreement to these values the residual resistivity measurements of Ref. [8] yield a much lower activation enthalpy of less than 3 eV.

Recent Monte Carlo studies indicate that in the ordering/disordering processes the annihilation/creation of nearest neighbour antisite *pairs* plays a dominant role. It is the aim of the present paper to report an *ab initio* approach studying quantitatively defect formation and defect migration within a suitable thermodynamic model which gives evidence for a *cooperative* motion of Ni and Al atoms during order-order relaxation.

CALCULATION OF DEFECT FORMATION AND MIGRATION ENTHALPIES

After cross-checking carefully its reliability against the all-electron full-potential linearized augmented plane wave (FLAPW) method [9] the calculations were carried out using the Vienna *ab initio* simulation package VASP [10] which is based on ultrasoft pseudopotentials. Exchange-correlation was treated within the generalised gradient approximation (GGA); this is the best choice as the ground state properties of Ni_3Al in contrast to the usual overbinding effects of the local density approximation (LDA) are obtained in very good agreement with experimental data of the lattice parameter (a_{exp}=3.572 Å, a_{GGA}=3.576 Å) and bulk modulus (B_{exp}=1.75 Mbar, B_{GGA}=1.77 Mbar).

Single vacancies and antisite defects for both sublattices and both atomic species were modelled by supercells of 32 atomic sites for which the structural relaxation and total energies could be handled by *ab initio* methods; tests were made also for larger supercells. Full ionic relaxation for the defects in equilibrium were allowed. For the calculation of defect formation enthalpies in equilibrium non-interacting defects embedded in a Ni_3Al infinite reservoir were assumed within a grand canonical statistical ensemble to conserve the overall stoichiometry.

The migration of atoms was described by a static displacement of Ni and Al atoms relaxing locally the ionic positions of the surrounding atoms for successive fixed positions of the migrating atoms. Relaxation effects are of a strong influence for the derivation of migration barriers leading to a reduction of up to 50% when compared to the unrelaxed case.

NUMERICAL RESULTS

The results of the calculated defect formation enthalpies for vacancies on the Ni and Al sublattices and for both types of antisites are presented in table I. The values are derived for stoichiometric Ni_3Al but also for the weakly off-stoichiometric case of $Ni_{76}Al_{24}$. All values refer to 1000K but depend rather weakly on temperature.

The calculated migration energy profiles for Al jumping into a nn Ni vacancy and Ni jumping into an Al vacancy are shown in figure 1 for the unrelaxed (dashed line) and the relaxed case (full line). Whereas the migration barrier results symmetric for Ni jumps a markedly asymmetric jump barrier is obtained for Al jumps and backjumps: to climb the barrier starting from a regular site the Al atom needs about 1 eV but only 0.2 eV to jump back. This is a most striking result which means that there is a very low probability for the creation of Al antisites if the possibility for a backjump into the regular position is given.

Table I. Defect formation enthalpies H_f for Ni_3Al and $Ni_{76}Al_{24}$ at 1000K. Ionic positions: frozen and relaxed. Values as derived from GGA calculations

1000K	H_f [eV]	V_{Ni}	V_{Al}	Al_{Ni}	Ni_{Al}
Ni_3Al	frozen	1.38	2.35	1.01	1.01
	relaxed	1.50	2.01	0.51	0.51
$Ni_{76}Al_{24}$	frozen	1.54	1.88	1.63	0.40
	relaxed	1.54	1.91	0.63	0.39

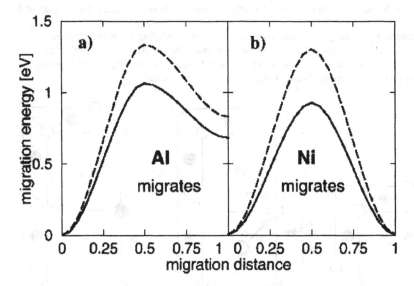

Figure 1. Energy barriers of Al jumping to a Ni vacancy (a) and of Ni jumping to an Al vacancy (b). Full lines: local atomic surroundings of the migrating atom relaxed; dashed lines: without local relaxation. Distance in units of nn-distance.

DISCUSSION

It is well-known from the literature that small deviations from stoichiometry are accommodated by the generation of antisite atoms rather than by structural vacancies [11] and that thermal vacancies are generated predominantly on the Ni-sublattice [12]. Due to the calculated asymmetry of the jump barrier for Al atoms Al antisites should not be stable and therefore not contribute to changes of the long-range order parameter. However, changes of long-range order have clearly been observed experimentally. The answer to this puzzling fact is intriguingly simple. As already concluded from Monte Carlo simulations of ordering kinetics [13] Al backjumps must be prevented by a *cooperative* motion of Ni atoms into the Al vacancies. This is sketched in figure 2. If the Al vacancy generated by the Al atom jumping to a Ni vacancy is 'blocked' by a Ni atom jumping into this antisite position a nn pair of antisites is generated. If this is not the case no antisite at all is produced. Out of a multitude of possible cooperative jump paths we select two representative cases: (i) Ni makes a complete jump into the Al vacancy left behind by Al having moved to the Ni vacancy site as sketched in figure 2; (ii) a cooperative move of Ni towards the Al site at first makes the migration barrier symmetric in order to enable a reasonable dynamic equilibrium between forward and backward jumps of Al. In figure 3 the migration processes and their barriers are illustrated for this second case (ii). When Ni has moved towards the Al site by 24% of its total distance to Al (panel a) a rather symmetric barrier of about 3 eV for Al migration follows (panel b). Al then moves to the vacant Ni site over this barrier. A final jump of Ni to the now vacant Al site (panel c) also costs about 3 eV.

Extensive first principles calculations were performed modelling a cooperative atom migration as in cases (i) and (ii) by quasistatically moving Ni and Al thereby scanning the enthalpy hypersurface of $E_{mig}(r_{Ni}, r_{Al})$ as a function of the positions (r_{Ni}, r_{Al}) of the migrating atoms. In general, a minimum barrier value $E_{mig} \approx 3$ eV was derived.

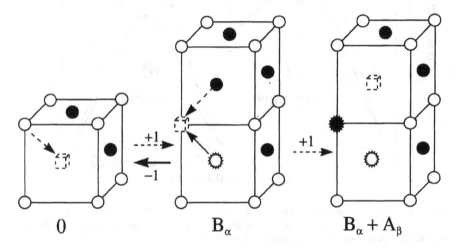

Figure 2. *Sketch of single and cooperative disordering jumps in an A_3B compound (here Ni_3Al) of $L1_2$ structure (full circles: A atoms, open circles: B atoms). α, β denote lattice sites of A and B atoms in the perfectly ordered compound. Leftmost cube: initial defect structure is an A vacancy (dashed cube). Atom B jumps to A vacancy creating (+1) an antisite defect B_α. Central doublecube: initial defect structure: B_α antisite and a B vacancy. Competition of jumps of B_α (ragged open circle) back to B vacancy (dashed cube, full arrow) annihilating (−1) the antisite and jump of A atom to B vacancy creating (+1) a second antisite A_β as visualized in the rightmost doublecube by the ragged full circle. Finally, two antisites and an A vacancy have been generated.*

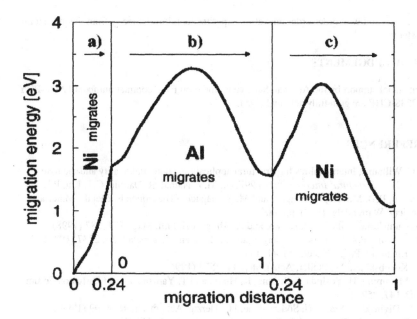

Figure 3. Energy barriers as determined from migration enthalpy hypersurface for a chosen cooperative jump. Local surroundings relaxed during migration. (a) Ni migrates towards Al position by 24% of Ni-Al nn-distance. Al remains fixed. (b) Ni is fixed and Al now jumps completely with practically symmetric barrier. (c) For fixed Al the displaced Ni atom of (a) completes its jump to the Al vacancy left behind when process in (b) is completed. The resulting defect structure correponds to the rightmost doublecube in figure 2.

Adding up the calculated formation enthalpy of a Ni vacancy of 1.5 eV and the migration barrier for the cooperative motion of about 3 eV results in a total activation enthalpy of 4.5 eV for order-order relaxations close to the experimentally observed value of 4.6 eV [5].

CONCLUSIONS

It turns out that changes in the degree of long-range order i.e. changes in the concentration of Ni and Al antisite atoms in Ni_3Al require *cooperative jumps* of the involved atoms. An activation enthalpy considerably higher than the usual Ni* tracer diffusion but consistent with

Al* diffusion and with order-order relaxation experiments follows directly from the present *ab initio* study.

ACKNOWLEDGEMENTS

Financial support by the Austrian Science Foundation FWF contract numbers 12420-PHY and 12538-CHE are gratefully acknowledged.

REFERENCES

1. J.C. Williams, Intermetallics for structural applications; potential, reality and the road ahead, *Structural Intermetallics 1997*, ed. M.V. Nathal, R. Darolia, C.T. Liu, P.L. Martin, D.B. Miracle, R. Wagner, and M. Yamaguchi (The Minerals, Metals, Materials Society, Warrendale, 1993), pp3-8.
2. H. Numakura, T. Ikeda, M. Koiwa, and A. Almazouzi, Phil. Mag. **A77**, 887 (1998).
3. R.W. Cahn, P.A. Siemers, J.E. Geiger, and P. Bardhan, Acta metall. **35**, 2737 (1987).
4. R. Kozubski, Progr. Mater. Sci. **41**, 1 (1997).
5. R. Kozubski and W. Pfeiler, Acta mater. **44**, 1573 (1996).
6. Y. Minamino, H. Yoshida, S.B. Jung, K. Hirao, and T. Yamane, Defect Diffusion Forum **143-147**, 257 (1997).
7. S.V. Divinski, S. Frank, U. Södervall, and C. Herzig, Acta mater. **46**, 4369 (1998).
8. C. Dimitrov, X. Zhang, and O. Dimitrov, Acta mater. **44**, 1691 (1996).
9. H. Schweiger, E.G. Moroni, W. Wolf, W. Püschl, W. Pfeiler, and R. Podloucky, Mat. Res. Soc. Symp. 552, KK5.15.1 (1998).
10. G. Kresse and J. Hafner, J. Phys. Cond. Matter **6**, 8245 (1994).
11. K. Aoki and O. Izumi, phys. stat. sol. (a) **32**, 657 (1975).
12. C.-L. Fu and G.S. Painter, Acta Mater. **45**, 481 (1997).
13 P. Oramus, R. Kozubski, M.C. Cadeville, V. Pierron-Bohnes, and W. Pfeiler, Solid State Phenomena **72**, 209 (2000).

Mat. Res. Soc. Symp. Proc. Vol. 646 © 2001 Materials Research Society

Ni₃Al Thin Foil by Cold Rolling

Toshiyuki Hirano, Masahiko Demura, Kyosuke Kishida and Yozo Suga[1]
National Research Institute for Metals, 1-2-1 Sengen, Tsukuba, Ibaraki 305-0047, Japan
[1] Nippon Cross Rolling Co. 697 Mobara, Chiba 297-0026, Japan

ABSTRACT

Thin foils of stoichiometric Ni₃Al below 100 µm in thickness were successfully fabricated by cold rolling of the sheets which were sectioned from directionally solidified ingots. Maximum rolling reduction in thickness amounted to 96%, irrespective of the initial orientation or the existence of columnar grains in the starting sheets. The as-rolled foils were characterized in terms of microstructures, textures and dislocation structures. The deformation microstructures were of a dual banded structure composed of two different {110} textures in the case of <001> rolling direction, while a rather homogeneous structure with a single {110} texture resulted in the case of <112> rolling direction. TEM observation revealed homogenous dislocation structures in either case without cell formation, accompanied by very fine grained-regions at higher reduction.

INTRODUCTION

Considerable efforts have been made in recent decades to use Ni₃Al as high-temperature structural materials in a bulk form [1]. In contrast, we are focusing on a plan to use it in a foil form, e.g. honeycomb structure known as a lightweight and high-rigidity structure. However, at present it is impossible to cold-roll to thin foil below 100 µm in thickness, because of severe intergranular brittleness [2]. Even with the beneficial effect of boron addition which Aoki and Izumi discovered [3], the ductility is not enough to reduce the thickness below 800 µm [4,5].

Alternatively, we found that directional solidification (DS) using a floating zone (FZ) method provides us a significant ductility improvement for Ni₃Al without any boron additions [6,7]. The DS materials with columnar-grained structure show very high tensile elongation, more than 70% in ambient air [7,8]. The high ductility is ascribed to the large fraction of low angle and low-Σ value boundaries [9]. Using the same technique we have succeeded in growing single crystals of binary stoichiometric Ni₃Al [10,11]. As is well known, single-crystalline Ni₃Al has substantial ductility, more than 100% elongation [2]. Taking advantage of the high ductility of the DS materials, we fabricated thin foil by cold rolling in this study. We present the details of the results.

EXPERIMENTAL

Four rods, designated as Nos. 31-2, 41-1, 42-2, and 47-1, of boron-free binary stoichiometric Ni₃Al were grown by a FZ method in the same way as previously described [6]. Optical microscopic observation showed that Nos. 31-2 and 47-1 were mostly single crystal but contained columnar grains with low angle boundaries in places, while Nos. 41-1 and 42-2 were fully single crystals. Table 1 summarizes the Al contents of the grown rods.

The grown rods were sectioned into sheets along the growth direction by electric discharge machining. The initial rolling direction (RD) or the growth direction and normal direction (ND) of the sheets were determined by the Laue X-ray back reflection method as summarized in Table 1. The sheets were cold-rolled without intermediate annealing or lubricant by using four-high

Table 1 Al contents and initial orientation of the rolling sheets

Sample No.	Al content (at%)	Rolling plane	Rolling direction	Structure
31-2	24.4	{0.2 0.1 1.0}	<1.0 0.0 0.2>	Columnar-grained
47-1	24.6	{3.0 0.1 4.9}	<0.0 1.0 0.0>	Columnar-grained
41-1	24.7	{3.9 1.0 5.2}	<4.0 9.0 4.8>	Single-crystalline
42-2	24.8	{2.0 1.0 4.3}	<1.1 2.0 1.0>	Single-crystalline

mills and cemented carbide rolls. The rolling texture was measured by X-ray Schultz back reflection method. The as-rolled microstructures were examined by optical microscopy and transmission electron microscopy (TEM).

RESULTS

It turned out to be possible to cold-roll the starting sheets to thin foil less than 100 μm in thickness without intermediate annealing. Figure 1 shows a 91μm-thick, 10 mm-wide and 1 m-long foil of No. 41-1. In this case the total reduction in thickness amounts to 96%. The surface is crack-free and smooth, with little fluctuation in thickness along the rolling direction. The foil is heavily work-hardened, with a Vickers hardness of about 620, nevertheless it is possible to make a coil, as shown in Fig. 1. Also, the foil is ductile and can be bent plastically, and hence

Figure 1. 91 μm-thick, 10 mm-wide and about 1 m-long foil of sample No. 41-1 cold-rolled to 96% reduction.

Table 2 Summary of cold rolling

Sample No.	Thickness/μm Before	After	Reduction (%)	Rolling texture	Structure
31-2	959	315	67		
		73	92	{110}<-113>+{110}<1-17>	Banded
		57	94		
47-1	1789	384	78	{110}<-114>+{110}<1-14>	Banded
		96	95	{110}<-112>+{110}<1-12>	
41-1	2043	91	96	{110}<-113>	
42-2	1907	319	83	{110}<-114>	

honeycomb structures are possible.

Similarly high rolling ductility was obtained in other samples, Nos. 31-2, 42-2 and 47-1, which had different initial orientations. The results are summarized in Table 2. Samples Nos. 41-1 and 42-2 are of single crystals, and hence the results may be somewhat expected because Ni_3Al is known to be ductile in a single crystal form [2]. Still, it is worth noting that such high-quality thin foil was fabricated by cold-rolling boron-free, binary Ni_3Al, which is considered a brittle intermetallic compound. Samples Nos. 31-2 and 47-1 had some columnar grains, but this did not hinder cold rolling. As we previously reported [7], the columnar-grained polycrystals grown by the FZ method are ductile because most of the boundaries are low-angle and low Σ types [9].

Two typical types of deformation microstructures were observed in the etched as-rolled foils as shown in Fig.2. In the case of <112> RD, fine slip traces lie on the surface, being inclined about 60° to the rolling direction ((No. 42-2) in Fig. 2(a)). Also, coarse and wavy lines, which are regarded as shear bands [12], are observed clearly on the longitudinal section. In the case of <001> RD, however, the deformation structure is composed of repeated double shear bands with differently oriented slip traces (No. 47-1 in Fig. 2 (b)). The width of the bands ranges from 20 to 100 μm parallel to the rolling direction. There is no significant difference in the morphology of the shear bands between the two types.

Figure 3 shows the {220} pole figures of the foils cold rolled to 83% reduction: (a) No. 42-2 and (b) No. 47-1. In all the foils cold rolled over 83% reduction, the {220} pole has the highest intensity peak at the ND and surrounding high intensity peaks about 60 degrees away from the ND, showing well-developed {110} texture. However, there is some difference in RD among the samples (Table 2). Samples Nos. 41-1 and 42-2 whose initial RD is <112> possess a single {110} texture, while No. 31-2 and no.47-1 whose initial RD is <001> consists of two different {110} textures. The difference observed in the texture corresponds well to the deformation microstructures, which is related to the initial RD of the sheet. The {110} texture is thought to be stable in the rolling deformation of Ni_3Al whose slip system is <110> on {111}. Crystal changes its orientation by operating two {111}<110> slips under compressive deformation, which ends up in the most stable orientation <011>. We consider that this compressive deformation led to {110} texture evolution as discussed elsewhere [13].

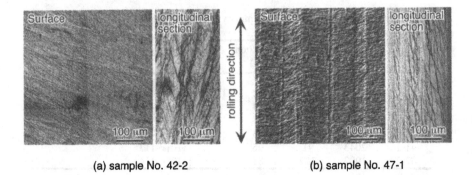

(a) sample No. 42-2   (b) sample No. 47-1

Figure 2. *Optical microstructures observed on the etched surface and longitudinal section of cold-rolled foils (83% reduction): (a) sample No. 42-2 and (b) No. 47-1.*

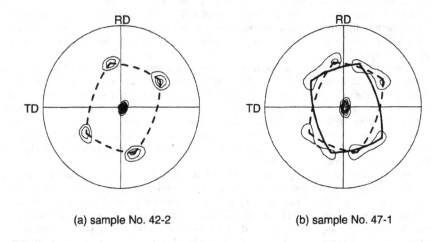

(a) sample No. 42-2   (b) sample No. 47-1

Figure 3. *The {220} pole figures of the cold-rolled foils: (a) sample No. 42-2 (83% reduction) and (b) No. 47-1 (78% reduction). Open and solid squares represent {110}<-113> and {110}<-117> textures, respectively.*

Figure 4 shows TEM micrograph of the as-rolled No. 47-1 with a high density of dislocations. Similar to the fcc metals with low values of stacking fault energy [14], the dislocations are not arranged in cell structures or subgrains as previously reported [15], instead they are rather uniformly distributed. When the amount of cold reduction becomes larger than 90%, very fine grained-structure with about less than 50 nm in diameter is developed. This structure probably corresponds to the shear bands.

DISCUSSION

Metallic thin foils have been manufactured for ordinary metals by cold-rolling, for example aluminum, copper, titanium, stainless steel and so on. However, there exist no thin cold-rolled foils of intermetallic compounds. The reason for our success may be mainly because our starting materials were single-crystalline or near single-crystalline alloys. Even so, it is worth considering why Ni_3Al, a species of intermetallic compounds, can be cold-rolled to such a thin foil without premature fracture.

In order to achieve large rolling reduction, crack nucleation must be suppressed during deformation as much as possible. This tendency can be seen on the tensile stress-strain curves of both columnar-grained and single-crystalline binary stoichiometric Ni_3Al which exhibit more than 70% uniform elongation in ambient air [7,8]. In addition to the low yield stress, the tensile stress-strain curves were accompanied by large linear work hardening rate [8,10], which induces uniform elongation by preventing local intensive deformation. In fact, work softening, which is due to dislocation rearrangement or annihilation such as cell formation, was not observed. Fracture occurred abruptly without necking in the late linear work hardening range [8,10], indicating little local deformation. These tensile characteristics, which must originate in the

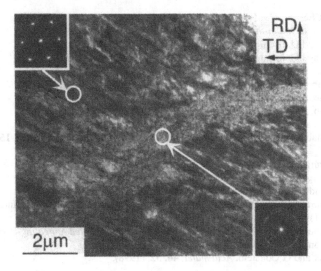

Figure 4. TEM micrograph of sample No. 47-1 cold-rolled to 92% reduction

properties of dislocations in Ni_3Al, indicate that crack nucleation was significantly suppressed until the applied stress reached a fracture value.

Microscopically, there were indications of heterogeneous deformation such as shear bands and banded structures in the as-rolled foils (Figs. 2 and 4). Further deformation would eventually cause crack nucleation along them. However, these heterogeneous deformation microstructures are commonly observed in ductile fcc metals such as aluminum, copper and brass after heavy cold reduction [14]. Therefore, it is better to consider that the development of the heterogeneous microstructure shows how heavily the alloys cold-rolled. In other words, our Ni_3Al can be deformed without premature cracking until these heterogeneous microstructures develop.

CONCLUSIONS

Thin foils of stoichiometric Ni_3Al below 100 μm in thickness with crack free and smooth surfaces were fabricated by cold rolling without intermediate annealing. Starting sheets, which were sectioned from directionally solidified rods by using the floating zone method, were of single crystals or nearly single crystals with a few columnar grains. In all the sheets, high rolling ductility was obtained. Maximum reductions in thickness amounted to 96%, irrespective of the initial orientation and existence of columnar grains. The deformation microstructures showed a dual banded structure with two different {110} textures in the case of <001> RD, and a rather homogeneous structure with a single {110} texture in the case of <112> RD. No cell formation was observed in either case, accompanied by very fine grained-regions at higher reduction.

ACKNOWLEDGEMENT

We would like to thank E. P. George at Oak Ridge National Laboratory for his helpful discussions.

REFERENCES

1. N. S. Stoloff, *Int. Mater. Rev.*, **34**, 153(1989).
2. K. Aoki and O. Izumi, *Trans. JIM*, **19**, 203(1978).
3. K. Aoki and O. Izumi, *Nihon Kinzoku Gakkai Shi*, **43**, 1190 (1979).
4. C. T. Liu and V. K. Sikka. *J. Metals*, **38**, 19(1986).
5. A. I. Taub, S.C. Huang and K. M. Chang, *Metall. Tran.s A*, 15A, 399(1984).
6. T. Hirano, *Acta metall. mater.*, **38**, 2667(1990).
7. T. Hirano, *Scripta metall. mater.*, **25**, 1747(1991).
8. T. Hirano and T. Kainuma, *ISIJ International*, **31**, 1134(1991).
9. T. Watanabe, T. Hirano, T. Ochiai and H. Oikawa, *Materials Science Forum*, **157-162**, 1103 (1994).
10. M. Demura and T. Hirano, *Phil. Mag. Letters*, **75**, 143(1997).
11. D. Golberg, M. Demura and T. Hirano, *J. Crys. Growth*, **186**, 624(1998).
12. J. Ball and G. Gottstein, *Intermetallics*, **1**, 171(1993).
13. M. Demura, Y. Suga, O. Umezawa, K. Kishida, E. P. George and T. Hirano, Intermetallics, **9**, 157(2001).
14. F. J. Humphreys and M. Hatherly, *Recrystallization and Related Annealing Phenome*na, (Pergamon, 1995)pp.11-56.
15. C. Escher and G. Gottstein, *Acta mater*, **46**, 525(1998).

Mat. Res. Soc. Symp. Proc. Vol. 646 © 2001 Materials Research Society

Transient creep behaviour of Ni₃Al polycrystals

Tomas Kruml, Birgit Lo Piccolo and Jean-Luc Martin
Département de Physique, Ecole Polytechnique Fédérale de Lausanne (EPFL)
CH 1015 Lausanne, SWITZERLAND

ABSTRACT

Repeated creep tests were used for measuring various constant strain-rate deformation parameters. The results are consistent with those of repeated stress relaxations, although the precision is lower for creep in the present case. The small yield point observed in reloading after the transient is directly related to the amount of exhausted mobile dislocations, i.e. it originates from multiplication processes. During the transient test (180s total), the total exhaustion rate of mobile dislocations can be as high as 99%. It exhibits a maximum at the same T (about 500 K) as the work hardening. This supports the validity of a model which considers the work-hardening peak temperature to correspond to the stress under which incomplete Kear-Wilsdorf locks yield.

INTRODUCTION

The technique of repeated stress relaxations is well established and has been used frequently since the sixties for the determination of some material parameters, above all the activation volume of the microscopic deformation mechanisms [1-4]. The repeated creep technique was proposed recently as an alternative [5-7] and its abilities are shown below.

EXPERIMENTAL DETAILS

Polycrystalline binary Ni₃Al rods with a nominal composition of 24 at.% aluminium and 76 % nickel were kindly provided by Dr. T. Khan (Onera, Paris). The material was homogenised at 1583 K for 48 hours. The metallurgical inspection revealed a single-phase material with large equiaxed grains of mean size about 800 micrometers.

Parallelipipedic compression specimens with a length of 7 mm and a gauge section of 3.5 x 3.5 mm² were cut by a diamond saw. The Schenck RMC 100 machine was used for the compression tests. These were performed between room temperature and 800K under an inert helium atmosphere. A nominal strain rate of 5×10^{-5} s^{-1} was kept constant.

THEORY OF REPEATED CREEP TESTS

When the machine is switched from the constant strain rate mode to the constant force mode, a logarithmic dependence of plastic strain on time is experimentally observed :

$$\Delta \gamma_p = (kT/MV_c)\ln(1 + t/C_c) \qquad (1)$$

T is the absolute testing temperature, M is the elastic modulus of the machine-specimen assembly, V_c has the dimension of a volume, t is the time, C_c is a time constant. The Orowan equation is:

$$d\gamma_p / dt = \alpha \rho_m bv \tag{2}$$

ρ_m being the mobile dislocations density, v their average velocity, b the Burgers vector and α a geometrical parameter. The applied stress τ is decomposed according to:

$$\tau = \tau^* + \tau_i \tag{3}$$

the effective stress τ^* depending only on temperature and strain-rate, τ_i being the internal stress. The velocity of mobile dislocations is a function of the effective stress:

$$v = \nu d \exp\left[-\Delta G(\tau^*)/kT\right] \tag{4}$$

ΔG is the activation free enthalpy and ν and d have the respective dimensions of a frequency and a distance. The microscopic activation volume V is:

$$V = -\partial \Delta G / \partial \tau^* \tag{5}$$

Combining (1), (2) and (4) yields a power law between ρ_m and v:

$$\rho_m / \rho_{m0} = (v / v_0)^\beta \tag{6}$$

where v_0 and ρ_{m0} refer to the onset of relaxation. The coefficient β can be determined from the experiment which enables to calculate the mobile dislocation density after the time t of the creep test:

$$\rho_m / \rho_{m0} = \left[C_c /(C_c + t)\right]^{\beta/(1+\beta)} \tag{7}$$

RESULTS AND DISCUSSION

As seen in figure 1, the sequence followed for the repeated creep procedure consists of 1) loading the specimen under constant strain rate, 2) performing the first creep for a selected time interval, 3) switching to the constant strain rate mode and increasing the stress by a certain amount, 4) performing the second creep test for the same time duration etc.

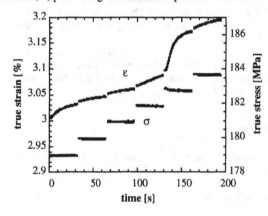

Figure 1. *Repeated creep test. Duration of one creep = 30 s. Stress increase between two subsequent creep tests = 1 MPa. T = 490 K.*

The creep strain is a logarithmic function of time over 30 s time intervals. These results agree with previous ones of conventional or transient tests at low or intermediate temperatures, which were interpreted in terms of exhaustion of octahedral glide through cross-slip pinning [8,9]. It has been shown that creep can be reinitiated by a stress increment through the start of a new cycle of dislocation motion [10].

Sometimes a sudden increase of the strain rate is observed during the test (figure 1). Such instabilities were observed at all temperatures and different stress levels. This behaviour looks similar to the appearance of inverse creep on intermediate temperature creep curves of Ni_3Al single crystals [9]. The latter observation was interpreted in terms of bowing and glide on the cube cross-slip plane of the Kear-Wilsdorf locks (KWL) formed during primary creep [9]. Electron microscope observation are underway to test this assumption in the present conditions.

The validity of the procedure of repeated creep test was checked in the following way (see figure 2). Several transient tests were performed along one stress-strain curve (repeated stress relaxation and creep tests alternately) and the activation volume V was calculated. It is expressed in b^3 units, b being the Burgers vector of one superpartial dislocation (b = a/2 <110> = $2.54 \cdot 10^{-10}$ m). The repeated relaxations R1, R2 and R3 and repeated creep tests C1, C2, C3 will be studied in more details in the following. Figure 2 shows that the agreement between the volumes calculated from relaxation and creep tests is rather good since their variation as a function of strain follows a single monotonic curve.

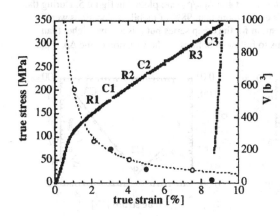

Figure 2. *Stress-strain curve and the microscopic volume as a function of strain calculated from repeated stress relaxations (o) and repeated creep tests (●). T=490 K. Transient test are performed where the stress-strain curve is interrupted.*

A close inspection of the stress-strain curve reveals a small yield point in reloading after each transient test, as shown on figures 2 and 3. The height of this multiplication yield point is directly connected with the exhaustion of mobile dislocations during the transient [6].

The coefficient β (equation 6) can be calculated for both transients (see figure 4). β changes substantially along the stress-strain curve. It increases systematically with plastic strain from 1 to 4.5 and it is larger for creep tests than for stress relaxations. A larger value of β results in higher exhaustion for the same drop of velocity (see eq. 6), i.e. the rate of exhaustion depends on the microstructure. Not surprisingly, the larger the plastic strain (i.e. the forest dislocation density), the easier the exhaustion.

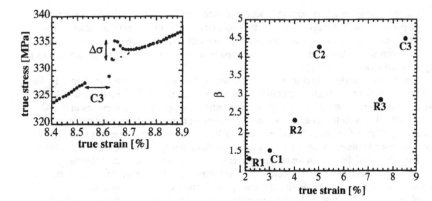

Figure 3. *Yield point after the repeated creep test C3.*

Figure 4. *Coefficient β for the transients of figure 3.*

β being known, it is possible to determine mobile dislocation exhaustion rates $\Delta\rho_m/\rho_{m0}$, where $\Delta\rho_m = \rho_{m0} - \rho_m$. The two parameters $\Delta\sigma/\sigma$ and $\Delta\rho_m/\rho_{m0}$ are plotted in figure 5. During the series of 6 stress relaxations (30 seconds each), more than 98% of mobile dislocations were exhausted. This value varies more with strain for the creep series but it is always higher than 95%. The height of the yield point seems to follow in both cases the variation of the $\Delta\rho_m / \rho_{m0}$ parameter.

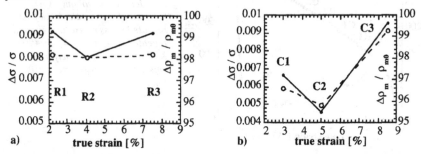

Figure 5. *Relative mobile dislocations exhaustion (o) and the yield point height (●) after series of a) stress relaxations, b) creep tests.*

Values of this parameter for a given plastic strain are shown on figure 6a), determined using stress-relaxation test and creep test. On figure 6b), the amplitude $\Delta\sigma$ of the yield point at reloading after a creep series (figure 3), normalised to stress is also plotted as a function of temperature. The comparison of figure 6a) and b) shows parallel trends for both parameters $\Delta\rho_m/\rho_{m0}$ and $\Delta\sigma/\sigma$ as a function of temperature : as T increases, they both increase to a peak temperature close to 500K after which they decrease.

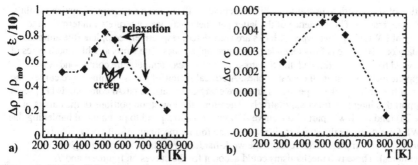

Figure 6. *Mobile dislocation exhaustion corresponding to a plastic strain rate decrease by a factor of 10 a), yield points in reloading after creep tests b). Constant plastic strain = 5%.*

This fair correlation can be understood as follows: when dislocation exhaustion is intense during the transient, an excess of stress is necessary to activate new dislocation sources so that the imposed strain rate can be achieved by the crystal. The curve of the work hardening coefficient θ as a function of temperature also exhibits a shallow maximum in the vicinity of 500K [6]. Therefore a maximum of exhaustion during the transient corresponds to a maximum of θ at imposed strain-rate.

The work hardening peak temperature has recently received some attention [11]. Figure 7 schematically illustrates how the stress-strain curve evolves with temperature and the corresponding variation of work-hardening.

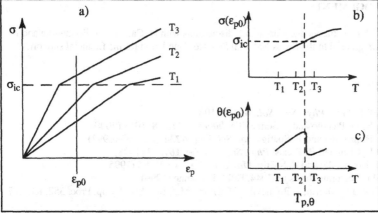

Figure 7. *Model for the interpretation of the work hardening peak temperature (from [12]) – see text.*

On figure 7a, the stress-plastic strain curve consists of two linear parts that respectively correspond to two distinct hardening mechanisms. At given ε_{p0}, figure 7b illustrates the positive variation of stress with temperature, while the curve θ (T) is shown on figure 7c, which exhibits a peak temperature. The existence of the latter parameter has been interpreted in terms of the stability of KWL [11]. Two types of locks are considered, respectively complete (the antiphase boundary on 010 plane is fully extended) and incomplete ones. The stress at which both types of KWL yield have been estimated and the stress σ_{ic} of the peak temperature corresponds to incomplete locks which are the weaker ones. A critical review of stress - temperature curves, the values of hardening peak temperatures and core energies in various Ni_3Al compounds [11] seem to support this interpretation quantitatively. Therefore, the low strain portions of the curves of figure 7a) (and the low T part of the curve of figure 7c) correspond to pronounced hardening due to KWL formation. As the stress σ_{ic} is reached, the former mechanism still operates but exhaustion of incomplete locks reduces the work-hardening (at large strains or higher temperatures). These two mechanisms could account for the curves on figures 6 and 7.

CONCLUSIONS

* Repeated creep tests can be used for measuring the activation volume, the parameter β and the mobile dislocation exhaustion.
* Mobile dislocation exhaustion can be estimated using the $\Delta\rho_m/\rho_{m0}$ and the $\Delta\sigma/\sigma$ parameters measured respectively during and after the transient test.
* The temperature of maximum exhaustion rate coincides with the maximum of work-hardening at given strain. This supports the validity of a model for the work-hardening peak temperature, based on the stability of incomplete Kear-Wilsdorf locks.

ACKNOWLEDGEMENTS

The authors wish to acknowledge stimulating discussions with D. Caillard , J. Bonneville and P. Spätig. They are grateful to the Swiss National Science Foundation for the financial support.

REFERENCES

1. F. Guiu and P.L. Pratt, *Phys. Stat. Sol.*, **6**, 111 (1964)
2. J.L. Martin, B. Lo Piccolo and J. Bonneville, *Intermetallics*, **8**, 1013 (2000)
3. G. Saada, J. Bonneville and P. Spätig, *Mat. Sci. Eng*, **A234-236**, 263 (1997)
4. W. Pichl, D.L. Chen and B. Weiss, *Phys. Stat. Sol. (a)*, **167**, 43 (1998)
5. A. Orlova, J. Bonneville and P. Spätig, *Mat. Sci. Eng.*, **A 191**, 85 (1995)
6. B. Lo Piccolo, Doctorate thesis no 2044, EPFL Lausanne (1999)
7. B. Matterstock, J.L. Martin, J. Bonneville, T. Kruml, Mat. Res. Soc. Symp. Proc. 552, KK5.17 (1998)
8. P.H. Thorton, R.G. Davies and T.L. Johnson, *Metall. Trans.* **1**, 207 (1970)
9. K.J. Hemker, M.J. Mills and W.D. Nix, *Acta metall. mater.* **39**, 1901 (1991)
10. M.D. Uchic and W.D. Nix, Mat. Res. Soc. Symp. Proc. 460, 437 (1997)
11. D. Caillard, Proc. of ICSMA13, *Mat. Sci. Eng.* A, in print
12. D. Caillard, private communication

Mat. Res. Soc. Symp. Proc. Vol. 646 © 2001 Materials Research Society

Slip trace characterisation of Ni₃Al by atomic force microscopy.

Christophe Coupeau, Tomas Kruml[1] and Joël Bonneville
Université de Poitiers, LMP, UMR-CNRS 6630, SP2MI,
F-86962 Futuroscope Cedex, FRANCE.
[1] Ecole Polytechnique Fédérale de Lausanne (EPFL), DP-IGA,
CH-1015 Lausanne, SWITZERLAND.

ABSTRACT

We examined by atomic force microscope the slip traces produced on Ni_3Al single crystals pre-deformed up to nearly 1% plastic strain at three temperatures in the anomaly domain: 293K, 500K and 720K. It is observed that, whatever the deformation temperature, the slip traces essentially belong to the primary octahedral slip system. The lengths of the slip lines become shorter and shorter with increasing temperature, while the number of dislocations that constitutes the lines is approximately constant. These results are interpreted in terms of a decreasing mean free path of the mobile dislocations when the temperature is raised. The implications of these results in the understanding of the flow stress anomaly are underscored.

INTRODUCTION

Positive temperature dependence (PTD) of the flow stress of the Ni_3Al intermetallic compound has not yet received a satisfactory explanation. While several theoretical works succeed in explaining the increase of the flow stress with temperature, they generally fail to predict other characteristic parameters of the plastic deformation, such as the work-hardening rate (which also exhibits a PTD) and the strain-rate sensitivity of the flow stress (for a review see [1]). The proposed models usually consider that a thermally activated cross-slip process plays a key role in understanding the PTD of the flow stress, but, depending on the dislocation dynamics considered, cross-slip process leads with increasing temperature either to a decrease in the dislocation velocity or to a decrease in the mobile dislocation density.

Indeed, on the one hand, direct measurements of the stress and temperature dependences of the dislocation velocity have been carried out in the anomaly domain of Ni_3Al by the double etching technique [2,3]. These two studies have shown that, at a given stress, the screw dislocation velocity decreases when the temperature is raised. In addition, small activation areas, i.e. $A < 100 \, b^2$ (b being the Burgers vector of a superpartial dislocation), have been reported in both investigations. These latter results contrast with the very low strain-rate sensitivity of the flow stress obtained by deformation experiments performed either at different strain-rates (see for instance [4]) or by strain-rate changes [5,6]. On the other hand, variation in the density of mobile dislocations has been estimated by indirect techniques [7,8], based on transient tests performed during constant strain-rate experiments. These studies have demonstrated that, in the PTD range of the flow stress, high exhaustion of the mobile dislocations takes place, which is associated with large activation areas [9].

This study is aimed at quantifying the respective contributions of the mobile dislocation density and dislocation velocity to the anomalous temperature dependence of the flow stress of Ni_3Al intermetallic compounds. For this, we examined the slip traces produced on Ni_3Al single crystalline specimens by atomic force microscope (AFM). Preliminary results obtained at room temperature have already been published elsewhere [10]. We then examined specimens that have been plastically deformed at various temperatures in the anomaly domain. The overall features of the slip traces are analysed with particular attention to the temperature evolutions of cross-slip events, deviations of the slip traces from the primary octahedral slip plane and slip trace lengths. The results are discussed in terms of the models that have been proposed for explaining the yield stress anomaly.

EXPERIMENTAL DETAILS

Compression specimens with a length of 7 mm and a square cross-section of approximately 3.3mm x 3.3mm have been spark-eroded from a single crystalline rod with the $Ni_{75}Al_{24}Ta_1$ nominal composition, provided by David P. Pope at the University of Pennsylvania. The compression axis of the specimens was along the [$\overline{1}23$] crystallographic orientation, which particularities have already been presented in detail in [11]. This orientation has a high Schmid's factor on the (010) cube cross-slip plane, for promoting dislocation glide on this plane, while the Schmid's factor on the ($1\overline{1}1$) octahedral cross-slip plane is zero. The lateral faces are ± [$54\overline{1}$] and ±[$1\overline{1}1$] oriented. Consequently, the slip markings corresponding to the [$\overline{1}01$](111) primary octahedral glide system are visible only on two opposite specimen sides, which are the ± [$54\overline{1}$] oriented faces. Prior to deformation, the specimens were mechanically polished by alumina and silica powders of decreasing granularities. A description of the mechanical polishing procedure has been given elsewhere [10]. This procedure yields specimen surfaces that are suitable for AFM imaging, obviating the need of electro-polishing techniques.

The specimens have been deformed in compression at a nominal strain-rate of 1.3×10^{-4} s^{-1}. A complete description of the deformation set up can be found in [12]. Three deformation temperatures have been selected in the flow stress anomaly domain, 293 K, 500 K and 720 K. These three temperatures correspond respectively for the alloy investigated to the onset of the anomaly domain, the temperature at which the strain-rate sensitivity of the stress exhibits a sharp discontinuity and a temperature nearly 80 K below the stress peak temperature [12]. In order to minimise surface contamination the deformation chamber was evacuated by a repeated sequence of a primary vacuum pumping followed by filling with pure Ar gas. During the heating and deformation procedures a small flow of argon was maintained in the chamber. A radiant heating furnace has been used, which is monitored by an accurate PID controller. This type of furnace presents the advantage of having practically no thermal inertia and allows rapid heating and a short time for temperature stabilisation. The pre-deformation conditions are summarised in table 1. In table 1, deformation time is the elapsed time between yielding and the final achieved shear stress at the corresponding plastic strain amount at the end of the test. One may note that the increase in the deformation time with increasing temperature does not result from the small additional plastic strain (see table 1) but reflects, since the deformation experiments are performed at constant total imposed strain-rate, the increase in work-hardening with temperature. This also means that the plastic strain-rates are not exactly similar for all temperatures; they do not differ by more than 35%.

Table I. Predeformation conditions of the specimens observed by AFM.

Deformation temperature (K)	Yield stress (MPa)	Ultimate shear stress (MPa)	Plastic shear strain (%)	Deformation time (s)
293	~ 35	50	1.25	100
500	~ 80	127	1.37	150
720	~ 140	210	1.47	180

The deformed specimens were examined *post mortem* at room temperature by using an AFM. The slip line lengths were estimated by measuring the number of ending slip traces on scan areas of 10 μm x 10 μm. The size of the scan area results from a compromise between good lateral resolution for discriminating between parallel slip traces and safe observation of ending lines inside the investigated areas.

RESULTS AND DISCUSSION

The main features of the slip markings observed on the specimen surfaces as a function of temperature are shown in figure 1. The slip traces that are visible on the figures belong to the primary (111) octahedral glide plane, which is predominantly activated in the central part of the specimens. One general observation is that, whatever the deformation temperature, the slip traces appear to be rather straight, without exhibiting appreciable deviations from the (111) octahedral glide planes. This result is in agreement with previous observations obtained by optical

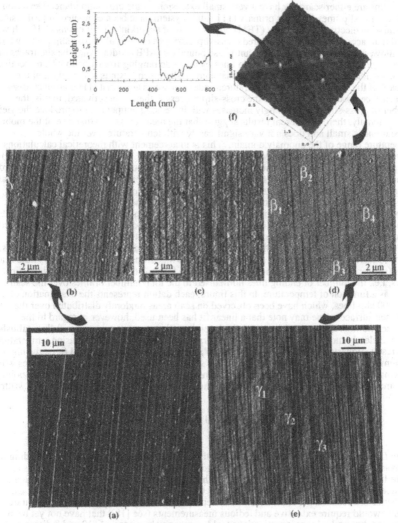

Figure 1. *Typical slip traces associated with the (111) primary slip plane observed after 1% plastic strain at three temperatures: 293 K, a and b) 500 K, c) and 720 K, d, e and f) in the anomaly domain of the Ni$_{75}$Al$_{24}$Ta$_{1}$ intermetallic alloy.*

microscopy [13,14], where the slip traces corresponding to the primary (111) glide system were also observed to be very rectilinear, even for deformation temperatures close to the stress peak temperature. AFM imaging at higher magnification indicates that the intimate structure of the slip lines consists of several finer lines on parallel planes, but in the limit of the apparatus resolution these slip lines are not connected by cross-slip events. This observation indicates that cross-slips are either scarce or have a very small extension on the cross-slip plane. Deviation angles of the slip lines from the primary (111) glide system have been estimated in [10], using transmission electron microscopy (TEM) observations of superkink distributions [1,15]. It was found that, according to the considered cross-slip distances on the cube cross-slip plane, which was allowed to vary from b up to the antiphase boundary (APB) width d_{APB}, the slip traces may deviate from the primary octahedral slip lines by an angle ranging from 1.5° to 27°, respectively. This does not invalidate the current idea that a cross-slip mechanism is at the origin of the increase of the yield stress with temperature. Indeed, if we assume that (1) no artefact arises from surface observations, and (2) the cross-slip process is thermally activated, that is, the number of cross-slip events strongly increases with increasing temperature to produce the yield stress anomaly, then, the present results suggest that the mean cross-slip distance of the mobile dislocations is small and does not vary significantly with temperature, over the whole temperature range of the anomaly domain. This is in agreement with theoretical calculations concerning the stability of Kear-Wilsdorf (KW) locks [16a&b]. It has been predicted that (1) complete KW locks, resulting from cross-slip distances on the cube cross-slip plane of the order of the APB width, require very high stresses for back cross-slipping on the primary octahedral slip plane, and (2) only incomplete KW resulting from a small cross-slip length on the cube cross-slip plane, i.e., of the order of b, can be driven by the applied stress. This result also supports the recent idea that incomplete KW locks certainly play a key role in the understanding of the PTD of the flow stress of Ni_3Al [17]. Therefore, high local internal stresses are certainly responsible for the large double cross-slip events that are sometimes observed [10].

Another striking feature of the slip lines concerns the high propensity of short slip lines, which is easily discernible on the AFM images by the numerous slip traces that end in the scan areas (see for instance figure 1d). We have plotted in figure 2 the percentage of ending slip traces, i.e., the number of ending lines normalised to the total number of lines over the scan areas, as a function of temperature. In this figure, each datum represents the examination of at least 400 slip lines, which have been observed on scan areas randomly distributed over the specimen surfaces. One may note that a linear fit has been used, however in regard to the restricted number of data, this temperature dependence is not yet considered as fully established. Figure 2 clearly indicates that this percentage drastically increases with increasing temperature. This result suggests that the slip trace length, which follows a reciprocal trend (i. e., the higher the number of ending slip traces the shorter the slip trace lengths), considerably decreases when the temperature is raised. Considering that an ending slip trace exhibits only one ending point per scan area, an empirical formula can be established for relating both quantities, which is written as

$$
L_m \approx \left(\frac{2}{p} \times 100 - 1 \right) \ell_s , \qquad (1)
$$

where L_m represents an 'average' slip line length associated with the percentage p of ending lines (expressed in percent) and ℓ_s is the size of the scan area in the direction of the slip traces (see figure 1). L_m values that have been calculated by using relation (1) are given in figure 2 as a function of temperature. We have also estimated by measuring the heights of the traces that the number of dislocations per slip line does not depend too much on temperature. An accurate statistic would require extensive and tedious measurements (see [18]) that have not yet been performed, but preliminary investigations yield as mean values nearly 5, 10 and 8 dislocations per slip line at 293 K, 500 K and 720 K respectively. One example of a step associated with a slip line resulting from the emergence of 8 dislocations at 720 K is shown in figure 1f.

Therefore, with increasing temperature, we observe (1) a nearly constant or even slightly increasing dislocation number per slip trace and (2) a decreasing length of the slip traces. For coherence, since all specimens have undergone an almost identical plastic strain, the number of slip traces per unit area must increase with increasing temperature. This is indeed observed and, for instance the number of slip traces (N) measured at room temperature on figure 1a) is $N = 46 \pm 2$ while at 720 K on figure 1d) we have $N = 63 \pm 3$. All these results clearly indicate that, in the anomaly domain of Ni_3Al, the distance travelled by the mobile dislocations becomes shorter and shorter with increasing temperature. The two interpretations that can be proposed to explain this feature are a large decrease either of the dislocation velocity, v, or of the dislocation mean free path, Λ (or a combination of both effects). A decrease of v may contribute to some extent to the shortening of the slip lines. However, if we consider that v exhibits in $Ni_3(Al, Ta)$ with temperature an anomalous behaviour, i. e. $(\partial v / \partial T)_\sigma < 0$, similar to the one of $Ni_3(Al,Ti)$ [3], a drastic change in $v = v(\sigma, T)$ is not expected for our deformation conditions, roughly 0.1 $\mu m.s^{-1}$ for temperatures and related final shear stresses reported in Table I. In particular, because the applied stress goes up with increasing temperature. Thus, the decrease in the slip line length with increasing deformation temperature must be ascribed essentially to the decrease of Λ. Since Λ is a measure of mean free path of the dislocations before they become definitively locked, its reciprocal value can be interpreted in terms of an exhaustion parameter. The decrease of Λ with increasing temperature therefore suggests an exhaustion mechanism of mobile dislocations that is more and more efficient when the temperature is raised. These results are in fair agreement with the variation of the density of mobile dislocations found with the transient tests (successive relaxations and repeated creep tests) performed during constant strain-rate experiments [9], with the mechanical spectroscopy technique [19] and with the increase in the dislocation density measured by TEM [13]. The exhaustion of ρ_m has to be compensated by a multiplication process, whose efficiency has to increase with temperature as well. The latter process usually requires an increase in stress with increasing temperature, which has been proposed and modelled in [20a&b] for explaining the origin of the PTD of the flow stress.

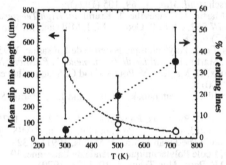

Figure 2. *Percentage of ending lines and associated mean slip line lengths as a function of temperature. Note that the error bars indicate the distribution of each parameter about its mean value.*

SUMMARY AND CONCLUSION

The slip traces of $Ni_3(Al,Ta)$ single crystalline specimens deformed at different temperatures in the range of the anomaly domain of the flow stress have been examined by using an AFM. The slip traces were observed to be highly rectilinear whatever the deformation temperature, which indicates that the movement of the mobile dislocations is fairly planar. From this, we conclude that the cross-slip distance of the moving dislocations is small, on the order of b, and does not depend too much on temperature. The length of the slip traces is found to decrease with increasing temperature, which can be reasonably interpreted as a decrease in the dislocation mean free path when the temperature is raised. This supports the idea of an exhaustion mechanism for the density of mobile dislocations that becomes more and more efficient with increasing temperature and which can be at the origin of the PTD of the flow stress [20a&b]. However, it is not yet possible to discriminate if the flow stress anomaly arises from the anomalous behaviour of the dislocation velocity or from the exhaustion of the mobile dislocation density. This requires *in situ* AFM observations using a special AFM device [21,22]. Such experiments are under progress.

ACKNOWLEDGEMENTS

The authors wish to thank Prof. J.-L. Martin (Ecole Polytechnique Fédérale de Lausanne) and Prof. J. Grilhé (Université de Poitiers) for the numerous and highlighted discussions on the subject. We are also indebted to F. Préau for his help in the AFM observations and to G. Beney for preparing high quality specimen surfaces.

REFERENCES

1. P. Veyssière and G. Saada, *Dislocations in solids*, ed. F.N.R. Nabarro and M.S. Duesbery (Amsterdam:Elsevier Science, 1996) p. 253.
2. E. M. Nadgorny and Y. L. Iunin, in: *High Temperature Ordered Intermetallic Alloys VI* **364**, ed. J. A. Horton, I. Baker, S. Hanada, R. D. Noebe and D. S. Schwartz (Materials Research Society, Pittsburgh, 1995) p. 707.
3. C. B. Jiang, S. Patu, Q. Z. Lei and C. X. Shi, *Philos. Mag. Letters* , **78**, 1 (1998).
4. S. Miura, S. Ochia, Y. Oya, Y. Mishima and T. Suzuki, in: *High Temperature Ordered Intermetallic Alloys III* **133**, ed. C. T. Liu, A. I. Taub, N. S. Stoloff and C. C. Koch (Materials Research Society, 1989) p. 341.
5. P.H. Thornton, R.G. Davies and J.L. Johnston, *Metall. Trans.*, **1**, 207 (1970).
6. S.S. Ezz and P.B. Hirsch, *Phil. Mag. A*, **69**, 105 (1994).
7. B. Matterstock, J.-L. Martin, J. Bonneville, T. Kruml. in: *High Temperature Ordered Intermetallic Alloys VIII* **552**, ed. E. P. George, M. J. Mills and M. Yamaguchi (Materials Research Society, 1999) p. KK. 5.17.
8. P. Spätig, PhD Thesis, Ecole Polytechnique Fédérale de Lausanne, 1995.
9. J. Bonneville and J. L. Martin, in: *Multiscale Phenomena in Plasticity Nato Science Series –* Vol. **367** ed. J. Lepinoux, D. Mazière, V. Pontikis and G. Saada (Kluwer Academic Publishers, London, 2000) p. 57.
10. C. Coupeau, J. Bonneville, B. Matterstock, J. Grilhé and J.-L. Martin, *Scripta Mat.*, **41**, 945 (1999).
11. A.E. Staton-Bevan and R.D. Rawlings, *Phys. Stat. Sol. A*, **29**, 613 (1975).
12. J. Bonneville, N. Baluc and J.-L. Martin, Proc. 6th Int. Symp. (JIMIS-6) *on Intermetallic Compounds*, ed. O. Izumi (The Japan Institut of Metals, 1991) p. 323.
13. N. Baluc, PhD Thesis, Ecole Polytechnique Fédérale de Lausanne, 1989.
14. C. Lall, S. Chin and D.P. Pope, *Met. Trans. A*, **10**, 1323 (1979).
15. A. Couret, Y.Q. Sun and P.B. Hirsch, *Phil. Mag. A*; **67**, 29 (1993).
16. G. Saada and P. Veyssière, *Phil. Mag. A*, a) **66**, 1081 (1992); b) **70**, 925 (1994).
17. D. Caillard and G. Molenat, Proc. 20th Risø Int. Symp. on Materials. Science: *Deformation-Induced Microstructure: Analysis and Relation to Properties*, ed. J. B. Bilde-Sørensen et al., Risø National. Laboratory. (Roskilde, Denmark 1999) p. 1.
18. C. Coupeau and J. Grilhé, *J. of Mat. Sc. Eng. A*, **271**, 242 (1999).
19. B. Cheng, E. Carreno-Morelli, N. Baluc, J. Bonneville and R. Schaller, *Phil. Mag. A*, **79**, 2227 (1999).
20. F. Louchet, a) *Phil. Mag. A*, **72**, 905 (1995) ; b) *Phil. Mag. A*, **77**, 761 (1998).
21. C. Coupeau, J.C. Girard and J. Grilhé, *J. of Vacuum Sci. Technology B*, **16** (4), 1964 (1998).
22. C. Coupeau, F. Cleymand and J. Grilhé, *Scripta Mat.*, **43**, 187 (2000).

Mat. Res. Soc. Symp. Proc. Vol. 646 © 2001 Materials Research Society

Microstructural Evolution and Mechanical Properties of
Al₃Ti-based Multi-Phase Alloys

Seiji Miura[1], Juri Fujinaka[2], Rikiya Nino[2] and Tetsuo Mohri

Division of Materials Science and Engineering,

Graduate School of Engineering, Hokkaido University,

Kita-13, Nishi-8, Kita-ku, Sapporo 060-8628, Japan.

[1] Corresponding author: miura@eng.hokudai.ac.jp

[2] Graduate Student, Graduate School of Engineering, Hokkaido University.

ABSTRACT

A preliminary study on the phase relations in Al-Mo-Ti-X quaternary systems in the vicinity of Ti-trialuminide phases is carried out with various additives X= Mn, Cr, Fe, Ni and Ag. In the Al-Mo-Ti ternary system, a bcc-phase field extends from the Ti-Mo edge to high Al region at high temperatures and it equilibrates with a $D0_{22}$-Al₃Ti phase containing a large amount of Mo. It is found that, by additions of X= Mn, Cr, Fe or Ni, an $L1_2$-(Al, X)₃Ti phase appears near the two-phase region composed of the $D0_{22}$-Al₃Ti and bcc phases in the Al-Mo-Ti ternary system. By heat treatment at 1223 K, the bcc phase of quaternary alloys decomposes into the A15-Mo₃Al, $D0_{22}$, $L1_2$ and/or σ phases, and no voids are observed. The mechanical properties of these alloys are also investigated by Vickers hardness.

INTRODUCTION

A Ti-trialuminide ($D0_{22}$-Al₃Ti phase) and its derivative, a $L1_2$-(Al,X)₃Ti phase, have been investigated as potential high temperature materials because of their low density and high oxidation resistance [1, 2]. Ternary $L1_2$-(Al,X)₃Ti phases with the addition of Mn, Cr, Cu, Fe, Ni and Ag have been expected to have a better ductility due to a simpler crystal structure than the $D0_{22}$-Al₃Ti phase has, but are still brittle [3]. One of the reasons is voids formed in $L1_2$-(Al,X)₃Ti based alloys during homogenization. HIP is effective to suppress void formations. An introduction of secondary phase is another solution to form void-free microstructures [4, 5]. It suggests that the solid-solid transformation process is an effective way for suppressing the void formation. We attempted to apply a eutectic decomposition from a sound and homogeneous single phase to form tri-aluminide-based multi-phase microstructures.

In a previous study we found that a lamellar structure composed of an A15-Mo₃Al phase and an mC22-Mo₃Al₈ phase has better toughness with an inter-lamellar spacing of a few μm than the same lamellar structure with a narrower inter-lamellar spacing [6]. Such a dependence on the inter-lamellar spacing of toughness is attributed to the interaction between the dislocations and crack-tips in the A15 phase. The previous studies also suggest that an introduction of a phase in which dislocations can move more easily improves toughness of the lamellar structure with proper inter-lamellar spacings.

By a broad search on ternary phase diagrams, it was found that in the Al-Mo-Ti system a ternary bcc-phase decomposes in the manner of eutectoid reaction into the DO_{22}-Al_3Ti and A15-Mo_3Al phases as shown in Figure 1 [7]. The mechanical properties of the fine lamellar structure are investigated and also explained in terms of dislocation-crack tip interaction [6, 8]. In this study further attempts were made to find multi-phase structures composed of the $L1_2$-$(Al,X)_3Ti$ and A15-Mo_3Al phases formed by a eutectoid decomposition of high-temperature bcc phase by adding Mn, Cr, Fe, Ni and Ag to the Al-Mo-Ti ternary. Present results will be a basis for a further study on the effect of microstructure on the toughness of a multi-phase alloy.

EXPERIMENTAL PROCEDURES

Alloy ingots were prepared by arc-melting under an Ar atmosphere. The nominal compositions of specimens are listed in table I. Each alloy has a composition between pure Mo and 65Al-25Ti-10X in at.% which is a typical composition of the $L1_2$-$(Al,X)_3Ti$ phase in various Al-Ti-X ternary systems.

A differential thermal analysis (DTA) was carried out to evaluate the phase transformation temperatures and/or melting points. Some of the specimens sealed in evacuated silica tubes were heat-treated at 1623 K for 24 hours and/or 1223 K for 18-48 hours. Microstructure was observed by SEM and micro-vickers hardness are measured with a load of 2.94N or 300g. An

Table I. Nominal composition of alloys investigated.

at.%	Al	Mo	Ti	X
Mn-alloy	52	20	20	8
Cr-alloy	52	20	20	8
Ni-alloy	52	20	20	8
Fe-alloy	52	20	20	8
Ag-alloy	52	20	20	8

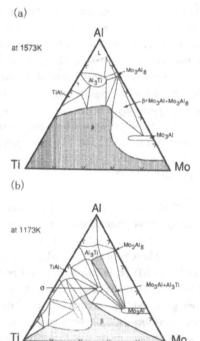

Figure 1. Isothermal sections of Al-Mo-Ti ternary system at 1573 K and 1173 K [7].

X-ray diffractometry (XRD) and a Wave-length dispersive X-ray spectroscopy (WDS) are performed to identify the phases existing.

RESULTS AND DISCUSSION

Figure 2 shows the microstructure of an alloy with Mn. After the heat treatment at 1623 K, a bcc single phase was attained. After the heat treatment at 1223 K, the bcc phase decomposes into several phases. From the results of WDS, a bright phase with high-Mo composition is thought to be the σ-phase and needle-like precipitates are the DO_{22}-Al_3Ti phase. The gray bcc matrix still remains in the alloy.

Figure 3 shows the microstructure of an alloy with Cr. As shown in (a), the bright bcc phase appears in an as-cast ingot as a primary phase with needle-like precipitates. XRD revealed that the dark and gray precipitates are the Ll_2-$(Al,Cr)_3$Ti and DO_{22}-Al_3Ti phases, respectively. DTA revealed that the eutectoid transformation temperature for this alloy is about 1500 K, almost the same with that in the Al-Mo-Ti ternary system. After the heat treatment at 1223 K, below the eutectoid temperature, most of the bcc phase decomposes into several phases as shown in Fig. 3 (b). It is also noteworthy that no voids are observed. Although the grains and precipitates are too fine to identify their phases from their compositions by WDS, XRD revealed that the bright $A15$-Mo_3Al phase appears and both the gray bcc and dark DO_{22}-Al_3Ti phases co-exist in the microstructure.

By the addition of Fe, the bright bcc phase also appears as a primary phase in Fig. 4 (a). The gray DO_{22}-Al_3Ti phase covers the bcc phase, and a dark Ll_2-$(Al,Fe)_3$Ti phase exsists between them. Needle-like precipitates are also observed in the bcc phase. DTA revealed that there are several transformation reactions in this alloy between 1350 K

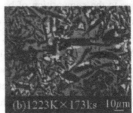

Figure 2. SEM images of microstructure of the Mn-added alloy. (a) An ingot heat-treated at 1623 K and (b) an ingot heat-treated successively at 1223 K.

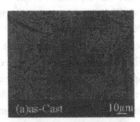

Figure 3. SEM images of microstructure of the Cr-added alloy. (a) An as-cast ingot and (b) an ingot heat-treated at 1223 K.

and 1500 K. After the heat treatment at 1223 K, the bcc phase decomposes into several phases as shown in Fig. 4 (b). Voids are also not observed. XRD revealed that the DO_{22}-Al_3Ti phase appears in the microstructure. A bright phase with a high volume fraction in the decomposed area is thought to be the σ-phase. Dark regions are the DO_{22}-Al_3Ti phase.

Figure 5 shows the microstructure of an alloy with Ni. The bcc phase appears as a primary phase as bright phases with a dark $L1_2$-$(Al,Ni)_3Ti$ and gray DO_{22}-Al_3Ti phases in Fig. 5 (a). Similarly with the Cr addition, DTA revealed that the eutectoid transformation temperature for this alloy is about 1400 K, rather lower than that in the Al-Mo-Ti ternary system. After the heat treatment at 1223 K, the bcc phase also decomposes into several phases. XRD revealed that the bright σ-phase appears and the gray DO_{22}-Al_3Ti and dark $L1_2$-$(Al,Ni)_3Ti$ phases co-exist in the microstructure as shown in Fig. 5 (b).

Figure 6 represent schematic drawings of an Al-Mo-Ti-X quaternary system with Mn, Cr, Fe and Ni as X deduced from the result of these microstructural observations. At high temperature the bcc region has a large solubility area in the diagram. At lower temperature, it shrinks and is replaced by multi-phase microstructures composed of the A15, DO_{22}, $L1_2$ and/or σ phases. Difference in the combination of phases may be due to the composition of bcc in the alloys and to the difference of the equilibrium compositions of each phase in each quaternary system. It is noteworthy that the high temperature bcc phase is easily quenched in upon air-cooling. It enables various heat treatments for a formation of various microstructures.

Figure 7 shows the microstructure of an as-cast alloy ingot with Ag. Although the gray bcc phase appears as a primary phase, a bright δ-Ag_2Al phase is observed. The melting point of the δ-Ag_2Al phase is 999 K in binary Ag-Al system [9] and a corresponding peak

Figure 4. SEM images of microstructure of the Fe-added alloy. (a) An as-cast ingot and (b) an ingot heat-treated at 1223 K.

Figure 5. SEM images of microstructure of the Ni-added alloy. (a) An as-cast ingot and (b) an ingot heat-treated at 1223 K.

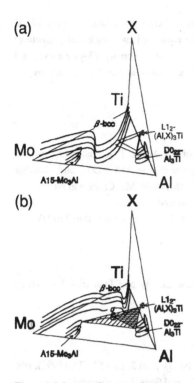

(a)

(b)

Figure 6. *Schematic drawings of an Al-Mo-Ti-X quaternary system with Mn, Cr, Fe and Ni as X at (a) high temperature and (b) low temperature.*

appears at 997 K in the DTA profile of the quaternary alloy specimen. The dark DO_{22}-Al_3Ti phase also co-exists in the as-cast structure. Figure 8 indicates a schematic drawing of the Al-Mo-Ti-Ag quaternary system deduced with the present result. The δ-Ag_2Al phase equilibrates with the DO_{22}-Al_3Ti and Mo-lean bcc phases, but not with the $L1_2$-$(Al,Ag)_3Ti$ phase.

Micro-Vickers hardness tests were performed on alloys with Cr, Mn and Fe, heat-treated at 1223 K. Hardness values are 600 HV, 620 HV and 770 HV, respectively. These values are higher than that for a DO_{22}-Al_3Ti single phase alloy with 10 at.% of Mo, which was 500 HV [8]. It is due to a fine microstructure with harder phases such as the A15 or σ phases. These values are comparable with those for Al-Mo binary lamellar structure or Al-Mo-Ti ternary lamellar structure composed of the A15, mC22 or DO_{22} phases [6, 8].

It is already reported that a lamellar structure composed of the A15 and DO_{22} phases is formed in the Al-Mo-Ti ternary alloys by a continuous cooling from the bcc phase region, whereas an

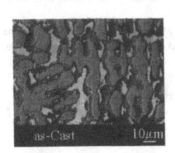

Figure 7. *SEM images of microstructure of the Ag-added as-cast alloy.*

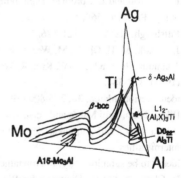

Figure 8. *A schematic drawing of an Al-Mo-Ti-Ag quaternary system at high temperature.*

equi-axed two-phase microstructure of the same alloy is formed by a heat-treatment of the quenched bcc phase at a certain temperature [7]. The equi-axed two-phase microstructures observed in some of the present alloys strongly suggest these quaternary alloys are expected to form a lamellar structure composed of the A15 and DO_{22} and/or $L1_2$ phases by the continuous cooling from the bcc region.

CONCLUSIONS

1. A bcc solid solution appears at the center region of the Al-Mo-Ti-X quaternary systems with Mn, Cr, Fe, Ni and Ag. The bcc phase decomposes into an A15, a DO_{22}, an $L1_2$ and/or a σ phases by low temperature heat treatment of the alloys with Mn, Cr, Fe and Ni.
2. After the heat treatment at 1223 K, no voids were observed.
3. Hardness of the alloys with multi-phase microstructure is ranging from 600 to 770 HV.

ACKNOWLEDGEMENT

This research is partially supported by NEDO (New Energy and Industrial Technology Development Organization).

REFERENCES

1. M. Yamaguchi and H. Inui, "Intermetallic Compounds", vol.2, pp.147-173, (1995), eds. by J. H. Westbrook and R. L. Fleischer, John Weily and Sons, Ltd., England.
2. T.Matsubara, T. Shibutani, K. Uenishi and K. F. Kobayashi : J. Intermetallics, **8**, 815-822 (2000).
3. R. A. Varin and L. Zbroniec, High-Temperature Ordered Intermetallic Alloys VII, MRS Symp. Proc. vol.**460**, C. C. Koch, C. T. Liu, N. S. Stoloff & A. Wanner eds., MRS, Pittsburgh, (1997), pp.121-126.
4. J. Y. Park, M. H. Oh, D. M. Wee, S. Miura and Y. Mishima : Proc. Intnl. Symp. Gamma Titanium Aluminides, Y-W. Kim, R. Wagner and M. Yamaguchi eds., TMS, Warrendale, (1995), 377-384.
5. idem: Scripta Mater., **36**, 795-800 (1997).
6. Rikiya Nino, Seiji Miura and Tetsuo Mohri : J. Intermetallics, **9**, 113-118 (2001).
7. Rikiya Nino, Juri Fujinaka, Seiji Miura and Tetsuo Mohri : to be submitted to J. Intermetallics.
8. idem : to be submitted to J. Intermetallics.
9. H. Okamoto, "Phase Diagrams for Binary Alloys", ASM International, Materials Park, Ohio, (2000).

Mat. Res. Soc. Symp. Proc. Vol. 646 © 2001 Materials Research Society

Hydrogen Pulverization in Intermetallic-based Alloys

Satoshi Semboshi, Naoya Masahashi and Shuji Hanada,
Institute for Materials Research, Tohoku University,
2-1-1 Katahira, Aoba-ku, Sendai 980-8577l, JAPAN

ABSTRACT

Pulverization behavior and microstructure evolution with hydrogenation in hydrogen absorbing Ta-Ni intermetallic-based alloys, such as Ta_2Ni with Ta solid solution (Ta_{ss}), $TiMn_2$ with TiMn and Nb_3Al with Nb solid solution (Nb_{ss}), are investigated to elucidate the mechanism of the hydrogen pulverization. Ta-10at.%Ni consisting of Ta solid solution (Ta_{ss}) and Ta_2Ni Laves phase is pulverized to coarse powder over 100 μm in hydrogenation. Crack propagation occurs preferentially in the brittle Ta_2Ni phase rather than in the ductile Ta_{ss} phase. When the volume fraction of brittle Ta_2Ni increases with increasing Ni content, hydrogen pulverization is enhanced. The lattice parameter of Ta_{ss} increases by hydrogenation, while it does not change in Ta_2Ni. In addition, nano-sized regions with Moiré patterns are produced in Ta_{ss} and Debye rings corresponding to tantalum hydride β−TaH appear in the diffraction pattern. These features are very similar to those of $TiMn_2$ based alloy and Nb_3Al based alloys in the literature. Based on the present results along with those in the literature it is concluded that the hydrogen pulverization is attributable to (1) the absorption of a large amount of hydrogen in constituent phase(s), (2) the large strain introduced by lattice expansion and the hydride formation, and (3) the ease of crack nucleation and propagation in brittle constituent phase(s).

Introduction

Intermetallic-based alloys as hydrogen absorption materials are pulverized in a hydrogen atmosphere, during cycle hydrogenation [1,2]. From the viewpoint of hydrogen absorption property the pulverization should be avoided, because it causes the degradation of hydrogen absorbing capacity with hydrogenation [2]. It is necessary to suppress the hydrogen pulverization to improve the cyclic property of hydrogen absorbing capacity. On the other hand, the hydrogen pulverization is expected for a new powder fabrication process to produce fine powder with high quality and low cost, especially for refractory metals like Nb and Ta based alloys [3-5]. Therefore, there are different reasons for investigating the mechanism of hydrogen pulverization for intermetallic-based alloy; one is the improvement of the cyclic property in hydrogen absorption material by pulverization suppression, and the other is the application to fine powder fabrication of intermetallic-based alloys.

In this paper, hydrogen pulverization behavior and microstructure evolution in hydrogen absorbing Ta-Ni alloys consisting of Ta_2Ni and Ta solid solution (Ta_{ss}) are investigated. Based on the obtained results the mechanism of hydrogen pulverization is discussed, referring to the results on the hydrogen pulverization in $TiMn_2$ based and Nb_3Al based alloys, which has been already reported [2-5].

EXPERIMENTAL PROCEDURE

Button ingots of Ta -5at.%Ni, Ta -8at.%Ni, Ta -10at.%Ni and Ta -15at.%Ni were arc melted in an argon atmosphere containing 5 % hydrogen, using 99.9 % purity tantalum and 99.9 % purity nickel as raw materials, with four times remelting to obtain homogeneity. After arc melting the atmosphere was replaced by 0.1 MPa hydrogen (99.99999 % purity) in the arc-melting chamber to hydrogenate Ta-Ni alloys without exposure to air. The pulverization behavior of Ta-Ni alloys with hydrogenation was monitored through a small window of the chamber. The structures of the alloys before and after hydrogenation were identified by X-ray diffraction (XRD) analysis of conventional 2θ scans using MoK_α radiation. The microstructures of as-cast Ta-Ni alloys and the powders pulverized by hydrogenation were investigated with a scanning electron microscope (SEM). The microstructure of hydrogenated powders was also observed using a high-resolution transmission electron microscope (HRTEM). For HRTEM observation the sample powders pulverized by hydrogenation were used. These powders were put on a copper grid coated with carbon for HRTEM observation.

RESULTS

The pulverization of Ta-Ni alloys by hydrogenation

After the melted buttons were sufficiently cooled on a water-cooled copper hearth, high purity hydrogen was introduced to 0.1 MPa into the chamber, and the hydrogen pulverization of Ta -(5-15)at.%Ni alloys was monitored. The pulverization quickly occurred from the button surfaces in the two phase alloys of Ta_{ss} and Ta_2Ni, Ta -8at.%Ni, Ta -10at.%Ni and Ta -15at.%Ni, slowed down a few minutes after hydrogenation and then stopped in less than an hour. However, no pulverization occurred in Ta_{ss} single phase alloy of Ta -5at.%Ni. This suggests that the hydrogen pulverization takes place in two-phase alloy rather than in single phase alloy.

The SEM images of Ta -8at.%Ni, Ta -10at.%Ni and Ta -15at.%Ni powders obtained by the hydrogen pulverization are shown in Fig.1 (a)-(c), respectively. SEM observation of the powder shape of Ta -10at.%Ni and Ta -15at.%Ni alloys reveals that flake-like powders are mostly formed by hydrogenation. The Ta -8at.%Ni button is seen to delaminate from the surface, suggesting that the hydrogen pulverization occurs by peeling off surfaces. SEM observation at a higher magnification was conducted to examine the pulverization process, as shown in Fig.2 (a)-(c). In Ta -10at.%Ni and Ta -15at.%Ni, cracks are observed to propagate preferentially in the brittle Ta_2Ni phase (dark contrast) rather than in the ductile Ta_{ss} phase (bright contrast). On the other hand, the crack propagation in Ta -8at.%Ni appears not to be influenced by ductile Ta_{ss} or brittle Ta$_2$Ni. The volume fractions of Ta$_{ss}$ and Ta$_2$Ni are evaluated by analyzing the SEM micrographs and they are listed in Table 1. The volume fraction of Ta$_2$Ni phase increases with Ni content, implying a decrease in fracture toughness. In a sample with low fracture toughness, cracks are easily propagated by hydrogenation, suggesting that the hydrogen pulverization is closely related to the fracture toughness of an alloy.

Figure 1. Hydrogen pulverization in (a)Ta-8at.%Ni, (b)Ta-10at.%Ni and (c)Ta-15at.%Ni.

Figure 2. Cracks of hydrogen-pulverized powder in (a)Ta-8at.%Ni, (b)Ta-10at.%Ni and (c)Ta-15at.%Ni.

Table 1 Volume fractions of Ta$_{ss}$ and Ta$_2$Ni analyzed from point counting in SEM images.

Composition	Ta -5at.%Ni	Ta -8at.%Ni	Ta -10at.%Ni	Ta -15at.%Ni
Volume fraction of Ta$_{ss}$/ %	100	90	80	66
Volume fraction of Ta$_2$Ni/ %	0	10	20	34

Structural analysis by XRD and HRTEM

Fig.3 shows XRD patterns of (a) as-cast and (b) hydrogenated Ta -10at.% Ni alloy. The XRD pattern of as-cast Ta -10at.%Ni represented in Fig.3 (a) is indexed on the basis of Ta$_{ss}$ and Ta$_2$Ni, indicating that as-cast Ta -10at.%Ni consists of two phases, Ta$_{ss}$ and Ta$_2$Ni. Comparing the XRD patterns of as-cast and hydrogenated Ta -10at.%Ni, the peak positions of Ta$_{ss}$ are shifted to lower angles, while those of Ta$_2$Ni are not changed. The volume change in Ta$_{ss}$ analyzed by XRD is calculated to be 10% expansion after hydrogen pulverization, suggesting that a large amount of hydrogen is absorbed at interstitial sites in Ta$_{ss}$ after hydrogenation, which causes lattice expansion. This introduces interfacial strain generation between the two phases. Similar results were obtained in Ta -8at.%Ni and Ta -15at.%Ni which were pulverized by hydrogenation.

Fig.4 shows HRTEM micrographs of (a) Ta_{ss} and (b) Ta_2Ni with the associated electron diffraction patterns of hydrogenated Ta -10at.%Ni. After hydrogenation, nano-sized regions with random orientations are observed in Ta_{ss} phase as shown in Fig.4 (a), and both diffraction spots from Ta_{ss} and Debye rings are visible. Precise measurement of distances between (000) and Debye rings revealed that the rings correspond to the formation of β-TaH. Therefore, the nano-sized regions observed in Fig.4 (a) are considered to correspond to β-TaH, in spite of the lack of corresponding spectra in XRD patterns. The detection of β-TaH only by the selected area diffraction pattern in TEM is probably due to the small fraction of β-TaH. In hydrogenated Ta_2Ni phase, its lattice image is unchanged as shown in Fig.4 (b), indicating that hydrogen is not readily absorbed or not retained in Ta_2Ni even if absorbed.

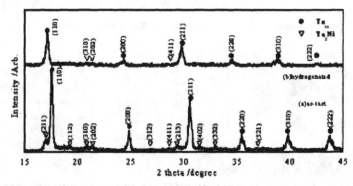

Figure 3. XRD patterns of (a)as-cast and (b)hydrogenated Ta-10at.%Ni.

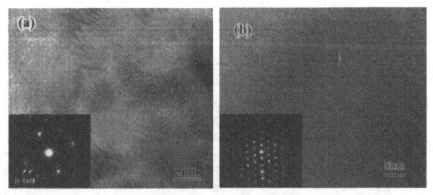

Figure 4. HRTEM micrographs and diffraction patterns of (a)Ta_{ss} and (b)Ta_2Ni in Ta-10at.%Ni after hydrogenation.

DISCUSSION

The pulverization of intermetallic-based alloys by hydrogenation

Semboshi et al. have demonstrated that two phase alloys consisting of Nb_{ss} and Nb_3Al and of Nb_3Al and Nb_2Al are likely to be pulverized, as compared with single phase Nb_3Al and Nb_{ss} [3,5]. The precise measurement of lattice parameters by XRD indicated that there are differences in lattice expansion among Nb_3Al, Nb_{ss} and Nb_2Al after hydrogen pulverization as listed in Table 2, suggesting that large strain is induced between the two phases. In addition, when the volume fraction of brittle Nb_3Al increased in the two phase alloy of Nb_{ss} and Nb_3Al, the average particle size of powder produced by hydrogenation became small. This means that the hydrogen pulverization can be correlated with the toughness of an alloy.

Ti -60 at.% Mn consisting of $TiMn_2$ Laves phase and TiMn have been reported to be pulverized to fine powder under 30 μm in hydrogenation [2]. Crack propagation parallel to the sample surfaces was observed throughout both phases after hydrogenation. The lattice parameter in $TiMn_2$ increases after hydrogenation, while it does not change in TiMn as shown in Table2. Nano-sized regions with Moiré patterns corresponding to titanium hydride δ-TiH are produced in $TiMn_2$.

Table 2 Volume expansions of constituent phases in two phase alloys after hydrogen pulverization.

Alloy	Ta-10at.%Ni		Nb-16at.%Al		Nb-28at.%Al		Ti-60at.%Mn	
Phase	Ta_{ss}	Ta_2Ni	Nb_{ss}	Nb_3Al	Nb_3Al	Nb_2Al	$TiMn_2$	TiMn
Volume expansions / %	10.2	0	10.9	8.6	8.6	1.5	2.2	0

The mechanism of the hydrogen pulverization

From the present results as well as the references described above, common features are recognized for the hydrogen pulverization in intermetallic-based alloys. Hydrogen pulverization occurs in two phase alloys when,

\# the alloy surface is not oxidized,

\# hydrogen is absorbed in constituent phase(s) so as to induce large lattice expansion,

\# hydride or nano-sized region forms in a phase absorbing a large mount of hydrogen,

\# the alloy is composed of brittle intermetallic(s),

And hydrogen pulverization proceeds with the following behavior.

* pulverization starts at alloy surface, and progresses by delamination from surface,

* the shape of the powders is flake-like,

* crack propagation occurs preferentially throughout a brittle phase.

Based on the above features, the mechanism of the hydrogen pulverization in intermetallic-based alloys is assumed to occur as follows . Hydrogenation processing of an alloy causes hydrogen absorption from its surface, thereby leading to a gradient of hydrogen content.

According to Fick's law, hydrogen content exponentially decreases as a function of distance from the surface. The increased hydrogen content in regions near the surface generates large strain. Further, in two phase alloy, the difference of microstructure evolution with hydrogenation in the respective phases induces local strains at the interface between phases. When a brittle intermetallic phase exist in the regions, even if it absorbs hydrogen or not, cracks will nucleate at the brittle phase. The observed delamination fracture would be due to the large gradient of hydrogen content from surface. Once a surface layer is spalled, fresh surfaces emerge, which creates a large gradient of hydrogen content from the new surfaces. As a result, delamination fracture will be repeated and flake-like powders will be produced. From these considerations, it could be deduced that a single phase alloy with high toughness is desirable to suppress hydrogen pulverization, while multi-phase alloys with low toughness are desirable for fine powder fabrication.

CONCLUSIONS

It is concluded that the hydrogen pulverization of two phase alloys is attributable to (1) the absorption of a large amount of hydrogen in constituent phase(s), (2) the large strain introduced by lattice expansion and the hydride formation, and (3) the ease of crack nucleation and propagation in brittle constituent phase(s).

ACKNOWLEDGMENTS

The authors are grateful for E. Aoyagi, Y. Hayasaka for technical assistants on HRTEM. This work was supported by Japan Society for the Promotion of Science, and partly supported by a Grant-in-Aid for Scientific Research (11305048) from the Ministry of Education, Science and Culture, Japan.

REFERENCE

1. T. Kastrissios, E. Kisi and S. Myhra, *J. Materials Science*, **30**, 4979, (1995)
2. S. Semboshi, N. Masahashi and S. Hanada, *Acta mater.*, **19**, 927, (2001)
3. S. Semboshi, H. Hosoda and S. Hanada, *J. Japan Inst. Metals*, **61**, 1132 (1997)
4. S. Semboshi, T. Tabaru, H. Hosoda and S. Hanada, *Intermetallics*, **6**, 61 (1998)
5. H. Hosoda, T. Tabaru, S. Semboshi and S. Hanada, *J. Alloy Comp.*, **281**, 268 (1998)

Mat. Res. Soc. Symp. Proc. Vol. 646 © 2001 Materials Research Society

High Temperature Compressive Properties and Deformation Microstructure of Fe-25Al-10Ti Intermetallic Alloy

Su-Ming Zhu, Kazushi Sakamoto, Makoto Tamura, and Kunihiko Iwasaki
Japan Ultra-high Temperature Materials Research Institute, Ube, Yamaguchi 755-0001, Japan

ABSTRACT

The compressive properties of Fe-25Al-10Ti intermetallic alloy were studied as a function of temperature and strain rate. Optical microscopy and transmission electron microscopy (TEM) were used to examine the deformation microstructure. The alloy exhibited strong tendency of strain hardening at low temperatures. A positive temperature dependence of strength was observed in the 673 - 873 K range. The mechanical behavior was interpreted in terms of dislocation structures. Based on the analysis of the stress-strain response, strain rate sensitivity and deformation microstructure, possible hot working range was proposed.

INTRODUCTION

Iron aluminides based on Fe_3Al have attracted great attention in the past decades because of their availability of raw materials, conservation of strategic elements (Cr as an example), relatively low density (as compared with Fe and Ni-based alloys), and excellent oxidation and corrosion resistance [1]. Though many scientific studies have been devoted to the physical properties, mechanical properties and corrosion resistance, commercialization of these materials has been very limited due to their fabrication difficulty and poor mechanical properties both at ambient and elevated temperatures. The best application of Fe_3Al to date is the use in porous hot-gas filters [2]. Recent studies by Hawk et al. [3,4] demonstrated that additions of Ti to Fe_3Al are effective in improving the tribological properties. For example, the alloy with 10 at. % Ti substituting for Fe in Fe_3Al shows 40% decrease in volume wear compared with Fe_3Al. This inspiring result suggests that iron aluminides may find their application in some tribological circumstances, especially where the oxidation or sulfidation is also a major concern.

Relatively little work has been reported in literature on the mechanical properties and deformation behavior of Ti-alloyed Fe_3Al. Diehm and Mikkola [5] first investigated the effect of Ti and/or Mo addition on yield strength and work hardening rate of Fe_3Al. They reported that the Ti addition significantly improved the high temperature compressive yield strength. Agarwal et al. [6] examined the corrosion and tensile properties of Fe_3Al alloys with Ti addition. Their results showed that the addition of Ti led to improved room temperature ductility but poor yield strength. They attributed the improvement in ductility to the increase in passivity and the change in fracture mode. However, Viswanathan et al. [7] reported that Ti addition tended to decrease the ductility of Fe_3Al-based alloys. Thus, there is still a need to study the mechanical behavior of Ti-alloyed Fe_3Al. In this work, the compressive behavior and deformation microstructure of Fe-25Al-10Ti alloy was studied in a temperature range of 298 to 1273 K, with an aim of identifying the underlying deformation mechanisms. It is also a goal of the present study to define a temperature/strain rate domain in which good workability could be achieved.

EXPERIMENTAL

Button ingots of Fe-25Al-10Ti (at. %) alloy were prepared by non-consumable electrode arc melting on a water-cooled copper hearth under a purified Ar atmosphere. The ingots were remelted several times to ensure homogeneity. After melting, the ingots were homogenized in vacuum (10^{-4} Pa) for 24 h at 1323 K, followed by furnace cooling. Rectangular compression specimens with dimensions of 3 mm × 3 mm × 6 mm were sectioned from the ingots by electro-discharge machining (EDM), with the long axis parallel to the solidification direction. These specimens were polished with 1000 grit emery paper and then tested in air at temperatures ranging from room temperature to 1273 K. The crosshead speed employed was 0.5 mm/min, corresponding to a nominal strain rate of 1.4×10^{-3} s^{-1}. At high temperatures, strain rate change tests were also conducted. To minimize friction effect, boron nitride powder was used for lubrication in the compression tests. Optical microscopy and transmission electron microscopy (TEM) were employed to characterize microstructures of the alloys before and after compression testing. Metallography samples were etched using an etchant containing 33% HNO_3, 33% CH_3COOH, 33 % H_2O and 1 % HF. TEM foils were prepared by electrolytic thinning at 228 K in a solution containing 30 % nitric acid and 70 % methanol, and were examined in a JEOL JEM-4000FX transmission electron microscope operated at 400 kV.

RESULTS AND DISCUSSION

Initial microstructure

Fe$_3$Al-based alloys generally show a cast microstructure consisting of coarse columnar grains along the solidification direction. However, relatively fine and equiaxed grains are observed in the present Fe-25Al-10Ti alloy (Fig. 1). The grain size is determined to be less than 100 μm. This confirms the previous study by Viswanathan et al. [7] that Ti additions tend to refine the microstructure of Fe$_3$Al-based alloy. Moreover, the Ti addition also leads to the precipitation of second phase particles in the matrix. Preliminary wave dispersive spectroscopy (WDS) analysis shows that the precipitates have much higher Ti content than the matrix.

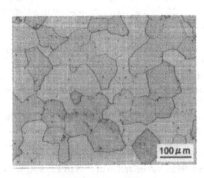

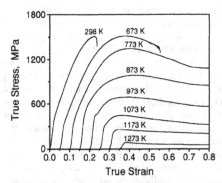

Figure 1. Optical micrograph of Fe-25Al-10Ti alloy homogenized at 1323 K for 24 h.

Figure 2. Compressive stress-strain curves at various temperatures (strain rate = 1.4×10^{-3} s^{-1}).

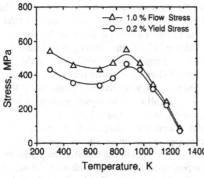

Figure 3. Temperature dependence of 0.2 % yield strength and 1.0 % flow stress.

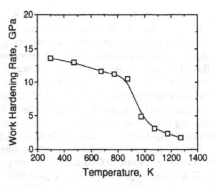

Figure 4. Temperature dependence of work hardening rate in Fe-25Al-10Ti alloy.

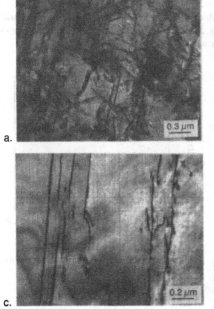

a.

b.

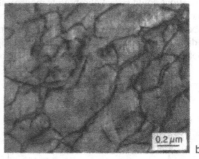

Figure 5. TEM micrographs showing the dislocation substructure developed after compression testing at (a) 873 K, (b) 1073 K and (c) 1273 K.

c.

Mechanical properties

Figure 2 shows the compressive stress-strain curves of the alloy at various temperatures. Yielding is generally followed by a stage of strong strain hardening at temperatures lower than 1073 K. At room temperature, the strain hardening continues throughout the whole testing period. As a result, an ultimate strength as high as 1.5 GPa is attained. It is also noted that the alloy exhibits considerable compressive ductility (0.23). With increasing temperature, the strain hardening is gradually confined to low strain regime and followed by a fall in stress; but the compressive ductility is increased. At temperatures above 1173 K, the yielding is immediately followed by a stress saturation or fall. Figure 3 shows the temperature dependence of yield strength and flow stress. The alloy exhibits a pronounced strength anomaly (i.e. yield strength increases with temperature) in 673 - 873 K range. From Fig. 3, work hardening rate ($\theta = \partial\sigma/\partial\varepsilon$) is calculated and plotted against temperature in Fig. 4. The alloy shows significantly high work hardening rate at low temperatures, and the work hardening rate decreases slightly with temperature up to 873 K. Above 873 K, a rapid drop in work hardening rate is observed.

Figure 5 presents the dislocation substructure developed in specimens after compressive deformation to a strain of 0.08 at different temperatures. The specimens tested at temperatures of 873 K or below generally show a high density of superlattice dislocations with a <111> vector (Fig. 5(a)). These superlattice dislocations are considered to be responsible for the high work hardening rate of the alloy at low temperatures. For the specimen deformed at 1073 K, analysis shows that most dislocations have a <110> vector (Fig. 5(b)). The curved nature of these dislocations could reflect climb-glide movement of dislocations around or over obstacles. In some areas, dislocation networks are also observed (not shown), indicating some recovery of dislocations. A low density of dislocations is observed in the 1273 K deformed specimen, and the dislocations exhibit a planar distribution (Fig. 5(c)). The formation of subgrain boundaries is not evidenced, probably due to the small deformation strain.

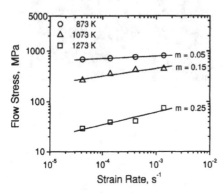 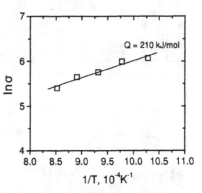

Figure 6. Dependence of flow stress (at 0.2 true strain) on strain rate in Fe-25Al-10Ti alloy.

Figure 7. Plot of ln σ against $1/T$ for the determination of activation energy.

High temperature deformation behavior

The high temperature deformation behavior was further studied by strain rate change testing and by optical observation of the deformed microstructure. Figure 6 shows the strain rate dependence of flow stress (at 0.2 true strain) at various temperatures. From the plots, the strain rate sensitivity (m) is determined. We can see that the strain rate sensitivity increases with increasing temperature and the value is 0.25 at 1273 K. Generally, m value of above 0.3

indicates the possibility of superplastic deformation [8]. For the present alloy, the low m values at temperatures below 1073 K imply poor hot workability at these temperatures.

The activation energy for flow deformation, Q, can be determined by the following equation [9]

$$Q = n \cdot R \cdot \frac{\partial \ln \sigma}{\partial (1/T)}, \tag{1}$$

where n is the stress exponent ($n = 1/m$), R is the universal gas constant, σ is the flow stress and T is the absolute temperature. Based on this equation, $\ln \sigma$ is plotted against $1/T$ for the present alloy around the temperature of 1073 K in Fig. 7. The calculated flow activation energy is 210 kJ/mol. This value is much lower than the creep activation energy of 300-350 kJ/mol for Fe_3Al-based alloys [10], suggesting that the flow deformation is not controlled by the lattice diffusion, but by grain boundary related mechanisms.

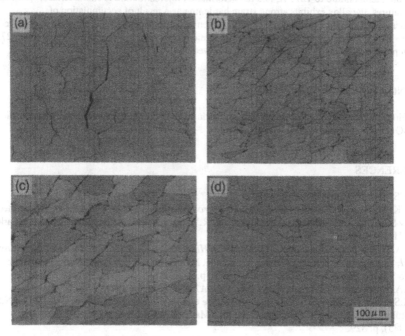

Figure 8. Optical micrographs showing the deformed microstructure after compression at (a) R.T., (b) 873 K, (c) 1073 K and (d) 1273 K.

Examination of the deformed specimens reveals that the alloy tends to exhibit surface cracking at temperatures below 1073 K. The cross sectional deformation microstructures are shown in Fig. 8. The alloy fails in an intergranular manner at room temperature. Grain boundary cavitation is dominant in specimens deformed at 873 K and 1073 K. Such cavitation occurs preferentially at grain boundaries, especially at their triple junctions, indicating that grain boundary sliding plays an important role in the deformation process. The occurrence of grain boundary cavitation should account for the observed flow softening at 873 and 1073 K. Formation of new grains is observed in the 1273 K deformed specimen. These newly formed grains are very fine, indicating the occurrence of dynamic recrystallization. It is well known that dynamic recrystallization causes not only grain refinement but also flow softening [8]. The flow softening at 1273 K has been evidenced in Fig. 2. The above results reveal that the hot working temperature for this alloy should be above 1273 K.

CONCLUSIONS

The mechanical behavior of Fe-25Al-10Ti intermetallic alloy is studied under compression from room temperature to 1273 K, and the deformation microstructure is characterized. At low temperatures, the alloy exhibits strong strain hardening as a result of accumulation of high-density superlattice dislocations. Positive temperature dependence of strength is observed in 673 - 873 K range. The alloy undergoes dynamic recrystalization at temperatures above 1273 K, where successful hot deformation can be achieved.

ACKNOWLEDGEMENTS

Financial support by the New Energy and Industrial Technology Development Organization (NEDO), Japan, is greatly acknowledged.

REFERENCES

1. N.S. Stoloff, Mater. Sci. Eng. **A258**, 1 (1998).
2. V.K. Sikka, in *Nickel and Iron Aluminides: Processing, Properties, and Applications*, edited by S.C. Deevi, P.J. Maziasz, V.K. Sikka, and R.W. Cahn, (ASM International, Materials Park, OH, 1997), p. 361.
3. J.A. Hawk, D.E. Alman, and R.D. Wilson, in *High Temperature Ordered Intermetallic Alloys VI*, edited by J.A. Horton, I. Baker, S. Hanada, R.D. Noebe, and D.S. Schwartz, (Mater. Res. Soc. Symp. Proc. **364**, Pittsburgh, PA, 1995), p. 243.
4. J.A. Hawk and D.E. Alman, Mater. Sci. Eng. **A239-240,** 899 (1997).
5. R.S. Diehm and D.E. Mikkola, in *Temperature Ordered Intermetallic Alloys II*, edited by N.S. Stoloff, C.C. Koch, C.T. Liu, and O. Izumi, (Mater. Res. Soc. Symp. Proc. **81**, Pittsburgh, PA, 1985), p. 329.
6. A. Agarwal, M.J. Akhtar, and R. Balasubramaniam, J. Mater. Sci. **31**, 5207 (1996).
7. S. Viswanathan, V.K. Andleigh, and C.G. McKamey, in *High Temperature Ordered Intermetallic Alloys VI*, edited by J.A. Horton, I. Baker, S. Hanada, R.D. Noebe, and D.S. Schwartz, (Mater. Res. Soc. Symp. Proc. **364**, Pittsburgh, PA, 1995), p. 97.
8. J.W. Edington, K.N. Melton, and C.P. Cutler, Progr. Mater. Sci. **21**, 61 (1976).
9. D. Lin, A. Shan, and M. Chen, Intermetallics 4, 489 (1996).
10. D.H. Sastry and R.S. Sundar, in *Nickel and Iron Aluminides: Processing, Properties, and Applications*, edited by S.C. Deevi, P.J. Maziasz, V.K. Sikka, and R.W. Cahn, (ASM International, Materials Park, OH, 1997), p. 123.

Mat. Res. Soc. Symp. Proc. Vol. 646 © 2001 Materials Research Society

Mechanical Properties of Fe$_3$Al Intermetallic Matrix Composites

B.G. Park, S.H. Ko and Y.H. Park
Tohoku National Industrial Research Institute,
4-2-1 Nigatake, Miyagino-ku, Sendai, 983-8511, Japan

ABSTRACT

Iron aluminides are of considerable interest due to their low cost, relatively high melting point, relatively low density, and excellent resistance to oxidation, sulfidation and molten salts. However, poor ductility and fracture toughness at room temperature hinder their use as a structural material. Refining of the microstructure is known to be one method to increase the room temperature ductility. Mechanical alloying (MA) is an easy way to obtain fine microstructures. In addition, pulse discharge sintering (PDS) is a new technology which suppresses grain growth during sintering because the sparks generated during sintering break the surface oxide layer of the powder particles and thus speed up the sintering process. Therefore, the combination of MA and PDS processes results in final products with very fine microstructures. Fe-28at.%Al alloy and its composite reinforced with 5vol.% of TiB$_2$ particles were fabricated by the MA-PDS process. The mechanical properties of these materials were improved significantly as compared to conventionally processed materials.

INTRODUCTION

Iron aluminides have long been a strong candidate for a high temperature structural material because of their low cost, relatively high melting point, relatively low density, and excellent resistance to oxidation, sulfidation and molten salts. However, the main drawback of those materials are poor ductility and fracture toughness at room temperature. To overcome this disadvantage, several methods such as ternary element additions [1], grain size refinement [2] and fine particle dispersions [3-5] were attempted.

Since Benjamin [6] fabricated oxide dispersion strengthened alloy using mechanical alloying (MA), it became one of the most important methods to refine microstructures, that is, grain refinement and/or fine particle dispersions. Colliding balls impart sufficient energy to the powders to repeat the processes of cold welding and fracture of raw powders and fine microstructures can achieved in this way. Several researchers reported successful fabrication of nano-structured Fe-Al alloys by MA [7,8]. On the other hand, pulse discharge sintering (PDS) has strong potential to be an effective sintering method [9]. The advantage of this process is that the sparks generated between the particles during sintering break the surface oxide layer and clean the surface of the powders resulting in much faster sintering. The reduced sintering time can suppress grain growth of the final product. Therefore, the combination of MA and PDS processes can result in very fine microstructures of the final product.

This research is preliminary work exploring the possibility to improve mechanical properties in Fe$_3$Al alloys and their composite system using the MA-PDS process. Fe-28at.%Al and its composite reinforced with 5 vol.% of TiB$_2$ were fabricated by an *in-situ* process. The room temperature mechanical properties were investigated.

EXPERIMENTAL PROCEDURE

Binary Fe_3Al and a $Fe_3Al\text{-}TiB_2$ composite were fabricated using the MA-PDS process. Starting materials were pure iron (99.9%, 5μm), aluminium (99.9%, 10μm), titanium (99.9%, 10μm) and amorphous boron (97%) powders. The composition of the binary alloy was Fe-28at.%Al. The composite was formulated to contain 5% volume of TiB_2 particles as reinforcement.

Mechanical alloying (MA) was performed at room temperature for 100 hours with a high energy vibratory mill. A stainless steel container and 25.4mm diameter chrome steel balls were used. The volume fraction of the ball charge in the container was 0.6. The container was charged with a total of 62g of elemental powders and the ratio of ball to powder mass was 50:1. The milling container was vibrated with a gyratory motion that was driven by rotation of unbalanced weights. The amplitude of the gyratory motion was 5mm and its frequency was 25Hz. To avoid oxidation during milling, the container was sealed in the glove box filled with high purity argon gas.

Pulse Discharge Sintering (PDS) was carried out to fabricate the iron aluminide and the *in-situ* composite. The mechanically alloyed powders were packed in a graphite mold and sintered at 1100°C for 10 minutes. During sintering, a pressure of 30MPa was continuously continuously applied. The sintered samples were 30mm in diameter and 5mm in thickness. From the sample discs tensile specimen was cut by electric discharge machining. The gauge length of the specimens was 10.5mm and their cross section was 1.5x2mm. Tensile testing was performed at room temperature using an Instron testing machine at an initial strain rate of $1.6\times10^{-4}s^{-1}$

A Field Emission Gun SEM (Philips XL30S) equipped with EDX was used to observe microstructures, fracture surfaces and compositions of the MA powders and the sintered samples. XPS was used to confirm the formation of TiB_2. The density was measured using the Archimedes principle by weighing in water and air.

The elastic moduli were measured using ultrasonic pulse echo method and calculated by the following equation [10],

$$E = \frac{(3V_L{}^2 - 4V_T{}^2)V_T{}^2}{1000(V_L{}^2 - V_T{}^2)} \times \rho$$

where V_L and V_T (km/s) are the sound velocity in the longitudinal and transverse directions, respectively and ρ (kg/m^3) is the density of the material. The Vicker's hardness was measured using a micro Vickers hardness tester with 100g load and 10 second indentation time.

Figure 1. SEM micrographs which shows the shape and size distribution of milled powders, (a) binary alloy and (b) composite

RESULTS

Figure 1 shows the shape and size distribution of the mechanically alloyed powders used to fabricate the specimens. The shape of the milled powder for the binary alloy tended to be angular while that for the composite appeared to be more spherical.

The microstructures of sintered specimens are shown in Fig. 2. Even at high magnification pores could not be found. Density measurement confirmed that full densification was obtained. The micrograph of the composite indicated areas of different contrast. The light gray areas were an *in-situ* particle rich zone whereas the dark gray areas were an *in-situ* particle depleted zone. The average size of the particle rich zones was nearly the same as that of the milled powder. Therefore, the particle depleted zones are associated with the surfaces of the powders prior to the sintering. *In-situ* formation of TiB_2 was confirmed by X-ray photoelectron spectroscopy (XPS) analysis. The TiB_2 particles were about 100nm in size. Because the particles were very fine and because of the particle depleted zones the volume fraction of the *in-situ* particles was not checked by normal quantitative microscopy analysis. XPS analysis result did not show any peaks other than those for TiB_2. Therefore, about 5% of TiB_2 was assumed to be formed. The matrix was mainly Fe_3Al with a small amount of Al_2O_3.

Figure 2. SEM micrographs which shows microstructures of sintered samples, (a) binary alloy and (b) composite

Table I. Mechanical and physical properties of sintered samples

Specimen	Y.S. (MPa)	U.T.S. (MPa)	Ductility (%)	E (GPa)	Hv (kgf/mm^2)	Density (g/cm^3)
Binary	1192	1412	6.8	173	466	6.51
Composite	1294	1562	7.2	185	527	6.42

Mechanical and physical properties are summarized in Table I. Typical tensile strengths of iron aluminides are in the range 500-900MPa and elongations are 4-12% [11]. In this material, the strength was significantly improved without sacrificing elongation. The average grain size estimated from a TEM micrograph was about 2μm. The increase of strength without loss in elongation loss is mainly due to the refined grain size. In the case of composite, TiB$_2$ particles produced during fabrication gives another improvement in the mechanical strength.

Figure 3 shows the tensile fracture surfaces of the binary and the composite specimens. The fracture surfaces of the specimens show a ductile fracture mode. It is well known that the fracture mode of iron aluminides is a mixed intergranular/cleavage failure [12]. However, this material exhibited no intergranular fracture and only little cleavage fracture. This means that the grain boundaries were strengthened and hydrogen embrittlement was suppressed by the microstructure produced by the MA-PDS process. As far as binary alloy Fe-Al alloys are concerned, the only difference as compared to normal Fe$_3$Al alloy is a fine grain size. Therefore, the fine grain size is assumed to have a significant effect in this study.

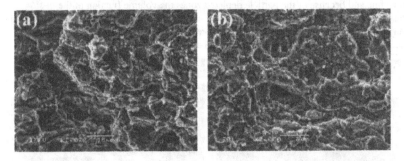

Figure 3. SEM micrographs of fracture surfaces, (a) binary alloy and (b) composite

CONCLUDING REMARKS

Fe-28at.%Al and its composite reinforced with 5vol.% of TiB_2 were fabricated using the MA-PDS process. Both yield and tensile strengths were significantly improved while maintaining good ductility. The fracture mode was ductile rather than intergranular or cleavage. Because the most important difference of the microstructure in the binary alloy used in this study is its fine grain size, as compared to conventionally processed iron aluminides, the fine grain size is thought to be the reason for the significant ductility of the MA-PDS processed iron aluminide and its composite.

REFERENCES

1. C.G. McKamey and P.J. Maziasz, Intermetallics **6**, 303-314 (1998).
2. D.J. Gaydosh, S.L. Draper, R.D. Noebe and M.V. Nathal, Mat.Sci.&Eng. **A150**, 7- 20 (1992).
3. S. Strothers and K. Vedula, Proc. Powder Metallurgy Conf., vol. 43, 597-610, Metal Powder Industry Fedration, Princeton, NJ (1987).
4. R. Baccino and F. Moret, 29th Int. Symp. On Automotive Technology & Automation, edited by D. Roller, 501-506, Automotive Automation Limited, Florence, Italy (1996).
5. R.G. Baligidad, U. Prakash, A. Radhakrishna and V. Ramakrishna Rao, Scripta **36**, 667-671 (1997).
6. J.S. Benjamin, Metall. Trans. **1**, 2943-2951 (1970).
7. D. Oleszak and P.H. Shingu, Mat. Sci. & Eng. **A181/A182**, 1217-1221 (1994).
8. G. Valdre, G.A. Botton and L.M. Brown, Acta Mater. **47**, 2303-2311 (1999).
9. Z.M. Sun, H. Hashimoto, Y.H. Park and T. Abe, Mater. Trans. JIM **40**, 879-882 (1999).
10. T. Abe: Doctorate Dissertation, Tohoku University, Sendai, Japan (1987).
11. J.H. Schneibel, 'Processing, Properties, and Applications of Iron Aluminides', edited by J.H. Schneibel and M.A. Crimp, 329-342, TMS society (1994).
12. M.G. Mendiratta, S.K. Ehlers, D.K. Chatterjee and H.A. Lipsitt, Metall. Trans. 18A, 283-291 (1987)

Mat. Res. Soc. Symp. Proc. Vol. 646 © 2001 Materials Research Society

Effect of E2₁-Fe₃AlC Precipitation on Mechanical Properties of γ-Austenite Stabilized by Addition of Mn and Co

Seiji Miura[1], Hiroaki Ishii[2] and Tetsuo Mohri
Division of Materials Science and Engineering,
Graduate School of Engineering, Hokkaido University,
Kita-13, Nishi-8, Kita-ku, Sapporo 060-8628, Japan.
[1] Corresponding author, miura@eng.hokudai.ac.jp
[2] Graduate Student, Graduate School of Engineering, Hokkaido University.

ABSTRACT

Effects of Co addition on the phase relation and mechanical properties of an E2₁-Fe₃AlC phase in the Fe-Mn-Al-C system are investigated. The relation between the γ-austenite and E2₁ phases is the same with Co-free alloys. Two kinds of precipitation processes of E2₁ phase from γ-austenite matrix are observed: a fine E2₁ phase precipitation and a cellular structure formation from grain boundaries. The cellular structure formation is explained in relation to the lattice mismatch evaluated by XRD.

The lattice mismatch is estimated to be ⁓3%, rather higher than that of γ/γ' in Ni-based superalloys, and the volume fraction of the E2₁-Fe₃AlC phase precipitated in the γ-austenite phase is as large as 30%. The micro-vickers hardness increases with increasing Al or C, caused by the increase in the volume fraction of the E2₁-Fe₃AlC phase.

INTRODUCTION

An E2₁-Fe₃AlC phase (κ) has been investigated as a potential high temperature material with the L1₂-like fcc-based structure [1, 2]. However, no sound single-phase Fe₃AlC alloys can be fabricated by a conventional melting and casting method because of a peritectic reaction associated with the formation of a graphite phase. In the Fe-Al-C ternary system, the E2₁-Fe₃AlC phase equilibrates with the graphite phase and a B2-FeAl phase or a D0₃-Fe₃Al phase at low temperature region [3-5]. On the other hand it was reported that the Fe₃AlC phase equilibrates with a γ-austenite when a large amount of Mn is added [6], and it has been attempted to introduce the "ductile carbide" E2₁ as a precipitate into γ-austenite-based alloys for high temperature use [7]. Figure 1 shows the lattice constants of various E2₁-T₃MC phases [8]. Additives such as Mn, Co, Ni are expected to substitute for Fe in the E2₁-Fe₃AlC phase because Mn₃AlC, Fe₃AlC, Co₃AlC and Ni₃Al form a continuous solid solution each other [9, 10]. Judging from these data, the addition of Mn to Fe₃AlC is expected to increase the lattice constant of the E2₁ phase, resulting in the larger lattice mismatch between the γ-austenite. On the other hand, the addition of Co is anticipated to decrease the lattice constant of the E2₁ phase.

In the present study, effects of Co on the stability of the E2₁-Fe₃AlC phase and the lattice mismatch in the Fe-Mn-Al-C system are investigated. Their mechanical properties are also evaluated by vickers tests at room temperature.

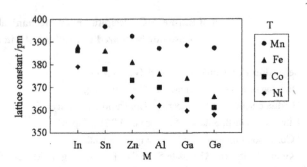

Figure 1. *The lattice constants of various E2₁-T₃MC phases [8]*

EXPERIMENTAL PROCEDURES

Master Fe-C alloy ingots containing carbon of about 20 at.% were prepared by melting high purity Fe in a high purity graphite crucible by an induction heater under an Ar atmosphere. Alloy specimens were then prepared by arc-melting with addition of high purity Mn, Co and Al under an Ar atmosphere. In order to compare with the Fe-Al-C ternary system, one-tenth of the atomic fraction of Fe is replaced by Mn and Co, respectively. Carbon concentrations in alloys were chemically analyzed and the

Table I. Composition of alloys investigated.

at.%	Fe	Mn	Co	Al	C
1	bal.	8.4	9.0	10.8	8.2
2	bal.	8.2	8.4	11.2	9.2
3	bal.	7.1	8.4	10.8	11.5
4	bal.	7.6	8.2	11.7	9.9
5	bal.	7.7	8.0	13.8	9.8
6	bal.	6.6	7.8	12.0	11.5
7	bal.	7.5	8.3	13.8	9.2
8	bal.	7.4	7.8	15.6	8.8
9	bal.	6.9	8.0	14.5	10.1
10	bal.	8.0	8.6	15.0	11.5

concentrations of other elements were confirmed by WDS. The nominal compositions of specimens are listed in table I.

Specimens in evacuated silica tubes were homogenized at 1473 K for 24 hours, followed by water quenching. Additional heat treatment were conducted at 1273 K or 1073 K for various periods up to 24 hours. Microstructural observations were carried out by SEM and micro-vickers hardness values were measured with a load of 2.94N or 300g. XRD were performed to evaluate the lattice mismatch between the γ-austenite and E2₁ phases in the sample heat-treated at 1073 K.

RESULTS AND DISCUSSION

Figure 2 shows the microstructure of several alloys with various heat treatments. As both Mn and Co substitute for Fe in the E2₁ phase, 80Fe-10Mn-10Co is treated as an element M to indicate the Fe-Mn-Co-Al-C quaternary system in the form of ternary system in Figure 2. The

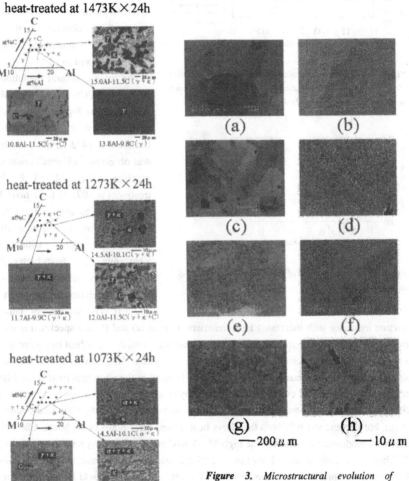

Figure 2. The microstructure of several alloys with various heat treatments. M corresponds to (80Fe-10Mn-10Co).

Figure 3. Microstructural evolution of Fe-7.6Mn-8.2Co-11.7Al-9.9C alloy heat-treated at 1473 K for 24 hr((a) and(b)), at 1073 K for 1 hr((c) and(d)), for 5 hr((e) and(f)),and for 24 hr((g) and(h)).

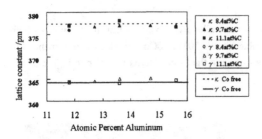

Figure 4. *The lattice constant of γ-austenite and E2₁ phase observed in the samples heat-treated at 1073 K for 24 hours in the two-phase region of the Fe-Mn-Co-Al-C system.*

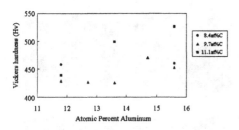

Figure 5. *Vickers hardness of the Alloys as a function of Al concentration.*

E2₁ phase equilibrates with the γ-austenite in the Fe-Mn-Co-Al-C system at various temperatures. The γ-austenite has a large single-phase region at 1473 K. At lower temperature, the E2₁ κ phase and α-ferrite precipitate. It is the same with that observed in the Fe-Mn-Al-C quaternary system with low amount of Mn.

The microstructural evolution of Fe-7.6Mn-8.2Co-11.7Al-9.9C alloy is indicated in Figure 3 as a typical example. By a homogenization treatment at 1473 K, the γ-austenite was observed with small amount of graphite in (a) and (b). By heat treatment at 1073 K for 1 hour, E2₁ phase starts to appear from grain boundaries in the manner of a cellular structure with new γ-austenite phases from high carbon and high Al γ₀-austenite matrix with small precipitates as shown in (c) and (d). The amount of lamellar structure increases with increasing heat-treatment time. In (e) and (f), the specimen is almost covered with the cellular structure after 5-hours heat treatment. A longer heat treatment results in a coarsening and spheroidization of plate-like E2₁ phase. XRD revealed that the lattice mismatch between the γ-austenite and E2₁ phases is small in the alloy heat treated for 1 hour, and it is much higher and constant in both the alloys heat-treated for 5 hours or 24 hours. The XRD peak for both the γ-austenite and E2₁ phases are rather broad in the alloy heat treated for 1 hour, but become sharp in both the alloys heat-treated for 5 hours or 24 hours. Choo et al. reported a modurated structure in the high-Mn, low-C and Co-free alloy heat-treated at 873 K [7]. They found a continuous change in the lattice constants of the γ-austenite and E2₁ phases. However, in the present alloys the change in the lattice constants should be attributed to the discontinuous formation of the cellular structure from the high C and high Al γ₀-austenite matrix with fine κ precipitates. The cellular structure has been thought to form in the alloys to alleviate a large strain which originates from a lattice mismatch between two phases [11]. A sudden and large change in the lattice constant found in the present alloy can be explained in terms of the formation of the cellular structure. The volume fraction of the $E2_1$ phase is about 30 % in the cellular structure.

Figure 4 indicates the lattice constant of the γ-austenite and $E2_1$ phases observed in the samples heat-treated at 1073 K for 24 hours in the two-phase region of the Fe-Mn-Co-Al-C system as a function of Al and C contents. The lattice constants of a Co-free alloy are also indicated by lines. By the addition of Co, the lattice constant of the $E2_1$ phase slightly decreases, whereas that of the γ-austenite slightly increases. Therefore, it can be concluded that Co has a small effect to lower the lattice mismatch of this alloy system.

Vickers hardness were measured and a part of the results are shown in Figure 5 as a function of Al concentration. The hardness values of alloys heat-treated at 1073 K is ranging from 400 to 550 HV. As can be expected from the phase diagram, the amount of the $E2_1$ phase increases with increasing Al. The increase in hardness with Al is, therefore, due to the increased fraction of the $E2_1$ phase. At higher Al concentration, softer α-ferrite appears which may decrease the hardness as indicated in the figure. On the other hand, the hardness value of the specimens heat-treated at 1473 K is about 600HV regardless of the composition. This implies the high solid solution hardening caused by C. No crackings are observed around the indents.

CONCLUSIONS

1. The addition of Co doesn't change the phase relation in the Fe-Mn-Al-C system seriously and keeps the two-phase region composed of the γ-austenite and $E2_1$ phases to low temperatures.
2. By the heat treatment at 1073 K, the $E2_1$ phase precipitates in the γ_0-austenite matrix first, then a cellular structure composed of new γ-austenite and $E2_1$ phases starts to form from grain boundaries. The lattice mismatch between the γ-austenite and $E2_1$ phases is higher in the latter case and it results in the cellular structure formation.
3. The addition of Co decreases the lattice mismatch between the γ-austenite and $E2_1$ phases.
4. Hardness of the γ - $E2_1$ two-phase alloys increases with increasing the amount of the $E2_1$ phase.

ACKNOWLEDGEMENT

This research is partially supported by Kawasaki Steel 21st Century Foundation.

REFERENCES

1. W. Sanders and G. Sauthoff : Intermetallics, **5**, 361-375,(1997).
2. idem : ibid, 377-385, (1997).
3. Keizo Nishida and Michihiro Tagami : Nippon Kinzoku Gakkaishi **30**, 68-72, (1966). (in Japanese)
4. idem : ibid, 73-78, (1966). (in Japanese)
5. M. Palm and G. Inden : Intermetallics, **3**, 443-454, (1995).
6. Kiyohito Ishida, Hiroshi Ohtani, Naoya Satoh, Ryosuke Kainuma and Taiji Nishizawa : ISIJ international, **30**, 680-686, (1990).
7. W. K. Choo, J. H. Kim and J. C. Yoon : Acta Met., **45**, 4877-4885, (1997).
8. Makoto Kikuchi, Sigemaro Nagakura and Shigueo Oketani : Tetsu to Hagane, **57**, 1009-1053, (1971). (in Japanese)
9. Hideki Hosoda, Kensyo Suzuki and Shuji Hanada : High-Temperature Ordered Intermetallic Alloys VIII, MRS Symp. Proc., vol.**552**, E. P. George, M. J. Mills & M. Yamaguchi eds., MRS, Pittsburgh, (1999), KK8.31.1-KK.8.31.6.
10. idem : ibid, (1999), KK8.32.1-KK.8.32.6.
11. G. Qin, J. Wang and S. Hao, Intermetallics, **7**, 1-4, (1999).

Mat. Res. Soc. Symp. Proc. Vol. 646 © 2001 Materials Research Society

MICROSTRUCTURES AND MECHANICAL PROPERTIES OF THE γ / κ TWO PHASE Fe-Mn-Al-C (-Cr) ALLOYS

Kazuyuki Handa*, Yoshisato Kimura and Yoshinao Mishima
Tokyo Institute of Technology, Materials Science and Engineering,
4259 Nagatsuta, Midori-ku, Yokohama 226-8502, Japan; *Graduate student

ABSTRACT

In view of the possibility of the E21-type intermetallic compound as a strengthener in austenitic alloys at high temperatures, mechanical properties and microstructures of the γ / κ two-phase alloys in the Fe-Mn-Al-C system were investigated. Alloy compositions with more than 1.5wt% C were selected in order to stabilize the γ / κ two-phase microstructure in the high temperature regime exceeding 873K. During furnace cooling after the homogenization treatment, κ phase precipitates on the γ grain boundary, which results in the poor tensile ductility. In these alloys, tensile yield strength at 873K and 1073K increases with increasing κ phase volume fraction. Applying ageing treatment at 1073K after the solid solution treatment followed by water quenching causes the formation of a uniform γ / κ lamellar microstructure, and the remarkable improvement of ductility was achieved. The addition of Cr, which is expected to improve the oxidation resistance, reduces the phase stability of κ phase, and leads to the formation of Cr_7C_3 carbide and β-Mn phase with more than 5.5wt.% Cr addition.

INTRODUCTION

The L12 type ordered intermetallic compound (γ' phase) plays a major roll as a strengthening precipitates in Ni-base superalloys with the highest strength among the high temperature structural materials currently used. In the case of Fe-base heat resisting alloys, strengtheners are usually carbides, and γ' phase can be used only at very high Ni content because γ' phase never exists in the Fe-Al binary system. Extending to the Fe-Al-C ternary system, an fcc-base intermetallic compound Fe_3AlC_x (κ phase) with E21-type ordered structure is available as a strengthener. The crystal structure of the E21-type is quite similar to that of the L12-type, and the only difference between them is that E21 has an interstitial carbon atom on body-center site of the L12. In view of the structural similarities of these two phases, E21 κ phase is expected to have desirable properties as a strengthener of Fe-base alloys for high temperature structural applications. The fcc γ-Fe phase is selected as the matrix rather than bcc α-Fe since the E21-type structure is based on fcc, which is similar to the case of the well-known γ / γ' microstructure of Ni-base superalloys. Austenitic Fe-Mn-Al-C alloys with the various carbon contents, mainly at around 1wt%, have been widely investigated for substituting conventional Fe-Ni-Cr stainless steels [2]. It is reported that the fine precipitation of κ phase provides relatively high strength through aging treatment at 673-973K after solid solution treatment in γ-Fe single phase field. As the stability of γ-Fe increases with temperature [4], the phase boundary between γ single phase and γ / κ two-phase regions at about 1 wt% C is situated at around 1000 K, just above the aging temperature. Hence, it is not possible to apply these alloys at higher temperatures than these aging temperatures [5].

The present study is focusing on the γ / κ two-phase Fe-Mn-Al-C quaternary alloys with high C content from the viewpoint to stabilize the γ / κ two-phase microstructure at high temperatures over 1000 K and to increase the volume fraction of κ phase. The aim of Mn addition is twofold: one is the austenite former and the other is the stabilizer for E21 κ phase. The E21-type T3AlC (T: transition metals) exists in many ternary systems such as Fe-Al-C, Co-Al-C, and Mn-Al-C, and it is reported that they form a continuous solid solution [1]. The

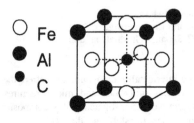

○ Fe
● Al
• C

Fig.1 The unit cell of the E21 type
ordered crystal structure.

Table 1 Nominal compositions of the alloys
investigated in the present study in wt%.

Alloy	Fe	Mn	Al	Cr	C
A	59.5	29.3	8.6	0	2.6
B	60.8	29.9	7.3	0	2.0
C	60.0	29.5	8.5	0	2.0
D	59.2	29.1	9.7	0	2.0
E	58.3	28.6	11.0	0	2.1
F	60.8	30.0	7.7	0	1.5
G	60.1	29.5	8.9	0	1.5
H	59.2	29.1	10.2	0	1.5
2.2Cr	58.5	28.8	8.5	2.2	2.0
5.5Cr	56.3	27.7	8.5	5.5	2.0
8.8Cr	54.1	26.6	8.5	8.8	2.0
12.1Cr	51.9	25.5	8.5	12.1	2.0

γ / κ two-phase alloys are less costly and also beneficial in resource securities compared to current austenitic stainless steels and Fe-base superalloys, because the present alloys do not contain Ni which is an expensive and strategic alloying element. The objective of the present study is to evaluate the relationship between microstructures and tensile properties of the γ / κ two-phase alloys in the Fe-Mn-Al-C quaternary system. Moreover, the variations of microstructures by the addition of Cr, which probably improves the oxidation resistance, were investigated as well.

EXPERIMENTAL PROCEDURES
          Nominal compositions of the alloys investigated in the present work are shown in Table 1. The C contents are more than 1.5wt% and Al contents are 7-9wt%. These compositions are required to stabilize κ phase at temperatures above 873K. The Mn contents are fixed at the ratio of Fe:Mn in at% is 2:1. As a consequence Mn contents in wt% are between 28.6 and 30.0%, which is rich enough to stabilize γ phase to avoid the coexistence of bcc-α phase and poor enough to avoid the formation of brittle β–Mn phase. Alloys were prepared by induction melting under an Ar atmosphere. The ingots were approximately 250g in weight and 30mm diameter x 50mm height in size. Wet chemical analysis performed on the alloy A revealed that the deviations of chemical compositions from nominal composition of each element were less than 5%. All the alloys were homogenized at 1323K in air for 1h and then furnace cooled to room temperature. Hereafter, alloys after this heat treatment are denoted as "homogenized alloy". To improve the microstructure, a specific heat treatment involving solution treatment, hot rolling and ageing treatment were carried out on some alloys with 2.0wt% C. Solution treatment condition was at 1323 K in the γ single phase field for 0.5h. Hot rolling was conducted at 1323 K with the reduction in area of about 40% followed by immediate water quenched. Ageing treatment at 1073K, which is higher than assumed utilizing temperature of austenitic heat resistant alloys, was conducted. These alloys are denoted as "aged alloy" hereafter. Microstructural observation was performed by back-scattered electron images of scanning electron microscopy (SEM-BEI). The constituent phases were identified by X-ray diffractometry (XRD) and electron probe microanalysis (EPMA). Tensile specimens with 15mm of gage length, 3mm of width and 1mm of thickness were cut off from the ingots by electro-discharge machining. Mechanical polishing to remove electro-discharged layer and electrolytic machining to remove work-hardened layer were conducted. Tensile tests were carried out at room temperature, 873, 973, and 1073K under vacuum conditions of 1.0×10^{-4} torr and strain rate of 1.1×10^{-4} s^{-1}. Fracture surface and sub-cracks distribution near the fracture surface were observed by SEM.

RESULTS AND DISCUSSION

Microstructures and tensile properties of homogenized alloys

All the Fe-Mn-Al-C quaternary homogenized alloys have the γ / κ two-phase microstructure. Typical microstructures of these alloys are shown in Fig. 2. The configuration of κ phase precipitation mainly depends on the C content. In Alloy A with 2.6wt% C, coarse κ grains supposed to form by solidification are collocated on the phase boundaries of the γ matrix, and also a large amount of fine κ phase precipitates in the γ matrix. Regions like precipitates free zone (PFZ) is formed around the coarse κ grains. In Alloy C with 2.0wt% C, the coarse κ grain and PFZ are not observed. Relatively thick κ phase film exists on the γ matrix grain boundaries and a small amount of fine κ phase with cross shape is observed. These are both supposed to precipitate during furnace cooling after homogenization treatment. Almost the same microstructures are observed in alloys B, D, and E with 2.0 wt% C. In alloy F with 1.5wt% C, κ precipitates inside γ grain are very fine and a small amount of κ phase decorates the grain boundaries of γ matrix. Almost the same microstructures are observed in alloys G and H with 1.5wt% C.

Temperature dependence of the tensile yield strength (YS), which is determined by the 0.2% flow stress, is shown for alloys A, C, and F in Fig. 3. These alloys fracture before the onset of yielding and show no ductility at room temperature. Yield strength of the alloys increases with increasing volume fraction of κ phase at 873 and 1073K. This result suggests that κ precipitates seem to be effective as a high temperature strengthener in the Fe-Mn-Al-C system at least up to 1073K. Crack propagations and sub-crack distribution in the vicinity of the fracture surface of alloys A and F tested at room temperature are shown in Fig. 4. In alloy A cleavage fracture takes place in the coarse κ grains and phase separation at γ / κ interfaces are observed as well. In alloy C cracks propagate along the grain boundaries of γ matrix. Thus, it is necessary to prevent κ phase from precipitating on the γ grain boundaries in order to improve the room temperature ductility of homogenized alloys.

Fig.2   Back scattered electron images of (a) alloy A, (b) alloy C and (c) alloy F after homogenization.

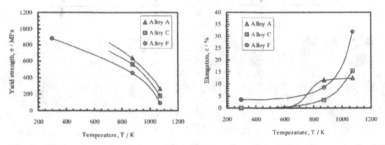

Fig. 3   Temperature dependence of yield strength and ductility of homogenized alloys.

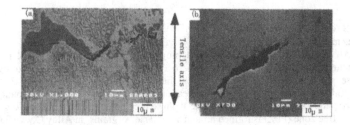

Fig. 4   Back scattered electron images in the vicinity of the fracture surface of
(a) alloy A and (b) alloy F tested at room temperature.

Microstructure modification and ductility improvement by ageing treatment
        Heat treatment sequence consisting of solution treatment, hot rolling, water
quenching, and aging at 1073K applied to alloy C is shown in Fig. 5(1). The alloy exhibits
single-phase γ microstructure after hot rolling followed by water quenching. Microstructural
variation of the alloys during ageing at 1073K is shown in Fig. 5(2). In the alloy aged for
0.5h, a small amount of fine κ phase precipitates from the γ grain boundaries. In the alloy
aged for 6hs, lamellar region seems to proceed toward the interior of γ grain with advancing
interface. Lamellar microstructure consisting of γ / κ two-phase is formed for 120h aging, and
the microstructural feature is completely different from that of homogenized alloys. The
interspacing of lamellar is about 1 μm. Judging from microstructures aged for 0.5h and 6h,
precipitation mechanism seems to be discontinuous precipitation. These lamellar
microstructure forms only by the ageing after water quenching from γ-Fe single phase field.
        Temperature dependence of strength and ductility of aged alloy C having lamellar
microstructures compared to homogenized alloys is shown in Fig. 6. The circles represent the
aged alloy, and the triangle, the homogenized alloy. The solid line indicates the yield strength
(YS) and the dotted line, the ultimate tensile strength (UTS). It is noteworthy that remarkable
improvement of ductility has been achieved by the microstructure modification at whole test
temperatures, especially at room temperature from zero to 12% plastic strain. On the other

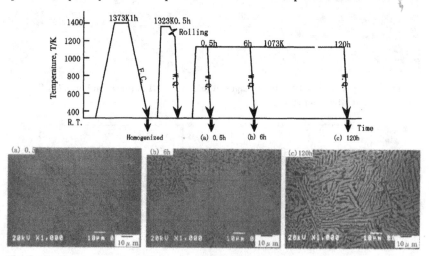

Fig. 5   (1) Heat treatment profile conducted for microstructural modification, and
(2) microstructures during aging treatment : (a) 0.5h, (b) 6h and (c) 120h.

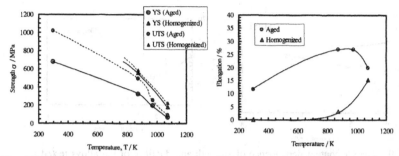

Fig. 6 Temperature dependence of yield strength and ductility of aged and homogenized alloys.

hand, YS and UTS of aged alloys are lower than those of homogenized alloys. Aged alloys maintain relatively low yield ratio from room temperature to 873K. Strength of aged alloy at 1073K is considerably lower than that of homogenized alloys. The difference in ductility between homogenized and aged alloys at 1073K is small, and it is suggested that in this temperature regime κ precipitates in the γ matrix effectively work as a strengthener. Figure 7 shows crack propagations of aged alloy tested at room temperature and 873K. Cracks seem to be initiated at κ phase of the lamellar microstructure and propagation is prevented at the γ / κ interfaces. At 873K, prevention of crack propagations by γ phase works more effectively than at ambient temperature.

Variation of microstructures by Cr addition

In the present work, Cr addition from 2.2 to 12.1wt% was performed based on the alloy 2.0C. Typical microstructures of these alloys after ageing are shown in Fig. 8. Heat treatment sequence is the same as aged alloys. The alloy 2.2Cr has γ / κ lamellar microstructure, which is almost the same as alloy 2.0C. In the alloy 5.5Cr, microstructure consists of γ / κ two phases and κ precipitates are coarser than those in the alloy 2.2Cr. In the alloy 8.8Cr, a large amount of Cr7C3 carbide forms in the γ matrix, while the volume fraction

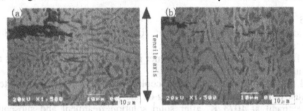

Fig. 7   Back scattered electron images of alloy C tested (a) at room temperature and
(b) at 873K.

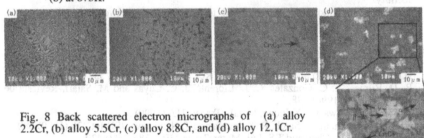

Fig. 8 Back scattered electron micrographs of   (a) alloy
2.2Cr, (b) alloy 5.5Cr, (c) alloy 8.8Cr, and (d) alloy 12.1Cr.

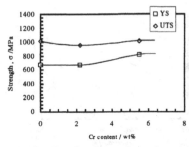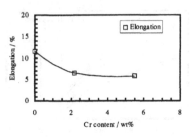

Fig. 9   The Cr content dependence of strength and ductility of the alloys tested at
room temperature.

of κ phase decreases due to a lowering of carbon content in the matrix. This means phase
stability of Cr7C3 is higher than that of κ phase. In the alloy 12.1Cr, there exists no κ phase,
and brittle β-Mn phase with the brightest contrast is seen. Close-up of the inset shows that the
microstructure is composed of α-Fe, γ-Fe, β-Mn and Cr7C3. The electron probe
microanalysis (EPMA) revealed that Al is not soluble in Cr7C3 carbide. Formation of β-Mn
and α-Fe is attributed to the enrichment of Al content in matrix caused by the increase of
Cr7C3 carbide. This is in a good agreement with the Fe-Mn-Al ternary phase diagram
information. Suitable amount of Cr addition must be up to about 5 to 6 wt%, if we want to
use this material with the γ / κ two-phase microstructure.

Figure 9 shows the Cr content dependence of strength and ductility at room
temperature. Strength of alloy 2.2Cr is comparable to that of Cr-free 2.0C alloy while alloy
5.5Cr shows slightly higher strength than the other two alloys. It is obvious that the addition
of Cr lowers the ductility of alloys, but 6 to 7 % plastic deformability is still good ductility.
Coarse κ precipitates in alloy 5.5Cr seem to be favorable in strength but not in ductility
compared with the lamellar microstructure of aged alloy C. Thus it may be possible to
balance the strength and ductility as we desire through the control of the morphology of κ
phase precipitates by the fifth element addition.

CONCLUSIONS

The following conclusions are drawn about the γ / κ two-phase alloys in the
Fe-Mn-Al-C(-Cr) system:
1. The κ phase effectively works as a strengthener in the present γ-Fe base alloys at elevated
   temperatures up to 1073 K.
2. Homogenized alloys followed by furnace cooling exhibit a poor ambient temperature
   ductility because of κ precipitates decorating on the γ grain boundaries.
3. An ageing treatment at 1073 K, after solid solution treatment and subsequent water
   quenching, results in uniform precipitation of γ phase and remarkable improvement of
   ductility.
4. Addition of Cr is possible up to 5.5 wt% to maintain the γ / κ two-phase microstructure
   without sacrificing much of ductility.

REFERENCES

1. H. Hosoda, K. Suzuki and S. Hanada, Mat. Res. Soc. Symp. Proc., 552 (1999) KK8.31.1.
2. B. D. Bilmes, A. C. Gonzales, C. L. Llorente, J. C. Cuyas and M. Salari, Reviesta de
   Metalurgia, 30 (1994) 298.
3. K. Sato, K. Tagawa and Y. Inoue, Metall. Trans., 21A (1990) 5.
4. K. Ishida, H. Ohtani, N. Satoh, R. Kainuma and T. Nishizawa, ISIJ Int., 30 (1990) 680.
5. K. S. Chan, L. H. Chen and T.S. Lui, Mater. Trans., JIM, 38 (1997) 420.

Mat. Res. Soc. Symp. Proc. Vol. 646 © 2001 Materials Research Society

Deformation Twins in Ni$_3$Nb Single Crystals with D0$_a$ Structure

Kouji Hagihara, Takayoshi Nakano and Yukichi Umakoshi
Department of Materials Science and Engineering, Graduate School of Engineering,
Osaka University, 2-1, Yamada-Oka, Suita, Osaka 565-0871, Japan

ABSTRACT

Plastic deformation behavior and substructures in Ni$_3$Nb single crystals were examined focusing on the operative deformation twin systems. Three twinning systems were operative depending on crystal orientation. The twinning behavior was discussed comparing with that in pure hcp metals. Microcracks initiated at the twin intersection. A large number of twin intersections were analyzed and classified into several types.

INTRODUCTION

Ni$_3$Nb with the D0$_a$ structure has been of considerable interest for a strengthening phase in unidirectionally solidified eutectic alloys of Ni(γ)/Ni$_3$Al(γ') -Ni$_3$Nb(δ) (for example, see [1],[2]). Recently, our group reported the deformation mode and plastic behavior in Ni$_3$Nb single crystals [3]. The D0$_a$ structure of Ni$_3$Nb is crystallographically classified as an orthorhombic system because Nb atoms are arranged in rectangular fashion on the close-packed (010) plane. However, since the atomic arrangement and stacking sequence of the (010) plane are identical to those on the (0001) basal plane in the hcp lattice, the D0$_a$ structure can be also regarded as an ordered structure based on the hcp lattice. As pyramidal slip (c+a dislocation in the hcp notation) is not operative, deformation twins are essential to provide strains along the b-axis in Ni$_3$Nb. On the other hand, deformation twins are known to induce fracture under some circumstances. The role of deformation twins on plastic behavior in pure hcp metals has been extensively investigated (for example, see [4]), but little information on intermetallics is available [5]. In this study, plastic deformation behavior and substructures in Ni$_3$Nb single crystals were examined focusing on the operative deformation twin systems.

EXPERIMENTAL PROCEDURE

Master ingots with a composition of Ni-25.0 at.% Nb were prepared by melting high purity Ni and Nb in a plasma arc furnace in high purity Ar gas. Single crystals were grown by the floating zone method using our NEC SC-35HD furnace at a growth rate of 2.5 mmh^{-1} under an Ar gas flow. Rectangular specimens for compression tests were prepared by the electro-discharge machining from the as-grown single crystals. Detailed preparation method should be referred to in our previous paper [3]. Six orientations of [302]*, [205]*, [1̄11]*, [270]*, [011]* and [012]* were chosen for the loading axes, where asterisk (*) indicates the reciprocal space representation. Compression tests were performed on an Instron-type testing machine at a nominal strain rate of 1.7×10^{-4} s^{-1} at temperatures between room temperature (RT) and 1000°C in a vacuum. Deformation markings were observed using an optical microscope with Nomarski interference contrast. Deformation substructures were observed in a transmission electron microscope (TEM, Hitachi H-800) operated at 200kV.

RESULTS AND DISCUSSION

Crystallographic aspect of three twinning systems in Ni_3Nb single crystals

Three twinning systems of $\{011\} < 0\bar{1}1 >$, $\{211\}<1\bar{0}\,7\,13>$ and $\{012\} < 0\bar{2}1 >$ were reported to be operative in Ni_3Nb single crystals [3]. Since the DO_a structure can be regarded as an ordered structure based on the hcp lattice, characters of the twins in Ni_3Nb are discussed comparing with those reported in pure hcp metals. The ordering arrangement of atoms in Ni_3Nb strongly influences the type of twin, especially the shear direction.

$\{011\}$ twin in Ni_3Nb was firstly confirmed by Grossiord et al. [6]. The twinning system of $\{011\} < 0\bar{1}1 >$ corresponds to $\{10\bar{1}2\} < 10\bar{1}\,\bar{1} >$ in the hcp notation. This type of twin was reported in many pure hcp metals [4]. In this case, ordering symmetry between twin and matrix is conserved with the same shear vector as that in pure hcp metals in Ni_3Nb. The morphology of $\{011\}$ twin is lenticular and the twin widely propagates, similar to that in pure hcp metals.

The twinning planes of $\{211\}$ and $\{012\}$ twins in Ni_3Nb correspond to $\{10\bar{1}1\}$ in the hcp notation. This $\{10\bar{1}1\}$ twin was reported in some pure hcp metals such as Mg, Ti and Zr [4]. The shear direction for $\{10\bar{1}1\}$ twin is known to be $< \bar{1}012 >$.

The shear direction for $\{211\}$ twin was proposed by Grossiord et al. as an irrational index near to $<\bar{10}\,7\,13>$ [1,6]. This nearly corresponds to $< \bar{2}3\bar{1}1 >$ in the hcp notation, which was different from $< \bar{1}012 >$ shear direction in pure hcp metals. This gives a relatively small magnitude of shear (g=0.40) without disturbing the ordered symmetry. It should be noted that this type of shear was reported even in the pure metals, however, as transformation twins which accommodate the internal strain during the bcc-to-hcp martensitic transformation in Li, Ti and Zr [7]. The detailed information was given by computer simulation [8].

Shear direction for $\{012\}$ twin is considered to be $< 0\bar{2}1 >$, which corresponds to $< 10\bar{1}\bar{2} >$ in the hcp notation. The direction is opposite to that in the pure hcp metals in order to conserve the ordered symmetry, which resulting in large magnitude of shear (g=1.07).

Plastic behavior by twinning deformation in Ni_3Nb single crystals

The asymmetry of twinning shear influences activation of the twins in tension and compression. Figure 1 shows the stereographic projections of possible activation areas for the twinning systems in Ni_3Nb single crystal in compression. A large number of twinning systems may be activated depending on loading axis. At low temperatures, however, (010)[100] slip was mainly operative in the wide range of crystal orientation because of its low CRSS. To examine plastic deformation behavior for the twinning systems, compression tests were conducted at some orientations where Schmid factor (SF) for (010)[100] slip is zero or relatively small.

Figure 2 shows temperature dependence of CRSS for three twinning systems in Ni_3Nb single crystals. The CRSS was evaluated from 0.2% proof stress in the stress-strain curves. All the three twinning systems were operative even at RT and their CRSS showed relatively high values. The CRSS rapidly decreased with increasing temperature around 200°C for $\{011\}$ twin and around 400°C for $\{211\}$ twin. The CRSS for $\{012\}$ twin at high temperatures could not be determined because single operation of $\{012\}$ twin did not occur at any loading orientation.

The observed twinning systems were in good agreement with those predicted in Fig.1. However, it was revealed the CRSS for the twinning systems showed orientation dependence and the Schmid law was not valid in Ni_3Nb. An example is the absence of $(01\bar{2})$ twin which was predicted from the Schmid factor consideration in the [011]*-oriented specimens deformed at low temperatures. The strong orientation dependence of CRSS and the violation of Schmid

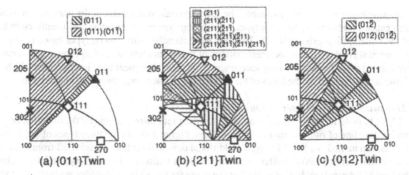

Figure 1 Stereographic projections of possible activation areas for the twinning systems in Ni₃Nb single crystal in compression. (a) {011} twin, (b) {211} twin, (c) {012} twin

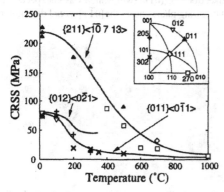

Figure 2
Temperature dependence of critical resolved shear stress (CRSS) for three twinning systems in Ni₃Nb single crystals.

law for twinning were reported in α-Ti with the hcp lattice [9].

Fracture induced by twinning

Tensile elongation of [110]*-oriented specimens is closely related to activation of twins. Although more than 100% elongation was obtained by (010)[100] slip at −196°C, elongation rapidly decreased to few percent at RT and cleavage-like fracture occurred. The fracture surfaces were macroscopically consisted of several twinning planes. This suggests that deformation twins induced fracture in an early stage of deformation. Stress concentration at the twin intersection may initiate microcracks and then the cracks easily propagate along twin boundary in Ni₃Nb. The reason is not clear yet but high cleavage sensitivity of {211} and {012} twin boundary was reported [3,10,11]. Hence the ductility of Ni₃Nb single crystal is strongly controlled by twinning deformation.

Intersection of twins

At the intersection of twins, residual strain is often accommodated by the additional deformation within the crossed twin and/or in the matrix. However, if an appropriate accommodation mode is not available, the residual strain will remain at the intersection, which may cause a microcrack. As eight types of twinning system can be operative, twenty-eight kinds of twin intersection exist in Ni_3Nb. But they were classified into several types based on the geometrical condition. In some combinations of twins, incident twin often propagated across the pre-formed twin. In the following sections, several types of typical twin intersections will be shown and discussed on the basis of the mechanism for accommodating residual strain.

{211} twin - {211} twin intersection

{211} twin-{211} twin intersection was classified into three patterns by the direction of intersecting line of two crossing twinning planes. Figure 3 shows the three types of {211} twin intersection named A-, B- and C-type. Additional deformation traces originated from twin intersection were observed. Different additional deformation modes were operated depending on the type of intersection. It is obvious that the plane for the additional modes was selected to have a common intersecting line of two crossing twinning planes. As shown in fig. 3(a), for example, the additional mode observed at A-type intersection was (011) twin, which was unexpected from the Schmid factor consideration. This (011) twinning plane shears the $[0\bar{1}1]$ zonal direction of the A-type intersection.

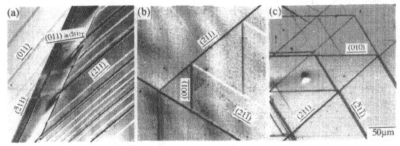

Figure 3 *Three types of {211} twin-{211} twin intersections observed in specimen deformed with [270]* orientation at 500°C. Different additional deformation modes (ADM) were observed depending on the type of intersection. (a) A-type intersection, zonal direction of intersection (ZD) // $<0\bar{1}1>$, ADM; {011} twin, (b) B-type intersection, ZD // $<\bar{1}20>$, ADM;(001) slip, (c) C-type intersection, ZD // $<\bar{1}02>$, ADM; (010) slip.*

Penetration of twin was frequently observed at C-type intersection, while most of twins were suppressed to propagate at A- and B-type intersection. Figure 4 shows a bright-field micrograph of a C-type intersection. (211) twin propagated across $(\bar{2}1\bar{1})$ twin. A large number of faults bounded by dislocations are seen especially in $(\bar{2}1\bar{1})$ twin. According to the tilting experiment and the g·b contrast analysis, faults lie on (010) and the Burgers vector for the bounding dislocations was parallel to [001]. These results are in accordance with the previous report [2].

Highly densed (010) faults and dislocations existed at the intersections. The calculated angle between [001] shear direction of (010) faults in $(\bar{2}1\bar{1})$ twin and $[\bar{10}\,7\,13]$ shear direction of (211) twin in matrix was very small to be 6.5°. In addition, (010) plane in $(\bar{2}1\bar{1})$ twin was also nearly parallel to (211) twinning plane of crossing twin and the deviation angle was 6.9°. Therefore, shear strain of crossing (211) twin may be transferred into the crossed $(\bar{2}1\bar{1})$ twin by the movement of dislocation on (010), mainly by [001] dislocations at C-type intersection.

Figure 4
Bright-field micrograph of a C-type intersection. (211) twin propagated across ($\bar{2}1\bar{1}$) twin. Specimen was deformed with [270]* orientation at 500°C. Beam direction is [$\bar{1}$02], which is parallel to zonal direction of intersection.

Similar geometrical relationship was obtained between (012) twin -(01$\bar{2}$) twin intersection. For example, the twinning plane and shear direction for (012)[0$\bar{2}$1] are approximately corresponding to those for (010)[001] slip system in (01$\bar{2}$) twin. Penetration of twins was actually observed (see fig.8 of ref. [3]), but it was revealed by TEM observation that additional operation of {011} twin was always accompanied at intersection with the (010)[001] faults in this case [12]. The {011} twin may be activated to accommodate the strain by the propagation of {012} twin with a large magnitude of shear.

Twin penetration was hardly observed at A- and B-type intersections since there does not exist any appropriate deformation modes for strain accommodation. Stress concentration, therefore, may occur at the intersection and it induces initiation of microcracks, which become a trigger for fracture.

{211} twin - {011} twin intersection
Figure 5 shows the intersection between ($\bar{2}$11) twin and (011) twin observed in specimen deformed with [111]* orientation at 700°C. Smooth intersections were often observed in this combination of twins, as previously reported [6]. Intersecting mechanism for this type of intersection was firstly explained by Cahn [13]. He proved that if the crossed twin belongs to the first kind, the only required condition for the perfect intersection is that shear direction of

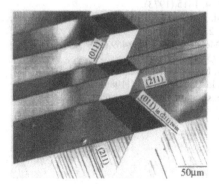

Figure 5
Optical micrograph of typical twin intersection between ($\bar{2}$11) twin and (011) twin. Zonal direction of intersection // [0$\bar{1}$1]. Specimen was deformed with [111]* orientation at 700°C.

crossing twin is to be parallel to the intersecting line of twins (i.e. the zone axis of intersection). The intersection between $(\bar{2}11)$ twin and (011) twin satisfied this criterion, that is, $[0\bar{1}1]$ shear direction for (011) crossing twin lies on $(\bar{2}11)$ twinning plane of crossed twin.

Such a smooth intersection did not occur at the intersection between $(\bar{2}11)$ twin and $(0\bar{1}1)$ twin because the intersecting criterion was not satisfied in this case.

CONCLUSIONS

(1) Three kinds of twinning systems of $\{011\} < 0\bar{1}1 >$, $\{211\}<\bar{10}\,7\,13>$ and $\{012\} < 0\bar{2}1 >$, were identified in Ni_3Nb single crystals. They were operative even at RT and the CRSS showed high values. The CRSS rapidly decreased with increasing temperature around 200°C for $\{011\}$ twin and around 400°C for $\{211\}$ twin.

(2) A large number of twin intersections were analyzed and classified into several types. In some cases, twin often propagated across the pre-formed twin. In the other cases, penetration was prevented because of the lack of deformation mode for the strain accommodation. At the intersection, stress concentration may promote initiation of microcracks.

ACKNOWLEDGMENTS

This work was supported by a Grant-in-Aid for Scientific Research Development from the Ministry of Education, Science, Sports and Culture of Japan. K. Hagihara would like to thank the Japan Society for the Promotion of Science (JSPS) for a research fellowship.

REFERENCES

1. C. Grossiord, G. Lesoult, and M. Turpin, in *Electron Microscopy and Structure of Materials*, ed. G. Thomas, (University of California Press, Berkeley, 1972), pp.678.
2. P.R. Bhowal and M. Metzger, Metall. Trans. A, **9A**, 1027 (1978).
3. K. Hagihara, T. Nakano and Y. Umakoshi, Acta Mater., **48**, 1469 (2000).
4. M.H. Yoo, Metall. Trans. A, **12A**, 409 (1981).
5. J.W. Lee, S. Hanada and M.H. Yoo, Scr. Metall. Mater., **33**, 509 (1995).
6. C. Grossiord and M. Turpin, Metal. Trans., **4**, 1415 (1973).
7. A.G. Crocker, in *Deformation twinning*, ed. R.E. Reed-Hill, J.P. Hirth and H. C. Rogers, (Gordon and Breach, New york, 1964), pp.272.
8. A.Serra, R.C. Pond and D.J. Bacon, Acta Metall. Mater., **39**, 1469 (1991).
9. A. Akhtar, Metall. Trans A., **6A**, 1105 (1975).
10. K. Hagihara, T. Nakano and Y. Umakoshi, Intermetallics, **9**, 239 (2001).
11. M. Fukuchi and K. Watanabe, Trans. Jpn. Inst. Met., **27**, 434 (1986).
12. Y. Umakoshi, K. Hagihara and T. Nakano, submitted to Intermetallics.
13. R. W. Cahn, Acta Metall., **1**, 49 (1953).

Mat. Res. Soc. Symp. Proc. Vol. 646 © 2001 Materials Research Society

Processing and Properties of Gamma+Laves Phase in-situ Composite Coatings Deposited via Magnetron Sputtering

Feng Huang[*], William S. Epling[**], John A. Barnard[*], Mark L. Weaver[*]
[*] Department of Metallurgical and Materials Engineering, The University of Alabama, Box 870202, Tuscaloosa, AL 35487-0202
[**] Department of Chemical Engineering, The University of Alabama, Box 870203, Tuscaloosa, AL 35487-0203

ABSTRACT

Recent research efforts have established that Laves phase reinforced gamma titanium aluminides (i.e. γ + Laves) offer significant potential as oxidation resistant coating in high-temperature structural applications and as wear-resistant coatings for cutting tools. In this study, TiAlCr coatings were magnetron sputtered from a Ti-51Al-Cr alloy target onto various substrates. The microstructure, hardness, and stress behavior of the as-deposited and annealed coatings have been investigated.

INTRODUCTION

In recent years, a series of γ+Laves phase (i.e., γ+Ti(Cr, Al)$_2$) based alloys have been developed for use as oxidation resistant coatings for γ-base alloys [1-7]. These alloys, which are capable of forming stable protective aluminum oxide scales, offer the potential for extending the application temperatures of γ-TiAl-base and conventional Ti-base alloys up to 1000 °C in air. This high temperature oxidation resistance additionally makes them potentially applicable as wear resistant coatings for cutting tools or as protective overcoats for magnetic storage media. Such applications call for high resistance to oxidation and corrosion, good adherence to the underlying substrate or media, coupled with high hardness and resistance to wear. In this paper, γ + Laves coatings have been deposited and characterized at relatively low temperatures with an emphasis on determining the potential of these coatings for application as protective overcoats for non-traditional applications such as magnetic storage media or cutting tool surfaces.

EXPERIMENTAL PROCEDURES

All Ti-Cr-Al alloy films were dc magnetron sputtered from a Ti-51Al-12Cr alloy target at ambient temperature using argon plasma. Substrates used include oxidized Si (100) and (111) wafers and glass pieces. The substrate to target distance was 60 mm. The base pressure was better than 4×10^{-7} Torr. Except for films used for deposition rate determination, all other films were deposited at 200 W and 3 mTorr (0.4 Pa) with 2 nominal thicknesses, 500 and 1500 nm, respectively. The deposition rates under various processing conditions were determined through the X-ray reflectivity (XRR) patterns [8]. Specular reflectivity measurements were performed on a Philips X'pert diffractometer utilizing Cu K_α radiation in the line focus mode. A graded parabolic X-ray focusing mirror was used to transform a divergent X-ray beam into a quasi-parallel yet intensive incident beam with angular divergence of about 0.05°.

To study the thermal stability of coating structures, coating specimens were air annealed at temperatures of 400, 600, 700, 800, and 1000 °C respectively for 1 hour, followed by air cooling to room temperature. The temperature variation was controlled within ±5 °C. Structural analysis was made with a Rigaku D/Max-2BX x-ray diffractometer with a thin film goniometer using Cu K_α radiation in the $\alpha - 2\theta$ mode. The glancing angle was fixed at $5°$ for all GAXRD scans.

A commercially available nanomechanical test system (Hysitron TriboScope®) equipped with a sharp Berkovich diamond tip was utilized to obtain the hardness (H) and modulus of the coatings through nanoindentation technique.

The stress in the films, σ_f, was determined by measuring the deposition- or annealing-induced curvature change of the substrates through the modified Stoney equation [9]:

$$\sigma_f = \frac{1}{6} \frac{E_s t_s^2}{(1-\nu_s)t_f R} \tag{1}$$

where E_s is the Young's modulus of the substrate, ν_s the Poisson's ratio for the substrate, t_f the film thickness, t_s the substrate thickness, and $1/R$ the substrate curvature. For the 2-inch oxidized Si (100) wafers used in current stress determination, $E_s/(1-\nu_s)$ is 180.5 GPa [10]. The intrinsic stress was determined at ambient conditions using a Flexus F2300 Stress Measurement system with in situ annealing capabilities, while the stress development with temperature was determined in situ during annealing the films. The coatings were first heated to 400 °C and held for 30 minutes, followed by heating to 500 °C and holding for another 30 minutes. The heating and cooling rates were both 5 °C min^{-1}.

RESULTS AND DISCUSSION

Film deposition rate

Table I presents the film deposition rates as a function of deposition power and pressure, which increase almost linearly with power for a fixed pressure. The deposition rates, as determined by the XRR technique, were found to decrease very slightly with increasing pressure for a fixed power. The influence of pressure on the deposition rates is negligible compared with those reported for pure Al films [11], indicative of the complexity in modeling alloy sputtering. It appears that the decreased bias voltage and increased gas scattering due to increased pressure do not affect the deposition rate significantly, contrary to what was observed for Al sputtering [11].

Structural analysis

Figure 1 presents the results of XRD analysis of as-deposited and selected annealed films. The as-deposited films are amorphous and remained stable even after annealing at temperatures up to 700 °C for 1 hour. However, after annealing at 800 °C for 1 hour, XRD peaks corresponding to rutile, γ-TiAl, and Laves phases were identified. Crystalline phases indicative of rutile, Al_2O_3, and γ+Laves phases were identified after the samples were annealed at 1000 °C for 1 hour. The coarsening of rutile grains is also evident. The stability of amorphous structure at intermediate temperatures might be beneficial for oxidation resistant coatings, because the absence of crystalline defects extending over distances comparable to the film thickness can effectively eliminate the fast diffusion paths for oxygen [12].

Hardness

To obtain accurate description of the coating hardness (H), the indentation force should be large enough to minimize the surface effect (roughness, oxidation layer, etc.), yet small enough to minimize the substrate effect. To evaluate these two effects, the peak indentation force P_{max}

Table I. Deposition rates as a function of deposition power and pressure (nm/s)[*]

	3 mTorr	5 mTorr	8 mTorr
100 (W)	0.56 (437 V)	0.54 (401 V)	0.53 (385 V)
200 (W)	1.08 (486 V)	1.05 (438 V)	1.02 (411 V)
400 (W)	2.23 (534 V)	2.20 (490 V)	2.19 (444 V)

[*]The number in the parenthesis following the deposition rate is the target bias voltage.

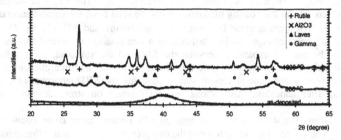

Figure 1. The GAXRD patterns for TiAlCr coatings after various heat treatments ($\alpha = 5^\circ$).

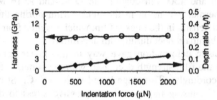

Figure 2. The hardness and contact depth versus indentation force plot for an as-deposited film.

was adjusted to obtain a $H - P_{max}$ plot for each film, as shown in Fig. 2 for 500 nm coatings deposited on glass substrates. As can be seen, the hardness remained essentially constant over the range of $P_{max} = 750 - 2000$ μN. The ratio of contact depth h_c to film thickness t was also presented in Fig. 2 for comparison. For indentation with a Berkovich tip, generally speaking, $h_c / t < 0.1$ indicates that the substrate effect on the hardness can be ignored. In our current studies, h_c was calculated using the method of Oliver and Pharr [13]. The essentially constant hardness for $P_{max} > 750$ μN can be attributed to the extremely fine structure of the sputtered films, in which case, the strains associated with indentation exerted no significant influence on the hardness [14]. Similar phenomena were also observed in films annealed at 400 and 600 °C for 1 hour. To evaluate the evolution of hardness with annealing conditions, the hardness data obtained under an indentation force of 1000 μN are taken, in which cases the h_c / t ratios were below 0.1. The hardness was found to depend on the heat treatment conditions. The hardness of as-deposited films was 8.8 GPa with a standard deviation of 0.2 GPa (8.8±0.2 GPa). It was increased by over 20% (10.7±0.4 GPa) after annealing at 400 °C for 1 hour, and was further raised to 14.4±0.4 GPa after annealing at 600 °C for 1 hours. This hardness increase might be ascribed to densification of the as deposited amorphous films as a result of excess vacancy annihilation. Although the as-deposited amorphous microstructure was retained after annealing at up to 600 °C, as confirmed by XRD analysis, the reduction in point defect density could be expected. The higher the annealing temperature was, the more significant the resulting reduction. This annealing-induced hardening is very beneficial in wear resistant applications at intermediate temperatures.

Stress

The intrinsic stress in the as-deposited films was compressive and independent of the coating thickness. The intrinsic stress was -267±20 MPa. The stress development with temperature depends on the coating thickness, as can be seen from the stress-temperature curve ($\sigma - T$) for the first thermal cycle between room temperature and 500 °C (Fig. 3). For films annealed in air, stress change can be attributed to a combination of several factors, including mismatch in the thermal expansion between the film and the substrate, point defect motion and/or annihilation, and the stresses generated due to oxidation and like phenomena although different factors may dominate at various stages. Because the XRD amorphous microstructure was retained after annealing, dislocations and their effect on stress development are not considered here.

Upon heating, the stress in both coatings initially became more compressive with rising temperature. The observed linear increase in the compressive stress indicates that deformation in the films was purely elastic as a consequence of mismatch in thermal expansion. The slopes of this stage for the 500-nm and 1500-nm coating are -2.32 and -2.16 MPa °C^{-1}, respectively, indicative of a smaller thermal expansion in the thicker film during this stage. This can be attributed to a lower defect density in the thicker film, which resulted from the larger temperature rise induced by elongated bombardment during film deposition.

At about 150 °C, elastic deformation gave way to non-linear deformation, followed by plastic yielding. Because the film strength decreased with temperature, the stress withstood by the coatings decreased accordingly. This stage lasted up to 400 °C. Stress relaxation occurred during the hold at 400 °C causing the overall compressive stress to decrease. Such stress relaxation likely resulted from a reduction in point defects during the elongated hold at intermediate temperatures. Film oxidation was expected at 400 °C, which would result in the generation of a compressive stress. It is speculated that this compressive stress was overwhelmed by the tensile stress resulting from vacancy annihilation. This speculation is supported by the stress development observed during the hold at 400 °C (see Fig. 3). It appears suitable to infer that the oxidation at 400 °C is not severe.

The strengths of polycrystalline films are generally believed to depend on the film thickness [15-17]. Such behavior was not observed in our current study of amorphous TiAlCr films, as can be seen from the close similarity between the $\sigma - T$ curves during heating. This could be ascribed to the amorphous film structure, in which the effect on yielding might be very different from those in polycrystalline structures.

The stress development during the hold at 500 °C for 30 minutes varied significantly between films, with significantly larger compression in the thinner coating. This oxidation-induced compression resulted from the significantly larger molar volumes of oxides compared with those of the corresponding elements.

With the presence of an oxide layer, the measured stress in fact reflects the contributions from both the oxide and the TiAlCr layers. In this case, the stress measured at the initial point of the hold at 500 °C, σ_f, is:

$$\sigma_f = \frac{\sigma_1 t_1 + \sigma_2 t_2}{t_1 + t_2} \tag{2}$$

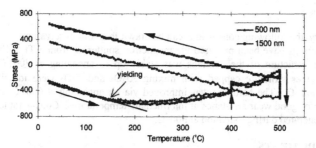

Figure 3. *The stress versus temperature profiles for 500 nm and 1500 nm coatings. The solid arrows indicate the direction of stress development, of which the vertical ones show the stress development in the two isothermal holds at 400 °C and 500 °C, respectively.*

where the subscript 1 indicates the oxide layer and subscript 2 the TiAlCr layer. The hold relaxed the stresses σ_1 by $\Delta\sigma_1$, and σ_2 by $\Delta\sigma_2$. And the oxidation changed the thicknesses t_1 by Δt_1, and t_2 by Δt_2. Then the overall stress change $\Delta\sigma_f$ during the hold is:

$$\Delta\sigma_f = \frac{(\sigma_1+\Delta\sigma_1)(t_1+\Delta t_1)+(\sigma_2+\Delta\sigma_2)(t_2+\Delta t_2)}{(t_1+\Delta t_1)+(t_2+\Delta t_2)} - \frac{\sigma_1 t_1+\sigma_2 t_2}{t_1+t_2}. \tag{3}$$

Noting that Δt_2 is negative since the TiAlCr layer was eroded, and that $\Delta t_1 + \Delta t_2$ is negligible compared with t_1+t_2 in thick TiAlCr coatings, Eq. (3) gives:

$$\Delta\sigma_f = \frac{\sigma_1\Delta t_1+\Delta\sigma_1 t_1+\Delta\sigma_1\Delta t_1+\sigma_2\Delta t_2+\Delta\sigma_2\Delta t_2}{t_1+t_2} + \Delta\sigma_2\left(\frac{t_2}{t_1+t_2}\right). \tag{4}$$

Again, noting $t_2 \gg t_1$ in thick TiAlCr coatings, and that $\Delta\sigma_2$ can be further assumed to be negligible compared with $\Delta\sigma_f$ because of the previous relaxation at 400 °C for 30 minutes. Eq. (4) is reduced to:

$$\Delta\sigma_f = \frac{\sigma_1\Delta t_1+\Delta\sigma_1 t_1+\Delta\sigma_1\Delta t_1+\sigma_2\Delta t_2+\Delta\sigma_2\Delta t_2}{t_1+t_2}. \tag{5}$$

Because the heating processes for these two coatings were almost identical, the numerator of the term on the right side of Eq. (5) can be assumed to be the same for both. Thus $\Delta\sigma_f$ is only a function of t_1+t_2 for both coatings. Although the exact t_1+t_2 values are hard to find presently, $\Delta\sigma_f$ can be roughly inversely proportional to the as-deposited coating thickness, provided the overall oxidation-induced thickness enlargement was comparatively insignificant in thick coatings. This assumption is supported by the observed stress change in the two coatings: $\Delta\sigma_f$ in the 500-nm coating ($\Delta\sigma_f \approx -360$ MPa) is exactly three times of that for the 1500 nm coating ($\Delta\sigma_f \approx -120$ MPa).

The linear and almost parallel stress-temperature developments in cooling both coatings indicate purely elastic deformation as a result of mismatch in thermal expansion. The slopes (-1.85±0.05 MPa °C^{-1}) are significantly smaller than that in the initial heating stage, due to the densification of the films. It also shows that the microstructure of the TiAlCr coatings and the overlapping oxide layers are essentially the same in these two coatings.

CONCLUSIONS

In summary, the following conclusions can be drawn from our observations:

TiAlCr coatings have been prepared via magnetron sputtering from a Ti-51Cr-12Al target. The amorphous structure of the coatings was stable up to annealing at 700 °C for 1 hour. Composite γ+Laves structure capped with mixing Al_2O_3 and TiO_2 were identified after air annealing at 1000 °C. The hardness was improved via air annealing at 600 °C by as much as 60%, indicative of good wear resistance at intermediate temperatures. Considerable compressive stress can be generated during oxidation at 500 °C.

ACKNOWLEDGMENTS

This work was supported by the U.S. Army Research Office under Grant No. DAAD 19-99-1-0152 and acknowledges the use of the facilities supported by the MRSEC Program of the NSF under Award No. DMR-9809423.

REFERENCES

[1]   M.P. Brady, W.J. Brindley, J.L. Smialek, and I.E. Locci, *JOM* **48**(11), 46 (1996).
[2]   R.L. McCarron, J.C. Schaeffer, G.H. Meier, D. Berztiss, R.A. Perkins, and J. Cullinan, in *Titanium '92*, edited by F.H. Froes and I. Caplan (The Minerals, Metals and Materials Society, 1993) pp. 1971.
[3]   M.P. Brady, J.L. Smialek, and F. Terepka, *Scripta Metall. Mater.* **32**, 1659 (1995).
[4]   R.A. Perkins and G.H. Meier, in *Proceedings of the Industry-University Advanced Materials Conference II*, edited by F. Smith (Advanced Materials Institute, 1989) pp. 92.
[5]   Z. Tang, F. Wang, and W. Wu, *Ox. Met.* **48**, 511 (1997).
[6]   C. Leyens, M. Schmidt, M. Peters, and W.A. Kaysser, *Mater. Sci. Eng. A* **239-240**, 680 (1997).
[7]   C. Leyens, J.-W. van Liere, M. Peters, and W.A. Kaysser, *Surf. Coatings Technol.* **108/109**, 30 (1998).
[8]   T.C. Huang, R. Gilles, and G. Will, *Thin Solid Films* **230**, 99 (1993).
[9]   W.D. Nix, *Metall. Trans. A* **20**, 2217 (1989).
[10] W.A. Brantley, *J. Appl. Phys.* **44**, 534 (1973).
[11] J.C. Helmer and C.E. Wickersham, *J. Vac. Sci. Technol. A* **4**, 408 (1986).
[12] H.P. Kattelus and M.A. Nicolet, in: *Diffusion Phenomena in Thin Films and Micro-Electronic Materials*, edited by D. Gupta and P.S. Ho (Noyes Publications, Park Ridge, NJ, 1988) pp. 432.
[13] W.C. Oliver and G.M. Pharr, *J. Mater. Res.* **7**, 1564 (1992).
[14] W.D. Nix, *Mater. Sci. Eng. A* **234-236**, 37 (1997).
[15] R. Venkatraman and J.C. Bravman, *J. Mater. Res.* **7**, 2040 (1992).
[16] C.V. Thompson, *J. Mater. Res.* **8**, 237 (1993).
[17] E. Artz, *Acta Mater.* **46**, 5611 (1998).

Mat. Res. Soc. Symp. Proc. Vol. 646 © 2001 Materials Research Society

Anomalous Strain Rate Dependence of Tensile Elongation in Moisture-Embrittled L1$_2$ Alloys

T. Takasugi, Y. Kaneno and H. Inoue
Department of Metallurgy and Materials Science, Graduate School of Engineering,
Osaka Prefecture University, Gakuen-cho 1-1, Sakai, Osaka 599-8531, Japan

ABSTRACT

The effect of strain rate on tensile ductility of moisture-embrittled L1$_2$-type Co$_3$Ti and Ni$_3$(Si,Ti) ordered alloys was investigated at ambient temperatures (298K~423K) by tensile test and SEM fractography. The anomalous increase of tensile elongation and ultimate tensile stress was observed in a low strain rate region and also at high temperatures, accompanied with an increased area fraction in ductile transgranular fracture pattern. The anomalous strain rate dependence of tensile ductility was shown to become more evident with decreasing grain size. As a process counteracting to the hydrogen decomposition from moisture in air, oxidation process on the alloy surface was suggested.

INTRODUCTION

L1$_2$-type Co$_3$Ti and Ni$_3$(Si,Ti) ordered alloys have some attractive mechanical and chemical properties as structural materials, e.g. a positive temperature dependence [1,2] of flow strength and strong corrosion resistance at ambient temperature. However, many L1$_2$-type ordered alloys including Co$_3$Ti [3] and Ni$_3$(Si,Ti) [4] have been shown to be embrittled at ambient temperature by hydrogen decomposed from moisture in air. Therefore, the prevention of a so-called environmental embrittlement is a crucial issue to develop them as structural materials. It has been shown that the environmental embrittlement of L1$_2$-type ordered alloys substantially depends on material conditions as well as testing conditions. As the testing conditions, atmosphere, temperature and strain rate have been investigated, and found to greatly influence tensile ductility and fracture mode of L1$_2$-type ordered alloys in air [5-7]. For example, the tensile elongation of L1$_2$-type ordered alloys in air decreases in a manner of ductile-brittle transition (DBT) as strain rate decreases. Corresponding to this behavior, fracture mode changes from ductile transgranular fracture to brittle intergranular fracture. In the previous study [8], the tensile elongation of Co$_3$Ti alloy in the embrittled condition (i.e. in a slow strain rate range) was low and insensitive to the strain rate, or, slightly increased with further decreasing strain rate, depending on the grain size. This behavior appears to be somehow distinct from the results observed in conventional metals and alloys; the hydrogen embrittlement becomes more severe as deformation rate decreases, i.e. approaches toward the so-called *delayed fracture* region.

In this study, the effect of strain rate on the moisture-induced embrittlement of L1$_2$-type Co$_3$Ti and Ni$_3$(Si,Ti) ordered alloys is investigated at ambient temperatures by tensile test and fractography. An emphasis is placed on the anomalous increase of tensile ductility in a low strain rate region, i.e. <10^{-4}s^{-1}. It is characterized how constituent element, testing temperature and grain size affect the anomalous strain rate dependence of the tensile ductility.

EXPERIMENTAL PROCEDURE

Raw materials used in this study were 99.9 wt.% cobalt, 99.9 wt.% nickel, 99.999 wt.% silicon and 99.9 wt.% titanium. A Co$_3$Ti alloy with a nominal composition of Co$_{77}$Ti$_{23}$ and a Ni$_3$(Si,Ti) alloy with a nominal composition of Ni$_{79.5}$Si$_{11}$Ti$_{9.5}$ were prepared by arc melting in an argon gas atmosphere on a copper hearth using non-consumable tungsten electrode. Homogenization heat treatment for both ordered alloys was conducted in a vacuum at 1323K for

48h, followed by furnace cooling. Homogenized ingots were rolled at 573K (for Ni$_3$(Si,Ti)) and 773K (for Co$_3$T) in air, and then annealed at 1273K for 5h. This procedure was repeated several times until desired thickness is obtained. In a final stage, rolling was conducted at room temperature. Final thickness of the rolled sheets was approximately 1mm. Tensile test pieces with a gauge size of 1x2~2.5x10~10.5 mm^3 were cut from the rolled sheets using an electro-discharge machine (EDM). These tensile specimens were recrystallized in a vacuum to obtain microstructures with desired grain sizes. Optical micrographic observation was carried out for recrystallized specimens. Average grain sizes were determined by a linear intercept method. Twin boundaries were excluded from counting for grain size.

Tensile tests were conducted in air at ambient temperatures between room temperature (~298K) and 423K over a range of strain rate from 1.6x10^{-2} to 1.6x10^{-7} s^{-1}. Tensile test in a vacuum was also conducted if necessary. The humidity in laboratory air was kept between 40~70% through day and season. The fracture surfaces of deformed specimens were examined by a scanning electron microscope (SEM).

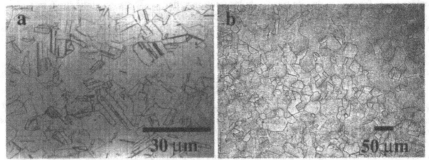

Figure 1 *Microstructures of (a) Co$_3$Ti and (b) Ni$_3$(Si,Ti) alloys annealed at 1373K for 24h and at 1173K for 4h, respectively.*

RESULTS

All specimens used in this study show fully recrystallized microstructures consisting of L1$_2$ equiaxed grains, e.g. as shown in figure 1. The grain sizes prepared for Co$_3$Ti alloy ranged from 4.6 to 19.8 μm while those prepared for Ni$_3$(Si,Ti) alloy ranged from 24 to 90 μm. Twin boundaries were more frequently observed in Co$_3$Ti alloy than in Ni$_3$(Si,Ti) alloy.

Figure 2 shows changes of tensile elongation with strain rate for Co$_3$Ti alloy with different grain sizes. The tensile elongation was generally high (~70%) in a high strain rate region (>10^{-1} s^{-1}), and then rapidly decreased with decreasing strain rate in an intermediate strain rate region (10^{-4}s^{-1}~10^{-1}s^{-1}). Thus, the decrease of the tensile elongation took place in a manner of a so-called DBT (Ductile-Brittle Transition). In a low strain rate region (<10^{-4}s^{-1}), the tensile elongation more or less tended to increase with further decreasing strain rate. As a result, a minimum in the tensile elongation vs. strain rate curve was observed in the strain rate region between 10^{-5}s^{-1} and 10^{-4}s^{-1}. In deformation in vacuum, the tensile elongation was high (~70%) and reported to be basically independent of the strain rate [3,8]. The ultimate tensile stress (UTS) behaved in a similar manner to the tensile elongation. The tensile elongation vs strain rate curve in Co$_3$Ti alloy was dependent on testing temperature and grain size. The DBT took place at a lower strain rate as the testing temperature increases and also as the grain size decreases. Also, the anomalous tensile elongation increase below 10^{-4}s^{-1} became more obvious as the testing temperature increases and also as the grain size decreases. As a result, the level of tensile elongation at a minimum was dependent on both the testing temperature and the grain size, and increased with decreasing grain size and with increasing testing temperature.

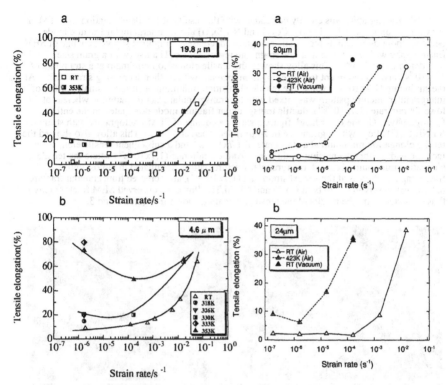

Figure 2 *Changes of tensile elongation with strain rate for Co₃Ti alloy with grain sizes of (a) 19.8 μ m and (b) 4.6 μ m. The tests were conducted at temperatures between RT (298K) and 353K.*

Figure 3 *Changes of tensile elongation with strain rate for Ni₃(Si,Ti) alloy with grain sizes of (a) 90 μ m and (b) 24 μ m. The tests were conducted at room temperature and 423K. Note that the filled marks are the data tested in a vacuum at room temperature.*

Figure 3 shows changes of tensile elongation with strain rate for Ni₃(Si,Ti) alloy with different grain sizes. Here, the data obtained from tensile tests in vacuum at a strain rate of 1.6x10⁻⁴ s⁻¹ were also included and denoted by filled marks. The tensile elongation in a vacuum is approximately 35% irrespective of testing temperature and grain size. In deformation in air, tensile elongation of Ni₃(Si,Ti) alloy deformed at a high strain rate was high and is almost the same as that of Ni₃(Si,Ti) alloy deformed in a vacuum. Below a strain rate of ~10⁻² s⁻¹, the tensile elongation rapidly decreases with decreasing strain rate, and then remains low, primarily insensitive to the strain rate. Again, the UTS behaved in a similar manner to the tensile elongation. Similar to the case of Co₃Ti alloy, the tensile elongation *vs* the strain rate curve in Ni₃(Si,Ti) alloy was very much dependent on the testing temperature and the grain size. The DBT took place at a lower strain rate as the testing temperature increases and also as the grain size decreases. However, the anomalous tensile elongation increase in a low strain rate region was observed only for Ni₃(Si,Ti) alloy with a grain size of 24 μm deformed at 423K.

SEM fractography was closely correlated with the results of tensile elongation (or UTS), as shown in figures 2 and 3. In both Co_3Ti and $Ni_3(Si,Ti)$ alloys, intergranular fracture mode becomes more dominant as the tensile elongation decreases. Figure 4 shows a change of SEM fractography with strain rate in a low strain rate region for Co_3Ti alloy (with a grain size of 4.6 μm) deformed at 353K. This alloy showed the tensile elongation minimum at a strain rate of $1.7 \times 10^{-4} s^{-1}$ and a subsequent tensile elongation increase with further decreasing strain rate. At the strain rate ($1.7 \times 10^{-4} s^{-1}$) where the tensile elongation minimum occurred, a small portion of intergranular fracture pattern was mixed with the transgranular fracture patterns, whereas at the lowest strain rate ($1.7 \times 10^{-6} s^{-1}$), ductile transgranular fracture mode dominated in the entire fractography. On the other hand, figure 5 shows a change of SEM fractography with strain rate for $Ni_3(Si,Ti)$ alloy (with a grain size of 24 μm) deformed at 423K. This alloy also showed the tensile elongation minimum at a strain rate of $1.7 \times 10^{-6} s^{-1}$ and a subsequent tensile elongation increase with further decreasing strain rate. At the strain rate ($1.7 \times 10^{-6} s^{-1}$) where the tensile elongation minimum occurred, intergranular fracture pattern dominated in almost entire fractography, whereas at the lowest strain rate ($1.7 \times 10^{-7} s^{-1}$), an increased transgranular fracture pattern was observed. In both Co_3Ti and $Ni_3(Si,Ti)$ alloys, the observed SEM fractography is thus consistent with the results of the tensile elongation shown in figures 2 and 3.

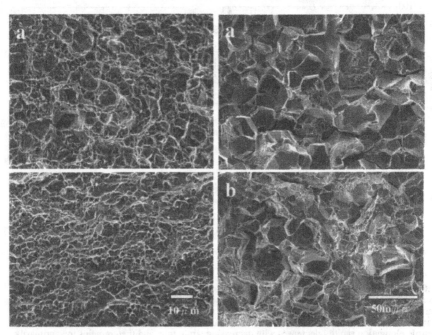

Figure 4 Change of SEM fractography with strain rate for Co_3Ti alloy (with a grain size of 4.6 μm) deformed at 353K in air. Observation was done at two strain rates of (a) $1.7 \times 10^{-4} s^{-1}$ and (b) $1.7 \times 10^{-6} s^{-1}$.

Figure 5 Change of SEM fractography with strain rate for $Ni_3(Si,Ti)$ alloy (with a grain size of 24 μm) deformed at 423K in air. Observation was done at two strain rates of (a) $1.7 \times 10^{-6} s^{-1}$ and (b) $1.7 \times 10^{-7} s^{-1}$.

DISCUSSION

It has been recognized that the moisture-induced embrittlement of $L1_2$-type ordered alloys strongly depends on strain rate. The tensile elongation (or the UTS) rapidly decreases with decreasing strain rate. It has been suggested that the moisture-induced embrittlement of $L1_2$-type ordered alloys including Co_3Ti [3] and $Ni_3(Si,Ti)$ [4] is caused by decomposition of moisture on alloy surface (or freshly exposed grain boundaries), and subsequent micro-processes of permeation into alloy, migration and condensation of atomic hydrogen to grain boundaries in front of a propagating micro crack [5-7]. When an applied strain rate is slow, the absorbed hydrogen has enough time to migrate and arrive at the grain boundary in front of the propagating intergranular microcrack, and results in the intergranular embrittlement. In both Co_3Ti and $Ni_3(Si,Ti)$ alloys, the reduction of the tensile elongation occurred in a manner of DBT.

With further decreasing strain rate, the tensile elongation however showed the minimum and then began to increase in the low strain rate region. Such an anomalous tensile elongation behavior reveals that a process counteracting to the embrittlement due to hydrogen operates in the low strain rate region. A possible process is an oxidation on alloy surface by which hydrogen decomposition from moisture (i.e. H_2O) can be suppressed. Reactive elements Ti, and Si and Ti are contained in Co_3Ti and $Ni_3(Si,Ti)$ alloys, respectively. On the alloy surface of both alloys, two reactions are suggested to occur:

$$\text{Ti (or Si)} + 2H_2O \rightarrow \text{Ti (or Si)}O_2 + 4H \qquad (1)$$

$$\text{Ti (or Si)} + O_2 \rightarrow \text{Ti (or Si)}O_2 \qquad (2)$$

By the reaction (1), atomic hydrogen is decomposed from moisture in air and then absorbed into alloy interior, leading to the embrittlement. On the other hand, on the reaction (2), oxygen molecule in air reacts with Ti (or Si) element and forms oxide surface film (product) by which the reaction (1) can be suppressed. The both reactions may occur at active points such as slip steps or grain boundaries freshly exposed during the deformation. The both reactions may compete with each other because the same active points are assumed to operate on the both reactions. If the reaction (2) is superior to (i.e. faster than) the reaction (1), the hydrogen production becomes less efficient because the pre-formed oxide surface layer makes the active points (relating to the reaction (1)) ineffective.

The reaction (1) is assumed to be very rapid and insensitive to deformation rate as well as temperature (around ambient temperature). On the other hand, the reaction (2) is assumed to be slow relative to the reaction (1), and sensitive to deformation rate as well as temperature. The formation of oxide surface film (or product) by reaction (2) more efficiently progresses as deformation rate decreases or temperature increases. Consequently, the net surface area (fresh area) available to the kinetics of reaction (1) is lowered and the elongation recovery takes place in low strain rate and high temperature regime because the amount of hydrogen generated becomes lower.

The anomalous tensile elongation increase in the low strain rate region was more obviously observed on Co_3Ti alloy than on $Ni_3(Si,Ti)$ alloy. If kinetics for oxide formation (i.e. reaction (2)) on the alloy surface is faster for titanium oxide (e.g. TiO_2) than for silicon oxide (e.g. SiO_2), a process counteracting to the embrittlement due to hydrogen is expected to more efficiently operate for Co_3Ti alloy than for $Ni_3(Si,Ti)$ alloy. In addition to such kinetics for surface oxide film formation, microstructure and thickness of the oxide film, and also adherence to the alloy substrate must be taken into the consideration.

In both Co_3Ti and $Ni_3(Si,Ti)$ alloys, the anomalous tensile elongation increase in the low strain rate region was more obviously observed on a fine-grained microstructure than a coarse-grained microstructure. Grain size effect affecting the present phenomenon can not be understood on the basis of the chemical process of decomposition (or its counteraction) because such a microstructural parameter is not involved in these chemical reactions. It appears that during deformation the concentration of hydrogen existing at grain boundaries is smaller in a fine grain microstructure than in a coarse grain microstructure because the former has larger grain boundary area per volume than the latter. It is easily assumed that the concentration of hydrogen existing at grain boundaries in a fine-grained microstructure is below the C_H^* value (where C_H^* is the critical hydrogen content at grain boundaries), below which the intergranular embrittlement does not occur. Consequently, the anomalous tensile elongation increase appears to effectively occur in a fine grain microstructure, combined with the reduction in the hydrogen absorption by the reaction (2). However, more detailed measurements for mechanical properties along with measurement for hydrogen kinetics are required to obtain a conclusive mechanism responsible for the anomalous strain rate dependence of the tensile ductility of Co_3Ti and $Ni_3(Si,Ti)$ alloys.

CONCLUSIONS

The effect of strain rate on tensile ductility of moisture-embrittled Co_3Ti and $Ni_3(Si,Ti)$ alloys was investigated at ambient temperatures (298K~423K) by tensile test and SEM fractography. The following results were obtained from the present study:
1. The anomalous increase of tensile elongation and ultimate tensile stress was observed in a low strain rate region, accompanied with an increased ductile transgranular fracture pattern.
2. The anomalous strain rate dependence of tensile ductility was shown to become more evident with decreasing grain size and also with increasing temperature.
3. Oxidation process on the alloy surface was suggested to operate as a process counteracting to the decomposition from moisture to hydrogen, and to result in an anomalous tensile ductility increase in a low strain rate region.

ACKNOWLEDGEMENT

This work was supported in part by the Functions of Hydrogen in Environmental Degradation of Structural Materials from the Special Coordination Funds for Promoting Science and Technology from the Japan Science and Technology Agency.

REFERENCES

1. T. Takasugi and O. Izumi, *Acta Metallurgica,* **33**, 39 (1985).
2. T. Takasugi, M. Nagashima and O. Izumi, *Acta Metallurgica,* **38**, 785 (1990).
3. T. Takasugi and O. Izumi, *Acta Metall.,* **34**, 607 (1986).
4. T. Takasugi, H. Suenaga and O. Izumi, *J. Mater. Sci.,* **26**, 1179 (1991).
5. T. Takasugi, *Critical Issues in the Development of High Temperature Structural Materials,* ed. by N. Stoloff, D.J. Duquette and A. F. Giamei, (The Minerals, Metals and Materials Society, Warrendale, PA, 1993) pp. 399-414.
6. C. T. Liu, *6th Int. Symp. Intermetallic Compounds – Structure and Mechanical Properties,* ed. by O. Izumi, (The Japan Inst. Metals, 1991) pp. 703-712.
7. C. T. Liu and E. P. George, *Proc. of Int. Symp. on Nickel and Iron Aluminide; Processing, Properties and Applications,* ed. by C. Deevi, V. K. Sikka, P. J. Maziasz and R. W. Cahn, (ASM, Metals Park, OH, 1997), pp. 21-32.
8. T. Takasugi, T. Tsuyumu, Y. Kaneno and H. Inoue, *Scripta Mater.,* **43**, 397 (2000).

Mat. Res. Soc. Symp. Proc. Vol. 646 © 2001 Materials Research Society

Tensile Properties of B2-Type CoTi Intermetallic Compound

Y. Kaneno and T. Takasugi
Department of Metallurgy and Materials Science, Graduate School of Engineering, Osaka
Prefecture University, Gakuen-cho 1-1, Sakai, Osaka 599-8531, Japan

ABSTRACT

B2-type CoTi intermetallic compound that was hot-rolled and recrystallized was
tensile-tested as functions of temperature and testing atmosphere. The tensile strength showed a
peak at intermediate temperature (~800K). The brittle-ductile transition (BDT) defined by
tensile elongation took place at about 800K, above which large tensile elongation was observed.
Corresponding to this transition, SEM fractography showed a change from cleavage-like pattern
mixed with intergranular fracture pattern to large cross-sectional reduction, i.e. necking of the
tensile specimen. Also, the observed mechanical properties were independent of heat-treatment
procedures, indicating that retained vacancies did not affect the mechanical properties of CoTi
intermetallic compound. However, the tensile elongation and UTS at room temperature were
dependent on testing atmosphere, indicating that moisture-induced embrittlement occurred in
CoTi intermetallic compound.

INTRODUCTION

B2-type CoTi intermetallic compound, which has its solid solution range from 50at.% to
55at.% Co, and is stable up to its melting point (-1600K) [1, 2], is one of intermetallic
compounds showing strength anomaly [3]. Deformation behavior of CoTi has been studied by
compression tests in both forms of polycrystal and single crystal [3-5]. Using CoTi polycrystals,
it was shown that an increase in yield strength with increasing temperature took place at an
intermediate temperature, and its behavior was significantly influenced by deviation from
stoichiometric composition [3]. Using CoTi single crystals, the strength anomaly has been
studied on the basis of deformed microstructures by a transmission electron microscope (TEM)
[5]. However, tensile property of CoTi has not been reported so far. Also, it is interesting to
know whether CoTi is susceptible to moisture-induced embrittlement because most intermetallic
compounds suffer from this phenomenon.

Many B2-type intermetallic compounds including CoTi have very low ductility at room
temperature but can be plastically deformed at elevated temperatures. To observe reliable
tensile properties of such a brittle material, specimens that consist of recrystallized microstructure
and have not solidified defects should be prepared, e.g. by adopting a thermomechanical
processing. In the present study, a stoichiometric CoTi was hot-rolled and then annealed. The
tensile tests were conducted as functions of temperature and testing atmosphere. SEM
fractography was also observed for tensile deformed specimens. Based on these results, the
temperature and atmosphere dependence of tensile properties and fracture mode of
polycrystalline CoTi was discussed.

EXPERIMENTAL PROCEDURES

A stoichiometric Co-50at.%Ti was prepared by arc melting in argon gas atmosphere on a
copper hearth using a non-consumable tungsten electrode. Homogenization heat treatment was
conducted in a vacuum at 1473K for 48h, followed by furnace cooling. Homogenized ingot was
sheathed with stainless steel, and then hot-rolled at 1273K in air to approx. 80% reduction in
thickness. Final thickness of the hot-rolled sheet was approximately 2mm. Using small pieces
cut from the hot-rolled sheet, recrystallization behavior in a vacuum was investigated at various
temperatures, followed by furnace cooling at a cooling rate of ~300 K/h. Optical micrographic

observation and microhardness measurement were carried out for these small pieces.

Tensile test pieces with a gauge dimension of $1 \times 2 \times 10.5$ mm^3 were cut from the hot-rolled sheet using an electro-discharge machine (EDM). After the surfaces of the tensile piece were mechanically abraded to 1mm thickness, the tensile pieces were annealed at 1323K for 10h, followed by furnace cooling. Tensile tests were conducted in a vacuum (6.7×10^{-4} - 9.3×10^{-3} Pa) in a temperature range between room temperature and 1223K at a stain rate of 1.6×10^{-3} s^{-1}. Also, some specimens were deformed in air at a stain rate of 1.6×10^{-3} and 1.6×10^{-4} s^{-1} in order to investigate the occurrence of moisture-induced embrittlement. Fracture surfaces of tensile deformed specimens were examined by a scanning electron microscope (SEM).

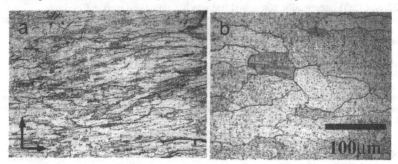

Figure 1. Optical micrographs of (a) as hot-rolled CoTi and (b) CoTi annealed at 1323K for 10h.

RESULTS

Microstructure and annealing behavior

Hot-rollig of CoTi at 1273K was successfully done and the material was crack free. The hot-rolled CoTi was heated up to 1223K, 1323K and 1423K in a vacuum, and held at each temperature for 10h, followed by furnace cooling. Microstructures of the as hot-rolled and annealed CoTi are shown in figure 1. From this figure, it is evident that the annealed CoTi consists of fully recrystallized microstructure, while the as hot-rolled CoTi appears to retain the deformed microstructure. Both microstructures consist of somewhat elongated grains. Grain size of the annealed CoTi increased as annealing temperature increased. Microhardness measurements were conducted for the hot-rolled and annealed CoTi, and then shown in figure 2. In this figure, the hardness of the CoTi annealed at 673K for 96h, preceded by annealing at 1323K for 10h is also included. This low-temperature and long-term annealing is aimed to eliminate thermal vacancies generated at a high temperature and/or retained during furnace cooling. For the annealed CoTi, the hardness slightly decreases with increasing temperature but their hardness reduction is not so large. According to these observation, it is suggested that the hot-rolled microstructure is recovered to a large extent or somewhat recrystallized during the hot-rolling. Also, it should be noted that the hardness of the CoTi annealed at 673K for 96h, preceded by annealing at 1323K for 10h is almost identical with that of the CoTi simply annealed at 1323K for 10h.

High-temperature tensile deformation

Observed stress-strain curves of the annealed CoTi can be divided into three temperature regions. At low temperature between room temperature and 773K, plastic flow was very small, and about 2% elongation was measured at room temperature. At intermediate temperature (between 773K and 1073K), moderate plastic flow was observed. At this temperature, a large elongation value is obtained, accompanied by moderate reduction in area. At high temperature above 1073K, a steady-state flow and a large reduction in area were observed, and about 50% elongation values were measured.

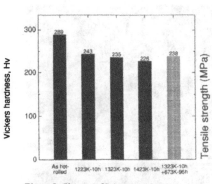

Figure 2. *Change in Vickers hardness by annealing the hot-rolled CoTi. The hardness of the CoTi annealed at 673K for 96h, preceded by annealing at 1323K for 10h is also included.*

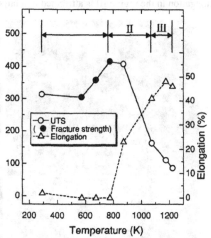

Figure 3. *Temperature dependence of tensile strength and elongation for the annealed CoTi.*

Tensile strength and elongation are plotted in figure 3 as a function of test temperature. In this figure, the fracture strength is plotted as tensile strength when the specimens fracture before yielding. On the other hand, the UTS is plotted as tensile strength when the specimens fracture after showing the UTS. The tensile strength at room temperature appears to be unchanged up to 573K. In the temperature range between 573K and 773K, the tensile strength increases with increasing temperature and then shows a peak at around 773K. A limited elongation value (~2%) is measured up to 773K. At the temperature range beyond 773K, the tensile strength decreases monotonously with increasing temperature. Above 773K, elongation rapidly increases with increasing temperature and shows a maximum value of 48% at 1173K. Here, it is noted that both high tensile strength and moderate value of elongation are obtained at an intermediate temperature (e.g. ~873K). From figure 3, the brittle-ductile transition temperature (BDTT) of the polycrystalline CoTi is defined as about 800K, and higher than that of NiAl single crystal (573K-773K) measured in a similar experimental condition [6].

SEM fractography of the tensile deformed CoTi is shown in figure 4 as a function of temperature. In the region I (figure 4(a)), brittle fracture surfaces are observed, corresponding to a limited elongation (figure 3). Although the predominant fracture mode in this region is quasi-cleavage fracture, intergranular fracture mode is also observed. However, the quasi-cleavage fracture mode was more dominant with increasing temperature. In the region II (at 873K), a mixed fracture mode of quasi-cleavage fracture with intergranular fracture changes to a single mode of quasi-cleavage fracture. In this region, the reduction of the fracture area is very significant and reaches a value more than 50%. Consequently, large elongation value is measured as shown in figure 3. However, a surprising point is that the fracture surface shows brittle fracture patterns, i.e. quasi-cleavage fracture mode quite similar to the fracture surfaces observed in the region I. As the fracture behavior in the region II, both pronounced necking at an early stage of deformation and a brittle fracture mode at a final deformation stage are emphasized. In the region III, a chisel edge type fracture takes place, as shown in figure 4(d). Consequently, the reduction in area becomes almost 100%. Detailed SEM observation indicates that fine voids are formed at the tip of the fracture surface, i.e. the chisel point. Thus, a large elongation in the region III is attributed to a large necking.

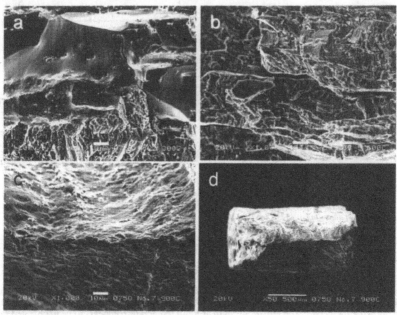

Figure 4. *SEM fractographies of the annealed CoTi. Observations were conducted on the CoTi tensile-deformed at (a) 573K, (b) 873K and (c,d) 1173K, respectively.*

Table 1 Summary of tensile test at room temperature

Strain rate	$\sigma_{0.2}$ (MPa)		Elongation (%)		UTS (MPa)	
	Air	Vacuum	Air	Vacuum	Air	Vacuum
1.6×10^{-3} s^{-1}	233	230	0.5	1.5	272	340
	230	195	0.2	1.9	234	314
1.6×10^{-4} s^{-1}	223	245	0.5	1.2	269	350
	222	237	0.3	1.2	241	355

Room temperature tensile deformation

At room temperature, the annealed CoTi was deformed in two testing atmospheres (of air and vacuum), and their differences were evaluated. Tensile tests were conducted at two strain rates of 1.6×10^{-4} s^{-1} and 1.6×10^{-3} s^{-1}, and summarized in table 1. Figure 5 represents the nominal flow stress-nominal flow strain curves for the annealed CoTi deformed at a strain rate of 1.6×10^{-3} s^{-1}. This figure indicates that the reproducibility of the stress-strain curves in each environmental medium is fairly good. As understood from table 1 and figure 5, the yield strengths were primarily insensitive to the environmental media used. However, it is evident that the elongation and the UTS depended upon the environmental media. Higher elongation and UTS values were observed in the CoTi deformed in vacuum than the CoTi deformed in air. Corresponding to this result, slightly higher fractional area consisting of quasi-cleavage fracture patterns was observed in the CoTi deformed in vacuum than the CoTi deformed in air. These results clearly indicate that a moisture-induced embrittlement takes place in CoTi, as has been observed in many intermetallic compounds such as B2-type FeAl.

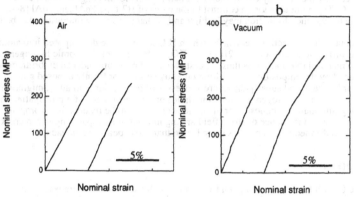

Figure 5. *Nominal flow stress-nominal flow strain curves for the annealed CoTi that were deformed at a strain rate of $1.6 \times 10^{-3} s^{-1}$ in (a) air and (b) vacuum.*

DISCUSSION

The observed temperature dependence of tensile property in CoTi is basically similar to that reported for NiAl [6,7]. In the region I, very limited elongation, an increasing strength with increasing temperature and quasi-cleavage fracture are characteristic. It has been observed that the active slip system of CoTi at low temperatures is {110}<001> [4,5]. Since high flow strength is required for the activation of <001> dislocation and the cleavage strength in tension is assumed to be low, the failure of CoTi may occur before a sufficient plastic deformation. The other reason why CoTi was so brittle at low temperatures may be attributed to insufficient slip systems. The number of slip systems guaranteeing von Mises' criterion is not satisfied only by the activation of <001> dislocations. Regarding the fracture mode in the region I, the quasi-cleavage fracture, i.e. more brittle fracture mode than the intergranular fracture, became more dominant with increasing temperature. This result means that the grain boundary strength in CoTi may be comparable to the quasi-cleavage strength at room temperature, and tends to become higher than the quasi-cleavage strength with increasing temperature in the region I.

In the region II, tensile ductility increases rapidly with increasing temperature. Here, the specimen deformed at 873K fractured by brittle quasi-cleavage though a considerable necking and a large elongation (23%) took place. A similar behavior has been observed in NiAl single crystal tensile-deformed at temperatures displaying an anomalously large elongation [6]. It is possible that the deformation of CoTi proceeds by movement of both <001> and <011> dislocations [5] at an early stage of deformation. Accumulation of dislocations occurs at grain boundaries or at points intersected between dislocations as the deformation proceeds further. However, the climbing of dislocations cannot compensate work-hardening arising from the accumulated dislocations because of an insufficient diffusion in the region II. Consequently, the stress enhanced at the triple junctions or at the grain boundaries is assumed to reach the fracture stress (i.e. the cleavage stress) and to result in crack initiation and propagation. As a result, both moderate ductility and high tensile strength can be achieved in the region II.

In the region III, the largest elongation (48% at 1173K) was measured, accompanied by no work-hardening. This implies that the climbing and annihilation of dislocations are very active in this temperature region. Also, there may be a possibility that dynamic recrystallization occurs in the region III, and results in low flow strength and large elongation.

The room-temperature hardness of CoTi was independent of annealing temperature and also was not affected by the vacancy elimination treatment. This result suggests that thermal vacancies in CoTi are not a factor, in contrast to the cases of B2-type FeAl and NiAl [8-10]. The enthalpy for vacancy formation may be high and/or that of vacancy migration may be low for CoTi.

CoTi contains Ti as reactive element by which H_2O in air can be decomposed into atomic hydrogen according to a reaction $Ti + 2H_2O \rightarrow TiO_2 + 4H$, and thereby atomic hydrogen can be released into material interior. It is thus suggested that the moisture-induced embrittlement of CoTi is induced by decomposition of moisture on alloy surface (or freshly exposed grain boundaries or lattices), and subsequent micro-processes of permeation into alloy, migration and condensation of atomic hydrogen to grain boundaries or lattices in front of a propagating micro crack. The grain boundary cohesion or the lattice cohesion can be reduced by hydrogen condensation under an influence of stress field arising in front of a propagating micro crack. Thereby, CoTi deformed in air becomes more brittle than CoTi deformed in vacuum.

CONCLUSIONS

B2-type CoTi intermetallic compound with recrystallized microstructure was tensile-deformed as functions of temperature and testing atmosphere. The following results were obtained from the present study;
(1) The tensile strength showed a peak at intermediate temperature (~800K). The brittle-ductile transition (BDT) defined by tensile elongation took place at about 800K, above which large tensile elongation was observed. Corresponding to this transition, SEM fractography showed a change from cleavage-like pattern mixed with intergranular fracture pattern to large cross-sectional reduction, i.e. necking of the tensile specimen.
(2) The observed mechanical properties were independent of heat-treatment procedures, indicating that retained vacancies did not affect the mechanical properties of CoTi intermetallic compound.
(3) The tensile elongation and UTS were dependent on testing atmosphere, indicating that moisture-induced embrittlement occurred in CoTi intermetallic compound.

REFERENCES

1. J. L. Murray, *Bull. Alloy Phase Diagr.*, **3**, 74 (1982).
2. T. Takasugi and O. Izumi, *Phys. Stat. Sol. (a)*, **102**, 697 (1987).
3. T. Takasugi and O. Izumi, *J. Mater. Sci.*, **23**, 1265 (1988).
4. T. Takasugi, K. Tsurisaki, O. Izumi and S. Ono, *Phil. Mag. A*, **61**, 785 (1990).
5. T. Takasugi, M. Yoshida and T. Kawabata, *Phil. Mag. A*, **65**, 29 (1992).
6. T. Takasugi, S. Watanabe and S. Hanada, *Mater. Sci. Eng. A*, **149**, 183 (1992).
7. A. G. Rozner and R. J. Wasilewski, *J. Inst. Metals*, **94**, 169 (1966).
8. P. Nagpala and I. Baker, *Metall. Trans. A*, **21**, 2281 (1991).
9. M. A. Morris, E. P. George and D. G. Morris, *Mater. Sci. Eng*, A, **258**, 99 (1998).
10. H. Xiao and I. Baker, *Acta metal. Mater.*, **43**, 391 (1995).

Mat. Res. Soc. Symp. Proc. Vol. 646 © 2001 Materials Research Society

Crystal Growth of RuAl-base Alloys

Sebastien Rosset, Rachel E. Cefalu, Louis W. Varner, and David R. Johnson
Materials Engineering, Purdue University, West Lafayette, IN 47907-1289

ABSTRACT

Different techniques to produce single crystals of RuAl were investigated. Processing from the melt is problematic due to the rapid loss of Al. However, very large grained RuAl specimens can be produced from the solid-state by arc-zone melting. Directional solidification of eutectic alloys is less problematic due to the lower melting temperature. Alloys of RuAl-Mo were found to be eutectic, and the eutectic composition was determined to be RuAl-51 wt% Mo.

INTRODUCTION

Intermetallics are often considered candidate materials for high temperature structural applications due to the high melting temperature and good oxidation resistance of many compounds. However, problems with brittleness have not been solved for intermetallic systems that are to operate above the use temperature of the superalloys. Ruthenium aluminide, having the CsCl (B2) type crystal structure, is unusual in this respect, as good room temperature toughness has been reported from limited and qualitative tests as first reported by Fleischer et al. [1]. More recently, Wolff et al. [2] have observed extensive room temperature plasticity from compression tests indicating a sufficient number of independent slip systems for polycrystalline deformation.

While prior work [1,2] has highlighted the promise of using RuAl for high temperature structural applications, ruthenium aluminide itself is likely to remain an exotic alloy regardless of its mechanical properties due to its great expense. The use in very small quantities may be acceptable such as in a multiphase alloy or as a coating material. Alloys of ruthenium aluminide have good high temperature oxidation resistance as reported by Fleischer et al. [3]. Furthermore, RuAl forms stable equilibria with a number of refractory metals. If multiphase microstructures are designed such that RuAl is the continuous phase, then the weight fraction of Ru can be significantly reduced and such materials may provide a tough coating for the refractory metals.

More important, though, is understanding the unusual deformation behavior of this material. Reasons for the good compressive ductility of polycrystalline RuAl are not well understood. The unusual room temperature toughness of this material appears to originate from the multiplicity of slip along the <110> and <100> directions which differs considerably from that of most other B2 compounds in which the common slip directions are either <111> or <100>. However, the observed polycrystalline plasticity in RuAl without grain boundary cracking precludes slip by <001> dislocations alone as the number of independent slip systems is less than that required by the von Mises criterion. Therefore, other slip systems must be operative during room temperature deformation. While slip parallel to <111> would allow for polycrystalline deformation, recent experimental evidence by Lu and Pollock [4] suggests that deformation by both <001> and <110> dislocations may occur thus satisfying the von Mises criterion.

The question of whether the observed dislocation structures are a result of multiple slip systems that are simultaneously active or the result of dislocation interactions has not been resolved. This is primarily due to difficulties in processing RuAl materials. Deformation studies using single crystals have not yet been performed, and the critical resolved shear stress corresponding to deformation on the possible slip systems and the single crystal elastic constants are unknown. The focus of this paper is to examine different crystal growth techniques to produce RuAl single crystals and multiphase alloys of controlled microstructure for future investigations of the fracture and deformation behavior.

PROCESSING OF RuAl

Processing single crystals of RuAl from the melt is a challenging task primarily because of the high vapor pressure of Al and the volatile Ru-base oxides that have a low melting point. The high melting temperature of RuAl precludes the use of most crucibles for containment. In this investigation, three different crystal growth methods were examined. These were directional solidification using a floating-zone image furnace, Czochralski method using a cold-crucible tri-arc furnace, and lastly, solid-state crystal growth by directional arc-melting.

The problems noted above are evident in producing the precursor ingots for the different crystal growth methods. Figure 1(a) shows the microstructure of an arc-melted RuAl ingot with a near stoichoimetric starting composition. However, significant amounts of interdendritic Ru-RuAl eutectic are clearly visible in the microstructure as a result of Al loss during melting. The resulting microstructure can be understood from the schematic Ru-Al phase diagram shown in Fig. 1(b) [5]. In order to avoid excess Al loss by evaporation, pre-alloyed Ru/Al powder blends with compositions of Ru-51 at% Al or Ru-52 at% Al were used. Pellets of 10 grams were pressed and melted in a single-arc furnace under a reducing atmosphere of 5%H$_2$/Ar. With the Al-rich powders, eutectic-free precursor ingots could be produced by arc-melting.

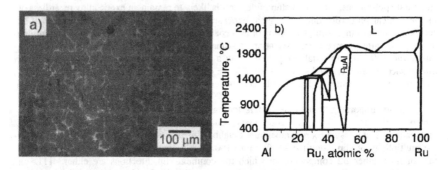

Figure 1. (a) Optical micrograph of a RuAl arc-melted ingot, and (b) schematic Al-Ru phase diagram (after Boniface and Cornish [5])

The float-zone processing was performed at Kyoto University using their optical imaging furnace. This same furnace has been used to process numerous transition metal silicides with melting temperatures similar to that of RuAl. However, when processing RuAl, metal vapor quickly coated the quartz containment cell decreasing the available input power. A stable melt of very small size could only be obtained for a slow growth rate of 5 mm/h. The resulting ingot (Fig. 2(a)) consisted of RuAl dendrites and Ru-RuAl eutectic as shown in Fig. 2(b).

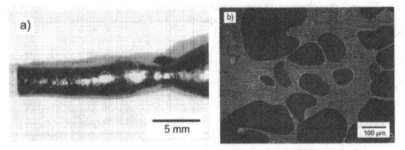

Figure 2. *(a) Photograph of a small RuAl sample that was directionally solidified within the floating zone furnace, and (b) an optical photomicrograph of the corresponding microstructure (longitudinal view).*

A similar problem was encountered when using a cold crucible tri-arc Czochralski furnace [6]. Very high temperatures are obtainable and there is not a problem in maintaining a melt pool. This process consists of melting a RuAl precursor ingot by three arcs under a 5%H_2/Ar atmosphere. A small sample is then pulled from the melt pool with typical pull rates between 60 to 100 mm/hr.

However, even with an Al-rich precursor ingot (Ru-52 at.% Al), the processed ingot contained a large amount of interdendritic eutectic as shown in Fig 3. The eutectic was found to be Ru-RuAl from analysis within a scanning electron microscope (SEM) equipped with an energy-dispersive spectrometer (EDS). Hence, due to Al loss, the melt composition quickly moves towards the Ru-rich side of the phase diagram during processing. A fast pull rate is desired such that a crystal can be pulled from the melt before Al evaporation changes the initial melt composition significantly. However attempts to pull crystals at rates much faster than 100 mm/h were unsuccessful. For a non-stoichoimetric melt composition (such as the case shown in Fig. 3), the simple constitutional supercooling criteria would predict a very slow growth rate to maintain a planar solid-liquid interface. Of course a slow growth rate only leads to greater Al loss; thus the practice is impractical.

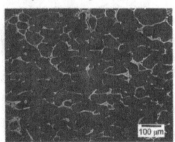

Figure 3. Optical photomicrograph of a RuAl specimen grown by the Czochralski method in a cold crucible tri-arc furnace at a pull rate of 100 mm/hr.

To better control the final composition, a procedure to grow RuAl crystals from the solid-state was investigated. The process employed was arc-zone melting (Fig. 4) which consists of pulling a "finger" shaped precursor ingot under a single plasma arc. Since only the top of the precursor ingot is melted, crystal growth is expected to take place in the solid state by grain boundary migration and solid-state diffusion [7].

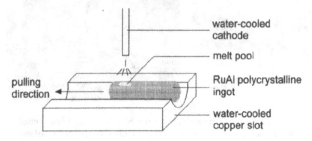

Figure 4. *Schematic of the arc zone-melting process*

This technique has been used to process alloys of Ru-52 at.% Al and Ru-50.5 at.% Al. The resulting ingot is not a single crystal but has a very large grain size and shows no eutectic formation. Grains of 5 mm diameter have been obtained and oriented using standard Laue back-reflection X-ray diffraction techniques. An example is shown in Fig. 5(a) for an oriented Ru-52Al crystal. The size and orientation was measured from two perpendicular surfaces as shown in Fig. 5(b). The perpendicular surfaces are needed to discern the orientation of the deformation marking observed after indentation testing. Indentation testing of the Ru-52Al and Ru-50.5Al crystals produced by this method is currently in progress.

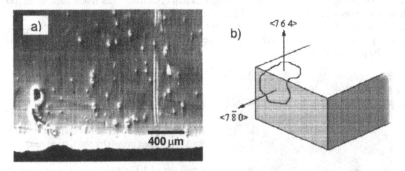

Figure 5. *(a) RuAl single crystal (52 at%Al) grown by arc-zone melting and a (b) schematic of the relative grain size and orientation.*

EUTECTIC ALLOYS

An alternate approach to producing single crystals is to produce in-situ composites from RuAl-base eutectic alloys. Eutectic alloys are congruently melting and thus can easily be processed due to the lack of a large mushy zone. Furthermore, the melting temperature is depressed below that of the constituent phases, which will greatly reduce the problems noted previously. Directional solidification can be used to align both phases parallel to the growth direction to produce "in-situ" or "natural" composites. As specific orientation relationships often exist between the phases and the growth direction, these ingots can then be used as a vehicle to study the deformation behavior of RuAl when constrained by the surrounding eutectic phase.

One system identified for composite growth is the RuAl-Mo eutectic. The eutectic composition was found to be RuAl-51 wt% Mo from the microstructure of arc-melted buttons. Analyses of dendritic button ingots on either side of the eutectic point (RuAl-26.5 wt% Mo and RuAl-70 wt% Mo) were used to roughly estimate the shape of the pseudo-binary phase diagram (Fig. 6(a)). The chemical compositions of the dendritic phases were determined by EDS analysis for arc-melted and heat treated (1350 °C, 5h) button ingots. After heat treating at 1350°C, the compositions of the respective solid solutions were RuAl-5.7 wt% Mo and Mo-9.3 wt% RuAl. The eutectic temperature was estimated from optical pyrometer measurements taken from the arc-melted buttons and referenced to melts of Ni, NiAl, and RuAl.

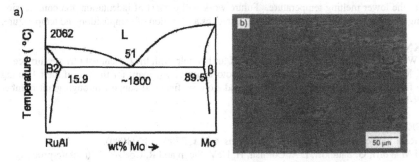

Figure 6. (a) Schematic of the RuAl-Mo pseudo-binary phase diagram, and (b) the fully eutectic microstructure of a RuAl-51wt% Mo Czochralski speciman (optical microscopy, longitudinal view).

Because of the lower melting temperature of the eutectic, the ability to processes this material is greatly improved. Small ingots can easily be pulled from the cold crucible tri-arc Czochralski furnace. A fully eutectic microstructure from such an ingot is shown in Fig. 6(b) in which lamellar RuAl-Mo eutectic colonies are aligned with the growth direction. The size and shape of the pulled specimen is shown in Fig. 7.

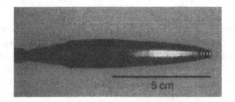

Figure 7. RuAl-51wt% Mo ingot pulled from a cold crucible Czochralski furnace.

CONCLUSIONS

Small single crystals of RuAl can be produced from the solid-state by arc-zone melting. The advantage of this method over growth by floating zone or Czochralski techniques is that crystal growth takes place in the solid-state and the desired stoichoimetry is maintained. Processing from the melt is problematic due to the rapid Al loss and the formation of a non-stoichoimetric melt composition. However, a RuAl-Mo eutectic can easily be processed from the melt primarily due to the lower melting temperature. Future work will consist of indentation and compression testing of oriented RuAl single crystals and eutectics as function of composition and temperature.

ACKNOWLEDGMENTS

We wish to acknowledge Patricia Metcalf for her help with the Czochralski tri-arc furnace, Prof. M. Yamaguchi and Prof. H. Inui of Kyoto University (Japan) for the use of their floating zone furnace and The National Science Foundation for financial support through grant number DMR-0076219.

REFERENCES

1.   R. L. Fleischer and R. J. Zabala, *Met Trans A* **21A**, 2709-2715 (1990).
2.   I.M Wolff, G. Sauthoff, L.A. Cornish, H. DeV. Steyn and R. Coetzee, "Structure-property-application relationships in ruthenium aluminide RuAl," *in Structural Intermetallics 1997*, edited by M.V. Nathal, *et al.*, (TMS, Warrendale, PA 1996), p. 815-830
3.   R. L. Fleischer and D. W. McKee, *Met. Trans. A* **24A**, 759-762 (1993).
4.   D.C. Lu, and T.M. Pollock, *Acta Mater.* **47**, 1035-1042 (1999).
5.   T. D. Boniface, and L. A. Cornish, *J. Alloys and Compounds* **234**, 275-279 (1996).
6.   D. W. Jones, "Refractory metal crystal growth techniques," in *Crystal Growth, Theory & Techniques*, (Plenum press, New York, 1974) p. 252.
7.   T.A. Lograsso, F.A. Schmidt, *J. of Crystal Growth* **110**, 363-372 (1991).

Mat. Res. Soc. Symp. Proc. Vol. 646 © 2001 Materials Research Society

Thermoelastic Properties of Super High Temperature Ru-Ta Shape Memory Alloy

Y.Furuya [1] and H.Zhirong [2]

[1] Department of Intelligent Machines and System Engineering, Hirosaki University, Hirosaki 036-8152, Japan

[2] Visiting Professor of Hirosaki University, and at present, Department of Material Science and Engineering, Shaanxi Institute of Technology, Shaanxi,China

ABSTRACT

Equiatomic Ru-Ta alloy can show a thermoelastic phase transformation at super high temperatures above 1000℃. The temperature of shape memory effect of this alloy seems to be remarkably higher by two times than those of the formerly reported high temperature NiMnAl alloy and it seems very interesting as one of the new high temperature SMAs, however, detailed thermoelastic properties of Ru-Ta system alloy have not been clarified. The present paper investigates the basic material properties of Ru-Ta alloy, i.e. each transformation temperature, temperature hysteresis range, heat absorption by DSC measurement and metallurgical compounds by X-ray diffraction that depend on alloy composition by changing the Ru content from 46-54at%. Effect of heat treatment at solution temperature and aging temperature on thermoelastic phase transformation points is investigated and the thermal cyclic stability is also studied for engineering applications as high temperature actuator/sensor materials.

INTRODUCTION

It is desirable to develop high temperature actuator/sensor materials which can respond to changes of temperature above 400~500℃ especially in the fields of controlling parts in electric power plants, chemical reaction plants and fire accident prevention etc.. As for a candidate of high temperature potential actuator/sensor materials, piezoelectric ceramics (PZT) and/or magnetostrictive materials has been studied, however, these are inevitably restricted and impossible for application temperatures above 300~400℃ due to their low Curie points. On the other hand, some shape memory alloys(SMAs) have their phase transformation and recovery temperature above 400℃,i.e. the NiAlMn alloy system has its inverse transformation, Af point near 600℃. Moreover, it was found by R.Fonda [1] and Y.Furuya [2] that the equiatomic Ru-Ta alloy can show a thermoelastic phase transformation at super high temperatures above 1000℃ as shown in **Fig.1**. This Af value associated with the shape memory effect is remarkably higher, by two times , than that of the formerly reported high temperature NiMnAl alloy. Therefore, Ru-based shape memory alloys will be applicable to ultra-high temperature turbine blades for developing higher efficiency jet engines and super heat-resistance components such as the tip material of the wings of aerospace vehicle due to the better ductility, toughness and mechanical strength than ceramics at super high temperature region over 1000℃. These features seem to be very attractive, however, detailed thermoelastic properties of Ru-Ta system alloy have not been clarified. In the present paper, we investigate the basic material properties of Ru-Ta alloy, each transformation temperature, temperature hysteresis range, heat absorption degree by DSC measurement and metallurgical phases by X-ray diffraction that depend on alloy composition by changing the Ru content from 46-54at%. Effects of heat treatments on thermoelastic phase transformation points are investigated. Thermal stability has been also studied by applying several thermal cycles for engineering applications as high temperature actuator/sensor materials. In consequence, it has been found that the thermoelastic phase transformation temperature is ranging from 900 to 1200℃ depending on Ta content, and that their thermal stability exists against aging and heat cycles.

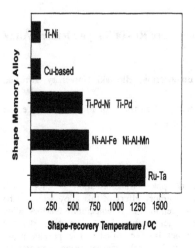

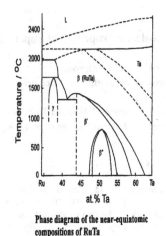

Phase diagram of the near-equiatomic
compositions of RuTa

Fig.1 Shape memory alloy map   Fig.2 Phase diagram of Ru·Ta alloy system

EXPERIMENTAL PROCEDURES

The phase diagram of the near-equiatomic compositions of binary Ru-Ta system is shown in Fig.2. Starting materials are Ru powder and Ta powder and they were mixed to control the composition from 46at%Ta to 54at%Ta. The mixed powders were arc-melted to make the Ru-Ta alloy bottom that was then annealed at 1400℃ for three days in vacuum to homogenize the composition. The melted alloys had the composition of Ru-Ta(46.2, 48.0, 49.4 50.8, 52.0,54.0 at%) respectively as verified by electron probe microanalysis (EPMA). The heat treatments, i.e. solution treatment and following aging treatment were carried out in evacuated quartz tubes, followed by quenching in water. The physical properties of thermoelastic phase transformation were measured by differential scanning calorimetry (DSC) ranging from room temperature(RT) to 1400℃, and the rate of heating and cooling rate was 10℃/min. Crystalline structures and precipitation were examined by X-ray diffraction methods whose target was CuKa at room temperature. Thermal expansion and contraction of Ru-Ta bulk rod was measured by dilatometer for the equiatomic Ru-Ta alloy. Thermal cyclic tests were performed during temperature ranging from 800℃ to 1400℃ for several cycles.

RESULTS AND DISCUSSON

After homogenization annealing at 1400℃,72hrs. an x-ray diffraction spectrum obtained from solution treated(a) and aged(b) Ru-Ta system alloys at room temperature is shown in Fig.3. It can be noticed that all samples of Ru=46at% to Ru=54at% contains the intermetallic compound of Ru(1):Ta(1)., and the intensity of Ru:Ta precipitation is strong near the equiatomic sample of Ru-50at%Ta. Fig.4 shows the effect of Ta content on transformation temperatures after heat treatments in Ru-Xat%Ta (X=46~54at%) under heating and subsequent cooling. It was found that near equiatomic Ru-Ta alloy, (i.e.Ta=49,50at%) undergoes two main phase transformations during heating and heating, i.e., the high-temperature M transformation (β (cubic) toward β '(tetragonal)) and the low-temperature P transformation (β "(monoclinic symmetry)). Martensite (M). After heat treatment, the A-M transformation (A:B2-ordered cubic(β),

M:tetragonal (β')) takes place in Ru-(46~54)Ta alloys during heating and cooling. With increasing Ta content, the Ms, Af and the hysteresis on the DSC curve of RuTa alloys decrease. The effect of Ta content on the transformation temperature of RuTa and the effect of Ta content on M transformation hysteresis(=Af-Ms) and absorption heat mass are shown in Fig.5 and Fig.6 respectively. From the narrower hysteresis and larger absorption heat-mass, near equiatomic Ru-Ta alloy seems to be the best selection as a high performance SMA.

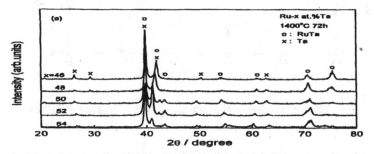

Fig.3   X-ray diffraction pattern obtained from solution treated Ru-xat%Ta (X=46~54) alloys

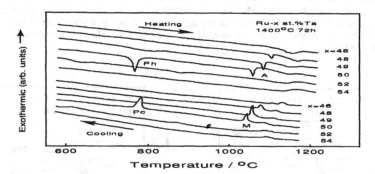

Fig.4   Effect of Ta content on DSC curves of RuTa alloys after solution treatment.

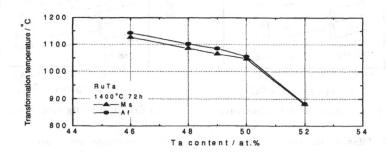

Fig.5   Effect of Ta content on the transformation temperature of RuTa SMA.

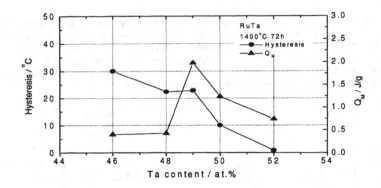

Fig.6   Effect of Ta content on M transformation hysteresis and heat of RuTa SMA.

Next, the effect of heat treatment on thermoelastic phase transformation is discussed. **Fig. 7** shows the effect of solution temperature on Af and Ms points in Ru-49at%Ta. DSC curve peak and Af,Ms points scarcely changed by changing heat-solution temperature from 600℃ to 1400℃. Effect of 24hr aging temperature after 1400℃ 24hr annealing on M and A phase transformation temperature is shown in **Fig.8**.   These features are typically shown in **Fig.9** which indicates the stability of each transformation temperature, Ms,Af in RuTa SMA alloy against thermal aging treatment at 900℃.

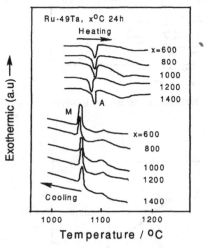

(a) Effect of solution temperature

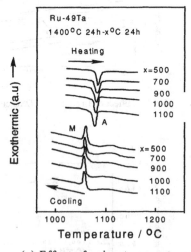

(c) Effect of aging temperature

Fig.7   Effect of solution temperature on   Fig.8   Effect of aging temperature on
DSC curve profile in Ru-49at%Ta .   DSC curve profile in Ru-49at%Ta

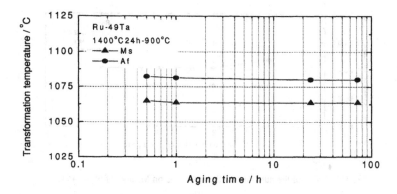

Fig. 9   Effect of aging time on Ms and Af of Ru-49Ta SMA.

Lastly, the thermal stability has been also studied by applying several thermal cycles. **Fig.10** shows the DSC curve for repeated heating and cooling in Ru-49at%Ta alloy which was heat-treated,1400℃ 24hr.   For as long as four thermal cycles between 800℃ and 1300℃, the phase transformation temperatures, Ms,Af did not change as shown in **Fig11.**
In future work, mechanical thermal-hysteresis, i.e. strain~stress curve in one thermal cycle, have to be studied, and the cyclic stress ~strain curves also will become important in using this SMA as a machinery actuator component.

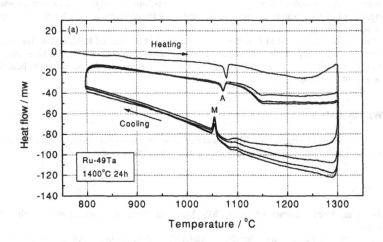

Fig. 10   DSC curves of the thermal cycles of Ru-49Ta SMA.

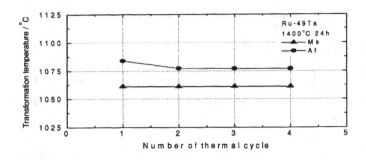

Fig.11   Effect of the number of   thermal cycle on Ms and Af of Ru-49Ta.

CONCLUSIONS

The results of this study are summarized as follows.

(1) The A-M transformation (A(parent-phase): B2- ordered cubic (β), M (martensite): tetragonal (β ')) took place in solution treated Ru-(46~54)Ta alloys during heating and cooling . The M transformation temperature of those alloys was very high, and the Ms of Ru-(46~50)Ta alloys was even higher than 1000°C. With increasing Ta content, the Ms, Af, and hysteresis decreased, and effects of aging on Ms, Af, and hysteresis of those alloys were very weak.

(2) Ru-50Ta alloy underwent two main phase transformations during cooling and heating, i.e. the high-temperature M transformation (β (cubic) - to - β ' (tetragonal)) and the low-temperature P transformation (β '- to- β " (monoclinic symmetry)) . During thermal cycling, the M transformation was very stable, and the P transformation was also stable except for the first thermal cycle.

(3) Ru-49Ta alloy had the largest M transformation heat (Q_M) in the studied RuTa alloys. The A-M transformation was very stable, and effects of the solution temperature, solution time, aging temperature, aging time, and thermal cycling on Ms, Af, hysteresis and transformation heat of this alloy were not strong.

ACKNOWLEDGEMENTS

The authors would like to thank Prof.T.Hirai and his laboratory members in Institute for Material Research (IMR),Tohoku University ,Sendai, Japan for many supports to do this experiment.

REFERENCES

1. R.Fonda et at, Final Abstract of ASM-TMS Materials Week'96 , (Oct.,1996, Cincinnati, USA),153(1996)

2. Y.Furuya,   Proc. 1st Japan-France Intelligent Materials and Structures Seminar (Sendai, '98) 113-122,(1998)

Mat. Res. Soc. Symp. Proc. Vol. 646 © 2001 Materials Research Society

First-Principles Calculations of Electronic Structure and Structural Properties for MoV, MoNb, and MoTa

R. de Coss, A. Aguayo, and G. Murrieta
Departamento de Física Aplicada, CINVESTAV-Mérida,
A.P. 73 Cordemex 97310, Mérida, MEXICO.

ABSTRACT

First-principles total-energy electronic structure calculations based on the full-potential linearized augmented plane wave (LAPW) method have been used to study the electronic and elastic properties of MoV, MoNb, and MoTa with the B2 (CsCl) estructure. From the calculated values for the bulk modulus we have determined the melting temperature using an empirical correlation. The chemical bond and the electronic structure around the Fermi level are analyzed. In particular, we found that MoTa which have the experimental determined highest melting point of the studied materials, present the largest bulk modulus and the highest degree of covalence bonding of these intermetallic compounds.

INTRODUCTION

Intermetallic compounds of bcc refractory transition metals (V, Cr, Nb, Mo, Ta and W) are of great technological interest because of their high hardness, high melting point and high corrosion resistance. The low ductility of the alloys with compositions in the central portion of the binary phase diagram, limits the application of these materials in the bulk [1]. However, recent studies show that also thin films of the molybdenum based alloys Mo-Cr [2], Mo-Ta [3], Mo-Nb [4] containing ~50% Mo give a high protection against corrosion. This discovery open up the possibility of applying this type of materials in a thin film configuration, were the low ductility is less of a problem. Structural characterization of the Mo-(Nb, Ta) alloys using x-ray diffraction shows that forms a bcc solid solution [3,4].

Purely experimental research, with the aim to understand and consequently improve the above mentioned technological properties, is however time consuming and expensive. It is possible to shorten this research with the aid of theoretical tools. Current *ab initio* density-functional calculations cannot adequately determine material properties at high temperature, but it is possible to compute the static lattice equation of state and elastic moduli of ordered binary compounds [5,6]. Known correlations between equilibrium properties and high-temperature properties such as the melting temperature (T_m) can then be used as a guide to steer the experimental research. For example, Friedel has shown that T_m of metals is strongly correlated with the cohesive energy and thus on the bonding properties in the ground state [7], while Fine et al. reported a correlation with the elastic moduli [8]. More recently, T_m of cubic metals was found to be correlated with the band energy [9].

As a first step to study the molybdenum based alloys, in this work we present a first-principles total-energy study of the electronic and structural properties for MoV, MoNb, and MoTa in the B2 phase (CsCl structure). From the results of our calculations we have obtained the heat of formation, the melting temperature, and an insight into the chemical bond of the intermetallic compounds in the ordered phase.

DETAIL OF CALCULATIONS

Self-consitent calculations of total energies and the electronic structure based on the scalar relativistic full-potential linearized augmented-plane-wave method (FP-LAPW) [10] were carried out using the WIEN97 code [11]. This is one of the most accurate schemes to solve the Kohn-Sham equations of Density Functional Theory. Here, exchange and correlations effects are treated by the generalized gradient approximation (GGA), which often leads to more accurate energetics and equilibrium structure than the local density approximation (LDA) [12]. The electron density is obtained by summing over all occupied Kohn-Sham orbitals and plays a key role in this formalism. The required precision in total energy ($\Delta E \leq 10^{-5}$ Ry) was achieved by using a large planewave cutoff of $RK_{MAX}=9.0$ and a k-point sampling in the Brillouin zone of 165 and 84 special k points in the irreducible wedge of the B2 and the bcc structures, respectively. The atomic sphere radii used were 2.0 a.u. for all the atoms (V, Nb, Mo, and Ta). For each of the studied metals and compounds, the full analysis was carried out at the calculated equilibrium volume based on the gradient generalized approximation for the exchange-correlation potential. We fitted the calculated total energies to the Murnaghan equation of state [13] to obtain the ground state energy, equilibrium volume, bulk modulus (B), and its pressure derivative (B').

The formation energy or heat of formation (H) of the MoV, MoNb, and MoTa was obtained by substracting the bulk total energies of the constituent elements (in bcc structure) from the total energy of the ordered compounds at equilibrium (in the CsCl phase). This procedure is justified since at zero temperature, the entropy contribution to the free energy becomes zero [14].

We have used our obtained bulk modulus to calculate the melting temperature for each compound. The melting temperature was calculated according an empirical correlation between the bulk modulus and the melting temperature described by Fine et al. [8]. For cubic materials they found that

$$T_m = 607 \text{ K} + (9.3 \text{ K/Gpa}) B \pm 500 \text{ K}, \qquad (1)$$

where B is the bulk modulus in Gpa and the error represent the standard deviation as determined from the fits of Ref. 8.

RESULTS AND DISCUSSION

In order to determine the accuracy of our calculations, we have calculated the structural properties of the elementary metals forming the ordered compounds MoV, MoNb, and MoTa. In the Table I we present the calculated lattice parameter and bulk modulus for V, Nb, Mo, and Ta in the bcc structure. From a comparison of the calculated values with the experimentals ones [15], we find root-mean-square (rms) errors of 0.55% and 5.43% for the lattice constant and bulk modulus, respectively. These values for the mrs errors in the lattice parameter and bulk modulus are around to that obtained in systematic studies for transition metals were also the GGA for the exchange correlation functional was used [16,17,18].

Table I Calculated and experimental values of the lattice parameter (a) and bulk modulus (B) for V, Nb, Ta, and Mo in the bcc structure, the experimental values are from Ref. 15.

Metal	a^{cal} (Å)	a^{exp} (Å)	B^{cal} (Gpa)	B^{exp} (Gpa)
V	2.99	3.03	187	162
Nb	3.31	3.30	169	170
Ta	3.31	3.30	204	200
Mo	3.16	3.15	262	272

In the Table II we present the calculated lattice parameter and bulk modulus for MoV, MoNb and MoTa in the B2 phase. The calculated lattice constants, which neglect zero-point motion and thermal expansion, are within 1% of the experimental lattice constant [19] for all of the compounds. We can not compare our calculated bulk modulus with experimental values, because these are not available, therefore our calculations are a prediction and can be used as a reference for future studies.

Table II Calculated values of the lattice parameter (a) and bulk modulus (B) for MoV, MoNb and MoTa in the B2 phase, the experimental values of the lattice parameter are from Ref. 19.

Material	a^{cal} (Å)	a^{exp} (Å)	B (Gpa)
MoV	3.073	3.092	217
MoNb	3.235	3.213	225
MoTa	3.232	3.234	234

The calculated heat of formation and melting temperature, and the experimental values of the melting temperature [19] are presented in the Table III. We can see that the difference between the predicted and experimental melting temperature is no greater in magnitude than the uncertainty. We also see a direct correlation between the melting temperature and the heat of formation of the alloys. These correlations are expected for an isoelectronic series, where the number of valence electrons in the compounds remains the same and the electronic structure are rather similar, in our case V, Nb, and Ta are isoelectronic therefore the three compounds have the same number of valence electrons.

Table III Calculated values of the heat of formation per formula unit (*H*) and melting temperature (*T_m*) for MoV, MoNb and MoTa in the B2 phase. The experimental values of melting temperature are from Ref. 19 and are also included for comparison.

Material	H (eV)	T^{cal} (K)	T^{exp} (K)
MoV	−0.30	2625±500	2623
MoNb	−0.33	2700±500	2833
MoTa	−0.46	2783±500	3008

In Fig. 1 we show the calculated Density of States (DOS) for the intermetallic compounds MoV, MoNb, and MoTa in the B2 (CsCl) phase, the zero of the energy axes is at the Fermi level E_r. The electronic structure are rather similar for the three compounds, which consists of two sets of peaks separated by a deep minimum (pseudogap) that lies above the Fermi level. This minimum separates the bonding from the antibonding states. This behavior is indicative of hybridization between the orbitals on the Mo site and the V (Nb, Ta) site.

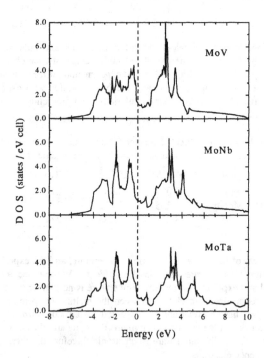

Figure 1. *Electronic density of states (DOS) for the MoV, MoNb, and MoTa intermetallic compounds in the B2 (CsCl) phase. The Fermi level is represented by the dotted vertical line.*

The similitude in the DOS for the three ordered compounds is a consequence of the fact that V, Nb, and Ta have the same number of valence electrons, as was mentioned above. However, if we look at the fine details, we can see that the bandwidth and therefore the pseudogap increases from MoV to MoTa. This behavior is a consequence of the gradual delocalization of the d-electrons from the $3d$-metal (V) to $5d$-metal (Ta), through the $4d$-metal (Nb). Also is interesting to note is that inside the pseudogap gradually appears a second minimum, which reaches the lower value in MoTa. This "quasigap" appearing in the DOS near E_r in MoTa indicates that there is a strong covalent contribution to the bonding and that the crystalline structure is very stable. This behaviour is often seen in compounds with a high stability, for example $MoSi_2$ [20]. The trends in the electronic structure and bonding properties of the studied compounds correlates with the calculated bulk modulus, heat of formation, and melting temperature (see Table II and III).

CONCLUSIONS

The calculated equilibrium lattice constants for MoV, MoNb, and MoTa are all within 1% of the experimentally determined values. We found that calculated melting temperatures obtained from the calculated bulk modulus using an empirical rule are in good agreement with the experimental values. The bandwidth and therefore the bonding strength increase in the studied compounds from MoV to MoTa. We find that MoTa present the more covalent bond of the studied intermetallic compounds.

ACKNOWLEDGMENTS

The authors would like to thank Alejandro Díaz-Ortiz for usefull discussions and Jaap Vente for critical review of the manuscript. Two of the authors (A.A. and G.M.) gratefully acknowledges a student fellowship from Consejo Nacional de Ciencia y Tecnología (CONACYT, México). Research supported by CINVESTAV through the JIRA program under Grant. No. 97/08 and CONACYT under Grant. No. 34501-E.

REFERENCES

1. L.I. van Torne amd G. Thomas, *Acta Mettalurgica* **14**, 621 (1966).

2. P.Y. Park, E. Akiyama, A. Kawashima, K. Asami, and K. Hashimoto, *Corros. Sci.* **37**, 1843 (1995).

3. P.Y. Park, E. Akiyama, A. Kawashima, K. Asami, and K. Hashimoto, *Corros. Sci.* **38**, 397 (1996).

4. P.Y. Park, E. Akiyama, H. Habazaki, A. Kawashima, K. Asami, and K. Hashimoto, *Corros. Sci.* **38**, 1371 (1996).

5. M.J. Mehl, B.M. Klein, and D.A. Papaconstantopoulos, in *Intermetallic Compounds: Principles and Practice*, ed- J.H. Weatbrook and R.L. Fleisher (Wiley, London, 1994), vol.1, chapter 9.

6. M.J. Mehl, J.E. Osburn, D.A. Papaconstantopoulos, and B.M. Klein, *Phys. Rev. B* **41**, 10311 (1990); erratum *ibid.* **42**, 5362 (1991).

7. J. Friedel, in *The Physics of Metals. I. Electrons*, ed. J.M. Ziman (Cambridge University Press, UK, 1969), p. 340.

8. M.E. Fine, L.D. Brown, and H.L. Marcus, *Scr. Metall.* **18**, 951 (1984).

9. R. de Coss, A. Aguayo, and G. Canto (to be published); A. Aguayo, M.Sc. thesis, Department of Applied Physics, CINVESTAV, México (1997).

10. D.J. Singh, *Planewaves, Pseudopotentials, and the LAPW Method* (Kluwer Academic, Boston, 1994).

11. P. Blaha, K. Schwarz, and J. Luitz, WIEN97, Vienna University of Technology 1997. [Improved and updated Unix version of the original copyrighted WIEN code, which was publisehd by P. Blaha, K. Schwarz, P. Sorantin, and S.B. Trickey, *Comp. Phys. Commun.* **59**, 339 (1990).]

12. J.P. Perdew, K. Burke, and M. Ernzerhof, *Phys. Rev. Lett.* **77**, 3865 (1996).

13. F.D. Murhagham, Proc. Natl. Acad. Sci. USA **30**, 244 (1944).

14. C. Ravi, P. Vajeeston, and R. Asokami, *Phys. Rev. B* **60**, 15683 (1999).

15. C. Kittel, *Introduction to Solid States Physics* (Wiley, New York, 1986).

16. V. Ozolins and M. Körling, *Phys. Rev. B* **48**, 18304 (1993).

17. A. Khein, D.J. Singh, and C.J. Umrigar, *Phys. Rev. B* **51**, 5105 (1995).

18. S. Kurth, J.P. Perdew, and P. Blaha, *Int. J. Quantum Chem.* **75**, 889 (1999).

19. W.B. Pearson, *The Crystal Chemistry and Physics of Metals and Alloys* (John Wiley & Sons, New York, 1972).

20. M. Alouani and R.C. Albers, *Phys. Rev. B* **43**, 6500 (1991).

Mat. Res. Soc. Symp. Proc. Vol. 646 © 2001 Materials Research Society

GROWING COBALT TRIANTIMONIDE USING VERTICAL BRIDGMAN METHOD AND EFFECTS OF POST ANNEALING

M. Akasaka, G. Sakuragi, H. Suzuki, T. Iida, and Y. Takanashi
Department of Materials Science and Technology, Science University of Tokyo 2641 Yamazaki, Noda-shi, Chiba 278-8510 Japan

S. Sakuragi
Union Material Inc., 1640 Oshido-jyoudai, Tone-Machi, Kitasouma, Ibaraki 300-1602, Japan

ABSTRACT

Crystals of $CoSb_3$ were grown using the vertical Bridgman method at growth rates that varied from 0.4 to 2.8 mm/h. Thermoelectric properties were analyzed for both as-grown and post-annealed samples. Polycrystalline $CoSb_3$ surrounded by Sb was obtained. Samples grown at the rate of 0.4 mm/h had larger $CoSb_3$ grains than samples grown at the 2.8 mm/h rate. For the as-grown samples, the Seebeck coefficient was smaller than 200 μ/K, which is a nominal value [1-3]. The presence of residual Sb resulted in a decrease in the Seebeck coefficient and an increase in the samples' electrical conductivity. A subsequent heat treatment at 800 °C for 20 h eliminated the residual Sb, resulting in a significant increase in the Seebeck coefficients (ranging from > 200 $\mu V/K$) in the annealed samples, as compared with the as-grown samples. The samples with a higher growth rate had larger Seebeck coefficients of ~500 μ/K after annealing.

INTRODUCTION

Cobalt triantimonide ($CoSb_3$), a peritectic-skutterudite compound, has been identified as a promising alloy for advanced thermoelectric materials [1-4]. For high conversion efficiency of thermoelectric devices with $CoSb_3$, a high thermoelectric property expressed by figure-of-merit $Z = S^2 \sigma / \kappa$ is needed, where S is the Seebeck coefficient, σ is the electrical conductivity and κ is the thermal conductivity. We used the conventional vertical Bridgman method, which is a promising method for growing high-quality crystalline bulk $CoSb_3$ [4-5]. In this paper, we describe the preparation of $CoSb_3$ crystals and refinement of the grown crystals by subsequent heat treatment. The quality of the grown crystals is expressed as a function of the growth parameters. The Seebeck coefficient increased to as high as 500 μ/K as a result of the heat treatment.

EXPERIMENTAL

The vertical Bridgman method was used to grow $CoSb_3$ crystals. The compound $CoSb_3$ exhibits a peritectical point at 876 °C and can be initiated from Sb-rich melt about 90 at% [6]. The source materials of 10 at% Co (99.99%) powder and 90 at% Sb (99.9999%) grains were charged into a quartz ampoule which was sealed under vacuum. Prior to loading the ampoule into the electric furnace, the parts were mixed at 1100 °C for 20 h to obtain a homogeneous melt. The growth rate varied from 0.4 to 2.8 mm/h with no seed. The axial temperature range was approximated by a linear gradient of 85 °C/cm near the melt-solid interface for a control-set point of 876 °C. The grown ingot was sliced, then polished to form wafers, 0.6 to 1.0 mm thick. Some samples were subsequently annealed in an ampoule sealed under vacuum at 800 °C for 20 h to remove the Sb from the as-grown samples. An x-ray-diffraction (XRD) analysis was performed using CuKa- radiation. The microstructure of the grown samples was analyzed using an optical microscope, and their composition was determined by electron-probe microanalysis (EPMA). The Seebeck coefficient and electrical conductivity at room temperature were analyzed.

RESULTS AND DISCUSSION

CRYSTAL GROWTH

The growth parameters observed in the as-grown samples are presented in Table 1. For sample C, the tip part of the ingot was $CoSb_2$, from which $CoSb_3$ was formed subsequently. At the interface, Sb and $CoSb_3$ were obtained for all the samples except sample C, with the lower part of the ingot corresponding to the skutterudite compound and the upper part corresponding to a Sb-rich eutectic. The size of the $CoSb_3$ grains increased when the diameter of the grown ingot was increased. The $CoSb_3$ alloy was clearly present. The EPMA results are shown in Figure 1. For samples A, B, and D, the lower part composition was close to the Co : Sb ratio of 1 : 3, although it contained some fragments of Co that came from the boundary between the $CoSb_3$ and Sb.

Table 1. Growth parameters.

Sample	A	B	C	D
Composition ratio (Co : Sb)	1.0 : 9.0	1.0 : 9.0	1.0 : 9.0	1.0 : 9.0
Growth rate (mm/h)	2.8	1.4	0.6	0.4
Axial temperature gradient (°C/cm)	85	85	85	85

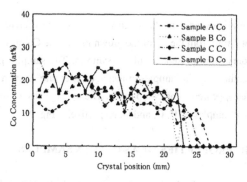

Fig. 1. Co concentration as a function of position in the crystal.

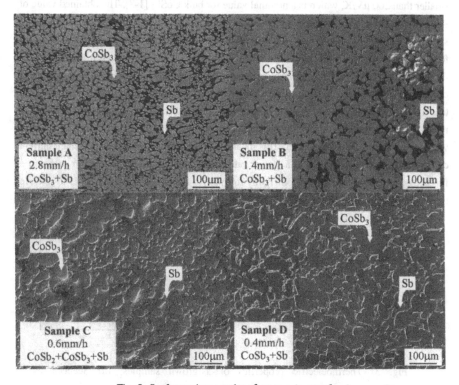

Fig. 2. Surface micrographs of as-grown samples.

Figure 2 shows surface micrographs of the as-grown samples at different growth rates. The samples were cut about 7 mm from the tip of the grown ingot. The $CoSb_3$ grains and Sb along the $CoSb_3$-grain boundary were present in all the samples. The samples grown at a slower rate had larger $CoSb_3$ grains than the samples grown at a faster rate. In terms of the amount of residual Sb, the samples appeared to be comparable. The average size of the grains was 4, 6, 12, and 21×10^{-3} mm^2 for samples A, B, C, and D, respectively (Fig. 2).

THERMOELECTRIC PROPERTIES OF AS-GROWN SAMPLES

Figures 3 shows the Seebeck coefficient and electrical conductivity at room temperature of the as-grown samples. For samples A, B, and C, the Seebeck coefficients were smaller than 200 µV/K, which is a nominal value for bulk $CoSb_3$ [1-3]. The obtained value of ~40 µV/K in these samples appeared to be affected by the residual Sb. Because the tip part of sample D had large $CoSb_3$ grains with a small amount of residual Sb, the Seebeck coefficient for sample D (90 µV/K) was larger than that in the other samples, although it was still lower than the nominal value. The presence of residual Sb resulted in a decrease in the Seebeck coefficient and an increase in electrical conductivity of the samples. In sample D, the size of the grains in the middle part was comparable to that of the grains in the tip part and was relatively larger than that of the grains of the other samples. However, the Seebeck coefficient was smaller than that in the tip part. This means that the residual Sb is a predominant current carrier path. Growing crystals at a slower rate might help curtail the formation of Sb in a furnace of the present design. Another way to improve the Seebeck coefficient is to remove the residual Sb by post-annealing.

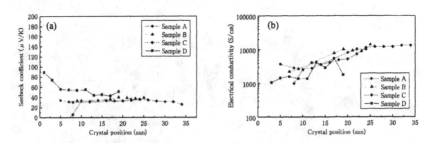

Fig. 3. Thermoelectric properties of as-grown samples; (a) Seebeck coefficient, (b) electrical conductivity.

POST ANNEALING

Figure 4 shows surface micrographs of sample A, both as-grown and annealed. During the heat treatment of the sample at 800 °C for 20 h in a vacuum ampoule, Sb evaporated from the sample, resulting in the formation of voids, and CoSb₃ grains agglomerated. The XRD revealed Sb peaks in the as-grown samples and no peaks in the annealed samples. The Co : Sb composition ratio with regard to the depth from the surface was determined by electron-probe microanalysis to be close to 1 : 3, with a slight shift towards Co, which can be explained by decomposition of a small amount of CoSb₃ near the boundary as a result of the annealing.

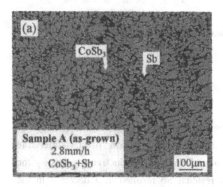 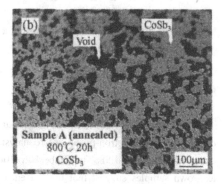

Fig. 4. Surface micrographs of annealed samples; (a) before annealing, (b) after annealing. Void was formed after Sb evaporation.

THERMOELECTRIC PROPERTIES OF ANNEALED SAMPLES

Figures 5 shows the Seebeck coefficient and electrical conductivity of the annealed samples at room temperature. Compared with the as-grown samples, all the annealed samples exhibited a significant increase in the Seebeck coefficients, which ranged from 200 to 500 µ/K. The samples grown at a higher rate had larger Seebeck coefficients (~500 µV) than the samples grown at a slower rate. These coefficients were also much larger than the nominal value of the bulk specimen. The Seebeck coefficient for the annealed sample A was more than twice as large as that for sample D, although the grain size of sample A in the as-grown condition was, on average, about one fifth that of sample D. It seems likely that such a large increase in the Seebeck coefficient is closely connected with the disappearance of the residual Sb and coalescence of the CoSb₃ grains as a result of the heat treatment. After the evaporation of the residual Sb, the current carrier path may have become dominated by the reformed CoSb₃. The electrical conductivity results showed that the post annealing at 800 °C for 20 h was not effective in removing the Sb in all the samples: it was still present in sample D after annealing. The lower Seebeck coefficient in the annealed sample D can be possibly explained by insufficient annealing that failed to remove the residual Sb surrounding the CoSb₃ grains and prevented grain coalescence.

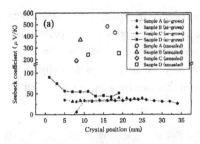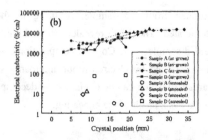

Fig. 5. Thermoelectric properties of annealed samples; (a) Seebeck coefficient, (b) electrical conductivity.

CONCLUSION

Crystals of $CoSb_3$ were grown using the vertical Bridgman method. X-ray diffraction and electron-probe microanalysis showed the formation of a skutterudite compound $CoSb_3$ and Sb. The Seebeck coefficient and electrical conductivity results for the as-grown samples suggest that the presence of residual Sb resulted in a decrease in the Seebeck coefficient and an increase in electrical conductivity of the samples. Post annealing removed the residual Sb and improved the Seebeck coefficient considerably.

REFERENCES

[1] H.Anno, T.Sakakibara, Y.Notohara, H.Tashiro, T.Koyanagi, H.Kaneko, and K.Matsubara, Proc. the 16 th Int. Conf. on Thermoelectrics, p.338 (1997)

[2] D.T.Morelli, T.Caillat, J.-P.Fleurial, A.Borshchevsky, J.Vandersande, B.Chen, and C.Uher, Phys. Rev. B, **51**, p.9622 (1995)

[3] D.Mandrus, A.Migliori, T.W.Darling, M.F.Hundley, E.J.Peterson, and J.D.Thomson, Phys. Rev. B, **52**, p.4926 (1995)

[4] J.-P. Fleurial, T.Caillat, A.Borshchevsky, J. Crystal. Growth. **166**, p.722-726 (1996)

[5] E. Monberg, Handbook of Crystal Growth, Vol.2, ed. D.T.J.Hurle, p.53 (1994)

[6] K.Ishiba and T.Nishizawa, Binary Alloy Phase Diagrams, Second Edition, ed. T.B.Massalski, P.1234 (1990)

Mat. Res. Soc. Symp. Proc. Vol. 646 © 2001 Materials Research Society

Microstructure and Plastic Deformation in Unidirectionally Solidified NbSi$_2$ (C40) /MoSi$_2$ (C11$_b$) Crystals

Takayoshi Nakano, Yasuhiro Nakai and Yukichi Umakoshi
Department of Materials Science and Engineering, Graduate School of Engineering, Osaka University, 2-1, Yamada-oka, Suita, Osaka 565-0871, Japan.

ABSTRACT

Microstructure and plastic deformation behavior in duplex-phase silicides composed of the C40 and C11$_b$ phases were examined using pseudo-binary (Nb,Mo)Si$_2$ crystals with a single set of lamellae. Single crystals of the C40 single-phase were grown from the master ingot with a composition of (Nb$_{0.15}$Mo$_{0.85}$)Si$_2$ by a floating zone method and duplex-phase microstructure containing a single set of lamellae was obtained by subsequent heat treatment at 1400°C for no less than 6h. During the heat treatment, the C11$_b$ phase was precipitated from the C40 matrix by satisfying the crystallographic relationship of $(0001)_{C40}//(110)_{C11b}$, $<1\bar{2}\bar{1}0]_{C40}//[1\bar{1}0]_{C11b}$ and $<10\bar{1}0]_{C40}//[001]_{C11b}$ at the lamellar boundary, while randomly oriented C11$_b$ grains also appeared at further annealing. As a result, the duplex-phase silicides with a single set of lamellae contained the C40 phase with a single orientation and three variants of the C11$_b$ phase. The lamellar spacing and the volume fraction of their phases depended strongly on annealing time.

In compression tests, yield stress and fracture strain of the duplex-phase silicides depended strongly on angle (ϕ) between the loading axis and lamellar boundaries, similar to TiAl-PST crystals. At $\phi=0°$, specimens fractured just after showing high fracture stress even at 1400°C. In contrast, at $\phi=45°$ where shear deformation in the C11$_b$ phase of lamellae occurred parallel to lamellar boundaries, low yield stress and significant plastic strain were achieved at 1400°C.

INTRODUCTION

Refractory metal silicides with the C11$_b$, C40 and C54 structures have been of great interest for high-temperature structural applications, but their low-temperature toughness and high-temperature strength should be improved before the industrial use [1]. MoSi$_2$ with the C11$_b$ structure is the most promising phase as a relatively more plastically deformable matrix as compared to C40 silicides due to operation of several slip systems even at low temperatures, but the high-temperature strength above 1200°C is not sufficient independent of addition of the third elements [2-5]. In contrast, NbSi$_2$ with the C40 structure has a limited slip system of $(0001)1/3<2\bar{1}\bar{1}0]$ due to its low crystal symmetry, but promotes anomalous strengthening around 1500°C by addition of the C11$_b$-stabilized elements [6]. As a result, a MoSi$_2$/NbSi$_2$ composite may, potentially, have a good balance of mechanical properties [7,8]. Moreover, a peculiar lamellar structure composed of the C40/C11$_b$ duplex-phase has been found in NbSi$_2$/MoSi$_2$ pseudo-binary system due to their similar atomic arrangement on $(110)_{C11b}$ and $(0001)_{C40}$ [8].

Similar lamellar structure was reported in Ni-base super-alloys [9], Ti-rich TiAl compounds [10] and so on. The lamellae were aligned by special techniques during crystal growth for industrial application in Ni-base superalloys and for understanding fundamental aspects of lamellar structure on mechanical properties in TiAl-PST crystals [11-13].

In this study, the C40/C11$_b$ duplex-phase silicides with a single set of lamellae were obtained by the floating zone method and subsequent heat treatment. This is the first report on microstructure and its related plastic deformation behavior in the silicides with oriented lamellae focusing on anisotropy of lamellar structure.

EXPERIMENTAL DETAILS

Crystals with a composition of $(Nb_{0.15}Mo_{0.85})Si_2$ were grown by a floating zone method at different rates of 2.5, 5.0 and 10.0mm/h under a high purity argon gas flow. Constituent phase and its lattice parameters were determined by the XRD method. The crystals grown at 2.5mm/h were annealed at 1400°C for various time to 168h. Microstructure was observed in a scanning electron microscope after polishing mechanically with diamond paste and subsequent etching in a 20vol.% nitric acid/ 40vol.% hydrochloric acid/ 40vol.% hydrofluoric acid solution.

Rectangular specimens with $2 \times 2mm^2 \times 5mm$ were cut from the annealed crystals. Loading axes were selected to be 0° and 45° from [10$\bar{1}$0] on [1$\bar{2}$10] zone in the C40 structure, which correspond to the angle (ϕ) between the loading axis and lamellar boundaries in duplex-phase silicides with a single set of lamellae as schematically drawn in Fig.4. Compression tests for specimens annealed at 1400°C for 1, 6, 24 and 168h were carried out at 1400°C at a strain rate of $1.67 \times 10^{-4}s^{-1}$. Thin foils for TEM observation were prepared from specimens before and after deformation and then observed in a Hitachi H-800 transmission electron microscope operated at 200kV.

RESULTS

Growth of C40 single-phase single crystals and subsequent formation of duplex-phase silicides with a single set of lamellae

Figure 1 shows XRD profiles from powders of $(Nb_{0.15}Mo_{0.85})Si_2$ silicides grown by the FZ method at different rates of 2.5, 5.0 and 10.0mm/h. The relative intensity at diffractions of the $C11_b$ phase to the C40 phase decreases with decreasing crystal growth rate and then the $C11_b$ phase disappears at 2.5mm/h. Duplex-phase silicides grown at 10.0 and 5.0mm/h contain the $C11_b$ primary phase in addition to the C40 single-phase grain and lamellar colony as polycrystalline form, while the C40 single-phase silicides obtained at 2.5mm/h have a single orientation in the entire rod. The solidification process at the different growth rates is discussed later in this article.

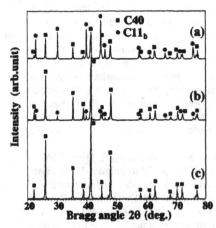

Figure 1
Variation in constituent phase of $(Nb_{0.15}Mo_{0.85})Si_2$ silicides grown by the FZ method at different crystal growth rates;(a) 10.0mm/h, (b) 5.0mm/h and (c) 2.5mm/h.

The C11$_b$ phase appears forming oriented lamellar structure in the C40 single-phase single crystals on the basis of phase transformation from the C40 to the C11$_b$ during annealing at 1400℃ for no less than 6h as shown in fig.2, while the specimen annealed at 1400℃ for 1h maintains the C40 single phase. The volume fraction of the C11$_b$ phase increases with annealing time, and finally saturated to about 57% around 24h. Each interspacing of the C40 and C11$_b$ phase in lamellae is 0.96μm and 1.32μm, respectively, in the duplex-phase silicides annealed for 24h and increases with annealing time due to growth of each phase. Almost all boundaries are parallel and flat, but colonies surrounded by rounded boundaries appear and its volume fraction increases with increasing annealing time.

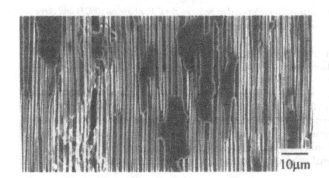

Figure 2
Oriented lamellar structure in (Nb$_{0.15}$Mo$_{0.85}$)Si$_2$ silicide grown by the FZ method and subsequent heat treatment at 1400 ℃ for 24h.

10μm

Crystallographic relationship of lamellar structure and C11$_b$ variants in the duplex-phase silicides with a single set of lamellae

According to analysis of selected area electron diffraction patterns (SAEDs) in the TEM, lamellar boundaries satisfy the following crystallographic relationship,

$$(0001)_{C40} // (110)_{C11_b}, \quad <\bar{1}2\bar{1}0]_{C40} // [1\bar{1}0]_{C11_b}, \quad <10\bar{1}0]_{C40} // [001]_{C11_b} \quad (1)$$

A similar relationship was reported on the lamellar boundary between the C40 and C11$_b$ in polycrystalline silicides in pseudo-binary systems of TaSi$_2$/MoSi$_2$ [7,14], CrSi$_2$/MoSi$_2$ [15], NbSi$_2$/MoSi$_2$ [8] and TiSi$_2$/MoSi$_2$ [16]. The atomic arrangement on $(0001)_{C40}$ and $(110)_{C11b}$ parallel to the lamellar boundary is quite same if the slight lattice distortion of the C11$_b$ structure is neglected, while periodicity of the stacking sequence on their planes is different; three-fold periodicity in the C40 phase and two-fold periodicity in the C11$_b$ phase. Therefore, the peculiar lamellar boundary must be produced during the phase transformation from the C40 to C11$_b$ accompanied by motion of ledges containing 4-types partial dislocations parallel to the interface [8].

Since the [110] axis in the C11$_b$ structure and the [0001] axis in the C40 structure are 6-fold and 2-fold rotation axis, respectively, three variants defined as variant 1 (V1), variant 2 (V2) and variant 3 (V3) of the C11$_b$ phase are produced from the C40 single crystal as follows.

$$V1: \quad (0001)_{C40} // (110)_{C11_b}, \quad [\bar{1}2\bar{1}0]_{C40} // [1\bar{1}0]_{C11_b}, \quad [10\bar{1}0]_{C40} // [001]_{C11_b} \quad (2)$$

$$V2: \quad (0001)_{C40} // (110)_{C11_b}, \quad [2\bar{1}\bar{1}0]_{C40} // [1\bar{1}0]_{C11_b}, \quad [0\bar{1}10]_{C40} // [001]_{C11_b} \quad (3)$$

$$V3: \quad (0001)_{C40} // (110)_{C11_b}, \quad [\bar{1}\bar{1}20]_{C40} // [1\bar{1}0]_{C11_b}, \quad [\bar{1}100]_{C40} // [001]_{C11_b} \quad (4)$$

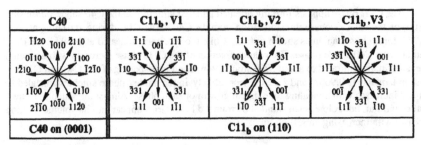

C40	C11$_b$, V1	C11$_b$, V2	C11$_b$, V3
C40 on (0001)	C11$_b$ on (110)		

Figure 3 *Crystallographic orientation relationship between the C40 phase on (0001) and three variants (V1, V2 and V3) of the C11$_b$ phase on (110) at lamellar boundaries.*

Figure 3 shows the above relation schematically. Since each variant is rotated by 120° around [110]$_{C11b}$ on the (0001)$_{C40}$ plane, the 120°-rotation boundary is produced at the C11$_b$/C11$_b$ lamellar boundary such as V1/V2, V2/V3 and V3/V1.

In contrast, the colony of the C11$_b$ phase does not satisfy equations (2)~(4) and develops accompanied by absorption of the C11$_b$/C11$_b$ lamellar boundary containing boundary dislocations. Since the phase transformation from the C40 single-phase to the duplex-phase induces internal strain near the lamellar boundary, the colony may be generated and then grow with annealing time to decrease the internal strain and lamellar boundaries. This suggests that additional elements for improving thermal stability of lamellar structure need to be assessed.

Anisotropy of plastic deformation behavior in silicides with a single set of lamellae

Plastic deformation behavior in the C40 single-phase single crystal and the C40/C11$_b$ duplex-phase silicides containing oriented lamellae was examined in compression at 1400°C. Figure 4 shows yield or fracture stress in (Nb$_{0.15}$Mo$_{0.85}$)Si$_2$ crystals grown at 2.5mm/h deformed at ϕ=0° and 45° at 1400°C as a function of annealing time. At ϕ=0° where the loading axis is parallel to lamellar boundaries, fracture occurs before macro-yielding, but the fracture stress is very high around 1000MPa. The fracture stress rapidly decreases when the C11$_b$ appears forming the lamellar after annealing for 6h. Thus, micro-yielding in the C11$_b$ phase may occur at a lower stress than the fracture stress of single-phase C40. However, lack of slip continuity across the C40/C11$_b$ interface may result in fracture, and hence the fracture stress of C40/C11$_b$ composites

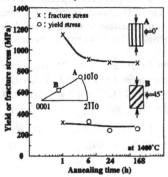

Figure 4

Orientation and annealing time dependence of yield or fracture stress in (Nb$_{0.15}$Mo$_{0.85}$)Si$_2$ crystals grown at 2.5mm/h deformed at ϕ=0° and 45° at 1400 °C.

is lower than that of single-phase C40.

At $\phi=45°$, fracture also occurred without yielding in the C40 single-phase single crystal annealed for 1h. The Schmid factor for $(0001)1/3<2\bar{1}\bar{1}0]$ slip corresponding to the easy mode in the C40 structure is 0.433, but no plastic strain was detected for the C40 single-phase single crystal.

In contrast, formation of lamellar structure makes the crystals deformable and significant plastic strain was obtained showing low yield stress around 300MPa in the duplex-phase silicides annealed for no less than 6h. This means that the precipitated $C11_b$ phase easily deforms in the duplex-phase at $\phi=45°$. Indeed, <100] and 1/2<111] dislocations which can operate in $MoSi_2$ at high temperatures were dominantly observed in the $C11_b$ variants of duplex-phase silicides [2,4].

DISCUSSION
Solidification process by the FZ method

In $TaSi_2$-$MoSi_2$ [7] and $NbSi_2$-$MoSi_2$ [8] pseudo-binary systems, the C40 phase near duplex-phase is well known to be formed by a peritectic reaction from the liquid phase and primary $C11_b$ phase, and the Nb concentration of the peritectic point is lower than that in $(Nb_{0.15}Mo_{0.85})Si_2$ as shown in fig.5. In general, formation of the primary $C11_b$ phase prevents the C40 single-phase from the growth of single crystal. The silicides grown at 5.0 and 10.0mm/h therefore were not single crystals because the temperature at the interface between the molten portion and the grown crystal is over the temperature for the peritectic reaction. In contrast, when the composition of molten portion is kept to be equilibrium with the bulk composition of $(Nb_{0.15}Mo_{0.85})Si_2$ by the FZ method and the difference in composition is maintained during the crystal growth, the C40 single-phase single crystals can be obtained as seen in the crystals grown at 2.5mm/h.

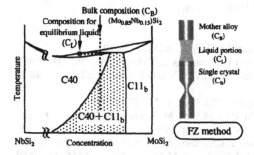

Figure 5

The schematic drawing showing equilibrium composition of the liquid and the C40 phase in $MoSi_2$-$NbSi_2$ pseudo-binary phase diagram during crystal growth by the FZ method.

Anisotropy of deformation in duplex-phase silicides in comparison with TiAl-PST crystals

The anisotropy of deformation behavior in the duplex-phase silicides is similar to that in TiAl-PST crystals with a single set of lamellae composed of six γ variants and the α_2 thin plate. When shear deformation parallel to lamellar boundaries occurs at $\phi=45°$, significant plastic strain is achieved accompanied by low yield stress in TiAl-PST crystals [11-13]. In contrast, crystals at $\phi=0°$ show the higher yield stress due to resistance of lamellar boundary to the dislocation motion.

At $\phi=0°$, moderate plastic strain appeared before failure in TiAl-PST crystals at room temperature, but fracture occurred without yielding in the duplex-phase silicides with oriented lamellae even at 1400℃. This is because the $\{10\bar{1}0\}1/3<1\bar{2}10]$ prism slip in the α_2 phase can be operative for maintaining the macroscopic strain continuity at the lamellar interface in the TiAl-PST crystals, but no slip system for the strain continuity at the C40/C11$_b$ interface appears in the C40 phase in the silicides. New four types of slip system of $\{10\bar{1}0\}1/3<1\bar{2}10]$, $\{10\bar{1}1\}1/3<1\bar{2}10]$, $\{10\bar{1}0\}<0001]$ and $\{11\bar{2}Y\}<11\bar{2}X>$ (XxY=-6) in addition to the easy slip mode of $(0001)1/3<2\bar{1}\bar{1}0]$ were recently confirmed near the crack path in NbSi$_2$ deformed at 1000°C [16]. Therefore, an alloying design to activate these systems should be necessitated for the further improvement of plastic behavior in duplex-phase silicides.

CONCLUSIONS

Lamellar structure and deformation behavior in duplex-phase silicides with a single set of lamellae were examined and the following conclusions were reached.

(1) The duplex-phase silicides with a single set of lamellae were obtained by the FZ method at a crystal growth rate of 2.5mm/h and subsequent heat treatment at 1400°C from the (Nb$_{0.15}$Mo$_{0.85}$)Si$_2$ crystal. The lamellar structure composed of the C40 phase with a single orientation and three variants of the C11$_b$ phase satisfies the crystallographic relationship of $(0001)_{C40}//(110)_{C11b}$, $<\bar{1}2\bar{1}0]_{C40}//[1\bar{1}0]_{C11b}$ and $<10\bar{1}0]_{C40}//[001]_{C11b}$ at the lamellar boundaries.

(2) Yield stress or fracture strain of the duplex-phase silicides at 1400°C depended strongly on angle (ϕ) between the loading axis and lamellar boundaries, similarly to TiAl-PST crystals. Specimens compressed at $\phi=0°$, fractured just after showing high fracture stress, while low yield stress and moderate plastic strain were obtained at $\phi=45°$.

ACKNOWLEDGMENT

This work was supported by a Grant-in-Aid for Scientific Research Development from the Ministry of Education, Science, Sports and Culture of Japan.

REFERENCES

1. J. J. Petrovic and A. K. Vasudevan, Mater. Sci. Engng., **A261**, 1 (1999).
2. Y. Umakoshi, T. Sakagami, T. Hirano and T. Yamane, Acta Metall. Mater., **38**, 909 (1990).
3. S. A. Maloy, A. H. Heuer, J. J. Lewandowski and T. E. Mitchell, Acta Metall. Mater., **40**, 3159 (1992).
4. K. Ito, H. Inui, Y. Shirai and M. Yamaguchi, Plil. Mag. A, **72**, 1075 (1995).
5. K. Ishikawa, H. Inui and M. Yamaguchi, PRICM 3, ed. M. A. Imam, R. DeNale, S. Hanada, Z. Zhong and D. N. Lee (Warrendale, PA,TMS, 1998) pp.2455.
6. T. Nakano, M. Kishimoto, D. Furuta and Y. Umakoshi, Acta Mater., **48**, 3465 (2000).
7. W. J. Boettinger, J. H. Perepezko and P. S. Frankwicz, Mater. Sci. Engng, **A155**, 33 (1992).
8. T. Nakano, M. Azuma and Y. Umakoshi, Intermetallics, **6**, 715 (1998).
9. E. R. Thompson and F. D. Lemkey, Transactions of the ASM, **62**, 140 (1969).
10. H. A. Lipsitt, D. Schechtman and R. E. Schafrik, Metall. Trans. A, **6**, 1991, (1975).
11. H. Inui, A. Nakamura, M. H. Oh and M. Yamaguchi, Acta Metall. Mater., **40**, 3095 (1992).
12. M. Yamaguchi, H. Inui, Structual Intermetallics, ed. M. V. Nathal, R. Darolia, C. T. Liu, P. L. Martin, D. B. Miracle, R. Wagner and M. Yamaguchi (Warrendale, PA,TMS, 1993) pp.305.
13. Y. Umakoshi and T. Nakano, Acta Metall. Mater., **41**, 1155 (1993).
14. F.-G. Wei, Y. Kimura and Y. Mishima, private communication.
15. Z. H. Lai, D. Q. Yi and C. H. Li, Scripta Metall. Mater., **32**, 1789 (1995).
16. T. Nakano, M. Azuma, S. Maeda and Y. Umakoshi, unpublished data.

Mat. Res. Soc. Symp. Proc. Vol. 646 © 2001 Materials Research Society

C11$_b$/C40 Lamellar Structure in a Ta-added Molybdenum Disilicide

Fu-Gao Wei, Yoshisato Kimura and Yoshinao Mishima
Department of Materials Science and Engineering, Tokyo Institute of Technology, 4259 Nagatsuta, Midori-ku, Yokohama 226-8502, Japan

ABSTRACT

C11$_b$/C40 fully lamellar microstructures, similar to the well-known TiAl/Ti$_3$Al lamellae, were obtained in Ta- and Nb-added MoSi$_2$ polycrystalline alloys in a previous work. In the present study, the crystallography of the lamellar structure is investigated in a MoSi$_2$-15mol%TaSi$_2$ pseudo-binary alloy after homogenized at 1400°C for 168h, in order to provide some useful parameters for microstructural control to improve mechanical properties. The orientation relationship between C11$_b$ and C40 phases and its three distinct variants were identified. Coherency of the lamellar interface is analyzed in comparison with the TiAl/Ti$_3$Al lamellae. Approach to modify the C11$_b$/C40 lamellar microstructure to increase its coherency is discussed based on the results obtained.

INTRODUCTION

The poor room temperature ductility and low high-temperature strength of molybdenum disilicide (MoSi$_2$) may be improved by making multi-phase. Advantage of selecting a C40-type disilicide as a counterpart lies in three major aspects: (1) both disilicides are thermodynamically stable up to about 1900°C, (2) addition of only a small amount of elements such as Nb, Ta and Ti, known to stabilize C40 structure, is sufficient to transform C11$_b$ structure into C40 structure, and (3) C11$_b$ and C40 structures, as shown in Fig.1, are closely correlated and may be designed to form a coherent C11$_b$/C40 lamellar structure. Significance of a coherent or zero-misfit two-phase alloy had been recognized in γ/γ' nickel-based superalloy [1], β/β' (NiAl/Ni$_2$AlTi) [2], and γ/α_2 (TiAl/Ti$_3$Al) [3] two-phase alloy because low misfit or high coherency generally gives a thermodynamically stable microstructure and a better performance in elevated temperature mechanical properties.

A recent investigation by the present authors [4] showed that besides the misfit factor, solidification process also plays an important role in forming a lamellar structure. Although CrSi$_2$ possesses a lattice misfit of less than 0.5% with MoSi$_2$, no lamellar structure is available in the MoSi$_2$-CrSi$_2$ system by normal solidification and heat treatment. The reason

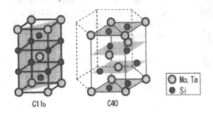

Fig.1 C11$_b$ and C40 crystal structures.

for the absence of lamellae is explained by the fact that $MoSi_2$ solidifies from the liquid and $CrSi_2$ forms subsequently via a peritectic reaction between $MoSi_2$ and the liquid; therefore no orientation relationship constraint applies on the two phases. Behavior of VSi_2 and $TiSi_2$ is similar to $CrSi_2$ and no lamellae were observed in $MoSi_2/VSi_2$ and $MoSi_2/TiSi_2$ two-phase alloys. In contrast, lattice misfits in $MoSi_2$-$TaSi_2$ and $MoSi_2$-$NbSi_2$ systems are somewhat higher than that in the $MoSi_2$-$CrSi_2$ system, but $C11_b/C40$ lamellar structure, even a fully lamellar structure, can be formed in the $MoSi_2$-$TaSi_2$ and $MoSi_2$-$NbSi_2$ pseudo-binary systems by selection of appropriate composition and heat treatment. In the latter two systems, C40 phase crystallizes in the liquid and $C11_b$ phase precipitates in the C40 matrix phase during subsequent heat treatment.

The present study attempts to characterize the $C11_b/C40$ lamellar structure in the $MoSi_2$-$TaSi_2$ system and to provide some crystallographic parameters that are helpful for microstructural control to improve the mechanical properties of $MoSi_2$-based disilicides.

EXPERIMENTAL PROCEDURE

The Ta-added $MoSi_2$-based disilicide used for investigation had a composition of $MoSi_2$-15mol%$TaSi_2$. The as-cast microstructure of arc-melted ingot consists of a large volume fraction of metastable $TaSi_2$-rich C40 phase and a small amount of $MoSi_2$-rich primary solid solution phase. Upon heat treatment at 1400°C for 168h, C40 phase decomposed into $C11_b/C40$ lamellae. Precise lattice misfit between $C11_b$ and C40 phases was measured by powder X-ray diffractometry (XRD). Microstructure was examined by means of scanning electron microscopy (SEM) in back-scattered electron image mode and transmission electron microscopy (TEM). Foils for TEM observation were prepared by twin-jet electropolishing method using 10% sulfuric methanol solution. TEM observation was conducted on a JEOL JEM-2011 transmission electron microscope operated at 200kV.

RESULTS AND DISCUSSION

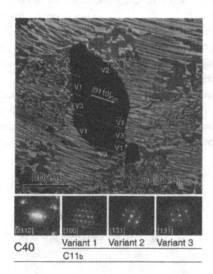

Fig.2 SEM micrograph of the lamellar microstructure around the perforation area of TEM foil and diffraction analysis of $C11_b$ variants. The C40 matrix is composed of a single crystal. Variants 1, 2 and 3 orient to [100], [131] and [13-1] respectively when C40 matrix is tilted close to [-2112] direction. Regions containing the same $C11_b$ variants are labeled V1, V2 and V3 for Variants 1, 2 and 3 respectively.

1. Orientation relationship between C11�b and C40 phases and its variants.

Fig.2 is an SEM micrograph taken from the perforation area of the TEM foil, showing a typical lamellar microstructure. The bright phase corresponds to C40 phase and the dark phase to $C11_b$ phase. Alternating plates of $C11_b$ and C40 phases construct the lamellar structure. Curvature and steps are also observed locally on the lamellar interfaces. The diffraction patterns from corresponding regions in the SEM microstructure are shown at the bottom of the figure, the area being composed of a C40 single crystal and three different variants of $C11_b$ phase. Orientation relationship (OR) between $C11_b$ and C40 phases obeys

$$(110)_{C11b} \ // \ (0001)_{C40}$$
$$[-110]_{C11b} \ // \ [2\text{-}1\text{-}10]_{C40}$$

for Variant 1, as expressed by stereographic projection as shown in Fig.3. The OR yields three variants, which is consistent with observation. This OR is kept valid even in the regions

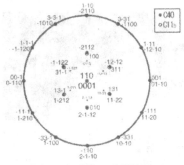

Fig.3 Stereographic projection expressing the orientation relationship between $C11_b$ and C40 phases in the lamellae. Three variants are expected.

Variant 1   (110)$_{C11b}$ // (0001)$_{C40}$
            [-110]$_{C11b}$ // [2-1-10]$_{C40}$

Variant 2   (110)$_{C11b}$ // (0001)$_{C40}$
            [-110]$_{C11b}$ // [-12-10]$_{C40}$

Variant 3   (110)$_{C11b}$ // (0001)$_{C40}$
            [-110]$_{C11b}$ // [-1-120]$_{C40}$

of irregular interfaces that seem to result from coarsening of lamellae as shown in Fig.2. The OR is consistent with that reported by Nakano et al. [5] in the $MoSi_2$-$NbSi_2$ system. In general, the $C11_b$/C40 lamellar microstructure is divided into several blocks; each block (labeled V1, V2 or V3 in Fig.2) consists of one of the three $C11_b$ variants (Variants 1, 2 and

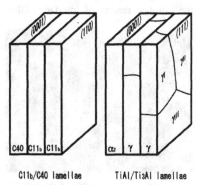

Fig.4 Comparison of the $C11_b$/C40 lamellae with the TiAl/Ti₃Al lamellae.

Cl1b/C40 lamellae   TiAl/Ti3Al lamellae

3). The number of variants within an individual block changes from place to place. The flat lamellar interfaces are basically parallel to the basal plane of C40 phase or the equivalent $C11_b$ (110) plane.

The structure of $C11_b$/C40 lamellae can be elucidated by comparison with the TiAl/Ti$_3$Al lamellae as schematically illustrated in Fig.4. Both lamellae are composed of a tetragonal phase and a hexagonal phase. However, they are different in configuration because TiAl/Ti$_3$Al lamellae consist of six variants of the tetragonal phase, which gives rise to the variant domains in the TiAl phase.

2. $C11_b$/C40 interfaces and $C11_b$/$C11_b$ boundaries.

A typical inclined $C11_b$/C40 interface is shown in Fig.5(a) when viewed along $[-2116]_{C40}$ direction. The relative low misfit between $(002)_{C11b}$ and $(01-10)_{C40}$ results in a

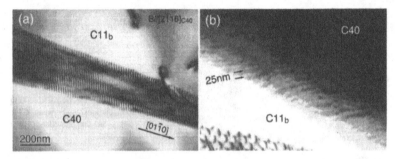

Fig.5 (a) Moiré fringes in $C11_b$/C40 interface and (b) interfacial dislocations. Micro-ledges on interface can be seen in (b).

strong oscillation of interface contrast due to strong interference of the two reflections, i.e. the so-called Moiré pattern. The period of the oscillating contrast is in agreement with the lattice misfit of $C11_b$ phase relative to C40 phase determined by XRD:

$$(110)_{C11b} \ // \ (0001)_{C40} \quad +4.05\%$$
$$[-110]_{C11b} // [2-1-10]_{C40} \quad -2.69\%$$

Although the Moiré pattern gives useful information about the interfacial coherency, it is usually an obstacle to direct observation of the intrinsic feature of interface. Tilting the

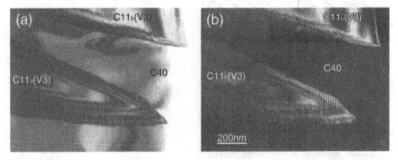

Fig.6 Appearance of the growth tip of a $C11_b$ variant under (a) bright-field and (b) dark field conditions.

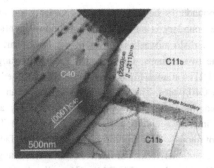

Fig.7 A macro-ledge on the C11$_b$/C40 interface.

sample to avoid the strong interference of reflections reveals a high density of misfit dislocations in the hetero-phase interface, as shown in Fig.5(b), with an interspacing of about 25nm. On this interface, micro-ledges are found to accommodate a slight deviation of interface plane from C40 (0001) orientation. High lattice misfit between C11$_b$ and C40 phases distinguishes the morphology of C11$_b$/C40 lamellae from TiAl/Ti$_3$Al lamellae. The TiAl/Ti$_3$Al lamellae have very low lattice misfit, so broad interfacial facets with rare misfit dislocations prevail in interfaces. As for the C11$_b$/C40 lamellae, as shown above, misfit is small to form a lamellar structure but is not small enough to build a lamellar interface with high coherency. High lattice misfit leads to a lot of ledges forming in the interfaces and makes macro-appearance of interfaces slightly deviate from C40 (0001) orientation. A growth tip of a C11$_b$ variant as shown in Fig.6 demonstrates a gradual change of interface orientation at the tip towards the C40 basal plane. However, facets are also found frequently at the growth tip of C11$_b$ variants. Fig.7 is an example of macro-facet in the lateral interface. The C11$_b$/C40 lateral interface aligns on (-2203)$_{C40}$ for the most part of it.

Larger misfit seems to account for the curved and thicker plates of the C11$_b$/C40 lamellae when compared to the TiAl/Ti$_3$Al lamellae. To increase the coherency of C11$_b$/C40 interface, addition of alloying elements may be the most effective. We found that only less than 0.5% misfit exists in MoSi$_2$-CrSi$_2$ system [4], indicating degree of coherency may be increased by addition of CrSi$_2$. However, low solubility of CrSi$_2$ in MoSi$_2$ and segregation of CrSi$_2$ during solidification have to be overcome.

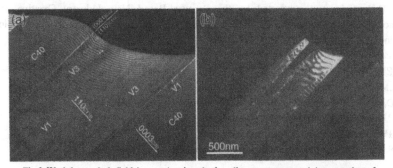

Fig.8 Weak-beam dark-field image showing the lamellar structure containing a packet of different type C11$_b$ variants. Interfaces are edge on in (a) and they are tilted inclined to the incident beam to show the boundary structure in (b).

Apart from hetero-phase interfaces, boundaries between two C11$_b$ variants are also present in the lamellar structure. Low angle boundary is built by collision of two identical variants, usually accompanying a network of dislocations, as shown in the lower part of Fig.5(b), due to slight misorientation of the two variants. On the other hand, high angle boundaries also form between two different C11$_b$ variants. Fig.8 shows an area composed of two different variants (V1 and V3) between C40 plates. The micrograph was imaged under weak-beam dark-field condition with the common reflection 0003$_{C40}$ for C40 phase and 110$_{C11b}$ for C11$_b$ variants. In Fig.8(a), all interfaces and boundaries align edge on. The contour fringes run across boundaries continuously but are disrupted at the interfacial dislocations. By tilting the specimen dislocation networks similar to those in low-angle boundaries are found in different C11$_b$/C11$_b$ boundaries (Fig.8(b)). Generally, however, packets of different C11$_b$ variants growing alternately in C40 matrix as shown in Fig.8 are not common in this alloy. Most of C11$_b$ variants grow individually but with the same type in the C40 matrix within a block across which different type variants appear, like the configuration identified in Fig.2.

CONCLUSIONS

The feature of C11$_b$/C40 lamellar structure in a 15mol%Ta-added MoSi$_2$ can be summarized as follows:

(1) Orientation relationship between C11$_b$ and C40 phases is found to be $(110)_{C11b}//(0001)_{C40}$, $[-110]_{C11b}//[2-1-10]_{C40}$. Lamellar interface prefers to align parallel to $(0001)_{C40}$ or $(110)_{C11b}$.

(2) Three distinct variants of C11$_b$ phase coexist in the lamellae. Most C11$_b$ variants form individually in the C40 matrix and neighboring variants are likely to take the same orientation, resulting in blocks of lamellae containing identical variants.

(3) High density of interfacial dislocations and numbers of micro-ledges are found to accommodate the relatively large misfit and the deviation of interface from $(110)_{C11b}$-$(0001)_{C40}$ matching plane.

ACKNOWLEDGEMENT

This study was supported by the Proposal-Based New Industry Creative Type Technology R&D Promotion Program from the New Energy and Industrial Technology Development Organization (NEDO) of Japan.

REFERENCE

1. I.L.Mirkin, and O.D.Kancheev, *Met.Sci.Heat Treat.*,**1/2**, 10(1967).
2. R.Darolia, D.F.Lahrman and R.Rield, *Scr.metall.mater.*,**26**, 1007(1992).
3. S.C.Huang, *Metall.Trans.*, **23A**,375(1992).
4. F.G.Wei, Y.Kimura and Y.Mishima, *JIM Symp. Proc.*, 1999.
5. T.Nakano, M.Azuma and Y.Umakoshi, *Intermetallics*, **6**, 715(1998).

Mat. Res. Soc. Symp. Proc. Vol. 646 © 2001 Materials Research Society

The β-FeSi₂ Formation and Thermoelectric Power of the FeSi₂+Co Base Alloys

Yoshisato Kimura, Kentaro Shindo* and Yoshinao Mishima
Tokyo Institute of Technology, Dept. of Materials Science and Engineering, 4259 Nagatsuta, Midori-ku, Yokohama 226-8502, Japan, *Graduate Student, now with Mistubishi Heavy Industries, Ltd., 5-717-1 Fukahori-machi, Nagasaki 851-0392, Japan.

ABSTRACT

Temperature dependence of thermoelectric power and resistivity was measured for the β-FeSi₂ based cast alloys on which n-type Co doping was performed with or without Cu addition. The Cu-free alloy shows a homogeneous eutectic microstructure consisting of metallic phases α-FeSi₂ and ε-FeSi, while the Cu-doped alloy has a quite heterogeneous microstructure dominated by coarse ε-FeSi dendrites in the α-FeSi₂ matrix. Adequate heat treatment condition for the semiconductor β-FeSi₂ phase formation has been evaluated as annealing at 923 K for less than 50 h, which is lower in temperature and shorter in duration than previously reported for alloys prepared by powder metallurgy. The resistivity measurement with the aid of microstructure observation has revealed that the β-FeSi₂ formation takes place and almost finishes at very early stage of annealing process. Addition of Cu effectively promotes the β-FeSi₂ formation rate of the present cast alloys.

1 INTRODUCTION

Thermoelectric material would give us a good solution to conserve the global environment and limit the use of fossil fuel, since the thermoelectric generation supplies electric power as a clean energy. The potential of thermoelectric materials is generally evaluated by the figure of merit, given as $Z = \alpha^2/\rho\kappa$, where α: thermoelectric power, ρ: resistivity and κ: thermal conductivity. Our special interest is focused on thermoelectric materials used as a component of the cogeneration system based on micro gas turbine engine. Some transition metal disilicides, especially β-FeSi₂, are attractive candidates because relatively large figure of merit with high thermoelectric power and excellent oxidation resistance at service temperatures can be expected [1-9]. Additionally, Fe and Si are less costly materials. The Fe-Si binary phase diagram is shown for Si-rich side in Fig. 1 [10]. There exist two "FeSi₂" phases: the high temperature phase is metallic α-FeSi₂ having a composition of 30Fe70Si (at%), often denoted as Fe₃Si₅, and the low temperature phase is intrinsic semiconductor β-FeSi₂ having the stoichiometric composition FeSi₂, 33.3Fe66.7Si. A proper heat treatment is required to obtain the semiconductor β-FeSi₂ phase since the as-cast microstructure is composed of metallic phases α-FeSi₂ and ε-FeSi. The transformation path for the β-FeSi₂ is complicated involving two major reactions, the peritectoid ε-FeSi+α-FeSi₂→β-FeSi₂ and the eutectoid α-FeSi₂→β-FeSi₂+Si, and a minor reaction, ε-FeSi +Si→β-FeSi₂. It was reported by Sakata [6] and Nishida et al. [7] that heat treatment for more than 200 h at relatively high temperatures around 1073 K are required for the alloys prepared by powder metallurgy. This is a practical problem of this material for a commercial use. To promote the β-FeSi₂ phase formation from α-FeSi₂ and ε-FeSi, Cu addition is known to be effective [11]. However, Yamauchi et al. [12-14] reported that heat treatment at 1073 K for 200 h is required for Cu and Mn doped β-FeSi₂ alloys.

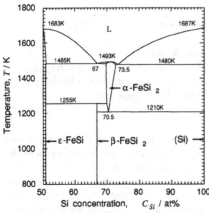

Fig. 1 The Fe-Si binary phase diagram.

Since the β-FeSi$_2$ is an intrinsic semiconductor, impurity element doping is necessary to enable the thermoelectric generation. For instance, Co is n-type and Mn and Al are p-type dopants for the β-FeSi$_2$ [1-5]. Note that the p-n junction based on the same material is advantageous to the fabrication of device. In the present work, 1at% doping of n-type Co was performed on the stoichiometric FeSi$_2$ alloy with or without Cu addition. We have investigated the temperature dependence of thermoelectric power and resistivity for the cast alloys annealed at temperatures from 873 K to 1073 K. The β-FeSi$_2$ formation was precisely observed with corresponding microstructural change during heating and holding stages of annealing process. The effect of Cu addition on promoting the β-FeSi$_2$ formation has also been studied.

2 EXPERIMENTAL PROCEDURES

Two FeSi$_2$-base alloys, FeSi$_2$-1Co and FeSi$_2$-1Co-1Cu (in at%), were prepared by arc melting under an argon gas atmosphere using highly pure raw materials. Hereafter the former is denoted as the Cu-free alloy and the latter as the Cu-doped alloy. Nominal alloy compositions are based on the stoichiometric β-FeSi$_2$, to which 1at% Co with or without 1at% Cu is added. Microstructures were observed by using scanning electron microscopy (SEM) using back scattered electron images (BEI). Thermoelectric power and resistivity measurements were conducted using a system manufacture in-house. Annealing conditions are at 873, 903, 923, 973, 1023 and 1073 K for 50 h. Thermoelectric power is measured using two sets of R-type thermocouples attached on a specimen by spot welding. The R-type thermocouple has an advantage that Pt is used as the standard reference for the absolute thermoelectric power evaluation. Temperature gradient is provided in the furnace so that the temperature difference between two junctions is at least greater than 1 K. Let T$_h$ be higher than T$_c$, then thermoelectric power α is given by V/(T$_h$-T$_c$). In the resistivity measurement system, Two couples of junctions are mechanically attached to the specimen: a couple made of pure Ag plates are for measuring current and the other made of Mo wires are for measuring potential. Both measurements were conducted under evacuation (about 10^{-3} Torr) in a quartz or Al$_2$O$_3$ tube to keep the temperature gradient stable and to protect a sample from oxidation.

3 RESULTS AND DISCUSSION

3.1 As-cast microstructure

Back scattered electron images showing typical as-cast microstructures of Cu-free alloy (a) and (b), and Cu-doped alloy (c) are shown in Fig. 2. The ε-FeSi is seen with bright contrast and the α-FeSi$_2$, dark contrast. The homogeneous eutectic microstructure consisting of ε-FeSi and α-FeSi$_2$ dominates the Cu-free alloy, and a small amount of primarily solidifying ε-FeSi dendrites is locally observed as well. Addition of 1at% Cu causes a remarkable change in as-cast microstructures. The Cu-doped alloy has a rather heterogeneous microstructure dominated by coarse primarily solidifying ε-FeSi dendrites, and there exists no eutectic microstructure. The third phase in needle like morphology seen with the brightest contrast is a Cu-rich phase.

3.2 The β-FeSi$_2$ formation and microstructure change

To investigate the β-FeSi$_2$ formation process, the resistivity measurement for Cu-free and Cu-doped alloys were conducted starting from as-cast state throughout annealing at 1073 K for 50h, including heating stage at the rate of 5 K/min. Changes in resistivity as a function of annealing time up to 200 min are shown with the annealing temperature profile in Fig. 3. Measuring resistivity is advantageous to evaluate the onset temperature of the β-FeSi$_2$ formation, because the semiconductor phase exhibits much larger resistivity than the metallic phase. Both Cu-free and Cu-doped alloys exhibit relatively low resistivity in the beginning of measurement due to the existence of metallic phases α-FeSi$_2$ and ε-FeSi. Then, the value of resistivity starts increasing rapidly according to the onset of the semiconductor β-FeSi$_2$ formation: at about 873 K, 90 min, in Cu-doped alloy, and at about 1003 K, 110min, in Cu-free alloy. Subsequently, the resistivity increases gradually after 110 min and at 1003 K in the Cu-doped alloy, and after 160 min and at 1073 K (including 30min hold at 1073 K) in the Cu-free alloy. It is noted that reactions regarding β-FeSi$_2$ formation seem to finish within the heating stage, i.e., at the very early stage of annealing. It is obvious from the figure that the onset temperature is lower and the formation rate of β-FeSi$_2$ is faster in Cu-doped alloy than in Cu-free alloy. This result indicates that Cu addition is effective to promote the β-FeSi$_2$ formation in the present cast alloys.

We have conducted water quenching immediately from temperatures on the heating and holding stages of annealing in order to observe microstructure change at each temperature shown in Fig. 3. Back scattered electron images of quenched alloys are shown in Fig. 4 (a) through (f)

Fig. 2   Back scattered electron images of as-cast microstructures: (a) and (b) Cu-free alloy and (c) Cu-doped alloy.

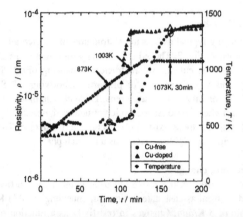

Fig. 3 Changes in the resistivity of Cu-free and Cu-doped alloys with temperature profile of annealing processs.

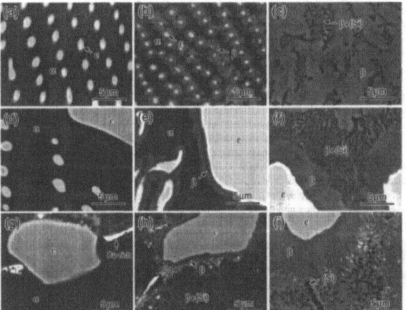

Fig. 4 Back scattered images of (a-f) Cu-free alloy and (g-i) Cu-doped alloy: water quenched (a, d, g) from 873 K, (b, e, h) from 1003 K, and (c, f, i) after holding at 1073 K for 30 min.

for Cu-free alloys, and (g) through (i) for Cu-doped alloys: (a,d,g) quenched from 873 K, (b,e,h) from 1003 K, and (c,f,i) after 1073 K for 30 min holding. There is no appreciable change observed in microstructures of both alloys at 873 K, compare (a), (d) and (g) with Fig. 2. The β-FeSi₂ formation starts right above 873 K in Cu-doped alloy, and above 1003 K in Cu-free alloy. Hence, it is clearly seen at 1003 K, in (c) and (f), for Cu-doped alloy and at 1073 K, in (h), for Cu-doped alloy that β-FeSi₂ forms surrounding ε-FeSi by the peritectoid reaction and β-FeSi₂ + (Si) lamellar microstructure also forms at former α-FeSi₂ matrix by the eutectoid reaction. The β-FeSi₂ formation further proceeds during holding at 1073 K for 30 min as shown in (i) that the volume fraction of β-FeSi₂ obviously increases and (Si) phase becomes spherical. The total amount of β-FeSi₂ phase formed for 50 h is greater in Cu-doped alloy than in Cu-free alloy.

3.3 Temperature dependence of thermoelectric power and resistivity

Temperature dependence of the thermoelectric power measured for specimens annealed for 50 h is shown for Cu-free alloy in Fig. 5 (a), and for Cu-doped alloy in (b). Note that negative thermoelectric power originates from n-type Co doping. The propensity of thermoelectric power is reflected by the a total amount of the β-FeSi₂ phase formed in each alloy annealing at each temperature for 50 h, hence the peak value of thermoelectric power indicates the efficiency of heat treatment conditions for the β-FeSi₂ formation. As a reference material, the alloy annealed at 1073 K for 200 h is shown together [12-14]. Cu-free and Cu-doped alloys exhibit almost the same tendency of temperature dependence of thermoelectric power and the peak temperature giving the maximum thermoelectric power appears at around 850 K. It is noteworthy that Cu-free alloy annealed at 923 K shows the highest maximum value of thermoelectric power exceeding the reference material. This result suggests that appropriate heat treatment condition for the present cast Cu-free alloys is annealing at 923 K for less than 50 h, which is lower in temperature and shorter in duration than 1073 K for 200 h. Homogeneity of as-cast α-FeSi₂/ε-FeSi eutectic microstructure allows much shorter atomic diffusion path even at relatively low temperature. The Cu-doped alloy annealed at 923 K also exhibits the highest maximum thermoelectric power, but never exceeds the reference material in this case. The Cu addition seems to be effective for the present cast Cu-doped alloys, however heterogeneous microstructure having coarse ε-FeSi dendrites is disadvantage for the β-FeSi₂ formation at low temperatures.

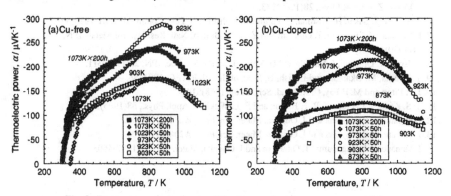

Fig. 5 Temperature dependence of the thermoelectric power measured for
(a) Cu-free alloy and (b) Cu-doped alloy.

4 CONCLUSIONS

The relationship between microstructures and thermoelectric power has been investigated on the β-FeSi$_2$ base alloys doped with the n-type element Co is doped with or without Cu addition. To seek for the proper heat treatment condition for the β-FeSi$_2$ formation in the cast alloys, extensive microstructure observation was conducted with the aid of resistivity measurements. The following conclusions are drawn from the present work:

1. In a comparison of as-cast microstructures, Cu-free alloy shows a relatively homogeneous α-FeSi$_2$/ϵ-FeSi eutectic microstructure while Cu-doped alloy exhibits a heterogeneous microstructure having coarse ϵ-FeSi dendrites in the α-FeSi$_2$ matrix.
2. Adequate heat treatment condition for the β-FeSi$_2$ formation in the present cast alloys deduced from thermoelectric power measurement is annealing at 923 K for less than 50 hours, which is lower in temperature and shorter in duration than previously reported for alloys prepared by powder metallurgy.
3. The resistivity measurements have revealed that the β-FeSi$_2$ formation rate is faster in Cu-Doped alloy than in Cu-free alloy. It means that Cu addition effectively promotes the β-FeSi$_2$ formation at its very early stage of annealing.

ACKNOWLEDGMENT

Authors are deeply grateful to Dr. Kazuhiro Hasezaki, Mitsubishi Heavy Industries Ltd., for his fruitful comments.

REFERENCES

1. T. Sakata and I. Nishida, JIM, 15(1976)11.
2. I. Nishida, Phys. Rev., B7(1973)2710.
3. R. M. Ware and J. J. McNeill, Proc. Inst. Electr. Eng., 111(1964)178.
4. U. Birkholz and J. Schelm, Phys. Stat. Sol., 27(1968)413.
5. J. Hesse, Z. Angew. Phys., 28(1969)133.
6. T. Sakata, Y. Sakai, H. Yoshida, H. Fujita and I. Nishida, J. Less-Comm. Met., 61(1978)301.
7. I. Nishida, M. Okamoto, T. Ohkoshi and Y. Isoda, Bull. of Natl. Res. Inst. for Metals JPN, 11(1990)145.
8. Y. Isoda, T. Ohkoshi, I. Nishida and H. Kaifu, J. Mater. Sci. Soc. JPN, 25(1989)311.
9. T. Kojima, M. Okamoto and I. Nishida, Proc. of the 5th ICTEC, 1984, 56.
10. J. P. Piton and M. F. Fay, C. R. Acad. Sci., C266(1968)514.
11. T. S. Kuan, J. L. Freeouf, R. F. Batson and E. L. Wilkie, J. Appl. Phys., 58(1985)1519.
12. I. Yamauchi, S. Ueyama and I. Ohnaka, Mater. Sci. Eng., A208(1996)101.
13. I. Yamauchi, S. Ueyama and I. Ohnaka, Mater. Sci. Eng., A208(1996)108.
14. I. Yamauchi, A. Suganuma, T. Okamoto and I. Ohnaka, Mater. Sci., 32(1997)4603.

Mat. Res. Soc. Symp. Proc. Vol. 646 © 2001 Materials Research Society

Microstructures and High-temperature Strength of Silicide-reinforced Nb Alloys

C.L. Ma, Y. Tan, H. Tanaka, A. Kasama, R. Tanaka, Y. Mishima[1] and S. Hanada[2]
Japan Ultra-high Temperature Materials Research Institute,
Ube 755-0001, Japan
[1]Department of Materials Science and Engineering, Tokyo Institute of Technology,
Yokohama 226-8502, Japan
[2]Institute for Materials Research, Tohoku University,
Sendai 980-8577, Japan

ABSTRACT

This article describes the phase stability, microstructures and mechanical properties of silicide-reinforced Nb alloys in Nb-Mo-W-Si quaternary system prepared by arc melting and heat treatment. There exists an equilibrium two-phase field of Nb solid solution (Nb_{ss}) and $\alpha(Nb,Mo,W)_5Si_3$ in a Nb-rich region of this quaternary system. Alloys in this region have a eutectic reaction of $L \rightarrow Nb_{ss}+\beta(Nb,Mo,W)_5Si_3$ during solidification. The $\beta(Nb,Mo,W)_5Si_3$ transforms to the stable $\alpha(Nb,Mo,W)_5Si_3$ at very high temperature. The cast and heat treated hypoeutectic alloys consist of dendritic Nb_{ss}, network-shaped Nb_{ss} matrix and $\alpha(Nb,Mo,W)_5Si_3$. These quaternary alloys exhibit excellent high-temperature strength, although the fracture toughness is still unacceptable for practical applications.

INTRODUCTION

Nb_{ss}/Nb-silicide in-situ composites have great potential to replace the Ni-based superalloys in future turbine engines, because they possess very high melting points and attractive balance of high-temperature strength and low temperature fracture toughness [1,2]. In these alloys, the silicide phase (Nb_5Si_3) is in equilibrium with the metallic phase (Nb_{ss}). The former provides the high-temperature strength and the latter provides low-temperature toughness. To produce Nb_{ss}/Nb-silicide composites with high performance, the microstructural and compositional optimizations were suggested to be the most important issue [3,4]. Great efforts have been devoted to this issue in recent years [5-7].

We have experimentally constructed the liquidus surface projections and isothermal section at 1700°C of Nb-Mo-Si [8] and Nb-W-Si [9] ternary systems. These diagrams show the existence of Nb_{ss}-$\alpha(Nb,Mo$ or $W)_5Si_3$ two-phase field at a Nb-rich composition zone. The most important feature of these phase diagrams is the existence of a eutectic reaction, $L \rightarrow Nb_{ss} + \beta(Nb,Mo$ or $W)_5Si_3$, in a wide Mo or W composition range. It is worth pointing out that such a eutectic reaction does not exist in Nb-Si binary phase diagram [10]. The brittle silicide phase was greatly refined by forming fine lamellar eutectic structure *via* the eutectic reaction during solidification [8,9]. This gives an indication that the microstructural optimization could be achieved by an appropriate amount of Mo or W addition. Studies on mechanical properties have shown that Mo or W addition to Nb-Si alloys not only strongly strengthens the Nb_{ss} phase, but also improves the room-temperature toughness [9]. The former is principally due to the solid solution hardening of Nb_{ss} by Mo or W, while the latter is mainly attributed to the distribution of fine silicide particles. It is evident from these results that the 'fine microstructure' offers an acceptable value of fracture toughness, even though the Nb_{ss} phase is embrittled by solid solution hardening. It is interesting to know whether or not this kind of microstructure could be produced in Nb-Mo-W-Si quaternary system. If so, the high temperature strength would be further improved by Mo or W addition without greatly degrading the room-temperature toughness.

The purpose of the present work is to develop Nb_{ss}/Nb-silicide in-situ composites with good balance of the mechanical properties in Nb-Mo-W-Si quaternary alloy system. In this paper, the relation among constituent phases, microstructural features and mechanical properties of several quaternary alloys will be described.

EXPERIMENTAL PROCEDURES

Alloy ingots, each about 25g, were prepared from high purity Nb, Mo, and Si elemental powders by arc melting in a water-cooled copper crucible under an ultra high-purity argon atmosphere. The ingots were remelted three times to ensure chemical homogeneity. Heat treatment was conducted at 1700°C for 48 h, followed by furnace cooling to room temperature. Cast and heat-treated microstructures were characterized by scanning electron microscopy (SEM), X-ray diffraction (XRD) and electron probe microanalysis (EPMA). The high-temperature strength was evaluated by compression tests using specimens with a dimension of 3 mm x 3 mm x 6 mm. Fracture toughness tests were performed on single-edge-notch bend specimens having dimension of 3 mm x 6 mm x 30 mm with a 3 mm notch in depth. Compression tests were carried out at a strain rate of $3 \times 10^{-4} s^{-1}$ in an argon atmosphere using an Instron-type machine, and were stopped when a plastic strain exceeded 16 %.

RESULTS AND DISCUSSION

A. Phase Stability and Microstructures in Nb-rich Zone of Nb-Mo-W-Si System

Based on the Nb-Mo-Si and Nb-W-Si ternary phase diagrams [8,9], a partial schematic quaternary Nb-Mo-W-Si phase diagram at 1700°C is proposed and presented in figure 1. There exists an equilibrium two-phase field of Nb_{ss} and $\alpha(Nb,Mo,W)_5Si_3$ at the Nb-rich corner of this diagram. Typical cast microstructures of alloys located in this field, Nb-16Si-10Mo-5W and Nb-16Si-10Mo-15W, are shown in figure 2 (a) and (b). In these back scattered SEM micrographs, the dendritic primary phase (bright) was identified to be Nb_{ss} by EPMA and XRD, and the interdendritic mixtures were Nb_{ss} and $\beta(Nb,Mo,W)_5Si_3$ (dark) produced from the eutectic reaction, $L \rightarrow Nb_{ss} + \beta(Nb,Mo,W)_5Si_3$. This demonstrates that the eutectic reaction $L \rightarrow Nb_{ss} + \beta(Nb,Mo,W)_5Si_3$ exists in the Nb_{ss}-$\alpha(Nb,Mo,W)_5Si_3$ two-phase region. $L \rightarrow Nb+Nb_3Si$ type eutectic reaction was also found in this region, but the eutectic composition range is very small (Mo<3 at% and/or W<2 at%). The eutectic $Nb_{ss} + \beta(Nb,Mo,W)_5Si_3$ exhibited fine lamella-like morphology in the low W content alloy (figure 2 (a)) and coarse in the high W content alloy (figure 2(b)). Similar phenomena have also been observed in Nb-W-Si ternary alloys [9]. In Nb-Mo-Si ternary alloys, on the other hand, the fine lamellar eutectic structure was almost not influenced by Mo content [2,8]. It is clear that W is more effective in affecting the cast microstructures than Mo.

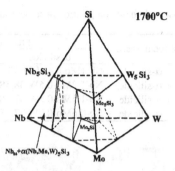

Figure 1. Schematic of the proposed Nb-Mo-W-Si pseudo-quaternary phase diagram at 1700 °C showing the equilibrium two-phase region of Nb_{ss} and $\alpha(Nb,Mo,W)_5Si_3$.

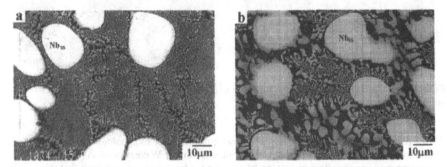

Figure 2. Cast microstructures of (a) Nb-16Si-10Mo-5W and (b) Nb-16Si-10Mo-15W alloys showing the dendritic primary Nb_{ss} phase and the interdendritic eutectic ($Nb_{ss}+\beta(Nb,Mo,W)_5Si_3$).

Heat treatment at 1700°C in Nb-rich alloys (3<Mo<20 at%, 2<W<25 at%) caused the transformation of the high-temperature phase of $\beta(Nb,Mo,W)_5Si_3$ to stable $\alpha(Nb,Mo,W)_5Si_3$. This suggests that the transformation temperature of β to α is higher than 1700°C in the composition range investigated. The constituent phases and their compositions of some heat treated alloys containing 16 at% silicon are shown in table I. It is evident from the table that the silicide is a line compound. The solubility of Mo and W in α-silicide is very low and there is strong Mo and W partitioning from silicide phase to Nb_{ss}. This is consistent with our previous results obtained in Nb-Mo-Si and Nb-W-Si alloys. It is expected, therefore, that Nb_{ss} is strongly strengthened by the Mo and W additions. Figure 3 (a) and (b) show the typical optical microstructure and SEM image of heat treated Nb-16Si-10Mo-15W alloy. The interdendritic eutectic becomes coarse and forms semi-continuous Nb_{ss} networks and α-silicide. The volume fraction of silicide phase in this alloy is about 45%. There is no clear evidence of any secondary

Table I. Constituent phase and phase composition of some heated Nb-Mo-W-Si alloys

| Alloys | Constituent phase | Phase composition | |
		α-silicide	Nb$_{ss}$
Nb-16Si-10Mo-5W	Nb$_{ss}$ + α-silicide	Nb-1Mo-0.4W-37.4Si	Nb-14.8Mo-7.4W-0.6Si
Nb-16Si-10Mo-10W	Nb$_{ss}$ + α-silicide	Nb-1.2Mo-0.5W-38.5Si	Nb-14.4Mo-15.5W-0.6Si
Nb-16Si-10Mo-15W	Nb$_{ss}$ + α-silicide	Nb-1.5Mo-0.8W-37.5Si	Nb-14.8Mo-25.3W-0.5Si
Nb-16Si-5Mo-15W	Nb$_{ss}$ + α-silicide	Nb-0.6Mo-0.9W-37.5Si	Nb-6.7Mo-25.7W-0.4Si

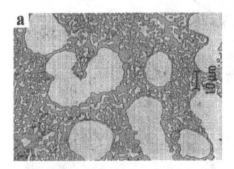 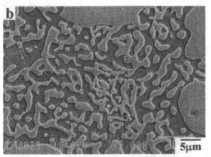

Figure 3. (a) Optical and (b) SEM micrographs of Nb-16Si-10Mo-15W showing the heat treated eutectic structure.

silicide phase precipitated from the primary metallic phase. These morphologic characters are similar to those of Nb-W-Si ternary alloys with the same Si and W concentration [9]. An undergoing study on the effect of heat treatment upon the microstructure indicates that this microstructure is very stable up to 1700°C.

B. Mechanical Properties of Alloys in the Nb$_{ss}$-α(Nb,Mo,W)$_5$Si$_3$ Two-phase Region

Plots of compressive yield strength *vs.* testing temperature for alloys listed in table I are shown in figure 4. Data of some binary [3,11], ternary [9] and multi-component [5] Nb/Nb-silicide in-situ composites are included for comparison. Though the compressive yield stress decreases with increasing temperature, the present alloys exhibit excellent yield strength at all the temperatures tested. It is seen that the high temperature strength increases with increasing W content in the quaternary alloys, suggesting the solid solution hardening by W alloying. The Nb-16Si-10Mo-15W alloy shows the highest compressive yield strength of ~796 MPa at 1500°C and ~550 MPa at 1600°C, which are higher than that of binary monolithic silicide Nb$_5$Si$_3$ (~380 MPa at 1600°C [11]). There are two plausible explanations for the result. One is that both the metallic and silicide phases in the present alloys are strongly strengthened by Mo and W additions. The other is that the data of Nb$_5$Si$_3$ cited here might not be intrinsic, because a great difference in strength would be risen when samples are prepared by different processes. The fabrication procedure of the monolithic Nb$_5$Si$_3$ was not described in the reference [11].

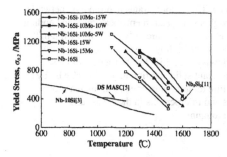

Figure 4. Changes of yield strength of some Nb-Mo-W-Si alloys as a function of testing temperature.

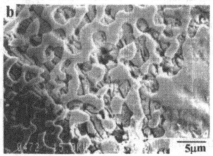

Figure 5. True stress - true strain behavior of Nb-10Mo-15W-16Si alloy at various high-temperatures.

Figure 6. SEM micrographs of Vickers indentation in Nb-5Mo-15W-16Si alloy, showing (a) outside- and (b) inside-view of the indentation.

Figure 5 shows true stress - true strain curves of Nb-10Mo-15W-16Si alloy tested at various temperatures. It can been seen that this alloy exhibits large plastic deformation above 1300°C. Deformation of the silicide phase is known to be difficult even at temperature above 1400°C [11]. Thus, the extensive plasticity above 1300°C is primarily due to the deformation of semi-continuous Nb$_{ss}$ networks in the alloy (see figure 3). A stress peak followed by a gradual decrease in stress is observed in all the stress-strain curves. This may be associated with dynamic recrystallization. Further studies are needed to clarify the deformation mechanisms at high temperature.

Figure 6 represents the SEM images of Nb-16Si-5Mo-15W alloy showing (a) the outside- and (b) inside-view of a Vickers indentation, where a 49N indentation-load was used. No large crack is observed around the indentation. However, a large number of cracks can be seen in the silicide phase inside the indentation. These cracks seem to nucleate in the silicide phase and propagate to Nb$_{ss}$-silicide interface. It is interesting to note that the cracks are blunted and arrested at Nb_{ss} phase or at Nb_{ss}-silicide interface. This demonstrates that the Nb_{ss} networks behave as a ductile phase. However, the fracture toughness measurements performed in bending test for alloys listed in table I indicated that the toughness values are only about 6~9 MPa·m$^{1/2}$, which are much lower than that of a complex Nb/Nb-silicide alloy (Nb-24.7Ti-8.2Hf-2.0Cr-1.9Al-16Si) reported by Bewlay et al. [5]. The brittleness of these quaternary alloys will be explained in terms of significant solid solution hardening of Nb_{ss} by Mo and W alloying, which results in a decrease of fracture toughness in spite of an increase of high-temperature strength. The existence of defects and microcracks in these cast and heat-treated alloys would also decrease the fracture toughness.

CONCLUSIONS

Phase stability, microstructures and mechanical properties of Nb-Mo-W-Si quaternary alloys were investigated to develop Nb/Nb-silicide in-situ composites for very high temperature structural applications. In this quaternary system, there exists an equilibrium two-phase field of Nb_{ss}-$\alpha(Nb,Mo,W)_5Si_3$, which is stable up to very high temperature. Moreover, a eutectic reaction $L \rightarrow Nb_{ss} + \beta(Nb,Mo,W)_5Si_3$ exists in this two-phase region. Microstructures of alloys located in this region can be changed by controlling the solidification processes and adjusting the alloy compositions so as to exhibit high performance in mechanical properties. These alloys have superior high-temperature strength as compared with some advanced high-temperature materials.

ACKNOWLEDGMENTS

This research is supported by the New Energy and Industrial Technology Development Organization (NEDO) of Japan.

REFERENCES

1. P. R. Subramanian, M.G. Mendiratta and D.M. Dimiduk, *JOM*, **48**(1), 33 (1996).
2. D. M. Shah, D. L. Anton, D. P. Pope and S. Chin, *Mater. Sci. Eng.*, **192/193A**, 658 (1995).
3. M. G. Mendiratta, J. J. Lewandowski, and D. M. Dimiduk, *Metall. Trans.*, **22A**, 1573 (1991).
4. S. Hanada, *Report of the 123rd Committee on Heat-Resisting Materials and Alloys*, JSPS, **38** 299 (1997).
5. B. P. Bewlay, M.R. Jackson and H.A. Lipsitt, *Metall. Trans.*, **27A**, 3801 (1996).
6. D.V. Heerden, A.J. Gavens, T.Foecke, T.P. Weilhs, *Mater. Sci. Eng.*, **261A**, 212 (1999).
7. C. L. Ma, A. Kasama, R. Tanaka, S. Hanada and M.K. Kang, *Transactions of Metal Heat Treatment*, **21**, 83 (2000).
8. C. L. Ma, Y. Tan, H. Tanaka, A. Kasama, R. Tanaka, S. Miura, Y. Mishima and S. Hanada, *Mater. Trans. JIM*, **41**, 1329 (2000).
9. C. L. Ma, A. Kasama, Y. Tan, H. Tanaka, R. Tanaka, Y. Mishima and S. Hanada, *Report of the 123rd Committee on Heat-Resisting Materials and Alloys*, JSPS, **40**, 335 (1999).
10. T. B. Massalski, H. Okamoto, P.R. Subramanian and L. Kacprzak: *Binary Alloy Phase Diagrams*, ASM, Metals Park, Ohio, 1992.
11. R. M. Nekkanti and D. M. Dimiduk, Mat. Res. Soc. Symp. Proc. **194**, 175 (1990).

Mat. Res. Soc. Symp. Proc. Vol. 646 © 2001 Materials Research Society

Tailoring the thermal expansion anisotropy of Mo_5Si_3

Joachim H. Schneibel, Claudia J. Rawn, Chong Long Fu
Metals and Ceramics Division, Oak Ridge National Laboratory,
Oak Ridge, TN 37831, U.S.A.

ABSTRACT

The silicide Mo_5Si_3, with the W_5Si_3 structure, exhibits a high anisotropy of its coefficients of thermal expansion (CTEs) in the a and c directions, namely, $CTE(c)/CTE(a)=2.2$. In order to determine whether the CTE anisotropy can be controlled, molybdenum was partially substituted with Nb. CTEs were determined by high temperature x-ray powder diffraction. Partial substitution with 44 at. % Nb reduced the value of $CTE(c)/CTE(a)$ to a value near 1. For higher Nb concentrations, $CTE(c)/CTE(a)$ increased again. When nearly all the Mo was replaced by Nb, the crystal structure changed to the Cr_5B_3 structure type and the CTE anisotropy decreased to a value near 1. Our thermal expansion results are interpreted in terms of the site occupation of Nb, the Nb-induced increase in the interatomic spacing of the Mo atom chains along the c-direction, and the reduction in the anisotropy of the lattice anharmonicity in the a and c directions.

INTRODUCTION

In the search for high-temperature structural materials, molybdenum silicides have frequently been considered as potential solutions [1]. Recently, Akinc and collaborators [2,3] have studied boron-containing molybdenum silicide alloys consisting of the phases Mo_3Si, Mo_5Si_3, and Mo_5SiB_2. These authors suggest that the Mo_5Si_3 phase is responsible for the high creep strength of these alloys. In order to further understand the Mo_5Si_3 phase Chu et al. [4] investigated Mo_5Si_3 single crystals. They found that the thermal expansion coefficients (CTEs) in the c and a directions of the tetragonal crystal structure of Mo_5Si_3 differed by more than a factor of 2. Unless the grain is sufficiently small (as it was in Akinc et al's experiments), such high CTE ratios result in micro-cracking and degrade the mechanical properties [5]. Fu et al. [6,7] performed ab-initio calculations of the CTEs of Mo_5Si_3. They obtained good agreement with Chu et al.'s experimental data and attributed the high CTE ratio of Mo_5Si_3 to the elastic rigidity of its basal planes and a high anharmonicity along the c-axis. The high anharmonicity along the c-axis was attributed to the presence of Mo-chains along the c-axis with unusually short atomic distances.

Recently, Ikarashi et al. [8] found that the CTE anisotropy of the tetragonal silicide Ti_5Si_3 could be significantly reduced by partial substitution of Ti atoms with larger Zr atoms. Whereas the CTE ratio, $CTE(c)/CTE(a)$, for Ti_5Si_3 was 1.68, the corresponding value for $Ti_2Zr_3Si_3$ was only 1.22. The purpose of the present work is to determine whether ternary alloying additions to Mo_5Si_3 can reduce the CTE anisotropy. It turns out that partial substitution of Mo atoms with (distinctly larger) Nb atoms is very effective in reducing the CTE anisotropy of Mo_5Si_3 type compounds.

EXPERIMENTAL DETAILS

Buttons of $(Mo_{1-x}Nb_x)_5Si_3$ alloys with several Nb concentrations were prepared from high-purity elements by repeated arc-melting in a partial pressure of argon. The buttons were homogenized by annealing in vacuum for 24 hours at 1600°C and crushed into –325 mesh (diameter ≤45 µm) powders. High temperature x-ray diffraction (HTXRD) measurements were conducted using a Scintag PAD X vertical θ/θ goniometer equipped with a modified Buehler HDK-2 diffraction furnace. The diffractometer utilized CuK$_\alpha$ radiation (45 kV and 40 mA) and a Si(Li) Peltier-cooled solid state detector. The data were collected as step scans, with a step size of 0.02 °2θ and a count time of 1 s/step between 24 and 65 °2θ. The sample temperature was monitored with a Pt/Pt-10%Rh thermocouple spot-welded to the Pt-30%Rh or Mo heater strip on which a thin layer of the powder was dispersed. Al_2O_3 (NIST SRM 676) was added as an internal standard. Data were collected in flowing He gas at room temperature, 100 °C and then up to 500°C in 100°C intervals. The lattice parameters were evaluated using the software package JADE 5.0 using at least 4 peaks each for the Al_2O_3 and the $(Mo_{1-x}Nb_x)_5Si_3$. The specimen displacement determined from the fit to the Al_2O_3 peaks was used to refine the $(Mo_{1-x}Nb_x)_5Si_3$ peaks. The coefficients of thermal expansion were calculated from regression lines to a plot of lattice parameter vs. temperature. To obtain an estimate of the error, data points deviating by ±esd (estimated standard deviation) from the lattice parameter values were added to the plots prior to fitting.

RESULTS AND DISCUSSION

Table I shows the CTEs and CTE ratios for Al_2O_3 and $(Mo_{1-x}Nb_x)_5Si_3$. For some of the alloys duplicate measurements were performed in order to assess the reproducibility of the results. The data for the Al_2O_3 standard exhibits some scatter - the error in the CTEs is typically 1 ppm/K and the standard deviation of the measured CTE ratios of Al_2O_3, CTE(c)/CTE(a), is 0.15. These errors provide an estimate for the scatter in the $(Mo_{1-x}Nb_x)_5Si_3$ data.

Figure 1 shows the CTE ratio of $(Mo_{1-x},Nb_x)_5Si_3$ as a function of the Nb concentration. The value for the binary compound, Mo_5Si_3, agrees well with the dilatometric data published by Chu et al. [4] (see also Table I). As more and more of the Mo is replaced by Nb the CTE ratio decreases until it reaches a value close to 1 near x=0.7. For x > 0.7 the CTE ratio increases again. Above x=0.9 the crystal structure changes from the structure of Mo_5Si_3 [W$_5$Si$_3$ type, D8$_m$, I4/mcm(140)] to that of Nb$_5$Si$_3$ [Cr$_5$B$_3$ type, D8$_l$, I4/mcm(140)] (see Table II). The change in the crystal structure is accompanied by a reduction of the CTE ratio to a value near 1. The $(Mo_{1-x}Nb_x)_5Si_3$ compound with the minimum CTE anisotropy may nevertheless have some advantages, such as the lower Nb concentration which may result in better oxidation resistance and is therefore worth further study.

As Mo in Mo_5Si_3 is partially substituted by Nb, the lattice parameters of the tetragonal unit cell increase (Fig. 2). Whereas the values of a vs the Nb concentration follow Vegard's law, the data for c appears to be non-linear with a decreasing slope for high Nb concentration. More precise lattice parameter measurements with internal standards will be needed to substantiate this result.

Table I. Coefficients of thermal expansion for $(Mo_{1-x}Nb_x)_5Si_3$ and Al_2O_3 determined from linear fits between 20 and 400 or 20 and 500°C. The estimated errors of the last digit are indicated in round brackets. In some cases two runs were performed for the same composition.

Nb, at. %	x	(Mo$_{1-x}$Nb$_x$)$_5$Si$_3$			Al$_2$O$_3$			Reference, Temperature range
		CTE(a) 10^{-6}/K	CTE(c) 10^{-6}/K	CTE(c)/ CTE(a)	CTE(a) 10^{-6}/K	CTE(c) 10^{-6}/K	CTE(c)/ CTE(a)	
					7.4	7.9	1.07	Ref. [9]
		5.2	11.5	2.21				Ref. [4]
0	0	6.4(2)	13.7(3)	2.14(8)	6.9(4)	9.1(4)	1.32(10)	20-500°C
15	0.240	6.5(2)	9.8(5)	1.51(9)	7.1(2)	7.7(1)	1.08(3)	20-500°C
30	0.480	8.6(4)	12.4(7)	1.44(11)	8.7(3)	8.1(4)	0.93(6)	20-500°C
30	0.480	6.2(2)	8.7(2)	1.40(6)	6.5(5)	7.1(4)	1.09(10)	20-500°C
37.5	0.600	5.6(5)	7.2(5)	1.29(15)	6.3(3)	6.5(3)	1.03(7)	20-500°C
44	0.704	8.5(8)	8.7(7)	1.02(13)	8.3(7)	7.9(2)	0.95(8)	20-400°C
44	0.704	6.0(5)	6.7(4)	1.12(11)	7.2(3)	7.1(3)	0.99(6)	20-500°C
50	0.800	5.0(4)	9.1(2)	1.82(15)	7.0(4)	8.1(1)	1.16(7)	20-500°C
50	0.800	5.4(5)	9.3(4)	1.72(18)	6.0(8)	7.0(4)	1.17(17)	20-400°C
56	0.896	6.0(4)	10.5(5)	1.75(14)	7.4(4)	8.2(2)	1.11(7)	20-500°C
62.5	1.000	7.8(2)	8.4(4)	1.08(6)	6.9(2)	8.1(2)	1.17(4)	20-500°C

Table II. Summary of the crystallographic details for Mo$_5$Si$_3$ and Nb$_5$Si$_3$.

Structure type	compound	Atom	Site*	x	y	z	lattice parameters (Å)
W$_5$Si$_3$	Mo$_5$Si$_3$	Mo	4b	0	1/2	1/4	a = 9.6425 c = 4.9096
		Mo	16k	0.075	0.224	0	
		Si	4a	0	0	1/4	
		Si	8h	0.165	0.665	0	
Cr$_5$B$_3$	Nb$_5$Si$_3$	Nb	4c	0	0	0	a = 6.570 c = 11.884
		Nb	16l	0.166	0.666	0.150	
		Si	4a	0	0	1/4	
		Si	8h	0.375	0.875	0	

*Wyckoff notation indicating how often a particular site (e.g., site b) occurs in the unit cell

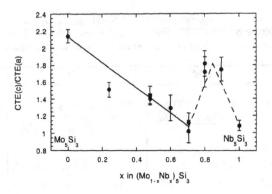

Figure 1. *Plot of the CTE(c)/CTE(a) ratio of (Mo,Nb)₅Si₃.*

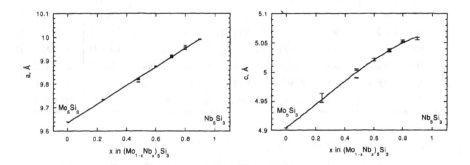

Figure 2. *Lattice parameters a and c of (Mo,Nb)₅Si₃.*

Figure 3 depicts the crystal structure of Mo_5Si_3. The Mo atoms occupy two different sites, namely, site *4b* and site *16k*. (see Table II). The Mo atoms on the *4b* sites form chains in the c-direction [see Fig. 3(a)]. The distance between these Mo atoms is $0.5 \times c = 2.452$ Å. This distance is significantly smaller than the smallest distance between the Mo atoms in the Mo bcc crystal structure, namely, $0.5 \times 3^{1/2} \times a_{Mo} = 2.725$ Å. The small spacing of the *4b* Mo atoms causes their interaction to be highly anharmonic and is therefore a major source of the high CTE value in the c-direction. Nb is a larger atom than Mo ($r_{Nb} = 1.44$ Å vs. $r_{Mo} = 1.37$ Å). When the Mo in Mo_5Si_3 is partially replaced by Nb the Nb is not likely to go to the *4b* sites since the Mo atoms are spaced very closely there. Instead, first principles calculations [10] support that the Nb goes to the *16k* sites. Since Nb is larger than Mo the lattice parameter will increase, in agreement with Fig. 2. At the same time, the spacing of the Mo atoms in the *4b* chains increases. For example, the spacing of the *4b* Mo atoms in Mo-44Nb-37.5Si (at. %) is 0.5×5.037 Å = 2.519 Å, which is

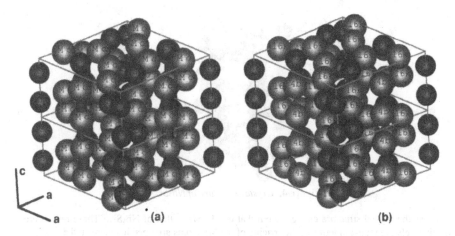

Figure 3. *Crystal structure of Mo₅Si₃ showing a stack of two unit cells with (a) the elemental occupation and (b) the Wyckoff sites indicated.*

almost 3% larger than the corresponding spacing in Mo_5Si_3, namely, 2.452 Å. This reduces the anisotropy in the anharmonicity [10], and therefore the CTE in the c-direction decreases (see Table I). If the Nb would fill up all the *16k* sites before starting to occupy *4b* sites, a transition in the site occupation would be predicted at a Nb concentration of 62.5× (16/20) at. % = 50 at. %. Once the large Nb atoms start to occupy the *4b* chains, the anisotropy of the anharmonicity is likely to increase again. This trend is qualitatively observed in Fig. 1. However, the minimum in the CTE ratio is observed around 44 at. % Nb, while the CTE ratio increases again for 50 at. % Nb (where it should be a minimum if the Nb fills all the *16k* sites first). It is therefore likely that the Nb starts to occupy *4b* sites somewhat before the *16k* sites are completely filled.

The CTE ratio of Nb_5Si_3 is close to 1. An inspection of the crystal structure of Nb_5Si_3 (see Fig. 4) shows (a) the absence of Nb-chains in the c-direction and (b) a relatively large minimum Nb-Nb separation, namely, 2.84 Å (calculated from the measured lattice parameter values $a=6.571$ Å and $c=11.892$ Å). Both the absence of Nb chains and the large spacing of the Nb atoms are consistent with the low CTE ratio of Nb_5Si_3 as compared to Mo_5Si_3. It should be noted that Mo_5Si_3 can be converted into the Cr_5Si_3 type structure not only by substitution of most of the Mo with Nb, but also by replacing 2/3 of the Si by B. Like Nb_5Si_3, the resulting ternary compound Mo_5SiB_2 has a CTE anisotropy ratio close to 1 [11]. Again, this result is attributed to the absence of Mo-chains.

CONCLUSIONS

Alloying of Mo_5Si_3 with Nb results in an unusual dependence of the thermal coefficient of expansion anisotropy on the Nb concentration. Below 44 at. % Nb, the CTE(c)/CTE(a) ratio decreases with increasing Nb concentration, and above 44 at. % it increases until it decreases

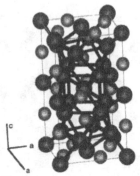

Figure 4. *Crystal structure of Nb₅Si₃.*

again as the crystal structure changes from that of Mo_5Si_3 to that of Nb_5Si_3. These results are qualitatively explained in terms of the spacing of the Mo atoms arranged in chains in the c-direction of Mo_5Si_3. They suggest that alloying additions which increase the Mo-Mo distance in the Mo-chains are effective in reducing the CTE anisotropy.

ACKNOWLEDGMENTS

This work was sponsored by the Division of Materials Sciences, Office of Basic Energy Sciences; and the Assistant Secretary for Energy Efficiency and Renewable Energy, Office of Transportation Technologies, as part of the High Temperature Materials Laboratory User Program, Oak Ridge National Laboratory. ORNL is operated by UT-Battelle, LLC, for the U.S. DOE under contract DE-AC05-00OR22725.

REFERENCES

1. A. K. Vasudevan and J. J. Petrovic, *Materials Science & Engineering* **A261**, 1 (1999).
2. M. K. Meyer, M. J. Kramer, and M. Akinca [sic], *Intermetallics* **4**, 273 (1996).
3. M. Akinc, M. K. Meyer, M. J. Kramer, A. J. Thom, J. J. Huebsch, and B. Cook, *Materials Science and Engineering* **A261**, 16 (1999).
4. F. Chu, D. J. Thoma, K. McClellan, P. Peralta, and Y. He, *Intermetallics* **7**, 611 (1999).
5. J. H. Schneibel, C. T. Liu, L. Heatherly, and M. J. Kramer, *Scripta Mater.* **38** [7], 1169 (1998).
6. C. L. Fu, X. Wang, Y. Y. Ye, and K. M. Ho, *Intermetallics* **7**, 179 (1999).
7. C. L. Fu and X. Wang, *Phil. Mag. Ltr.* **80**, 683 (2000).
8. Y. Ikarashi, K. Ishizaki, T. Nagai, Y. Hashizuka, and Y. Kondo, *Intermetallics* **4**, S141 (1996).
9. Y. S. Touloukian (Ed.), *Thermophysical Properties of High Temperature Solid Materials*, Macmillan, New York, 1967.
10. C. L. Fu, to be published.
11. C. J. Rawn, J. H. Schneibel, C. M. Hoffmann, and C. R. Hubbard , *The Crystal Structure and Thermal Expansion of Mo_5SiB_2*, in press.

Mat. Res. Soc. Symp. Proc. Vol. 646 © 2001 Materials Research Society

High-Temperature Compression Strength of Directionally Solidified Nb-Mo-W-Ti-Si *In-Situ* Composites

Hisatoshi Hirai[1], Tatsuo Tabaru[1], Jiangbo Sha[1], Hidetoshi Ueno[1], Akira Kitahara[1] and Shuji Hanada[2]
[1]Kyushu National Industrial Research Institute, Tosu, Saga 841-0052, Japan
[2]Institute for Materials Research, Tohoku University, Sendai, Miyagi 980-8577, Japan

ABSTRACT

Directionally solidified Nb-xMo-22Ti-18Si (x = 10, 20 and 30 mol%) and Nb-10Mo-yW-10Ti-18Si (y = 0, 5, 10 and 15 mol%) alloys were prepared. All of the alloys consisted of Nb solid solution and (Nb, Mo, (W,) Ti)$_5$Si$_3$ silicide. In compression tests performed on the Nb-xMo-22Ti-18Si alloys, a sample with x = 10 had the highest maximum stress at 1670 K, but it showed a sharp stress drop after reaching the maximum stress. Samples with x = 20 and 30 showed lower yield and lower maximum stress than that with x = 10, and they showed a small stress drop and an almost constant flow stress of 330 and 400 MPa, respectively. The sample with x = 20 showed a minimum creep rate ($\dot{\varepsilon}_m$) of 9.6 x 10^{-7} s^{-1} at 1670 K at a stress of 200 MPa. The yield stress of the Nb-10Mo-yW-10Ti-18Si alloys increased with increasing W content, although the stress decreased gradually after reaching the maximum stress and showed no steady state deformation. The $\dot{\varepsilon}_m$ of the sample with y = 15 was 1.4 x 10^{-7} s^{-1} at 1670 K at a stress of 200 MPa, which was much slower than that of Nb-20Mo-22Ti-18Si.

INTRODUCTION

Niobium (Nb) is known as one of the most promising refractory metals to be used as a base metal for future high-temperature structural materials. In view of its relatively low density (8.57 Mg/m^3 which is lower than that of Ni) and room temperature ductility, many studies have been performed in an effort to develop various Nb-base alloys since the 1960s [1]. By combining Nb-base alloys with intermetallics such as silicides and aluminides, their high-temperature strength can be further improved [2-8]. In our previous work [9], we prepared directionally solidified Nb-xMo-22Ti-18Si alloys with x = 0 to 30. We found that the samples with x = 10, 20 and 30 consist of Nb solid solution and Nb$_5$Si$_3$-base silicide, and these have excellent compressive strength at high temperatures. We also investigated the effect of added carbon on the mechanical properties of these alloys at room temperature and at high temperatures [10]. In the present study, tungsten (W) was added to directionally solidified Nb-Mo-Ti-Si alloys and their mechanical properties were investigated at elevated temperatures.

EXPERIMENTAL PROCEDURE

Specimens with nominal compositions of Nb-xMo-22Ti-18Si (x = 10, 20 and 30 mol%) and Nb-10Mo-yW-10Ti-18Si (y = 0, 5, 10 and 15 mol%) were arc-cast and then directionally solidified by the electron beam floating zone melting technique (EBFZ) and by the induction heating floating zone melting technique (FZ), respectively. EBFZ was done at a growth rate of 30 mm/h in vacuum, and FZ at a growth rate of 15 mm/h in a 0.1 MPa He atmosphere. Each specimen was annealed at 1870 K for 100 h in vacuum, and then the microstructure was examined by means of a scanning electron microscope (SEM). The composition of each phase in the specimen was analyzed by energy-dispersion X-ray spectroscopy (EDS) under the SEM. The bulk composition of the Nb-xMo-22Ti-18Si alloys (prepared by EBFZ) was analyzed by inductively coupled plasma spectroscopy (ICP). Compression tests were performed at 1670 and 1770 K in vacuum at an initial strain rate of 1 x 10^{-4} s^{-1}. Constant load compression creep tests were carried out at 1570, 1670 and 1770 K at initial stresses of 120, 200, 300 and 400 MPa in a 0.1 MPa Ar atmosphere. The dimensions of the compression and creep test pieces were 2 mm x 2 mm in cross-section and 5 mm in length. Each test piece was compressed parallel to the direction of growth.

During the EBFZ of the Nb-xMo-22Ti-18Si alloys, a relatively large weight loss was observed. The bulk composition of the specimens with x = 10, 20 and 30 mol% was Nb-10.4Mo-16.3Ti-19.8Si, Nb-21.0Mo-14.1Ti-21.2Si and Nb-31.6Mo-15.7Ti-20.4Si, respectively. Thus, the actual composition of each of these specimens is slightly different from the nominal composition, and this is mainly due to the evaporation of Ti during the EBFZ. We refer to each alloy by the nominal composition, however. As for specimens of Nb-10Mo-yW-10Ti-18Si alloys that were prepared by FZ, no weight loss was observed. For this reason, we assume that the bulk chemistry of these alloys did not change during processing.

RESULTS AND DISCUSSION

Microstructure and Composition of Each Phase

Back-scattered electron images of representative samples are shown in figure 1. All of the images were obtained from a cross-section parallel to the direction of growth. The bright phase is Nb solid solution (referred to as Nb$_{ss}$ hereafter) and the dark phase is (Nb, Mo, (W,) Ti)$_5$Si$_3$ silicide (referred to as 5-3 silicide hereafter). As seen in this figure, the microstructure is more or less oriented in the direction of growth, although it is not the microstructure typically observed in the case of directionally solidified materials. By adding a substantial amount of W, primary Nb$_{ss}$ appears and the microstructure becomes coarser (see figure 1 (d) for example). As seen in figure 1 (a) and (b), the addition of Mo results in a finer microstructure. As seen in figure 1 (a) and (c), Nb-10Mo-10Ti-18Si grown at 15 mm/h shows a finer microstructure than Nb-10Mo-22Ti-18Si grown at 30 mm/h. There is a possibility that the addition of Ti results in a coarser microstructure. Small precipitates of Nb$_{ss}$ are observed in the 5-3 silicide in the samples with coarser microstructure (see figure 1 (d)).

In each annealed sample, the Nb$_{ss}$ has an almost uniform composition. The 5-3 silicide has an almost constant Si concentration of 35 - 38 mol%, whereas the distribution of Ti, Nb and Mo is not homogeneous even in the same grain; the Ti-rich portion is poorer in Nb and Mo. The darker portion in the silicide in figure 1 is richer in Ti. The Ti concentration in the silicide is slightly higher than in the Nb$_{ss}$. On the other hand, Mo and W are more abundant in the Nb$_{ss}$ than in the 5-3 silicide. For example, the concentration of Mo and W in the Nb$_{ss}$ is about 20 and 36 mol%, respectively, whereas that in the 5-3 silicide is about 3.5 and 2.2 mol%, respectively, in the Nb-10Mo-15W-10Ti-18Si alloy. W is more abundant than Mo in the Nb$_{ss}$.

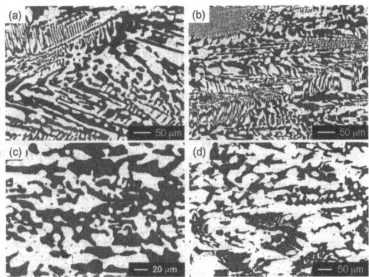

Figure 1. *Back-scattered electron image of annealed sample observed on a cross-section parallel to the growth direction (horizontal). Bright phase is niobium solid solution, dark phase (Nb, Mo, (W,) Ti)₅Si₃ silicide. (a) Nb-10Mo-22Ti-18Si. (b) Nb-30Mo-22Ti-18Si. (c) Nb-10Mo-10Ti-18Si. (d) Nb-10Ti-15W-10Ti-18Si.*

High-Temperature Compression Strength

The stress-strain curves of the Nb-xMo-22Ti-18Si alloys at 1470 and 1670 K are shown in figure 2 (a). At 1470 K, the 0.2 % yield stress $\sigma_{0.2}$ and the maximum stress σ_{max} of the Nb-xMo-22Ti-18Si alloy increases with increasing Mo content. After reaching the σ_{max}, however, all of

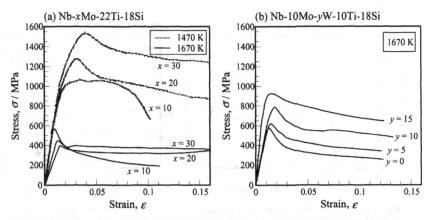

Figure 2. *Stress-strain curves of (a) Nb-xMo-22Ti-18Si alloys and (b) Nb-10Mo-yW-10Ti-18Si alloys at high temperatures and at an initial strain rate of 1 x 10⁻⁴ s⁻¹.*

the alloys show a stress drop, and steps caused by interphase boundary sliding and cracks were observed on the surface of heavily deformed test pieces. At 1670 K, the Nb-10Mo-22Ti-18Si alloy has the highest $\sigma_{0.2}$ and σ_{max}, but shows the largest stress drop and apparent cracking. On the other hand, the Nb-20Mo-22Ti-18Si and Nb-30Mo-22Ti-18Si alloys show only a small stress drop after reaching the σ_{max}, and show almost constant flow stresses of about 330 and 400 MPa, respectively.

The stress-strain curves of the Nb-10Mo-yW-10Ti-18Si alloys at 1670 K are shown in figure 2 (b). The $\sigma_{0.2}$ and σ_{max} of the Nb-10Mo-yW-10Ti-18Si alloy increase with increasing W content. The $\sigma_{0.2}$ and σ_{max} of the Nb-10Mo-10Ti-18Si alloy are almost the same as those of the Nb-10Mo-22Ti-18Si alloy, whereas those of the Nb-10Mo-15W-10Ti-18Si alloy reach as high as 890 and 920 MPa, respectively. The stress-strain curves of Nb-10Mo-yW-10Ti-18Si alloys have no constant flow stress regime, however, and the stress decreases gradually up to a strain of more than 0.1. The stress drop observed in the case of these alloys after reaching the σ_{max} is due to the cracking of the test piece during the compression test.

The $\sigma_{0.2}$ of each alloy at 1670 K is shown in figure 3. The effects of the microstructure and solid-solution strengthening are superimposed on the high-temperature strength of these alloys. In the case of the Nb-xMo-22Ti-18Si alloys, a specimen with a higher Mo content has a finer microstructure. For this reason, the Nb-10Mo-22Ti-18Si alloy with the coarsest microstructure shows the highest $\sigma_{0.2}$, and the effect of solid-solution strengthening of Mo becomes prominent in the case of the finer-grained Nb-20Mo-22Ti-18Si and Nb-30Mo-22Ti-18Si alloys (see figure 3 (a)). On the other hand, the addition of W coarsens the microstructure of the Nb-10Mo-yW-10Ti-18Si alloys. Thus, the effects of both microstructural change and solid-solution strengthening due to the added W make the alloy stronger at a higher W content (see figure 3 (b)). The effect of microstructure seems to be more conspicuous at higher temperature. Figure 3 (b) also shows the $\sigma_{0.2}$ of these alloys at 1770 K. The $\sigma_{0.2}$ of the Nb-10Mo-15W-10Ti-18Si alloy is still more than 600 MPa at this temperature.

Compression Creep Behavior

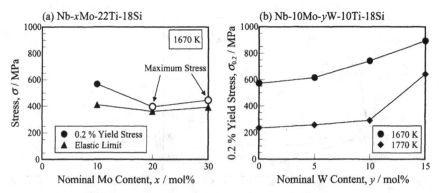

Figure 3. *High-temperature compression strength of alloys. (a) 0.2 % yield stress and elastic limit of Nb-xMo-22Ti-18Si alloys at 1670 K as a function of nominal Mo content. (b) 0.2 % yield stress of Nb-10Mo-yW-10Ti-18Si alloys at 1670 and 1770 K as a function of nominal W content.*

The creep curves of the Nb-20Mo-22Ti-18Si alloy at 1670 K at initial stresses of 120, 200 and 300 MPa are shown in figure 4 (a). The creep rate decreases with decreasing initial stress, and no specimen showed a steady state creep regime. The primary creep regime is relatively short and the minimum creep rate $\dot{\varepsilon}_m$ is obtained at a very small strain of less than 0.005, except for the result at an initial stress of 300 MPa, suggesting that the deformation mechanism of this alloy at 300 MPa is different from that at lower stresses. Figure 4 (b) shows the creep curves of the Nb-10Mo-yW-10Ti-18Si alloys at 1670 K at an initial stress of 200 MPa. The creep rate decreases with increasing W content. The primary creep regime is very short again, and the $\dot{\varepsilon}_m$ is obtained at a very small strain. Thus, the tertiary creep regime is dominant in the creep deformation of Nb-Mo-W-Ti-Si alloys. The creep in the Nb-10Mo-10Ti-18Si alloy is slower than the Nb-10Mo-22Ti-18Si alloy comparing at the initial stress of 200 MPa. This suggests that the creep resistance decreases with increasing Ti content, although the growth rates of these alloys are different.

Figure 5 summarizes the $\dot{\varepsilon}_m$ of the Nb-Mo-W-Ti-Si alloys as a function of the initial stress. The $\dot{\varepsilon}_m$ of the Nb-10Mo-5W-10Ti-18Si alloy is comparable with that of the Nb-20Mo-22Ti-18Si alloy, but the stress exponent, n, of $\dot{\varepsilon}_m$ is slightly lower than in the case of the latter. The n of the Nb-xMo-22Ti-18Si alloys is about 4.5, whereas that of the Nb-10Mo-5W-10Ti-18Si and Nb-10Mo-15W-10Ti-18Si alloy is 3.8 and 3.0, respectively. Although the reason why the n decreases with increasing W content is not clear, these n values indicate that the deformation mechanism of Nb-Mo-W-Ti-Si alloys is the dislocation creep mechanism (except for some cases at high initial stresses).

The high strength and creep resistance of these alloys at high temperatures are mainly attributable to the solid solution strengthening effect of Mo and W on Nb_{ss}. The size misfit parameter of the Mo and W atoms in Nb is known to be very large, -4.7 % and -4.3 %, respectively. It is also known that the diffusion rate of Mo and W in Nb is very slow [1]. For the above reasons, Mo and W are known as the elements most effective in solid-solution strengthening of Nb_{ss} [11]. The 5-3 silicide can be solid-solution strengthened by Mo and W also. The concentration of Mo and W are very small relative to those in Nb_{ss}, suggesting that the contribution of the solid-solution strengthening of the 5-3 silicide is not so large.

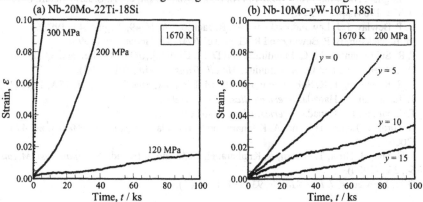

Figure 4. Creep curves of Nb-Mo-W-Ti-Si alloys. (a) Nb-20Mo-22Ti-18Si alloy at 1670 K and initial stress of 120, 200 and 300 MPa. (b) Nb-10Mo-yW-10Ti-18Si alloys at 1670 K and initial stress of 200 MPa.

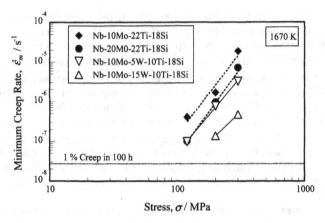

Figure 5. *Minimum creep rate of Nb-Mo-W-Ti-Si alloys at 1670 K plotted as a function of initial stress.*

CONCLUSIONS

In conclusion, the directionally solidified Nb-Mo-W-Ti-Si alloys have very high strength and creep resistance at high temperatures. Although we have to take into account the effect of microstructure on the mechanical properties of these alloys, it is obvious that Mo and W are elements quite effective in improving the high-temperature strength and creep resistance of Nb_{ss}/Nb silicide *in-situ* composites.

REFERENCES

1. C. English, *Niobium, Proc. of the International Symposium*, ed. H. Stuart (Metall. Soc. AIME, 1984) pp.239-324.
2. B. P. Bewlay, J. J. Lewandowski and M. R. Jackson, *JOM*, **49**, August, 44 (1997).
3. M. R. Jackson, B. P. Bewlay and R. G. Rowe, *JOM*, **47**, January, 39 (1996).
4. P. R. Subramanian, M. G. Mendiratta and D. M. Dimiduk, *JOM*, **48**, January, 33 (1996).
5. D. G. Mendiratta and D. M. Dimiduk, *Metall. Trans. A*, **24A**, 501 (1993).
6. E. S. K. Menon, P. R. Subramanian and D. M. Dimiduk, *Metall. Trans. A*, **27A**, 1647 (1996).
7. T. Tabaru and S. Hanada, *Intermetallics*, **6**, 735 (1998).
8. T. Tabaru and S. Hanada, *Intermetallics*, **7**, 807 (1999).
9. H. Hirai, T. Tabaru, H Ueno, A. Kitahara and S. Hanada, *J. Japan. Inst. Metals*, **64**, 474 (2000).
10. J.-B. Sha, H. Hirai, T. Tabaru, A. Kitahara, H. Ueno and S. Hanada, *J. Japan. Inst. Metals* **64**, 331 (2000).
11. G. D. McAdams, *J. Inst. Metals*, **93**, 559 (1964-65).

Mat. Res. Soc. Symp. Proc. Vol. 646 © 2001 Materials Research Society

Growth of The Mo_5SiB_2 Phase in A Mo_5Si_3/Mo_2B Diffusion Couple

Sungtae Kim[1], R. Sakidja[1], Z. F. Dong[1], J. H. Perepezko[1] and Yeon Wook Kim[2]
1 Department of Materials Science and Engineering, University of Wisconsin-Madison,
1509 University Ave, Madison, Wisconsin 53706, USA
2 Department of Materials Science and Engineering,
Keimyung University, Taegu, KOREA

ABSTRACT

The high melting temperature and oxidation resistance of the Mo_5SiB_2 (T_2) phase and multiphase microstructures incorporating the T_2 phase in the Mo-Si-B system have motivated further studies for applications in very high temperature environments. Since the long term microstructural stability is determined by diffusional processes, diffusion couples consisting of binary boride and silicide phases have been examined in order to evaluate the kinetics of T_2 phase development and the relative diffusivities controlling the kinetics. Long term annealing (500 hrs) of the Mo_5Si_3/Mo_2B diffusion couple yields the phase sequence of $Mo_5Si_3/Mo_3Si/T_2/Mo_2B$ at 1600°C. This indicates that the T_2 phase initiates and grows from the Mo_2B side to a thickness of about $32\mu m$ and the Mo_3Si phase initiates and grows from the Mo_5Si_3 side to a thickness of about $15\mu m$. Other annealing treatments allow for an analysis of the diffusion kinetics based upon the layer thickening and composition profile measurements. To identify the crystallographic growth direction of T_2 on Mo_2B, a wedge shaped TEM sample with very thin leading edge was prepared. Microstructure images indicate that the growth mode of the T_2 phase is columnar. There is a clear tendency for the growth of T_2 to be approximately normal to c-axis.

INTRODUCTION

The challenges of a high temperature environment (T>1400°C) impose severe material performance constraints in terms of melting point, oxidation resistance and structural functionality. Even though ceramic materials, intermetallic compounds and refractory metals with high melting temperature above 1500°C are available material choices, these materials as a single component rarely satisfy all the above requirements because of brittleness of ceramic materials and intermetallic compounds at low temperatures and oxidation problem and poor creep resistance of refractory metals at high temperatures. In order to address these issues multiphase designs such as composites or multicomponent alloys have been demonstrated to represent an effective approach. In this regard, the recent identification of refractory metal – refractory metal intermetallic compound combination based on Mo-Mo_5SiB_2 (T_2) two-phase equilibrium in the relevant ternary phase diagram [1,2], has attracted considerable interest for high temperature application. Based upon the electron probe microanalysis (EPMA) examination of phase compositions of the long-term annealed as-cast samples and rapidly solidified samples and x-ray diffraction determination of phase identity, the Mo-Si-B isothermal section at 1600°C is under continuing study. Figure 1 [2] shows the current determination for the Mo-MoB-Mo_5Si_3 region (excluding the two-phase equilibrium region of MoB-Mo_5Si_3) where the boundary lines

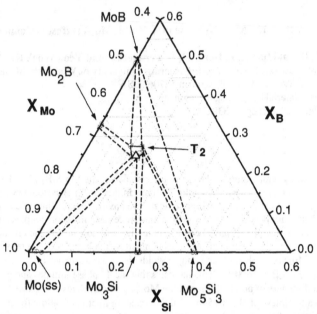

Figure 1. Ternary isotherm at 1600 °C of the Mo-rich Mo-Si-B system [2]

(dotted) are plotted on the basis of composition data for phases obtained using EPMA. The T_2 phase has an appreciable compositional homogeneity range around the stoichiometric composition and is in equilibrium with Mo_{ss}, Mo_2B, MoB, Mo_3Si and Mo_5Si_3. In order to confirm the phase stability in the Mo-Si-B system a diffusion study has been conducted to synthesize the T_2 phase during the diffusional reaction between binary boride and silicide phases. Knowledge of the diffusion kinetics is also essential in judging the microstructural phase stability. In this work a diffusion analysis has been conducted to examine growth of the T_2 phase between the Mo_2B and Mo_5Si_3 phases.

EXPERIMENTAL METHODS

Mo$_2$B and Mo$_5$Si$_3$ alloys were prepared by repetitive arc-melting of pure Mo and B, and pure Mo and Si in an atmosphere of high purity (oxygen gettered) argon and homogenized at 1600°C for 150 hours. In order to get smooth surfaces ingots were sectioned into 2-3 mm thick slices, ground with 14μm Al$_2$O$_3$ powder, and then rinsed with methyl alcohol in an ultrasonic cleaner and dried. The diffusion couples assembled from Mo$_2$B and Mo$_5$Si$_3$ slices were annealed at 1600°C in a flowing argon atmosphere. The diffusion couples were cut perpendicular to the interface using a diamond-impregnated low-speed saw. To identify the crystallographic growth direction of T_2 on Mo$_2$B by a transmission electron microscopy (TEM), a wedge shaped sample with very thin leading edge was prepared. Using a CAMECA SX50 EPMA, the compositions in a sectioned diffusion couple were line scanned. The EPMA analysis was carried out on the uncoated specimens at 7kV and a beam current of 30 nA, which resulted in decreased electron penetration and x-ray generation closer to the surface. This required a smaller absorption correction in the matrix correction. Serious interference problem of the first order Mo Mζ (Mz) line with the first order B Kα line was solved by a "Probe for Window" software [3].

RESULTS AND DISCUSSION

Annealing the Mo_2B/Mo_5Si_3 diffusion couple yields the diffusion pathway of either $Mo_2B/T_2/Mo_5Si_3$ or $Mo_2B/T_2/Mo_3Si/Mo_5Si_3$. Figure 2(a) shows the back-scattered electron (BSE) image from the cross-section of the diffusion couple annealed at 1600°C for 100 hrs and reveals the diffusion pathway of $Mo_2B/T_2/Mo_3Si/Mo_5Si_3$, which indicates that the Mo_3Si phase initiates and grows from the Mo_5Si_3 phase. The diffusion pathway of $Mo_2B/T_2/Mo_3Si/Mo_5Si_3$ that is drawn as a dotted line in Figure 3 crosses over an imaginary line between Mo_2B and Mo_5Si_3 to satisfy mass balance. Figure 2(b) shows the BSE image from the cross-section of the diffusion couple annealed at 1600°C for 400 hrs indicating a diffusion pathway of $Mo_2B/T_2/Mo_5Si_3$. In order to satisfy mass balance, the diffusion pathway that is drawn as a dashed line in Figure 3 is skewed to the Mo-rich side in the homogeneous T_2 phase region. The concentration profile as determined by EPMA measurement (Figure 5) supports the skewing of the diffusion pathway. The diffusion pathway of $Mo_2B/T_2/Mo_3Si/Mo_5Si_3$ has been found also in diffusion couples annealed at 1600 °C for 200 hrs. Observation of the diffusion pathway of $Mo_2B/T_2/Mo_3Si/Mo_5Si_3$ in 100 and 200 hrs annealed diffusion couples appears to provide a key consequence of changing diffusion pathway with annealing time, namely that the initial diffusion pathway is not the steady state path, but reflects a transient condition.

The T_2 grains grow with a columnar structure, which is identified from TEM analysis (Figure 4). The electron diffraction pattern on the T_2 grain indicates that the electron beam direction is [001] and normal to (001) plane. The concentration profile in Figure 5 reveals that Si atoms move from Mo_5Si_3 to Mo_2B and B atoms diffuse from Mo_2B to Mo_5Si_3. The diffusion pathway of $Mo_2B/T_2/Mo_3Si/Mo_5Si_3$ also supports that Si atoms move through the T_2 phase and

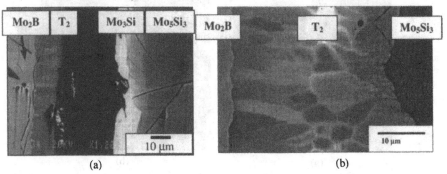

| Mo₂B | T₂ | Mo₃Si | Mo₅Si₃ | Mo₂B | T₂ | Mo₅Si₃ |

10 µm   10 µm

(a)   (b)

Figure 2. *BSE image of cross-section of the diffusion couple annealed at 1600 °C for (a) 100 and (b) 400 hrs*

react with Mo_2B to produce T_2. The T_2 (Mo_5SiB_2) phase interface with Mo_2B moves toward the Mo_2B phase side. In this regard, the T_2 phase initiates and grows in a columnar structure toward the Mo_2B phase side. The growth direction of the T_2 phase is shown in Figure 4(a). The back-scattered electron (BSE) image of the diffusion couple by the secondary electron microscopy (SEM) illustrates in Figure 2(b) that the T_2 phase appears to be growing toward both the Mo_5Si_3 and Mo_2B phases during the diffusion reaction. This indicates that there is a clear tendency for the growth of T_2 to be approximately normal to the c-axis of the T_2 structure.

The T_2 layer thickness in the Mo_2B/Mo_5Si_3 diffusion couple annealed at 1600°C for 400 hrs was ~38μm. Thickness measurements of the T_2 phase at other annealing times established that a diffusion control mechanism dominates the overall growth kinetics for the T_2 phase, which is influenced by the neighboring phases that change with time during heat treatment as the diffusion

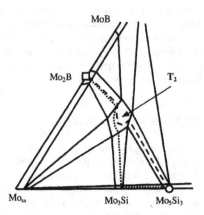

Figure 3. *Schematic diagram showing diffusion pathways of $Mo_2B/T_2/Mo_3Si/Mo_5Si_3$ (dotted path) and $Mo_2B/T_2/Mo_5Si_3$ (dashed path). Square and circle represent the diffusion couple end member compositions.*

pathway approaches steady state. In previous work, Dayananda et al. [4] proposed a method that determines the average interdiffusion coefficients from a single diffusion couple experiment. Kim et al. [5] suggested a method to determine the interdiffusion coefficient of the T_2 phase under assumptions of constant diffusion coefficients and narrow homogeneity ranges of composition. The Fick's first law in a ternary system is described by:

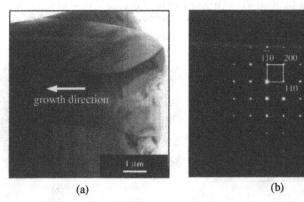

(a)   (b)

Figure 4. *(a) TEM image of the T_2 phase revealing a columnar growth and the growth direction (white arrow) and (b) electron diffraction pattern of the T_2 phase (001)*

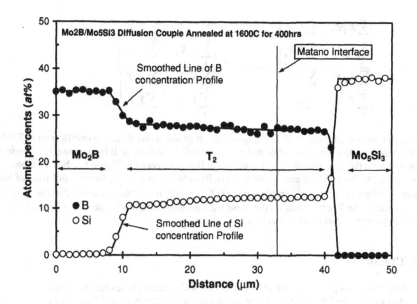

Figure 5. *Concentration profiles of Si and B in the Mo₂B/Mo₅Si₃ diffusion couple annealed at 1600 °C for 400 hrs.*

$$J_i(k) = -D_{i1}\left(\frac{\partial C_1}{\partial x}\right)_{x=x_k} - D_{i2}\left(\frac{\partial C_2}{\partial x}\right)_{x=x_k} \tag{1}$$

where i and j are component 1 or 2, $J_i(k)$ and $(\partial C_i/\partial x)_{x=x_k}$ are the atomic flow of i and concentration gradient of i at $x = x_k$, and D_{ij} is the interdiffusion coefficient related with the i atomic flow and concentration gradient of j. k represents the number of the T_2 composition data in Figure 5. In order to simplify equation (1), $J_i(k)$, $(\partial C_1/\partial x)_{x=x_k}$ and $(\partial C_2/\partial x)_{x=x_k}$ were substituted by Z_k, X_k and Y_k variables respectively. By applying the least squares data fitting method [6], which yields relationships between empirically obtained variables, with the T_2 composition data in Figure 5 the coefficients D_{i1} and D_{i2} are determined as (Table I):

Table I. *Evaluated interdiffusion coefficients of the T_2 phase in the Mo-Si-B system*

$D_{SiSi} = 2.56 \times 10^{-13}$ cm²/s	$D_{SiB} = 2.23 \times 10^{-13}$ cm²/s
$D_{BSi} = -1.93 \times 10^{-13}$ cm²/s	$D_{BB} = -1.70 \times 10^{-13}$ cm²/s

$$D_{i1} = - \frac{\begin{vmatrix} \sum_k X_k Z_k & \sum_k X_k Y_k \\ \sum_k Y_k Z_k & \sum_k Y_k Y_k \end{vmatrix}}{\begin{vmatrix} \sum_k X_k X_k & \sum_k X_k Y_k \\ \sum_k Y_k X_k & \sum_k Y_k Y_k \end{vmatrix}} \text{ and } D_{i2} = - \frac{\begin{vmatrix} \sum_k X_k X_k & \sum_k X_k Z_k \\ \sum_k Y_k X_k & \sum_k Y_k Z_k \end{vmatrix}}{\begin{vmatrix} \sum_k X_k X_k & \sum_k X_k Y_k \\ \sum_k Y_k X_k & \sum_k Y_k Y_k \end{vmatrix}} \tag{2}$$

Larger concentration gradient of Si compared the B gradient and diffusion coefficients in Table 1 yield a negative diffusion flow of Si and a positive diffusion flow of B by equation (1). These are in accord with the direction of Si and B diffusion flows observed in Figure 5, where Si atoms move from Mo_5Si_3 to Mo_2B and B atoms move from Mo_2B to Mo_5Si_3. The self diffusion of Mo in pure Mo (bcc) is 1.5×10^{-13} cm^2/sec at 1600°C [7], which is of the same order as the Si and B diffusivities in the T_2 phase and is consistent with the sluggish growth rate of the T_2 phase and the arrangement of metalloid atoms in the T_2 structure.

CONCLUSIONS

A diffusion couple consisting of binary boride and silicide phases has been examined in order to evaluate the kinetics of T_2 phase development and the relative diffusivities controlling the kinetics. The BSE and TEM analysis identified that the growth mode of the T_2 phase is columnar. The preferential growth direction of the T_2 phase in the diffusion couple is approximately normal to <001> direction. A new method to determine the interdiffusion coefficients from a single diffusion couple experiment indicates that the diffusion coefficients of the T_2 phase are in the order of 10^{-13} cm^2/s.

ACKNOWLEDGEMENTS

The support of the AFOSR (F49620-00-1-0077) is gratefully acknowledged. We thank Dr. J. Fournelle for expert guidance with EPMA measurements.

REFERENCES

[1] H. Nowotny, E. Dimakopoulou, H. Kudielka, Mh. Chem., **88**,180 (1957).
[2] S. Kim, R. Sakidja, J. Fournelle and J. H. Perepezko, to be published.
[3] J. H. Fournelle, J. J. Donovan, S. Kim and J. H. Perepezko, *Proceedings of 2nd Conference of the International Union of Microbeam Analysis Societies* (Kailua-Kona, Hawaii, Jul 9-13, 2000), pp. 425-426.
[4] M. A. Dayananda and Y. H. Sohn, Metall. Mater. Trans. A, **30A**, 535 (1999)
[5] S. Kim and J. H. Perepezko, to be published.
[6] K. Atkinson, *Elementary Numerical Analysis*, (John Wiley & Sons, New York, 1985) pp. 256
[7] K. Maier, H. Mehrer and G. Rein, Z. Metallkde., **90**, 271 (1979).

Mat. Res. Soc. Symp. Proc. Vol. 646 © 2001 Materials Research Society

Incommensurate Structure in Al-rich TiAl Alloys

S. Wang, D. Fort, I. P. Jones[1] and J. S. Abell
School of Metallurgy and Materials, The University of Birmingham, Birmingham, B15 2TT, England
[1]Also IRC in Materials for High Performance Applications, The University of Birmingham, B15 2TT, England

ABSTRACT

The microstructures of Ti-Al alloys in the range 56-62 at.% Al have been studied. These compounds consist of a basic $L1_0$ structure on which is superimposed an irrational modulation (Ti_3Al_5 related). This is incommensurate by occupation. It has two-dimensional modulations in which the wave vectors are $q_1 = \alpha(a^* + b^*)$ and $q_2 = \alpha(-a^* + b^*)$ where a^* and b^* are the $L1_0$ reciprocal basis and α is an irrational fraction. α decreases from 0.282 to 0.26 as the Al concentration increases to 60 at.% Al, i.e., the positions of the satellites change continuously with varying Al concentration. A third phase, $TiAl_2$, appears at a composition of 62 at.% Al which means that no Ti_3Al_5 (62.5 at.% Al) of stoichiometric composition can ever exist.

INTRODUCTION

TiAl single phase (γ phase, tetragonal $L1_0$ structure with $c_\gamma/a_\gamma \approx 1.02$ where a_γ and c_γ are lattice constants) has been found to exist over a wide composition range (49-56%Al at 700°C in Fig. 1). Another phase, Ti_3Al_5, was first reported to coexist with the $L1_0$ structure by Miida et al. [1]. This is a superstructure based on $L1_0$ and has tetragonal symmetry with the space group, P4/*mbm*, and lattice constants, $a_s = 2\sqrt{2}\, a_\gamma$ and $c_s = c_\gamma$. According to Miida et al. [1], the Ti_3Al_5 structure is commensurate and forms in the range 55-63at.%Al, the intensities of the additional reflections increasing with Al content. This phase has been confirmed to form precipitates at an alloy composition Ti-58at.%Al [2,3]. After studies of TiAl alloys over the composition range 56-63 at.% Al, Loiseau and Lasalmonie [4] reported weak irrational diffraction spots besides the main $L1_0$ reflections. It was originally proposed that these weak diffraction spots were caused by two phases with tetragonal and orthorhombic crystal structures [4], but this interpretation was questionable as the diffraction patterns could not be indexed in term of two such phases. Furthermore, dark field images using these superstructure reflections failed to reveal discrete precipitates, which should have been the case if other phases had been present. Subsequently, Loiseau et al. [5] re-interpreted these effects as being the result of islands surrounded by a phase having an intermediate composition, assumed to be Ti_7Al_{11} (a tetragonal structure with the lattice constants $a = 3a_\gamma$ and $c = c_\gamma$). However, the diffraction patterns do not support such an interpretation since no split satellite spots can be resolved, which would be the case if two intergrowth structures had formed. Indeed, high resolution electron microscopy failed to show such islands [6].

TiAl, and alloys based upon it, have received increasing attention in the recent two decades because of their low density and high strength up to 1000°C. On the other hand, their low

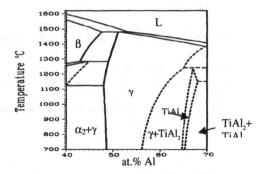

Fig. 1. The partial phase diagram for theTi-Al system by Murray (1986)

temperature ductility has limited their practical application. Existence of a peritectic reaction during solidification prevents crystal growth from the melt for the stoichiometric composition, and it is only possible to grow γ single crystals from Al-rich TiAl melts with an Al content larger than 54 at.%. Hence, deformation studies on single crystals have been limited to such Al-rich compositions. In fact the critical resolved shear stress (CRSS) especially for ordinary dislocations has been found to depend strongly on alloy composition [2].

Therefore, it is valid and interesting to examine the effect of Al concentration on the microstructure and plastic behaviour of TiAl.

EXPERIMENTAL DETAILS

Single-phase alloys with compositions Ti-56 at.% Al, Ti-58 at.% Al, Ti-60 at.% Al and Ti-62 at.% Al were prepared by arc-melting using high purity Ti (99.999 wt.%) and Al (99.999 wt.%) in an Ar atmosphere. To ensure homogeneity, the buttons were turned over and remelted 4 times. One as-cast alloy of composition 56 at.%Al was annealed at 1000°C for 48 hours. The buttons were then cut into thin slices using a spark machine and electropolished using a non-acid solution of 5.3g LiCl + 11.2g Mg(ClO$_4$)$_2$ in 500ml methanol and 100ml 2-butoxy-ethanol. TEM foils were examined using a JEOL 4000FX microscope or a Philips CM20 microscope both operating at 200 kV.

RESULTS

The γ phase TiAl has the L1$_0$ (CuAu) structure and the allowed reflections {hkl} have $h + k = 2n$ (n is an integer). Figs. 2a-2d show [001]$_\gamma$ diffraction patterns from Ti-56 at.% Al as cast, Ti-56 at.% Al as-cast followed by annealing at 1000°C/48hrs, Ti-58 at.% Al as cast and Ti-60 at.% Al as cast, respectively. The strong diffraction spots are produced by the γ phase; the weak spots, however, cannot be indexed according to the L1$_0$ structure (γ phase). They have the following characteristics: (1) the diffraction positions of the satellites varied with aluminium content; (2) the satellites intensified with increasing aluminium content; (3) there is

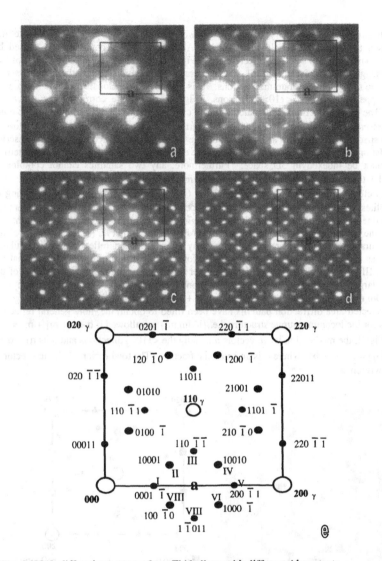

Fig. 2 [001]$_\gamma$ diffraction patterns from TiAl alloys with different Al contents
(a) Ti-56 at.%Al as-cast; (b) Ti-56 at.%Al as-cast after annealing at 1000°C for 48 hours;
(c) Ti-58 at.%Al as-cast; (d) Ti-60 at.%Al as-cast; (e) A schematic diagram. The open
and full circles represent the fundamental and the incommensurate structures, respectively

no 3D lattice translation symmetry; (4) the images were homogeneous when any of the satellites were chosen to form a dark-field micrograph. Accordingly, they could not be caused by precipitates, and the sharp satellite reflections (e.g. Fig. 2d) deny the possibility of islands of two phases as suggested by Loiseau et al [5]. These characteristics imply strongly that the diffraction spots are due to incommensurate modulations. This is a two-dimensional modulation since the satellites appear along two <110> directions. Fig. 2e shows the corresponding schematic diagram. The main reflections, designated by open circles, refer to the $L1_0$-structure. The other spots, represented by full circles, are defined in terms of five periodicities or basis vectors in reciprocal space. Three of them are the basis for the lattice of main reflections and two describe the modulations. By means of the incommensurate description, it is easy to index any satellites which cannot be indexed in 3D space. Furthermore, any two satellites cannot combine to produce the third since they originate from different reflection planes of the γ phase. For example, a cluster around the forbidden reflection $(100)_\gamma$ (position **a** in Fig. 2d), consisting of eight satellites, four strong and four weak, is produced from the five main reflections rather than from only one. (The first three indices show the origin of the fundamental reflections and the last two index the incommensurate modulations.) According to Janssen et al. [7], the intensity of the incommensurate reflections decreases rapidly away from the main reflections for modulated structures, as compared to the fundamental reflections. Consist with this, we observe that the satellites I, III, V and VII are weaker than the other four because the reciprocal spacings of the former are larger. From Fig. 2, the allowed reflections for satellites are: $\{h00m_1m_2\}$: $h + m_1 + m_2 = 2n$; $\{0k0m_1m_2\}$: $k + m_1 + m_2 = 2n$; $\{hk0m_1m_2\}$: $h + k + m_1 + m_2 = 2n$, where n is an integer. A series of selected area diffraction patterns have been tilted to obtain the more general reflection conditions for the incommensurate structure, i.e., $\{hklm_1m_2\}$ are allowed if $h + k + m_1 + m_2 = 2n$.

From Fig. 2, the modulation wave vectors are along the $<110>_\gamma$ directions and thus $\mathbf{q}_1 = \alpha(\mathbf{a}^* + \mathbf{b}^*)$ and $\mathbf{q}_2 = \alpha(-\mathbf{a}^* + \mathbf{b}^*)$ where α is an irrational fraction. The total reciprocal lattice vector **H** can be re-written as

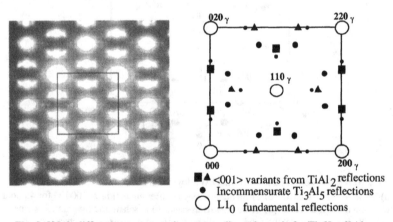

Fig. 3 $[001]_\gamma$ diffraction pattern and corresponding schematic for Ti-62at.%Al.

$$\mathbf{H} = h\mathbf{a}^* + k\mathbf{b}^* + l\mathbf{c}^* + m_1\alpha(\mathbf{a}^* + \mathbf{b}^*) + m_2\alpha(-\mathbf{a}^* + \mathbf{b}^*) \qquad (1)$$

α is determined to be 0.282, 0.272 and 0.260 for Ti-56at.%Al, Ti-58at.%Al and Ti-60at.%Al alloys, respectively.

Fig. 3 shows the $[001]_\gamma$ diffraction pattern for the Ti-62at.%Al alloy. Orthorhombic $TiAl_2$ with lattice parameters $a = 1.289$ nm, $b = 0.399$ nm and $c = 0.412$ nm, was present besides the Ti_3Al_5 incommensurate structure. The reciprocal vector $\mathbf{H}$ for the modulation structure can still be characterised by equation (1) with the slightly increased value of $\alpha = 0.269$, rather than a smaller value (the rational value of $\alpha = 0.250$ for the stoichiometric Ti_3Al_5 phase). This implies that the composition in Ti_3Al_5 domains is just over Ti-58at%Al where other Al-rich domains are $TiAl_2$.

DISCUSSION

Miida et al [1] reported a long-range ordered phase Ti_3Al_5, which had a tetragonal structure with the space group $P4/mbm$, in plasma-jet melted ingots with a nominal composition of Ti-63at.%Al followed by quenching from 1000°C and annealing below 800°C. Such a phase was observed as precipitates by Nakano et al. [2], at the composition of Ti-58at.%Al with as-cast and annealed samples. However, the present study and other work [4, 5, 8] show no long-range ordered Ti_3Al_5 phase, but instead an incommensurate structure with a modulation wave vector dependent on the Ti-Al composition. It is unclear what determines the structure as long-range order (precipitates) or incommensurate. It could be affected by the purity of the starting material or preparation with different levels of oxygen contamination, or the melting temperatures to which the order-disorder transformation is sensitive, or most importantly probably, the heat treatment.

The present work shows that the modulation irrational fraction, α, continuously decreases from 0.282 to 0.260 as Al content increases from 56 to 60at.%Al. The alloys have commensurate characteristics (precipitates) rather than short-range order (SRO) characteristics (e.g., Nakano et al 1996) since (1) no SRO clusters have been observed by HREM; (2) the spacing of the weak spots continuously changes with composition; (3) the satellites split well [9]. There are two ways of forming incommensurate structures in non-stoichiometric phases: either by atomic displacement or occupation (substitution). In the case of Al-rich Ti-Al alloys, excess Al atoms replace Ti atoms in the pure Ti plane of the γ structure [6] so that the incommensurate waves are caused by the occupation by excess aluminium of the Ti positions in the $L1_0$-structure. The incommensurability in the diffraction pattern results from the fact that spot positions are determined by the average spacing of occupation [10]. Such cases have been observed in other alloys by HREM, for example, a Cu-Zn-Al-Zr phase in a shape memory alloy [11].

Interestingly, Loiseau et al. [5] reported that Ti_3Al_5 incommensurate structures were unstable and disappeared after 1MeV electron irradiation. This could mean that the phase is unstable as suggested by these authors [5], or the order/disorder transformation may exist at a certain high temperature, like that in Cu-Zn alloys. Our studies (Figs. 3a and 3b) and those of Inui et al. [8], however, show that such structures are stable up to 1000°C, as shown in Fig. 3b.

CONCLUSIONS

In the present study, the structures of Al-rich TiAl alloys over the composition range from 56-62 at.%Al have been investigated by electron diffraction. The results show that the weak diffraction spots are related to an incommensurate Ti_3Al_5 structure with two-dimensional modulated composition waves. The modulation wavelength (inverse to the fraction α) increases with Al concentration up to 60 at.% and then drops as $TiAl_2$ appears. The modulation structure was caused by composition waves based on Ti_3Al_5, i.e., excess Al atoms substitute for Ti positions in the $L1_0$ Ti layer, rather than two composite phases as suggested by Loiseau et al. (1985) or short-range order. At 60at.%Al, the $TiAl_2$ phase coexists with incommensurate Ti_3Al_5.

ACKNOWLEDGEMENTS

This work is supported by the Engineering and Physical Sciences Research Council (Contract number GR/L 79960/01)

REFERENCES

1. R. Miida, S. Hashimoto and D. Watanabe, *Japanese J. Appl. Phy.*, **21**, L59 (1982).
2. T. Nakano, K. Matsumoto, T. Seno, K. Oma and Y. Umakoshi, *Phil. Mag. A*, **74**, 251 (1996).
3. T. Nakano, K. Hagihara, T. Seno, N. Sumida, M. Yamaoto and Y. Umakoshi, *Phil. Mag. Lett.*, **78**, 385 (1998).
4. A. Loiseau and A. Lasalmonie, *Acta Cryst.*, **B39**, 580 (1983).
5. A. Loiseau, A. Lasalmonie, G. Van Tendeloo, J. Van Landuyt and S. Amelincks, *Acta Cryst.*, **B41**, 411 (1985).
6. H. Inui, K. Chikugo, K. Kishida and M. Yamaguchi, *ICEM 14*, Cancun, Mexico,p175 (1998).
7. T. Janssen, A. Janner, A. Looijenga-Vos and P. M. de Wolff, *International Table for Crystallography*, Vol. C, Chaper 9.8. (1995).
8. H. Inui, M. Matsumuro, D. H. Wu and M. Yamaguchi, *Phil. Mag. A*, **75**, 395 (1997).
9. M. Sauvage and E. Parthé, *Acta Cryst.*, **A28**, 607 (1972).
10 P. M. De Wolff, *Acta Cryst.*, **A40**, 34 (1984).
11. C. Y. Chung, W. H. Zou, X. D. Han, C. W. H. Lam, M.Gao, X. F. Duan and J. K. L. Lai, *Acta Mater.*, **46**, 5541 (1998).

Mat. Res. Soc. Symp. Proc. Vol. 646 © 2001 Materials Research Society

Effect of surface defects on the fatigue behavior of a cast TiAl alloy.

M. Nazmy, M. Staubli, G.Onofrio[1] and V.Lupinc[1]
ALSTOM POWER, Baden CH
[1]CNR-TeMPE, Milano Italy

ABSTRACT

The effect of surface defects on the performance of TiAl-base alloys is an issue of importance in contemplating their application into engine components. Due to the relatively low ductility and low impact resistance of gamma alloys the validation of models for estimating economic life and for safe-life approaches employed for components becomes of great importance. Surface defects can be attributed to various sources during the manufacturing or handling of the components. In fact, little is known about the detrimental effects of surface defects on gamma alloys. In the present study, the effect of artificially introduced surface defects, on the high cycle fatigue behavior of the Ti-47Al-2W-0.5Si, will be investigated and correlated with the crack growth behavior at 700°C. The results are reported in the form of the Kitagawa diagram in which the safe and unsafe zones for crack advance and fracture are defined.

INTRODUCTION

In recent years significant effort has been put in attempt to use γ-based titanium aluminide alloys in gas turbine engine industry (1, 2). The low density and high stiffness of this class of materials relative to the more traditional nickel base superalloys and the mechanical properties retention up to temperatures of 700 – 750°C have made γ-TiAl an attractive candidate for use in the low pressure stage of gas turbine engines, both in large industrial turbines and in aeronautical engines, or as turbocharger for Diesel engines (3, 4).

At ALSTOM Power a proprietary gamma TiAl based intermetallic alloy named ABB-2 (Ti-47Al-2W-0.5Si nominal at. %) has been developed (5). Different heat treatments can produce duplex or nearly fully lamellar microstructure meant for different applications, e.g. turbochargers or large low pressure gas turbine blades. A number of investigations were carried out on this alloy to study several aspects of its mechanical properties as well as its oxidation behavior (6, 7).

In highly loaded applications adequate knowledge of the critical mechanical properties, e.g. creep strength, low and high cycle fatigue, and toughness of these materials, is essential. In particular high cycle fatigue has been recognized as one of the important parameters in design. The prediction of its dependence on defect size is of critical importance (8).

The fatigue strength estimation of the rotating components containing defects is, hence, of interest and in order to obtain an effective prediction of the fatigue limit the usual S-N curve approaches are of limited relevance. A more effective method consists in treating the defects like a crack and in calculating the limiting cyclic stress level below which this crack-like defect does not propagate.

The aim of this paper is to study the effect of artificially induced surface defects on the fatigue behavior of ABB-2 γ-TiAl alloy at 700°C, and correlate this behavior to the endurance limit and the crack growth threshold, i.e. determine the defect tolerance of the material.

These surface defects might come either from the manufacturing or from an incident during the performance of the component. Hence the importance and relevance of the aim of this investigation are obvious to the application (9).

MATERIAL

The 25.4 mm diameter bars, used in the present investigation, were cast, hot isostatically pressed at 1260°C and 172 MPa for 4 hours and heat treated at 1350°C for 1 h followed by gas fan cool, then at 1000°C for 6 h and furnace cooling by Howmet. The chemical composition of the material, measured by X-ray fluorescence using the relevant standards, is reported in Tab. 1.

Table 1 Chemical composition (wt %).

Al	Si	W	Fe	O	H	N	Ti
30.62	0.32	9.00	0.044	0.077	0.002	0.009	Bal.

The metallographic observation showed a duplex microstructure consisting of mainly lamellar γ/α_2 colonies and globular γ grains of 20 μm average grain size in the central part of the cast bar, Fig. 1, while in the columnar part of the cast bar the colonies were around 100 μm average size. As already reported (6), this kind of microstructure shows a good compromise of room temperature ductility and creep resistance.

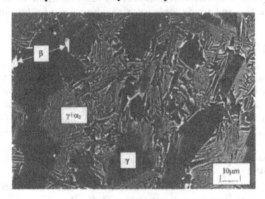

Fig 1. Microstructure of the central part of the ABB-2 alloy cast bar. Primary silicides are not observed in this micrograph.

TESTING

In order to evaluate the fatigue properties high cycle fatigue (HCF) and fatigue crack propagation rate (FCPR) tests have been performed at 700°C, a temperature of interest for application of the studied material and about 50°C below the brittle to ductile transition temperature as shown in Fig. 2.

The specimens were spark eroded from cast bars. The HCF testing was done at 40 Hz, R (= $\sigma_{min}/\sigma_{max}$) = 0 and temperature of 700°C on smooth specimens of 25 mm gauge length and 5 mm gauge diameter, as well as on specimens with small notches simulating a surface defect. These artificial surface defects were introduced using a special milling tool with a 30 μm radius diamond tip: the cracks were from 0.04 to 0.8 mm dip, covering both short and long crack sizes. The initial depth was determined by a microscope and confirmed after the fracture of the specimens. The HCF applied stress was calculated for each crack depth from

Eq. 1, assuming that the artificial crack in the specimen can be approximately considered as a surface crack in a specimen subjected to uniform tensile stress. The relationship for this case is (10):

$$\Delta K = F \, \Delta\sigma \, \sqrt{\pi a} \tag{1}$$

where the boundary correction factor F is about 1.1 for the present case.

The fatigue crack growth behavior and specially the crack growth threshold were measured using single edge notched tension (SENT) specimens of 5 mm x 12 mm cross-section and 1 mm deep starter notch, pre-cracked at room temperature. FCGR was measured at 700°C, a frequency of 10 Hz and R = 0.05. The following equation has been adopted for the stress intensity factor ΔK dependence on stress σ and crack depth a where W is the SENT specimen dimension in the crack growth direction (11):

$$\Delta K = (1-R) \, \sigma \, \sqrt{\pi a} \, \{1.12 - 0.23(a/W) + 10.6(a/W)^2 - 21.7(a/W)^3 + 30.4(a/W)^4\} \tag{2}$$

Tensile tests have been performed at 7×10^{-4} s^{-1} strain rate from room temperature to 900°C. The results on the yield strength, ultimate tensile strength and elongation percent of the specimens tested at different temperatures are given in Fig. 2. The excellent resistance retention of the alloy is observable up to 800°C with about 500 MPa yield and 600 MPa ultimate tensile stress. The brittle to ductile transition occurs in the 700-800°C range.

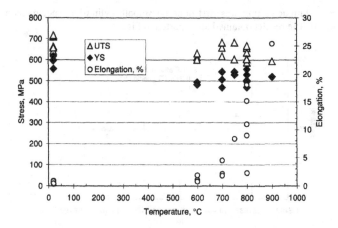

Fig.2 Tensile properties, as a function of the temperature, of the ABB-2 alloy.

RESULTS AND DISCUSSION

The fatigue crack growth behavior at 700°C, at the frequency of 10 Hz and R = 0.05 is shown in Fig. 3. The fatigue crack propagation test results have shown a significant resistance of this alloy to the fast defect propagation with a toughness value of about 40 MPa√m and, most important for this work, a threshold of about 8.5 MPa√m, not far from the value found in the literature (12,13) considering the different specimen geometry, different alloys and different microstructure.

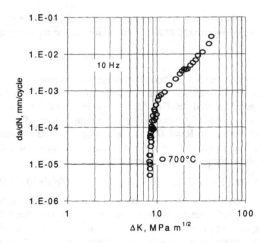

Fig. 3 Fatigue crack propagation rate of the ABB-2 alloy.

High cycle fatigue testing was first done on smooth cylindrical specimens at 40 Hz, 700°C, and R = 0. The results obtained are reported in Fig. 4.

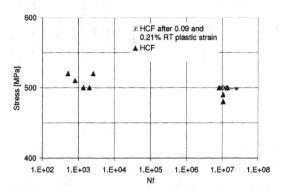

Fig. 4 High cycle fatigue behavior of the ABB-2 alloy.

A sizable number of cylindrical specimens were used for studying the effect of artificial surface cracks on the high cycle fatigue behavior. The applied stresses were calculated with Eq. 1 using ΔK = 8 and 9 MPa√m. The specimens tested at stresses corresponding to ΔK = 8 MPa√m did not show any crack growth after 10^7 cycles, i.e. run-outs. As the stresses were successively increased step-wise in order to increase the ΔK value from 8 to 9 MPa√m, most of the specimens did not exhibit any crack growth. At higher stresses, the specimens failed before reaching 10^7 cycles. These results, obtained at 700°C, are shown in Fig. 5 in the form of the so called Kitagawa diagram. The lines in the diagram correspond to the stresses for different crack depths and for ΔK = 8 and 9 MPa√m. The points below ΔK = 9 MPa√m line, represent non propagating cracks, i.e. run-out specimens, and the points above this line represent failed specimens. This behavior is in close agreement with the fatigue crack propagation $\Delta K_{threshold}$ value (=8.5 MPa√m) shown in Fig. 3. At crack lengths smaller than the critical value $a_c \approx 0.075$ mm the defect has no influence on the fatigue limit, as shown in Fig. 5. This critical size is consistent with an average size of microstructural features such as γ grains and lamellar colonies that range from 0.02 to 0.2 mm, being closer to the small values in the central part of the cast bar and to the large value in the columnar grain region.

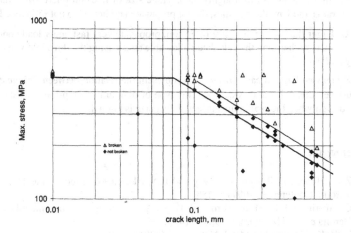

Fig. 5 Kitagawa plot for the ABB-2 alloy at 700°C. The two negative slope lines represent the ΔK values of 8 and 9 MPa √m.

The fractured surface of a specimen tested at a stress of 314 MPa with a = 0.42 mm is shown in Fig. 6. The stable crack propagation, followed by fast fracture, are clearly observable.

Two smooth specimens were pre-strained at room temperature to 0.09 % and to 0.21% plastic strain and then high cycle fatigue tested at 700°C and 500 MPa. Since, as shown in Fig.4, these specimens did not fail, it can be said that the applied pre-deformation had no effect on fatigue life.

fast fracture → | ←-FCP→ | ←-notch

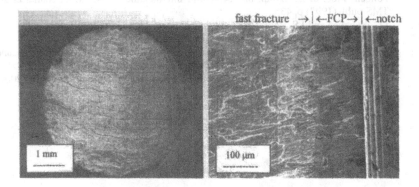

Fig. 6 Fracture surface after test with 0.47 mm notch simulating an artificial defect.

CONCLUSIONS

The conclusions of this work on the effect of surface defects on the fatigue behavior of ABB-2 alloy at 700°C can be synthesized as follows:

1) The fatigue life is strongly stress sensitive.

2) Cracks smaller than 75 μm, roughly the average size of relevant microstructural features such as γ grains and lamellar colonies, do not propagate and hence do not influence HCF life.

3) Cracks, larger than the critical size of 75 μm, propagate very fast under load control if the applied stress is higher than the calculated allowable stress using the threshold value of stress intensity factor.

4) The Kitagawa–diagram can satisfactorily represent the relation between fatigue endurance limit, surface defect size and $\Delta K_{threshold}$. This type of diagram can represent a useful tool for safe-life design when the maximum surface defect size is known.

REFERENCES

1   Z.W. Huang, P. Bowen, S. Davey and P.A. Blenkinsop, Proc. of the 4th Int. Parsons Turbine Conference (1997) p. 489.
2   C. Austin and T.J. Kelly, Proc. of 1st Symp. on "Structural Intermetallics", eds. R. Darolia et al., TMS publ. (1993) p. 1433.
3   Y.W. Kim, Journal of Metal, Vol. 46 (7) (1994) p. 30.
4   Y.W. Kim and D.M. Dimiduk, Proc. of 2nd Symp. on "Structural Intermetallics", eds. M.V. Nathal et al., TMS publ. (1997) p. 531.
5   M. Nazmy and M. Staubli, US Patent 5,207,982 & EP. 45505 B1
6   V. Lupinc, M. Marchionni, M. Nazmy, G. Onofrio, L. Rémy, M. Staubli and W.M. Yin , Proc. of "Structural Intermetallics 1997", eds. M.V. Nathal et al., TMS publ. (1997) p. 515.
7   A. Tomasi, S. Gialanella, P.G. Orsini and M. Nazmy, MRS Symposium Proceedings **Vol. 364** (1995) p. 999.
8   B.A. Cowles, Int. J. Fract., Vol. 80 (1996) p.147.
9   P.S. Steif, J.W. Jones, T. Harding, M.P. Rubal, V.Z. Gandelsman, N. Biery and T.M. Pollock, Proc. of 2nd Symp. on "Structural Intermetallics ", eds M.V. Nathal et al., TMS publ. (1997) p. 435.
10   J.C. Newman and I.S. Raju, NASA Technical Note, TP-1578 (1979).
11   W.F. Brown and J.E. Srowley, ASTM-STP 410 (1966) p.1.
12   A.L. McKelvey, K.T. Venkateswara Rao and R.O. Ritchie, Metallurgical and Materials Transactions A, Vol. 31A (2000) p.1413.
13   J.Lou, C. Mercer and W.O Soboyejo, MRS Symposium Proceeding, **Vol. 646** (2001) N1.7.1

Mat. Res. Soc. Symp. Proc. Vol. 646 © 2001 Materials Research Society

The generation of faulted dipoles and dislocation activity in γ-TiAl

Fabienne Grégori[1] and Patrick Veyssière
LEM, CNRS-ONERA, BP 72, 92322 Châtillon cedex, France.
[1]LPMTM, Institut Galilée, 99 av. J.B. Clément, 93430 Villetaneuse, France.

ABSTRACT

Faulted dipoles (FDs) nucleates at the tip of a <011] screw lock while the latter is unzipped by the connecting mixed segment. The transformation is favoured by the asymmetrical dissociation mode of the <011] dislocation, into two partials with 1/6<112] and 1/6<154] Burgers vectors bordering an intrinsic stacking fault. Where the parent <011] dislocation switches from a locked, 3D non planar to a fully coplanar core configuration at the connecting mixed segment, the 1/6<154] partial assumes a zonal form capable of expelling a partial dislocation with 1/6<112] Burgers vector. The latter transforms the fault from intrinsic to extrinsic and leaves a 1/3<121] partial with a zonal core on the other side of the fault. The presence of FDs reflects the propensity of <011] dislocations to form screw locks and thus attests to the operation of <011] dislocations during deformation.

INTRODUCTION

First reported by Shechtmann et al. [1], faulted dipoles (FDs) are a profuse debris in γ-TiAl alloys. They exhibit rather reproducible characteristics for deformation temperatures up to approximately 400°C. FDs consist of a hairpin-like partial dislocation elongated in a <011] direction and bordering a ribbon of extrinsic stacking fault (ESF). FDs are referred to as FD$_I$ or FD$_{II}$ according to whether they are bordered by a partial dislocation with 1/6<112] or 1/3<121] Burgers vector, respectively (Figure 1). The two displacement vectors are equivalent within a unit translation (i.e. 1/2<110], [2]). They are usually connected to a <011] or 1/2<112] dislocation; they may also assume the form of a closed faulted loop in places. When attached to a <011] dislocation, FDs are elongated in the screw direction of this dislocation. When attached to a 1/2<112] dislocation, they remain aligned along a <011] direction [2-5].

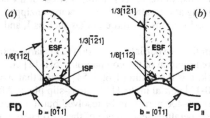

Figure 1. The two types of faulted dipoles that can be connected to a dislocation with <011] Burgers vector. (a) Type I. (b) Type II. The two types correspond to one another by interchanging the partials that border the extrinsic stacking fault (ESF) while the partial that borders the intrinsic stacking fault (ISF) on the perfect crystal side remains unchanged.

EXPERIMENTAL PROCEDURE

Single crystals of Al-rich γ-TiAl (54.3 at.% Al), pre-annealed for 100 h at 1200°C then 11 h at 900°C, were deformed along a near-[1 21 6] direction activating the [0-11](-111) slip system primarily.

Dislocations with 1/2[110] and 1/2[1-12] Burgers vectors nevertheless coexisted with the [0-11] dislocations in the (-111) plane (see "o" and "s" Figure 2). The presence of the 1/2[110] and 1/2[1-12] dislocations originates from the decomposition of [0-11] dislocations [6]. No slip planes other than (-111) were activated save for a slight cross-slip activity in order to form locks on [0-11] dislocations. Transmission electron microscope investigations were conducted on foils sectioned parallel to the slip plane, thus enabling a direct study of a possible correlation between FDs and operating dislocations.

OBSERVATIONS

As explained in the following, the mode of dissociation of [0-11] dislocations plays a crucial role in the generation of faulted dipoles. In the present experimental conditions, <011] dislocations are dissociated in two partials. Two alternative reactions may actually take place which, for a [0-11] Burgers vector, are written

$$[0-11] \rightarrow 1/2[0-11] + APB + 1/2[0-11] \tag{1}$$
$$[0-11] \rightarrow 1/6[1-12] + ISF + 1/6[-1-54] \tag{2}$$

referred to in the following as symmetrical and asymmetrical, respectively. We have determined that it is the asymmetrical configuration that prevails after deformation at room temperature (Figure 2, insets), [6]) but that the fraction of segments dissociated under the symmetrical mode increases beyond 400°C [7-8]. It was also demonstrated that the 1/6[1-12] and 1/6[-1-54] partials in reaction (2) assume the leading and trailing positions, respectively.

The key observation regarding the mechanism of formation of FDs is that of a local sub-splitting of [0-11] dislocations into a threefold configuration, always located at tips of screw locks (Figure 2, II_a, II_b and bottom left-hand side). The threefold configurations are not formed indifferently but sit at a specific extremity of the screw lock. This extremity is itself determined by the sense of displacement under stress of the connecting mixed segment. It is always sitting at the extremity that is being unzipped, as is indicated by the curvature of the [0-11] dislocations in Figure 2. FDs such as I_a and I_b are not aligned with screw locks.

The distance between the two partials of reaction (2) is large enough to ensure that the lock forms without recombination of the companion 1/6[1-12] and 1/6[-1-54] partials (Figure 3(a)). In the lock, it is the trailing 1/6[-1-54] partial that sub-dissociates forming an APB by glide of a 1/2[0-11] partial in the (111) cross-slip plane. After deformation at room temperature the APB strip is too narrow to be resolved but it is enough after deformation at 400°C. More than the temperature dependence of the APB energy, we believe that this difference in APB width between the two deformation temperatures reflects the friction exerted by the lattice on the 1/2[0-11] partial in the cross-slip plane.

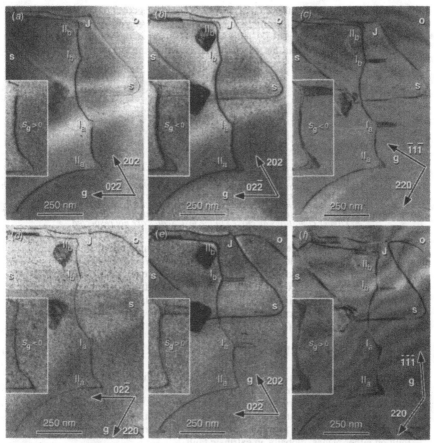

Figure 2. Properties relevant to the generation of FDs (Negative prints of weak-beam views). The contrast asymmetry under $g = 02\text{-}2$, $\pm s_g$ (figures (a) and (b)), together with the presence of one and two peaks under $g = \pm220$ and ±202, respectively, reveal a dissociation according to reaction (2) [6]. A reaction of decomposition into an ordinary dislocation (o) and a $1/2<112>$ dislocation (s) is visible at J. It is only in (c) that "o" is out of contrast, the weakness of its image in (a) and (b) is due to the relatively small value taken by $g.b$ (±1). Threefold configurations are present at the tips of screw locks at II_a and II_b, forming embryos of FDs of type II. The two FDs labelled I_a and I_b are of type I. Stacking faults are in contrast in (c) and (f). The difference in contrast between I_a and I_b and II_a and II_b stems from the position of the thickness fringes.

THE FORMATION OF FAULTED DIPOLES

In spite of the unusually long Burgers vector of the 1/6[-1-54] partial there is no documented evidence of its sub-splitting under the glissile form, i.e. in the non-screw orientation, which we then consider as compact or but slightly spread in the (-111) plane (this simplification is safe as far as the following reasoning is concerned). We are thus led to investigate what changes in configuration may occur between the cores of the locked, dissociated screw and of the glissile, undissociated non-screw segments, as this imposes a core constriction along the trailing 1/6[-1-54] partial where the lock terminates (Figure 3).

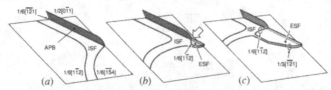

Figure 3. The unzipping of a portion of a screw lock by a mobile kink. (a) The configuration prior to unzipping. The mobile segment is asymmetrically dissociated into a leading Shockley and a trailing 1/6<154] partial (reaction (2)). The lock proceeds from the sub-dissociation of the latter in the cross-slip plane. (b) A Shockley partial is nucleated from the corner of the configuration, one plane above the habit plane of the ISF forming an ESF by glide (the location pointed with the open arrow is detailed in Figure 4). (c) The configuration expands upon stress generating an elongated FD under conditions discussed in the text. Scales are not respected.

Figure 4 is a formal representation of changes in core configuration at the interconnection between the lock and the mixed segment. Wherever the cross-slipped 1/2[0-11] partial assumes an exact screw orientation, its core relaxes slightly in a {111} plane. Although there is no experimental evidence for a splitting, one may always identify the core of the 1/2[0-11] partial as a pair of Shockley partials. As the 1/2[0-11] partial reaches the (-111) plane adjacent to the ISF plane, it may in addition undergo a slight spreading formally equivalent to

$$1/2[0-11] \rightarrow 1/6[1-12] + CSF + 1/6[-1-21] \tag{3}$$

Simultaneously, it reacts with the 1/6[-1-21] partial located in the ISF plane along a very short portion in the [0-11] direction. Therefore, the 1/6[-1-54] partial takes a zonal core that can be alternatively regarded as the superimposition on two adjacent (-111) planes
- of a 1/6[1-12] partial located on the upper plane that can escape from this core,
- of two 1/6[-1-21] partials forming a zonal 1/3[-1-21] partial that borders a stacking fault on either planes, globally an extrinsic fault.

This configuration corresponds to FD_{II} (Figure 1(b)). Now, in order to form FD_I, it is a non-zonal 1/3[-1-21] partial that should escape from the zonal 1/6[-1-54] partial on the upper plane forming an ESF. The other side of the ESF is then bordered by a 1/6[1-12] partial. As is classically postulated for ESFs in FCC crystals, this latter partial takes a zonal core according to the reaction

$$1/6[1-12] \rightarrow 1/6[211] + 1/6[-1-21] \tag{4}$$

It is noted that in both cases of a FD_I and of a FD_{II}, the total Burgers vector in each of the two adjacent planes is 1/2[0-11]. This implies that, formally, the two partials located on the two adjacent (-111) planes and forming the zonal dislocation are connected by an APB.

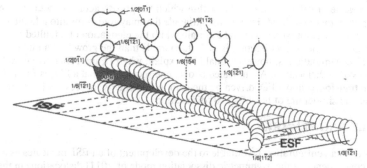

Figure 4. A schematic representation of the core configuration of the 1/6<154] dislocation in the region of transition between the screw lock and the connecting near-screw mobile segment (see open arrow in Figure 3(b)). The profiles associated with the vertical dashed lines represent sections at selected locations. Real scales are not accounted for.

DISCUSSION

Regarding the question of the formation of FD_I *versus* FD_{II}, it is worth keeping in mind that the latter can transform into the former but that the reverse reaction is totally unlikely. The reaction $FD_{II} \rightarrow FD_I$ corresponds to a series of transformations that greatly affect the whole configuration [2]. In the present case of a [0-11] dislocation in the (-111) plane, the transformation proceeds by emission of an ordinary 1/2[-1-10] dislocation from the tip of the faulted by a decomposition which is written

$$1/3[-1-21] \rightarrow 1/2[-1-10] + 1/6[1-12] \tag{5}$$

Atomic rearrangements between the ordinary dislocation and the first 1/6[1-12] partial that closes the hairpin may help form a 1/3[-1-21] partial with a stable core, that is, a faulted dipole of type I (Figure 1(a)). This is what we observe at I_a and I_b (Figure 2). However, since the ordinary dislocation exchanges but modest elastic interactions with 1/6[1-12] partials, it may also cross through these to reach the trailing [0-11] dislocation from which it may once again escape leaving a 1/2[1-12] dislocation. The latter actually look as though they trail FD_Is during deformation but there are reasons to believe that this interpretation is uncertain. We believe that 1/2<112> dislocations are all but mobile and that their presence reflects the decomposition rate of <011] dislocations, be it by the above manoeuvre or else by direct generation of an ordinary dislocation from a cusped <011] dislocation [6].

The growth of a faulted dipole upon unzipping of a screw lock is governed by several parameters such as the SF energies, γ, the applied stress, σ, and the angle α between the mobile (unzipping) mixed segment and the screw lock. The difference between the ESF and the ISF energies is probably modest and it should not influence the transformation. Mechanical work is however needed in order to create SF ribbons and this is where the SF energy intervenes and where the applied stress helps. The angle α determines the gain/loss in line energy and elastic interactions between segments. Numerical simulations show that there is a range of values of γ, σ and α within which FD growth actually occurs [9]. Within the simplified model utilised, i.e. a faulted triangle that may transform into a faulted parallelogram, it is found that beyond a certain width the elongation of a faulted parallelogram in the screw direction is more favourable than the growth of a faulted triangle. The simulation is, however, unable to explain why FD_I and FD_{II} may be trailed by the same <011] dislocation. As indicated above, one scenario is that a FD_{II} is formed first and then transforms into a FD_I driven by the gain in line energy that results from the emission and shortening of the 1/2<110] dislocation.

CONCLUDING REMARKS

In γ–TiAl, a configuration favourable to the development of an ESF nucleates as a natural consequence of the asymmetric dissociation mode of <011] dislocations in their glissile form and of the particular 3D configuration assumed by screw locks.

The microstructure involving faulted dipoles is consistent with the following scenario. Before all, the presence of FDs reflects the rate of formation of screw locks in the course of the expansion of <011] dislocations under stress. Transformation of FD_{II} into FD_I may occur by nucleation of an ordinary dislocation at the FD tip yielding configurations comprised of FD_{II}s and FD_Is all trailed by the same <011] dislocation. The ordinary dislocation may also be fully torn off the <011] dislocation leaving 1/2<112] dislocations interconnected with trailing-like FD_Is.

FDs are not expected to contribute to strain and it is rather difficult to envisage that they could take part to the brittle to ductile transition as this has been proposed in the past.

ACKNOWLEDGEMENTS

Philippe Penhoud (LEM) is gratefully acknowledged for his superb achievements in crystal growth, deformation tests and thin foil preparations. By offering us a generous access to his crystal growth facilities, Pr Yamaguchi (Kyoto University) has provided us with ideal working material.

REFERENCES

1. D. Shechtmann, M. J. Blackburn and A. Lipsitt, *Met. Trans.*, **5**, 1373 (1974).
2. B. Viguier and K. J. Hemker, *Phil. Mag. A*, **73**, 575 (1996).
3. G. Hug, A. Loiseau and A. Lasalmonie, *Phil. Mag. A*, **54**, 47 (1986).
4. G. Hug, A. Loiseau and P. Veyssière, *Phil. Mag. A*, **57**, 499 (1988).
5. G. Hug and P. Veyssière, *Gamma titanium aluminides*, Eds. Y.-W. Kim, R. Wagner and M. Yamaguchi (TMS, Warrendale, 1996) pp. 291.
6. F. Grégori and P. Veyssière, *Phil. Mag. A*, **80**, 2913 (2000).
7. F. Grégori, Ph. D. Thesis, University of Paris VI (1999).
8. F. Grégori and P. Veyssière, *Multiscale Phenomena in Materials - Experiments and Modeling*, Eds. I. M. Robertson, D. H. Lassila, B. Devincre and R. Phillips (MRS, Warrendale, PA, 2000), vol. 578, pp. 195.
9. F. Grégori and P. Veyssière, *Phil. Mag. A*, **80**, 2933 (2000).

Mat. Res. Soc. Symp. Proc. Vol. 646 © 2001 Materials Research Society

High temperature gas nitridation and wear resistance of TiAl based alloys

Bin Zhao, Jian Sun, Jiansheng Wu and Fei Wang
Key Laboratory of the Ministry of Education for High Temperature Materials and Tests
School of Materials Science and Engineering, Shanghai Jiao Tong Univ., Shanghai 200030, P. R. China

ABSTRACT

Gas nitridation of TiAl based alloys in an ammonia atmosphere was carried out in the present work. The nitride layers were characterized by X-ray diffraction (XRD) and scanning electron microscopy (SEM). The evaluation of the surface hardness and wear resistance was performed to compare with those of the non-nitrided alloys. It is concluded that the nitride layers are composed of Ti_2AlN as the inward-growing layer and TiN as the outward-growing layer. The nitridation temperature and time were two major factors influencing the thickness of the nitride layers of the alloys. The high temperature nitridation raised the surface hardness and the wear resistance of the TiAl based alloys markedly. The tribological behaviors of the nitrided alloys were also discussed.

INTRODUCTION

It is well known that the intermetallic compound TiAl has attractive comprehensive properties, especially high-temperature properties, such as high modulus, elevated temperature strength, good oxidation resistance and low density, which are particularly useful for applications in aircraft and automobile structures [1,2]. Therefore, in recent years more and more attention has been paid to the development of the engineering intermetallic alloys TiAl [3,4].

However, the TiAl alloy has two major drawbacks: very limited ductility at room temperature and poor surface properties at high temperatures. The former has been gradually enhanced by modification of the microstructure using ternary element addition [5] and thermo-mechanical processing, or powder metallurgy [6,7]. The latter can be improved to some extent by surface treatments. As we known, tribological problems are always involved in the structural materials, especially in case of a turbine engine or a turbocharger rotor. Some tribological events, such as adhesive wear, erosive wear, oxidation wear and fretting wear, could result in loss or damage of the parts. Therefore, the surface improvement of the intermetallic compounds by various surface modification methods [8-10] is being performed.

Due to its unique physical, mechanical and metallurgical properties, a thin titanium nitride (TiN) is effective in reducing tool wear so that the lifetime and cutting speed are considerably increased [11,12]. Because TiN is more thermodynamically stable than AlN, TiN layers can be formed on TiAl by direct gas nitridation in different atmospheres [13,14], implantation with N [10] and by ion nitridation [15]. Coatings of TiN applied by chemical vapor deposition (CVD) or physical vapor deposition (PVD) are often applied, but among various methods the direct gas nitridation is a less costly process. In the present work, high-temperature gas nitridation behaviors of TiAl based alloys in ammonia are investigated.

EXPERIMENTAL DETAILS

A conventional tungsten arc melting technique was employed to prepare titanium aluminide alloy. Experiments were performed with Ti-47Al-2Nb-2Cr-0.1Si (the compositions are given in at %) which had been cast and homogenized at 1050 °C for 100 h. Specimens with dimensions of 6 mm × 6 mm × 10 mm were cut from the homogenized ingots followed by surface polishing with emery paper up to No. 1000 for nitridation and the wear test.

After washed carefully in acetone and alcohol to remove grease, all specimens were hung in a high-temperature quartz reaction tube for nitridation. The tube was evacuated repeatedly and finally filled with argon. The specimens were heated to the desired temperature range of 800-940 °C with the emphasis at 940 °C. The nitridation time was 10 h, 30 h and 50 h, respectively. And then argon was replaced with ammonia flowing at the rate of 5-10 cm^3·s^{-1}. When the required period had been attained, the specimens were cooled down in argon to room temperature.

The wear test was performed on an Amsler test machine using block-on-ring setup. The counter ring was made of carbon steel containing 0.45 % C and surface electroplated with hard chromium with a hardness of HRC65. The load for the wear tests was 1.3 kg, and the sliding speed was 0.523 m·s^{-1} under unlubricated conditions. The sliding distance was 314 m.

X-ray diffraction (XRD) using Cu $K\alpha$ radiation and scanning electron microscopy (SEM) were used to study the microstructure and morphologies of the nitrided alloys. Scanning electron microscopy (SEM) and optical microscopy (OM) were employed to investigate the wear traces, the trace edge and the wear debris.

RESULTS AND DISCUSSIONS

Nitridation

The nitride layers were detected on the surface of the TiAl based alloys nitrided at all the temperatures for different periods. The typical XRD pattern of the nitride layers is shown in Figure 1. It has been found that besides the diffraction peaks of the constituents (TiAl and Ti$_3$Al) of the alloys, those of the TiN and Ti$_2$AlN occurred in the XRD pattern, and Ti$_2$AlN is the inner layer formed under the TiN outer layer [16]. AlN and other new Al-rich intermetallics, such as Al$_2$Ti and Al$_3$Ti in the subscale, were not found at the nitrides/matrix interface and only Al was found in the present work. It was indicated that Ti$_2$AlN formed more easily than TiN and their formation was dependent on temperature [15]. Figure 2 shows the cross-section back scattered electron images of the alloys nitrided at 940°C for 50 hours.

Generally, The nitriding behaviors of TiAl alloys in the ammonia atmosphere could be explained by the following steps: (i) the absorption of the activated nitrogen atoms to the alloy surface, (ii) the reaction of Ti and N to form TiN, (iii) the outward diffusion of Ti and the inward diffusion of

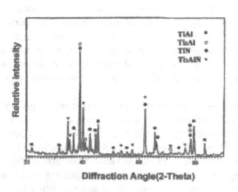

Figure 1. The X-ray pattern of TiAl based alloys nitrided at 940 °C for 50 h.

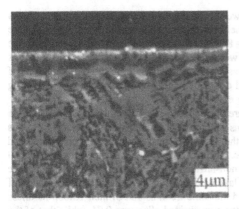

Figure 2. Cross-section micrograph of the TiAl based alloy nitrided at 940 °C for 50 h.

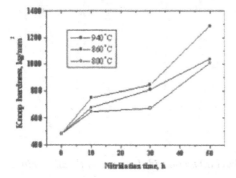

Figure 3. The Knoop hardness of the nitrded alloys before and after nitridation at the different temperatures.

Al, (iv) the formation of Al_2Ti and Al_3Ti and (v) the reaction of Al_2Ti or Al_3Ti with N to form Ti_2AlN. As to the details about the nitride layers characterization, please refer to the other paper [16].

Nitridation of the TiAl based alloys can be performed in either a nitrogen or an ammonia atmosphere. The advantage of using ammonia lies in that ammonia decomposition can provide the heated surface with more active atomic nitrogen than molecular nitrogen, which is beneficial to the nitridation processing. However, the hydrogen released by this decomposition will also diffuse into the substrate [17], and the effects of hydrogen on the mechanical properties of the nitrided alloys need further investigation.

Knoop hardness

The variations of the Knoop hardness of the nitrided TiAl based alloys vs. nitridation temperature and time were plotted in Figure 3. Compared with the non-nitrided alloy, the Knoop hardness of the nitrided alloys evidently increased with increasing nitridation temperature and time, which resulted from the hard surface layers of TiN and Ti_2AlN. As shown in Figure 2, the maximum thickness of the nitrided layers in the present work was about 4 μm. Thus, the apparent hardness of the nitrided layers was influenced by the soft substrate of TiAl based alloys, and was considered to be proportional to the thickness of the nitrides layers, which was related to the temperature and time of nitridation. In short, the composite of the nitride layers, the diffusing layer and the substrate formed by high temperature gas nitridation can greatly improve the surface hardness of the TiAl based alloys. When the nitriding time was prolonged to 50 hours at 940 °C, the Knoop hardness value was 1286 kg s^{-1}.

Wear resistance

Wear properties are generally related to the surface hardness of the alloys. Formation of the nitride layers increased the Knoop hardness and was also thought to be able to improve the wear resistance of the alloys. According to the optical micrographs of the TiAl based alloys in Figure 4, the wear traces of the non-nitrided specimens (Figure 4a) were wider than those of specimens nitrided at 940 °C for 50 h (Figure 4b). And its wear trace depth was also deeper. Figure 5 shows severe plastic deformation along the edge of the wear traces of the non-nitrided alloys (Figure 5a), compared with that of the alloys nitrided at 860 °C for 50 h (Figure 5b) which indicates the progressive wear. Two types of debris feature were observed. Figure 6 showed that the debris taken from alloys nitrided at 940 °C for 50 h was small, nearly round and smooth, while that of the non-nitrided alloys was inhomogeneous and exhibited signs of peeling and rupturing during wear testing which could be found from its wear edge (Figure 5a).

The comparison of the wear loss for different specimens is shown in Figure 7. As can be seen from it, the wear resistance of the specimens nitrided at different temperatures and time was better than that of the non-nitrided alloys. For the nitrided alloys, the wear resistance increased with increasing nitridation temperature and time. The wear loss of the specimen nitrided at 940 °C for 50 h was only half as large as that of the non-nitrided alloys. The superior wear resistance of the nitrided alloys is attributed to the nitride layers adherent to the substrate. The variation of the mean dynamic friction coefficients displayed the same trend as that of their Knoop hardness and wear resistance. The friction coefficient of the non-nitrided alloy was 0.74, while that of the alloy nitrided at 940 °C for 50 h was 0.53.

Wear mechanisms of TiN coating on steel in dry sliding contact using a pin-on-disk wear

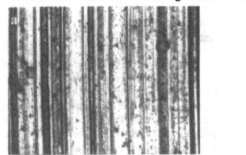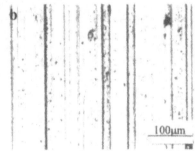

Figure 4. *The wear traces of the TiAl based alloys (a) non-nitrided alloys and (b) alloys nitrided at 940 °C for 50 h.*

Figure 5. *The edges of the wear traces (a) non-nitrided alloys and (b) alloys nitrided at 860 °C for 50 h.*

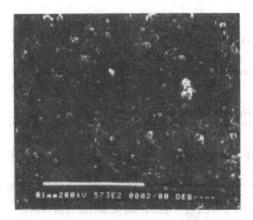

Figure 6. *The debris of the TiAl based alloys nitrided at 940 °C for 50 h.*

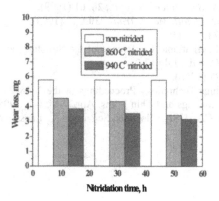

Figure 7. *Comparison of the wear loss of the nitrided alloys with that of the non-nitrided alloys.*

tester were investigated in the literature [18]. Three wear regimes were identified: (i) transfer and build up of oxidized pin debris on the coating at 20 N load, (ii) increased polishing damage and brittle spallation failure of the TiN at load in the 50-100 N range and (iii) a sharp transition to plastic deformation and microploughing of TiN at loads greater than 100 N. According to the present experimental observations, the wear mechanism of the nitrided alloys was similar to the second regime because of the more austere wear conditions than those in the literature [18]. As a result, the nitrded alloys show signs of increasing wear loss and damage of the nitride layers after the wear test. However, without the protection of the nitride layers, the non-nitrided alloys displayed severe abrasive wear and plastic deformation (see Figure 5a) just like in the regime three.

CONCLUSIONS

In the present work, the high-temperature nitridation behavior of the TiAl based alloys, their Knoop hardness and tribological behaviors were investigated. The nitride layers consisted of Ti_2AiN as the inward-growing layer and TiN as the outward-growing layer on the surface of the alloy and their thickness increased mainly with the nitridation temperature and time. Mechanical tests show that high-temperature nitridation can obviously increase the surface hardness and the sliding wear resistance of the TiAl based alloys. The Knoop hardness value was 1286 kg s^{-1} when the nitriding time was prolonged to 50 hours at 940 °C and its wear loss was only half as large as that of the non-nitrided alloys.

ACKNOWLEDGEMENT

This work is sponsored by the Science and Technology Commission of the Shanghai Municipal Government.

REFERENCES

1. R.L. Fleischer, *J. Materi, Sci.*, **22**, 2281 (1987).
2. D. Schectman, M.J, Blackburn and H.A. Lipsitt, *Metall. Trans.*, **5**, 1373 (1974).
3. Y.W. Kim, in L.A. Johnson, D.P. Pope and J.O. Stiegler (eds.), High Temperature Ordered Intermetallic Alloy, Mater. Res. Soc. Symp. Proc., Vol. 213. Pittsburgh, Pennsyvania, MRS, p.777.
4. D.M. Dimiduk, D.B. Miracle and C.H. Ward, *J. Mater. Sci. Technol.*, **8**, 3667 (1992).
5. K. Hashimoto, H. Doi, K. Kasahara, T. Tsujimoto and T. Suzuki, *J. Jap. Inst. Met.*, **54**, 539 (1990).
6. T. Degeawa and K. Kamata, Proc. Symp. Strength and Deformation of Intermetallic Compounds, JIM, Sendai, Sept., 1988, p 20.
7. H. Sugimoto, K. Ameyama, T. Inaba and M. Tokizane, *J. Jp. Inst. Met.*, **53**, 628 (1989).
8. Y. Wang, Z. Qian, X. Y. Li and K. N. Tandon, *Surface & Technology*, **91**, 37 (1997).
9. T. Nado, M. Okabe and S. Isobe, *Metarials Science & Engineering A*, A213, 157 (1996).
10. J. C. Privi, *Journal of materials Science*, **25**, 2743 (1990).
11. A. Matthews, *Surf. Eng.*, **1**, 93 (1985).
12. V. Murawa, *Heat. Treat. Met.*, **2**, 49 (1986).
13. S. Thongtem, T. Thongtem and M.J. Mcñallan. *Surf. Interface Anal.*, **28**, 61 (1999).
14. J. Magnan, G.C. Weatherly and M.C. Cheynet. *Metall. Mater. Trans.*, **30A** 19 (1999).
15. C.T. Chu and S.K.Wu. *Surf. Coat. Technol.*, **78**, 221 (1996).
16. J. Sun, J.S. Wu, B. Zhao and F. Wang. 5th International Conference of the Structural and Functional intermetallics. TMS. Vancouver, Canada. July 17-21.
17. K. Bungardt and K. Rudinger, *Z. Met.*, **47**, 577 (1956).
18. S. Wilson and A.T. Alpas, Surface & Coatings Technology Proceedings of the 1998 25th International Conference on Metallurgical Coatings and Thin Films, Apr.27-May 1, 1998 v108-109, n1-3, Oct 10 1998 San Diego, CA, USA, Elservier Science S.A. Lausanne Switzerland p369-376.

Mat. Res. Soc. Symp. Proc. Vol. 646 © 2001 Materials Research Society

Deformation of Ni₃Al Polycrystals at Extremely High Pressures

John K.Vassiliou[1], J.W. Otto[2], G. Frommeyer[3], A. J. Viescas[1], K. Bulusu[1] and H. Bellumkonda[1]
[1]Dept. Physics, Villanova University, Villanova, PA 19085, USA
[2]Joint Research Center of the European Commission, Brussels, Belgium
[3]MPI Eisenforschung, 40237 Dusseldorf, Germany
Corresponding author: John.Vassiliou@Villanova.Edu

ABSTRACT

The compression behavior in a multi-anvil apparatus of a foil of Ni₃Al embedded in a pressure medium of NaCl has been studied by energy-dispersive X-ray diffraction (EDX). At ambient temperature, the pressure and stresses, determined from line positions of NaCl, were constant throughout the sample chamber. Line positions and line widths of NaCl reflections were reversible on pressure release. Ni₃Al polycrystals, in contrast, undergo extensive (ductile) plastic deformation above 4 GPa due to the onset of high non-hydrostatic stresses and the introduction of stacking faults and dislocations. Plastic deformation due to stacking faults leads to a volume incompressibility followed by elastic compression of a fully plastically deformed state. The compression of a fully plastically deformed material is elastic and isotropic, independent of the presence and type of pressure medium. A discontinuity in the compressibility at the transition back from plastic to elastic compression is due to the yield strength of the plastically deformed material and corresponds to the Hugoniot elastic limit.

INTRODUCTION

Deviatoric stresses in high-pressure experiments can develop because of a non-isotropic macroscopic stress field set-up by the pressure device, substantial viscosity of the pressure transmitting medium or local stresses set up at the grain boundaries of a polycrystalline elastically anisotropic sample. For ductile fcc-based metals at ambient conditions, the critical resolved shear stress for the glide of dislocations in a single crystal, causing plastic deformation by slip, is of the order of only 10^{-4}-10^{-3} GPa . The yield stress in polycrystals is higher than this value by a factor of 3 for fcc and bcc based materials and may increase substantially with decreasing grain size[1]. Although the pressure effect on the yield stress is controversial[2], stresses of only a small fraction of the applied nominally hydrostatic pressure are required to cause plastic deformation of many materials. Such stresses may occur even in some liquid pressure media.

The general compression behavior of ductile elastically anisotropic polycrystals under increasing non-hydrostatic stress has recently been studied in detail with energy-dispersive X-ray diffraction of a foil of disordered Cu₃Au in a diamond anvil cell[3,4]. The sample was chosen to be a foil in order to illustrate the limiting case of non-ideal powders in which significant microstrains develop under compression. It was found that an initial region of elastic compression is followed by a pressure region over which the volume does not change, terminated by a discontinuity in the compression curve. Elastic compression sets in again at a higher pressure above the compressibility discontinuity. The onset of the elastic incompressibility and of plastic deformation was shown to shift to higher pressure and the pressure range of the elastic incompressibility was reduced with decreasing shear strength of the pressure media. A similar pressure shift of the discontinuity in the compressibility, using different liquid pressure media, had been noted by LeBihan[5] et al. Using a foil of ordered Ni_3Al as an example of a material with a higher yield strength than that of Cu_3Au, the features on the compression curve were demonstrated to shift to higher pressures[6], as expected. In particular, the region of incompressibility was extended. The discontinuity is interpreted as being due to the yield strength of the fully plastically deformed solid containing a maximum density of dislocations and stacking faults.

In the present study, the compression curve of a foil of Ni_3Al is investigated by choosing a multi-anvil apparatus. The cubic geometry of the anvils should make for a more hydrostatic macroscopic stress field than that in a diamond anvil cell. However, the cylindrical geometry of the sample container may impose a uniaxial stress component similar to that in a diamond anvil cell. The pressure medium used was NaCl.

Annealing experiments were carried out on NaCl and Ni_3Al to investigate the effects of restoring a hydrostatic macroscopic stress field on the microstrain in the sample by monitoring the position shift and broadening of line profiles.

EXPERIMENTAL

Polycrystalline foils of ordered Ni_3Al 30-50 μm thick were compressed in a multi-anvil apparatus (MAX80). The foils were embedded between two pre-compressed discs of NaCl. The crystallite sizes in the foils produced by splat-quenching were estimated from line broadening to be in the range 0.2-5 μm, providing for good powder averaging. Ni_3Al crystallize in an fcc superlattice structure ($L1_2$). The yield strength for plastically deformed ordered Ni_3Al composed of polcrystals of 1 μm in size is 2.2 GPa (for comparison, the tensile yield stress for single crystals of disordered Cu_3Au is 0.02 GPa) and the 0.2% flow stress for ordered polycrystalline Ni_3Al is 0.13 GPa [7]. Deformation results in introduction of stacking and twin faults on {111} planes and in the ordered state of anti-phase boundaries on {001} planes[7].

Experiments were carried out at beamline F2 (MAX80) at the Hamburg Synchrotron Radiation Laboratory using energy-dispersive x-ray techniques. The incident beam was parallel to the plane of the foil and diffraction was recorded at an angle $2\theta = 9^0$ above this plane. The crystallites probed are those with their diffraction vector almost perpendicular to the foil. Pressures were measured from the volume data of NaCl using the Decker equation of state[8] with state parameters B_0=24.008 GPa for the bulk modulus and B'_0=4.74 for its pressure derivative.

Analysis of lattice deformations with EDX

The stress field is analyzed to a hydrostatic component σ_p and a deviatoric stress component D_{ij} which is a function of the uniaxial stress t. The measured strain ε_m due to both stresses is given by the relation $\varepsilon_m = \dfrac{d_m(hkl) - d_0(hkl)}{d_0(hkl)}$ where d_m is the measured d-spacing between the (hkl) crystal planes and d_0 the spacing under zero hydrostatic pressure. The measured d-spacing is related to the uniaxial stress t, produced in the nonhydrostatic stress state, according the following relation[9,10]:

$$\varepsilon_m(hkl) = \varepsilon_p(hkl) + (1 - 3\cos^2\theta)\frac{t}{3}\{\alpha[2G_R(hkl)]^{-1} + (1 - \alpha)(2G_V)^{-1}\}$$

where $\varepsilon_p(hkl) = (S_{11} + 2S_{12})\sigma_p$ is the strain under pure hydrostatic pressure σ_p,

$$[2G_R(hkl)]^{-1} = [S_{11} - S_{12} - 3(S_{11} - S_{12} - \frac{1}{2}S_{44})\Gamma(hkl)],$$

$$[2G_V(hkl)]^{-1} = \frac{5}{2}\frac{(S_{11} - S_{12})S_{44}}{3(S_{11} - S_{12}) + S_{44}} \quad \text{and} \quad \Gamma(hkl) = \frac{(h^2k^2 + k^2l^2 + l^2h^2)}{(h^2 + k^2 + l^2)^2}$$

S_{ij} are the elastic compliances for a cubic material, θ is the scattering angle when the direction of the incident beam is perpendicular to the compression axis σ_p, G_R and G_V are the aggregate shear moduli calculated under the Reuss (isostress) or Voigt (isostrain) constrains. The parameter α varies between 0 and 1 and is determining the percentage of isosress or isostrain constraints. The product αt was obtained from the slope $d\varepsilon_m / d\Gamma$ and the known compliances under pressure[11].

The line profiles of a polycrystalline material contain information about size of the crystallites and the built in strains. When multiple orders of reflections are available or reliable the strain and the size contribution to the line width can be separated by the Williamson-Hall[13] or the Warren-Averbach[14] method. When muliple order reflection are not available the for a Williamson-Hall or Warren-Averbach analysis, an effective crystallite size L_{eff} is often estimated from the Lorentzian component and a strain ε from the Gaussian component to the line width of a single peak profile[15,16,17] using the Scherrer and Stokes-Wilson equations, respectively. The crystallite size and the strain were determined from the formulas[18] $\beta_C = \dfrac{6.199}{L_{eff}\sin\theta_0}$ and $\beta_G = 2\varepsilon E$ where β_C and

β_G is the Lorentzian and the Gaussian contributions to the line width due to crystallite size and stains respectively, θ_0 is the half diffraction angle and E is the energy corresponding to the reflection. Plastic deformation in Cu_3Au was shown to lead, indeed, to a strong line broadening and a dominant Lorentzian character of the peak profiles. Therefore, in fitting the line profiles we used the Voigt function which is a convolution of a Gaussian and a Lorentzian profile.

Plastic deformation introduces dislocations, stacking faults and twin faults on $\{111\}$ planes. The introduction of stacking faults in fcc metals leads to volume changes and a systematic shift in peak positions. The stacking fault probability P_{SF} can be estimated from the peak shifts following the analysis by Warren[12]. Since multiple orders of reflection don't exist or are not very reliable, the probability was estimated from the relative line shift $\Delta(E_{220} - E_{111})$. The stacking fault probability is given by the relation $P_{SF} = -\Delta(E_{200} - E_{111})\dfrac{4\pi}{\sqrt{3}}\left(\dfrac{E_{200}}{2} + \dfrac{E_{111}}{4}\right)^{-1}$. The resolution of EDX with synchrotron radiation requires a stacking fault probability of $\geq 1/200$, an effective crystallite size $\leq 0.2 \ \mu m$ -0.5 μm and strains $\geq 10^{-3}$ for detection.

RESULTS AND DISCUSSION

The pressure (at room temperature) was measured from the sample sections of pure NaCl, and NaCl in contact with the Ni_3Al foil. The pressure, uniaxial stress component and the strains were found to be constant over the length of the sample container. On pressure release, the strains and stresses determined from line widths and line positions, respectively, of NaCl were reversible within the experimental resolution. This confirms earlier findings of compression experiments in a diamond anvil cell[4].

Compression of a Foil of Ni₃Al

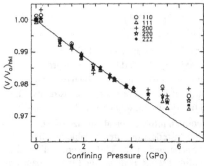

Figure 1. *Scaled volumes calculated from the (110), (111), (200), (220), (222) lattice spacing. Below 4 GPa there is good agreement with the equation of state*

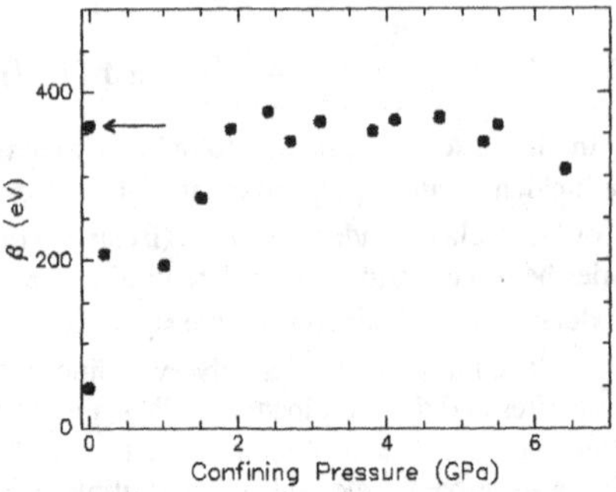

Figure 2. *Line width of (111) reflection of Ni₃Al. The arrow indicates the line width after the pressure was released*

Figure 1 shows the scaled volumes of Ni_3Al calculated from the d-spacing of (110), (111), (200), (220) and (222) reflections. This representation was chosen because a difference in lattice parameters (or volumes) obtained from multiple orders of a given reflection must be due to plastic deformation when pressure gradients are absent[4].

For elastic compressive strains, positive deviations from hydrostatic values of V/V_0 would occur in the order (100,200) ≥ (110,220) ≥ (111,222)[6]. The introduction of stacking faults on {111} planes, as expected during plastic deformation of fcc-based metals and alloys, would lead to a sequence (with increasing values of V/V_0 of (100) ≤ (111,220) ≤ (311) ≤ (222) ≤ (110,200) with (100) and (111,220) being shifted to lower values, and (311), (222) and (110,200) to higher values compared with those of unstrained Ni_3Al [6,12]. It is thus possible in principle to calculate the stacking fault probability free from contributions due to elastic stresses by using multiple orders of a single reflection and, conversely, to obtain the elastic stresses free from the contributions due to stacking faults by using pairs such as (111)-(220) and (110)-(200) in a Kennedy-Singh analysis. In this way, it is possible to distinguish in situ unequivocally between elastic strains and those plastic strains due to stacking faults. In the present experiments, the (100) reflection was not observed since it occurs in the energy range in which the radiation is strongly absorbed by the sample container.

Below 4 GPa, the lattice parameters calculated from the individual reflections agree very well (Figure 1). The values of the scaled average volume lie on the equation of state calculated from the ultrasonically determined elastic constants to 1.4 GPa[11]. In agreement with this observation, no deviatoric stresses up to 4 GPa were detected in Ni_3Al with a Kennedy-Singh analysis of line shifts.

Above 4 GPa, there is a rapid deviation in lattice parameters calculated from the individual reflections, especially as concerns multiple orders of a single direction {hkl}. The relative sequence in V/V_0 for the observed reflections and in particular for the multiple orders (111,222) and (110,220) indicates the presence of stacking faults[4,12]. The only exception is the relative sequence in V/V_0 calulated from (220) and (222); this is interchanged with respect to the sequence expected in the presence of stacking faults (and in the absence of uniaxial stresses). The relative shift due to stacking faults in the pair (111), (222) is rather small[12], and the pressure dependence of the (110) reflection could not be determined in the critical pressure range. The probability of stacking faults

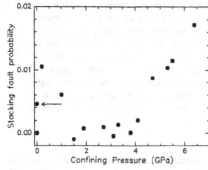

Figure3. *Stacking faults probability in Ni₃Al calculated from the shift of the relative separation of the (111) and (200) reflections. The arrow shows the stacking faults probability after the pressure was released*

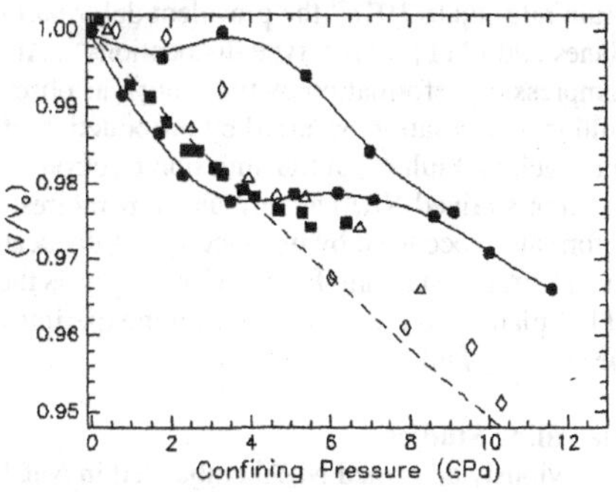

Figure 4. *Scaled average volumes of Ni₃Al in the present experiment (solid squares), in diamond anvil cell using NaCl as pressure medium (solid circles) [20], by Mauer et al. [19] in diamond anvil cell (triangles and rombs) using methanol-ethanol-water as a pressure medium. The dashed line is the ultrasonically derived equation of state[11].*

was thus calculated, not from multiple orders of a reflection, but from the relative shift in the separation of the (200)-(111) pair corrected for the shift due to hydrostatic pressure. The data clearly show the introduction of stacking faults starting at 4 GPa, with a maximum stacking fault probability at 6.4 GPa of 1 in about 60 layers (Fig.3). The average volume remains constant above 4 GPa up to the highest pressure reached. On pressure release, the stacking fault probability is around 0.005 (or 1 in 200 layers) indicating permanent deformation of Ni₃Al. Figure 4 compiles the data taken in a multi-anvil apparatus and in diamond cells[4,19]. The results of different experiments are in excellent agreement and in agreement with the ultrasonic data for pressures less than 4 GPa.

Line widths

Figure 2 shows the line widths of Ni₃Al. They remain constant to 1 GPa except for (111) and (220) which increase somewhat. Above 1 GPa, they increase rapidly and saturate at 2 GPa. The increase of the uncorrected Full Width Half Maximum is in the range 30% (for (110), (200) and (222)) to 50% (for (111), (220) and (311)). A decomposition of the Voigt profiles into Gaussian and Lorentzian components yielded line widths with limited scatter only for the (111) and (220) reflections. Large scatter in decomposed line widths was also found in previous studies even for strong reflections when the sample was embedded in NaCl[3,4] (rather than a glassy or liquid pressure medium) and it is due to the elastic anisotropy of the pressure medium. Little or no relaxation of

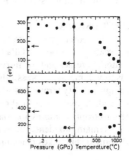

Figure 5. *Line widths of the Ni₃Al (111) (top) and (200) (botom) at 6.7 GPa. Arrows show data taken after the pressure release.*

linewidths ocurred on pressure release. The irreversibility of line widths on pressure release indicates that plastic deformation took place by the introduction of stacking faults. The superlattice reflections remained broadened by roughly 40% after the experiment.

The line profiles were symmetrical for all the pressure range, suggesting that no twin faults were introduced during the deformation[12]. In related studies of Ni_3Al at room temperature and at strain rates up to 10^4 s^{-1} the prevalent deformation modes were found to be stacking faults on {111} planes and {111} <110> type dislocations[21]. At very high strain rates such as in shock wave compression deformation by twinning was observed[22]. In materials, prepared by high-energy ball milling, deformation occurred by introduction of anti-phase boundaries, grain fracture, twinning and stacking faults[23]. In transmission electron microscopy studies of binary and boron doped[24] Ni_3Al or strained Ni_3Al at various temperatures[25] from 77 to 1023 K, it was observed that deformation occurred by introduction of dislocations in the form of anti-phase-boundary pairs or super intrinsic stacking faults. In our analysis the deformations are treated as stacking faults on {111} planes. At the resolution of our experiment we can not further characterize the variation of the stacking faults.

Annealing Studies
A previously deformed Ni_3Al embedded in NaCl was annealed in order to compare the temperature of the release of the stresses surrounding the foil to that of stress release within the foil. Release of the internal stresses in of NaCl at 6.7 GPa and recovery of the line widths started immediately on raising the temperature to 200 ^{0}C. The recovery of the line widths of the undeformed state was complete by 400 ^{0}C. On pressure release, there was minor broadening. The peak widths of Ni_3Al remained unchanged on compression (Fig.5) as expected for a fully plastically deformed material. In comparison to NaCl, the release of the internal stresses in Ni_3Al did not set in before 400 ^{0}C. The recovery of the line widths of the undeformed state was complete by 800 ^{0}C. This confirms the expectation that the temperature of release of the uniaxial stress in the pressure medium has no effect on an embedded sample which is plastically deformed (rather than elastically strained). Thus, in compression studies with annealing steps to ensure hydrostatic conditions it is important to make sure that it is not just the pressure medium which relaxes its internal stresses by annealing.

On pressure release from 5.3 GPa (resulting from temperature quenching the sample at 6.7 GPa), broadening in Ni_3Al was marked. The broadening may result from strains arising because the rather stiff interlocking grains cannot mutually accommodate changes in volume and shape. Again, this is a phenomenon expected in foils of elastically anistropic materials and needs to be taken into account in annealing studies of such materials.

CONCLUSIONS
The equation of state of Ni_3Al polycrystals and the line widths was measured to 6.4 GPa by x-ray scattering. Above 4 GPa the material undergoes plastic deformation due to the onset of deviatoric stresses and the introduction of stacking faults and dislocations. Plastic deformation leads to volume incompressibility and line widths broadening. The pressure range of the incompressibility terminates when the yield strength of the plastically deformed material is exceeded. Above this critical pressure the compression is elastic and isotropic. The critical pressure where isotropic compression starts depends upon the viscosity of the pressurizing medium. The line widths remain broadened upon release of the pressure. Recovery of the line widths of the undeformed state takes place upon annealing at 800 ^{0}C, which is a mach higher temperature than

the recovery temperature of the pressurizing medium. The resolution of our experiment does not allow further characterization of the variation of the stacking faults.

ACKNOWLEDGEMENT

The work has been carried under the proposals No. II-96-23 at HASYLAB and No. P621, NSF/CHESS at Cornell Synchrotron. JKV acknowledges support from the physics department at Villanova University.

REFERENCES

1. D. Hull and D.J. Bacon, *Introduction to Dislocations*, Pergamon Press, Oxford, 1984.
2. J. P. Poirier, *Creep of Crystals*, Cambridge Univ. Press, Cambridge, 1985.
3. J.W. Otto, J.K. Vassiliou and G. Frommeyer, *J. Synchr. Rad.* **4**, 155 (1997).
4. J.W. Otto, J.K. Vassiliou and G. Frommeyer, *Phys. Rev.*, B **57**, 3253 (1998); *ibido*, 3264 (1998).
5. T. Le Bihan, S. Heathman, S. Darracq, C. Abraham, J.M. Winand and U. Benedict, *High Temperatures-High Pressures*, **27/28**, 157 (1996).
6. J.W. Otto, J.K. Vassiliou and G. Frommeyer, *J. High Pressure Research* **16**, 45 (1998).
7. M. Yamaguchi and Y. Umakoshi, *Progress in Materials Science*, **34**, 149 (1990).
8. D.L. Decker, *J. Appl. Phys.*, **42**, 3239 (1971).
9. T. Uchida, N. Funamori and T. Yagi, *J. Appl. Phys.* **80**, 739 (1996).
10. A.K. Singh, *J. Appl. Phys.* **73**, 4278 (1993).
11. J. Frankel, J.K. Vassiliou, J.C. Jamieson, D.P. Dandekar and W. Scholz, *Physica*, **B139 & 140**, 198 (1986).
12. B. E. Warren, *X-ray diffraction*, Dover Reprint (1990).
13. G. K. Williamson and W. H. Hall, *Acta Metall.* **1**, 22 (1953).
14. B. E. Warren and B. L. Averbach, *J. Appl. Phys.* **21**, 595 (1950); **23**, 497 (1952).
15. J. L. Langford, *J. Appl. Crystallogr.* **11**, 10 (1978).
16. J. L. Langford, R. Delhez, Th. H. de Keijser, and E. J. Mittemeijer, *Aust. J. Phys.* **41**, 173 (1988).
17. Th. H. de Keijser, J. L. Langford, E. J. Mittemeijer, and A. B. P. Vogel, *J. Appl. Crystallogr.* **15**, 308 (1982).
18. A. R. Stokes and A. J. Wilson, Proc. Cambridge Phil. Soc. **38**, 313 (1942); *Proc. Phys. Soc.* London **56**, 283 (1944).
19. F. Mauer, R. G. Munro, G. J. Piermarini, B. C. Block and P. D. Dandekar, *J.Appl.Phys.* **58**, 3727 (1985).
20. J.W. Otto, J.K. Vassiliou and G. Frommeyer, *Rev. High Pressure Sci. Technol.,Vol.* **7**, 1511 (1998).
21. H. W. Sizek and G. T. Gray III, *Acta Metall. Mater.*, **41**, 1885 (1993).
22. D. E. Albert and G. T. Gray III, *Philosophical Magazine*, **A70**, 145 (1994).
23. L. Lutteroti, S. Gialanella and R. Caudron, *Materials Science Forum*, 228-231, 551 (1996).
24. K. J. Hemker and M. J. Mills, *Philosophical Magazine*, **A68**, 305 (1993).
25. I. Baker and E. M. Schulson, *Phys. Stat. Sol.* (a) **89**, 163-172, (1985).

Mat. Res. Soc. Symp. Proc. Vol. 646 © 2001 Materials Research Society

Long Term Oxidation of Model and Engineering TiAl Alloys

Ivan E. Locci[1], Michael P. Brady[2], James L. Smialek[1]
[1]NASA-Glenn Research Center, Cleveland, OH 44135, Ivan.E.Locci@grc.nasa.gov
[2]Oak Ridge National Laboratory, Oak Ridge, Tennessee 37831

ABSTRACT

The purpose of this research was to characterize the oxidation behavior of several model (TiAl, TiAl-Nb, TiAl-Cr, TiAl-Cr-Nb) and engineering alloys (XD, K5, Alloy 7, WMS) after long-term isothermal exposure (~7000 h) at 704°C, and after shorter time exposure (~1000 h) at 800°C in air. High-resolution field emission and microprobe scanning electron microscopy were used to characterize the scales formed on these alloys. Similarities and differences observed in the scales are correlated with the various ternary and quaternary microalloying additions.

INTRODUCTION

Titanium aluminides (TiAl) are of great interest for intermediate-temperature (600°C-850°C) aerospace, automotive, and power generation applications because they offer significant weight savings compared to today's nickel-based alloys. TiAl alloys are being investigated for low-pressure turbine (LPT) blade applications, exhaust nozzle components and compressor cases in advanced subsonic and supersonic engines [1-2], and exhaust valves in automobiles [2-4]. Significant progress has been made in understanding the fundamental aspects of the oxidation behavior of binary TiAl alloys [5-12]. However, most of this work has concentrated on shorter term (< 1000 hours), higher temperature (900°C-1000°C) exposures. Much less data is available in the literature regarding the oxidation behavior of the quaternary and higher order engineering alloys under the long term, low temperature conditions likely to be encountered in near term structural applications [5-6,11,13]. The present investigation was undertaken to characterize the long-term oxidation behavior of various model and advanced engineering titanium aluminides at 704°C in air. Some engineering alloys were also exposed to 800°C in air for 1000 h. Of particular interest for this study was the formation of nitrides, which have been linked to disruption of alumina scale formation in air [12,14], and the formation of brittle oxygen- and titanium- enriched, aluminum-depleted phase(s) at the metal scale interface, termed by various researchers as Z, X or NCP phase (new cubic phase) [14-18].

EXPERIMENTAL PROCEDURES

The alloys included in this investigation are listed in Table 1. All compositions presented in the paper are reported in atomic percent. Alloys were cast, hot isostatically pressed and heat treated to produce a duplex microstructure of γ-grains and $\alpha_2 + \gamma$ lamellae. Several alloys were isothermally exposed to 704°C for 7000 hours and 800°C for 1000 hours in static air. All samples had a 600-grit finish. Back-scattered electron (BSE) images and elemental maps of polished cross-sections were obtained by electron probe microanalysis (EPMA) equipped with a wavelength dispersive x-ray spectrometry (WDS) using pure element standards for Ti, Al, Cr, Nb, and a pure MgO standard for O. A high-resolution field emission electron microscope (FE-SEM) equipped with BSE and low angle secondary electron (SE) detectors and an energy dispersive x-ray spectrometer (EDX) was used to resolve and analyze the fine multiple layers observed in the metal/scale interface formed in these alloys. Total weight gain was measured before and after exposure using an analytical balance. The scale thickness was measured at three different locations in regions that were not in contact with the supporting alumina boat.

Results and Discussion

Oxidation at 704°C – 7000 h – In Air

Typical alloy compositions, the specific weight changes and the total scale thickness measured for each alloy after exposure to 704°C for 7000 h in air are included in Table 1. The weight gain results are presented in Fig. 1. The binary TiAl alloy is the only one where the scale was not adherent and tended to spall-off. The TiAl-2Cr alloy showed significant weight gain, the thickest scale of all the alloys studied, and was the only alloy that reacted with the supporting alumina boat during the exposure treatment. Therefore, the weight gain values for TiAl and TiAl-2Cr obtained may be inaccurate. The presence of Nb as a ternary or quaternary addition was extremely beneficial, minimizing the weight gain even when Cr was present in the alloy. A phosphoric acid surface treatment [19] on the Ti48Al-2Cr-2Nb alloy had a pronounced beneficial effect resulting in a further reduction in the weight gain and scale thickening. The advanced engineering alloys, K5, XD-TiAl, Alloy 7 and WMS showed the smallest weight gain. A description of the complex scales that formed in these alloys during the exposure follows.

Table I. Compositions, weight gain, and scale thickness for model and engineering TiAl alloys exposed to 700°C for 7000 hours in static air

Alloy Comp. (at.%)	Al	Ti	Cr	Nb	W	Mo	Mn	Wt. Change/area (mg/cm²)	Scale* Thickness (μm)
TiAl[3]	47.31	52.69	-	-	-	-	-	4.26[1]	20
TiAl-2Cr[4]	46.58	51.43	1.99	-	-	-	-	10.29[1]	150
TiAl-2Nb	47.67	50.36	-	1.96	-	-	-	1.43[1]	12
TiAl-2Cr-2Nb	48.19	47.95	1.94	1.92	-	-	-	2.04[1]	15
TiAl-2Cr-2Nb + Surf. Treat.	48.19	47.95	1.94	1.92	-	-	-	0.57[1]	6
K5	46.5	47.2	2	3	0.2	-	-	1.1	6
Alloy 7	46	48	-	5	1	-	-	0.41[1]	5
XD-TiAl	47	51	-	2	-	-	2	0.86	9
WMS	47	49	-	2	0.5	0.5	1	0.61[1]	4

[1]Average of 2 samples; [2]Average of 3 regions; [3]Including spall
[4]Reacted with Al₂O₃ boat

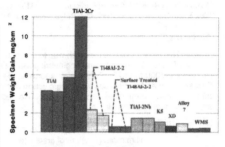

Figure 1. Specimen weight gain for model and advanced γ-TiAl alloys after isothermal exposure to 704°C for 7000 h in air.

Model Alloys: Ti-48Al -- A typical cross section of the scale (20 μm thick) formed and a series of microprobe elemental maps, which show the distribution of Ti, Al, N, and O in the different regions of the scale, are presented in Figure 2. The scale has fractured in the middle of the intermixed layer composed of TiO₂ and some Al₂O₃. Typically rutile (TiO₂) needles/platelets are observed on the scale surface, followed by a discontinuous Al₂O₃ layer and an intermixed Al₂O₃/TiO₂ region. Two other layers at the subscale, next to the metal, are clearly visible. The first layer, next to the metal is very rich in Al; microprobe analyses indicate that this rich phase has chemistry close to TiAl₂; the second layer is rich in Ti and N and Al. An Al-enriched zone below the scale has also been reported for Nb and Mo-containing alloys [5,20]. A Microprobe linescan that reflects the changes in chemistry for the multiple layers that formed at the metal/scale sub-surface is included in Fig. 2. It should be noted that such extensive nitride formation and Al enrichment at the metal/scale interface does not appear to occur under shorter term, higher temperature exposure conditions. For example, Dettenwanger et al. [14] report Al depletion and Z/X phase formation at the metal/scale interface for oxidation of Ti-50Al at 900°C in air, which suggests a change in oxidation mechanism between 900°C and 700°C for binary γ-TiAl alloys. On the other hand, Magnan et al., on nitridation of TiAl at 1000°C reported the formation of a surface nitride scale and an Al-enriched subscale [21].

TiAl-2Cr -- Figure 3 (a) shows an image and corresponding maps for the TiAl-2Cr sample after the 7000 hr exposure. Large blocky rutile crystals (30-40 μm thick) are observed on the outer part of the scale followed by a large area of porosity, probably left by the preferential outward growth of TiO₂. A discrete Al₂O₃ layer is detected, followed by finer porosity in a thin TiO₂ region. An extended region of alternating layers of Al₂O₃ / TiO₂ is beneath that layer. The appearance of this layered intermixed region is very atypical compared to the binary TiAl and other TiAl alloys, where a spottier intermixed region is generally observed. This intermixed region is followed by a bright Al-depleted (Ti-rich)/O-rich layer right at the metal/scale interface; its average microprobe chemistry is 58 Ti, 28 Al, 10.8 O and 2.3 Cr. This layer is speculated to be the Z/X phase described in [12-14]. No N-enrichment was detected at the subscale, although Haanappel et al. [8,22] reported the presence of Ti-nitrides after 150h at 700°C in Ti48Al-2Cr. Higher FE-SEM magnification images, shown in Figure 3 (b), provide some fine details of the scale/metal interface. Alternating vertical channels (pipes) of Al₂O₃/TiO₂ connect the bright Ti-rich/O-rich layer to the scale. A few Cr rich particles are also present in the channeled region. This channeled region morphology may be a form of internal oxidation and is reminiscent of that reported by Doychak et al. [23] for the accelerated oxidation of NbAl₃ and Brady et al. [24] for the rapid oxidation of a Ti modified σ-Nb₂Al alloy. Observation of the intermixed Al₂O₃/TiO₂ region suggests that the channel morphology occurred throughout the course of oxidation. The detrimental effects of small Cr additions on the oxidation of TiAl have generally been postulated to result from doping effects of Cr on TiO₂ growth rate [25]. However, the present results suggest that Cr may also promote this channel form

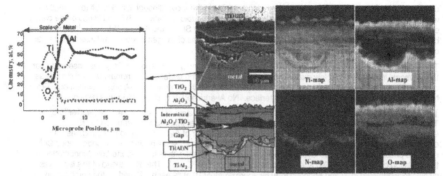

Figure 2. *Typical microstructure, microprobe elemental maps and linescans observed in a binary γ-TiAl alloy after exposure to 704 °C for 7000 hr in air.*

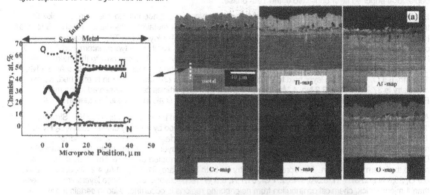

Figure 3. *(a) Typical microstructure, microprobe elemental maps and linescans observed in a TiAl-2Cr alloy after exposure to 704 °C for 7000 hr in air. (b) Higher magnification of the scale/metal interface showing the alternating vertical Al$_2$O$_3$/TiO$_2$ channels that connect the Ti-Al-O rich layer to the scale.*

of internal oxidation, possibly by increasing oxygen solubility in the alloy, although it is difficult to distinguish whether the channel morphology is a consequence of the more rapid oxidation with Cr additions or the cause of it. It is somewhat consistent with the mechanism forwarded by Shida and Anada [25], which interpreted beneficial or detrimental effects of ternary additives to TiAl in terms of oxygen solubility and internal/external oxidation and its effect on alumina morphology.

TiAl-2Nb -- A nearly continuous Al_2O_3 scale (~2 µm thick), containing a few spots of TiO_2 was observed near the gas/scale interface with a few discontinuous TiO_2 needles at the surface. The remainder of the scale as observed in the other alloys consisted of the intermixed TiO_2/Al_2O_3 richer in TiO_2. A thin discontinuous Ti(Al)-N rich layer was formed at the metal/scale interface. No Al-enriched or depleted layer was distinguished.

Ti48Al-2Cr-2Nb -- Detailed description of the scale formed in this system has been reported earlier [5,6]. Typically, the scale consisted of an outer layer of TiO_2 and an inner intermixed layer of Al_2O_3/TiO_2. Observations after 6000 h revealed that a nearly continuous layer of Al_2O_3 separated these regions. A continuous Ti(Al)-N rich layer nearly 1 µm thick was formed at the alloy/scale interface. Discontinuous Cr/Nb-enriched phases decorated this layer. These results are consistent with results reported by Sunderkotter for the oxidation of Ti-48Al-2Cr-2Nb after exposure for 150 h at 800 ºC [26]. The alloy ahead of this layer was significantly depleted in Ti and enriched in Al by approximately 10 at.% each. Based on the composition analyses, the N containing layer consisted of the TiN phase (with a minor amount of Al), and the Ti-depleted/Al-enriched layer consisted of the $TiAl_2$ phase.

Engineering Alloys: XD-TiAl -- The scale formed in the advanced engineering alloys is more complex than that observed on the model alloys with regards to the structure at the metal/scale interface. Figure 4 shows a typical cross section of the scale (~9 µm thick) formed on the XD-TiAl alloy. A few TiO_2 whiskers are observed at the surface followed by a nearly continuous Al_2O_3 layer. Next is the typical intermixed Al_2O_3/TiO_2 layer. The metal/scale interface consists of two fairly well defined layers. The one next to the matrix (white contrast) is very rich in Mn and also enriched in Nb; typical microprobe chemistry is 31.6 Ti, 34.5 Al, 5.4 Nb, 15.3 Mn, 6.5 O, 6.8 N. The layer next to the intermixed region (gray contrast) is Ti and N-enriched; typical chemistry is 37.7 Ti, 17 Al, 2.3 Nb, 4 Mn, 22 N, 17 O. Large TiB_2 particles can be observed in the matrix. A FE-SEM line scan showing the chemistry variation across the metal scale interface is included in Fig. 4.

Alloy K5 -- Figure 5 shows a typical cross section of the scale (~6 µm thick) formed on alloy K5. Similarly to the XD-alloy, a few TiO_2 whiskers are observed at the surface followed by a nearly continuous Al_2O_3 layer. The typical intermixed Al_2O_3/TiO_2 layer follows. The metal/scale interface is more complex. Higher magnification observations revealed at least two main layers. The layer next to the metal (white contrast) is Cr and Nb enriched. A Ti and N-enriched layer (gray contrast), interrupted by pockets of Al_2O_3, is observed next to the intermixed region. Microprobe chemistry for both layers are, 31.3 Ti, 30 Al, 4.2 Nb, 3.1Cr, 0.2 W, 6.9 O, 23.9 N and 37.7 Ti, 5.3Al, 1Nb, 2.1 Cr, 51.7 N, 0.3 W, respectively. Since these layers are much less than 1 micron thick, chemical contribution from neighboring regions is occurring. Also in certain areas (not shown in Fig. 5), a discontinuous Al-enriched layer was observed next to the metal. Typical microprobe chemical analysis for the matrix is 47.9 Ti, 47.4 Al, 2.6Nb, 1.9 Cr, 0.16 W, which is comparable to the bulk alloy composition.

Alloy 7 -- A typical cross section of the scale (~5 µm thick) formed on the Alloy 7 after exposure is presented in Fig. 6. This alloy, which contains 5 at.% of Nb, has very few TiO_2 whiskers at the surface and an almost continuous Al_2O_3 layer follows beneath. Next is an intermixed Al_2O_3/TiO_2 with apparently more Al_2O_3 compared to other alloys. A difficult to resolve thin Ti-rich nitride layer followed by a Nb-enriched region was observed. No Al-enriched or depleted layer at the metal/scale interface was detected.

Oxidation at 800°C-1000 hr-in Air

Ti48Al-2Cr-2Nb -- The appearance of the scale (~12 µm thick) formed on the Ti-48-2Cr-2Nb after the 800°C exposure is very similar to the long-term exposure at 704ºC. Main differences occurred at the interface where multiple layers have formed at the metal/scale interface. A high magnification FE-SEM image, shown in Figure 7, reveals four main layers, beyond the intermixed region. First, an alumina region containing small discrete particles is observed beneath the intermixed region. This is followed by a Ti-N region (white), a Nb-enriched region (grayish), and lastly (slightly darker) an Al-enriched region next to metal.

Alloy K5 -- The scale thickness formed on alloy K5 was approximately ~14 µm thick with features similar to the sample exposed to 704°C (Fig.5). Two well-defined bright regions, both enriched in Nb, were observed at the metal/scale interface (Fig.8 (a)). Oxygen was only detected in the Nb-enriched bright region next to intermixed region. A third region, enriched in Ti and N, separates the two Nb-enriched regions. No Al-enriched layer at the metal/scale interface was detected in the regions investigated.

Alloy 7 -- Alloy 7 developed a 6 µm thick scale during the 800°C exposure. Alumina nodules lightly coated with TiO_2 were observed at the surface. The nodules seem to be connected by a thin but continuous Al_2O_3 layer above the intermixed layer. Figure 8 (b) shows a high magnification of the metal/scale interface. A Nb-enriched layer next to the metal precede a brighter and thick Ti-N enriched region which is interrupted by pockets of Al_2O_3 ahead of the intermixed region. W-rich particles have concentrated near the metal/scale interface.

Conclusions

This study shows the complex nature of the scale that can form on TiAl alloys during high temperature exposure in air. None of the alloys form a true continuous alumina layer and nearly all formed nitrogen enriched zones at the alloy/scale interface. Alloy 7 and alloy K5 show excellent oxidation resistance at 704 and 800°C. With the exception of binary TiAl and TiAl-2Cr, all the TiAl alloys exhibit good long-term isothermal oxidation resistance at 704°C in air, the scale thickness only varying from 5 to 15 µm. The scales formed on the engineering γ-alloys are not identical and extremely complex, in particular with regards to the phases formed at the metal/scale interface. To some extent these results are at odds with generalizations made for protective behavior at higher temperatures. That is, very good behavior is observed here for 704°C oxidation of the engineering alloys, which do exhibit a nitride interlayer but not the brittle oxygen- and Ti-enriched, Al-depleted Z/X phase interlayer reported during oxidation above 900°C. While at high temperature, it appears that continuous alumina scales occur primarily when the Z/X phase is formed and the nitride layer and α_2-Ti_3Al layers are absent [27]. Much of the environmental concern for the TiAl engineering alloys stems from an embrittlement perspective - most do not form excessive scales, but may have compromised mechanical properties if a brittle surface layer forms. It is not yet clear whether the Z/X, nitride, or oxygen-saturated α_2-Ti_3Al phases are equally problematic.

Acknowledgments

Appreciation is expressed to J. W. Smith and T. R. McCue for dedicated and excellent microscopy work. The long-term oxidation study efforts of the engineering γ-alloys at NASA Glenn was initiated by Bill Brindley, with whom many useful discussions on the oxidation and embrittlement mechanisms in Ti-base alloys are gratefully acknowledged.

References

1. P.A. Bartolotta and D. L. Krause, NASA/TM-1999-209071, 1999.
2. H. Clemens, H. Kestler, Adv. Eng. Mater. **2** (9), 551 (2000).
3. D. Eylon, M. M. Keller and P. E. Jones, Intermetallics 6 (7-8), 703 (1998).
4. T. Noda, ibid., 709 (1998).
5. I.E. Locci, M.P. Brady, R.A. Mackay and J.W. Smith, Scripta Mat. **37** (6), 761 (1997).
6. M.P. Brady, W.J. Brindley, J.L. Smialek, I.E. Locci, JOM **48** (11), 46 (1996).
7. N.S. Jacobson, M. P. Brady, and G. M. Mehrotra, Oxidation of Metals **52**, 537 (1999).
8. V.A.C. Haanappel et al., Oxidation of Metals **48**, 263 (1997).
9. M. Nombela, V. Kolarik, M.Gross, H. Fietzek, and N. Eisenreich, Matl. High Temp. **17**, 49 (2000).
10. A. Rahmel, W.J. Quadakkers, and M. Schutze, Materials and Corrosion **46**, 271 (1995).
11. J.C. Schaeffer et al., in Gamma Titanium Aluminides, edited by Y-W Kim et al. (TMS,1995) p. 71.
12. J.M. Rakowski et al., Scr. Metall. Mater. **33**, 997 (1995).
13. M. Yoshihara, K. Miura and Y-W Kim, ibid ref. 11, p.93.
14. F. Dettenwanger et al., Matls & Corrosion 48(1), 23 (1996).
15. R.W. Beye and R. Gronsky, Acta Metall. Mater. **42** (4) 1373 (1994).
16. C.Lang and M.Schutze, Mater.Corros. **48**, 13 (1997).
17. W.J. Quadakkers, N. Zheng, A. Gil, E. Wallura, H. Hoven, High Temp. Corr. Prot. Mat. **4**, 187 (1997).
18. E.H. Copland, B. Gleeson, D.J. Young, Acta Mater. **47**(10), 2937 (1999).
19. W.B. Retallick, M.P. Brady, D.L. Humphrey, Intermetallics **6**, 335 (1998).
20. P. Perez, J.A. Jimenez, G. Frommeyer, and P. Adv. Matl. Sc. Eng. A **284**, 138 (2000).
21. J. Magnan et al., Met. and Mat. Trans. A **30**, 19 (1999).
22. V.A.C. Haanappel et al., Mat. High Temp. **14**(1), 19 (1997).
23. J. Doychak, S. V. Raj, I.E. Locci, M. Hebsur, NASA CP-10082, 18 (1991).
24. M.P. Brady, E.D. Verink, Jr., J.W. Smith, Oxidation of Metals **51** (5-6), 539 (1999).
25. Y. Shida, and H. Anada, Oxidation of Metals **45** (1-2), 197 (1996).
26. J.D. Sunderkotter et al., Intermetallics **5**, 525 (1997).
27. V. Shemet et al., Oxidation of Metals **54** (3-4), 211 (2000).

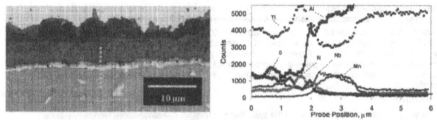

Figure 4. Typical microstructure, and FE-SEM elemental linescans observed in a XD-TiAl alloy after exposure to 704ºC for 7000 hr in air.

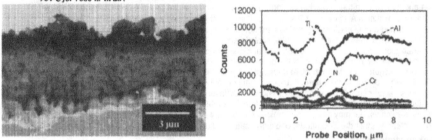

Figure 5. Typical microstructure, and FE-SEM elemental linescans observed in alloy K5 after exposure to 704ºC for 7000 hr in air.

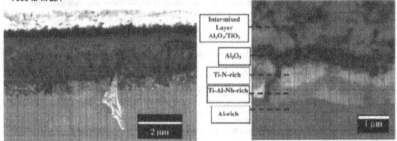

Figure 6. Typical microstructure observed in alloy 7 after exposure to 704ºC.

Figure 7. Typical microstructure observed at the metal/scale interface in TiAl-2Cr-2Nb after 800ºC.

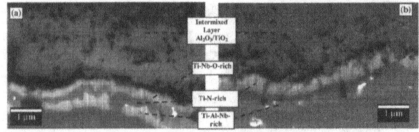

Figure 8. Typical microstructure observed at the metal/scale interface for (a) Alloy K5, (b) Alloy 7 after 800ºC -1000h.

Mat. Res. Soc. Symp. Proc. Vol. 646 © 2001 Materials Research Society

Interdiffusion and Phase Behavior in Polysynthetically Twinned (PST) TiAl / Ti Diffusion Couples

Ling Pan, David E. Luzzi
Department of Materials Science and Engineering, University of Pennsylvania
3231 Walnut Street, Philadelphia, PA 19104, U.S.A

ABSTRACT

Diffusion couples of pure Ti and polysynthetically twinned (PST) TiAl (49.3 at.% Al) were prepared by high vacuum hot-pressing, with the bonding interface perpendicular to the lamellar planes. Diffusion experiments were carried out by annealing the couples in the same furnace at 650, 700 and 850°C for various times. The cross-section of the couple was studied using scanning electron microscopy (SEM) and quantitative wavelength-dispersive x-ray spectroscopy (WDS). A reaction layer whose composition is close to that of the stoichiometric α_2–Ti_3Al phase formed along the PST TiAl / Ti bonding interface in PST TiAl side. Direct measurements of the thickness of the reaction zone were performed at different phase regions and various boundaries. By assuming the thickness of the reaction zone increases as $(Dt)^{1/2}$, where D is the diffusion coefficient and t is the annealing time, the diffusion coefficients at these temperatures were calculated. Composition profiles in the reaction zone, along the lamellae and at the lamellar interfaces were obtained by WDS analyses.

INTRODUCTION

Intermetallic compounds made of the light elements Ti and Al are promising candidates for aerospace, automotive and turbine power generation applications[1-5]. In the past decade, special interest has been paid to poly-synthetically twinned (PST) TiAl, composed of alternate lamellae of the γ-TiAl phase and the α_2-Ti_3Al phase with the orientation relationship $\{111\}_\gamma$ // $(0001)_{\alpha 2}$ and $<1\,\bar{1}0>_\gamma$ // $<11\,\bar{2}0>_{\alpha 2}$ [6]. PST TiAl exhibits low temperature ductility [7, 8] and higher toughness and high-temperature strength than TiAl alloys with other microstructures [9, 10].

At high temperatures the physical and mechanical properties of materials are generally associated with diffusion. In the Ti-Al alloy system, the formation and high-temperature stability of the lamellar structure are controlled by diffusion processes within the two phases and along the γ/γ and γ/α_2 lamellar boundaries. Moreover, diffusion is an important determinant of the creep resistance of the lamellar structure [11, 12]. Therefore, a fundamental understanding of the diffusion mechanisms in PST TiAl alloys is of great importance for the development of titanium aluminide alloys.

Current understanding of the diffusion processes in the Ti-Al system is largely based on the studies by Herzig et al. [13-16], who performed a series of tracer diffusion experiments, including self-diffusion, in polycrystalline γ-TiAl and α_2-Ti_3Al. The penetration profiles for diffusion in large grain size (> 1 mm) materials exhibit a $c \propto \exp(-x^2 / 4Dt)$ behavior for instantaneous sources and $c \propto \mathrm{erfc}\,[x / 2(Dt)^{1/2}]$ behavior for constant sources, as in semi-infinite materials. c is the average layered concentration of the diffusant and x is the penetration depth. In smaller grain size materials (350 - 500 μm), grain boundary diffusion is dominant in regions away from the surface (> 100 μm), where the concentration profile follows the numerical rule of $\log c \propto x^{6/5}$.

In the present paper, we present initial results on diffusion couples of Ti and PST TiAl crystal. This is the first experimental attempt to elucidate the diffusion and phase behavior in the two phases of a PST TiAl crystal, and will shed light on the interface diffusion along various lamellar boundaries in PST TiAl.

EXPERIMENTAL PROCEDURES

The composition of the master ingots used in the present study was Ti-49.3 at.% Al. 99.999% Ti and 99.99% Al were arc-melted at least four times to ensure homogeneity of the as-cast ingots. PST TiAl crystals were grown from the master ingots in an optical floating zone furnace under flowing argon gas with a growth rate of 3 mm/h. The back-reflection Laue X-ray diffraction technique was used to align the crystal along certain orientations. Slices parallel to the $\{1\,\overline{1}0\}$ planes, about 0.5 mm thick, which are also perpendicular to the lamellar planes, were cut from the as-grown PST crystals. The directions mentioned here and in the remainder of the paper are with respect to the γ-TiAl phase of the PST crystal. For diffusion-bonding experiments, the slices of PST crystals were mechanically and electrolytically polished in a solution of 6 vol.% perchloric acid (70%), 35 vol.% n-butyl alcohol and 59 vol.% methanol prior to diffusion bonding. Bulk Ti (99.999%) specimens were mechanically polished in parallel using an Allied High Tech MultiPrep polisher using silicon carbide paper to an ultimate finish of 1200-grit. Diffusion couples of PST TiAl and Ti were produced by diffusion bonding in a high vacuum furnace at 600°C for two hours. No extra mechanical stress was applied to the material during diffusion bonding except that from the thermal expansion of the graphite rams. Cross-sections of the as-bonded diffusion couples were cut perpendicular to both the bonding plane and the lamellar interfaces of the PST crystal. This cross-section was the observation surface in the microscope studies.

The as-bonded diffusion couples were subjected to diffusion anneals in the same furnace at three different temperatures under high vacuum conditions. Three diffusion couples were annealed at 650°C for 8 hours, 700°C for 8 hours, and 850°C for 2 hours, respectively. For the convenience of illustration, the diffusion couple annealed at 650°C for 8 hours is denoted as PST-Ti1, the one annealed at 700°C for 8 hours is denoted as PST-Ti2, and the one annealed at 850°C for 2 hours as PST-Ti3. SEM observations and quantitative WDS chemical analyses were carried out with a JEOL6400 scanning electron microscope operated at 15kV.

RESULTS AND DISCUSSIONS

Typical SEM back-scattered electron images of the as-bonded PST-Ti diffusion couple and the annealed PST-Ti1 are shown in figure 1. The upper white part at the very top of each image is the bulk Ti and the lower lamellar structure is PST crystal, with the lighter contrast vertical laths of α_2-Ti$_3$Al lamellae and darker γ-TiAl phase visible. A reaction zone with a wavy contour forms on the PST crystal side of the bonding interface and has a clear contrast difference with respect to both phases of the PST crystal. Especially, the image of the annealed diffusion couple (figure 1(b)) shows the reaction zone penetrates into the α_2 lamellae. Quantitative WDS chemical analysis indicated that the composition in the reaction zone is close to the stoichiometric α_2-Ti$_3$Al, while in the α_2 lamellae of the PST crystal the Al concentration is around 37 at. %, close to the expected equilibrium concentration of α_2 at 650°C for two-phase γ-α_2 alloys. These compositions are consistent with the contrast in the image.

The deeper penetration depth of the reaction zone into the α_2 lamellae than into the γ phase leads to an intuitive guess that Ti penetrates or diffuses into the α_2 lamellae at a faster rate. Considering the composition of the material and the morphology of the PST crystal, pure interdiffusion between the α_2 phase of the reaction zone and the α_2 as in a bulk diffusion couple is possible, whereas the γ-TiAl phase in the PST crystal must transform into the α_2–Ti$_3$Al phase prior to diffusion as the TiAl composition is at the Ti-rich phase boundary. However, the sharp contrast difference visible deep in the α_2 lamellae marking the boundary of Ti-rich Ti$_3$Al is not understood. The contrast seen in the SEM image of Figure 1(b) implies a sharp drop in average atomic number of the material at the point within the α_2 lamellae where the boundary between light and dark grey contrast is seen. Compositional analysis by WDS, within the approximately 1μm resolution limit of the SEM, confirms that this is associated with a sharp change in Ti concentration. The origin of this behavior will be explored through future TEM analysis.

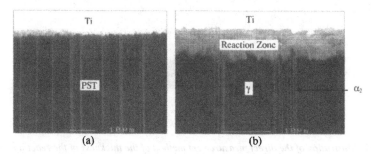

Figure 1. *SEM back-scattered electron images of (a) as-bonded PST-Ti diffusion couple and (b) PST-Ti1 annealed at 650 ℃ for 8 hours after bonding. The upper white part is Ti bulk and the lower lamellar structure is PST crystal, with the light contrast thinner lamellae α_2-Ti$_3$Al phase and the darker region γ-TiAl phase. The layer between Ti and PST along the bonding interface is what we call reaction zone.*

It was also found through SEM analysis that the penetration depth into the PST crystal of the reaction zone in any one diffusion couple varies. This variation is seen not only from differences in the penetration depth into the γ and α_2 phases, but also when comparing among individual γ lamellae or α_2 lamellae. The clear contrast difference in the SEM back-scattered images between the reaction layer and the two phases of the PST crystal allows a direct measurement of the penetration depth of the reaction zone as a function of position in each specimen. Hundreds of individual measurements of reaction zone thickness as a function of position where made using wide-area maps of each specimen created from many contiguous SEM images of the back-scattered electron signal. In order to create a common reference point, a line across the upper edge of the reaction zone through the images was drawn and regarded as the zero-depth level of the reaction zone, as shown in figure 2.

Measurements were divided into four categories. The penetration depths into α_2 lamellae were treated as one category. For the γ phase, the minima of the penetration depth were measured and labeled "γ". The maxima were labeled "γ/γ", with the assumption that the deepest penetration occurs at the points of the interfaces between the γ lamellae, which are invisible in the SEM. Finally, the thin γ lamellae sandwiched between α_2 lamellae, were placed in a separate category termed "γ in α_2". The thickness of the reaction zone, x, was measured vertically from the reference line to the point where the contrast difference marking the boundary of significant Ti penetration occurs, at the marked locations corresponding to each of the four categories as indicated in figure 2. Hundreds of measurements were made and then averaged for each category.

The distribution of the reaction zone thickness verified the deeper penetration into the α_2 lamellae than for the other three categories. For γ or α_2 lamellae with about the same thickness, the penetration depths usually are not the same. Hence, the thickness of the α_2 lamellae is not the dominant factor in the deeper penetration. This implies a faster penetration of Ti into the α_2 phase than into the γ phase.

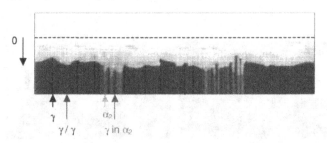

Figure 2. *Illustration of the direct measurement method of the thickness of the reaction zone.*

Under the approximation that the penetration depth, i.e. the thickness of the reaction zone, is approximately equal to $(Dt)^{1/2}$, where D is the diffusion coefficient and t is the annealing time, the diffusion coefficients for the four categories were calculated. The results are presented in table I. The values for the diffusion coefficients determined from the present measurements are 3-4 orders higher than the Ti self-diffusion coefficients in single phase γ-TiAl or α_2-Ti$_3$Al materials obtained from tracer diffusion experiments for diffusion through the bulk[14, 15]. It should be noted that the tracer diffusion experiments were on large grain, random polycrystals with a low density of interfaces.

Table I. Calculated results of diffusion coefficients for PST-Ti1, PST-Ti2 and PST-Ti3.

	t (hr)	T (K)	γ	γ/γ	γ in α_2	α_2
			D (m²/s)	D (m²/s)	D (m²/s)	D (m²/s)
PST-Ti1	8	923	4.15E-16	5.80E-16	5.46E-16	9.33E-16
PST-Ti2	8	973	1.81E-15	2.42E-15	2.33E-15	3.94E-15
PST-Ti3	2	1123	2.39E-15	3.21E-15	3.06E-15	6.00E-15

In order to study the details of the concentration gradient across the reaction zone and into the PST crystal, quantitative WDS analyses with ZAF corrections were carried out on PST-Ti2 using pure Ti and Al as standards. For these initial measurements and to examine the unexpected sharp contrast difference seen in the SEM images, the profile across the reaction zone and into the α_2 lamellae was studied. Two models were applied to analyze the composition profiles. One assumes that the concentration profile is determined by volume diffusion in the α_2 phase, i.e., the diffusion couple acts just like a bulk Ti/Ti$_3$Al diffusion couple. By applying the constant source condition and letting $c|_{x \to \infty}$ be the Ti concentration in the α_2 phase in PST TiAl, which is about 63 at.%, i.e.

$$c_1 = c|_{x<0} = 1, \quad c_2 = c|_{x \to \infty} = 0.63, \quad (1)$$

an analytical solution to Fick's diffusion equation for the case of two semi-infinite bulk diffusion couples is obtained:

$$\frac{c-c_2}{c_1-c_2} = \frac{1}{2}erfc\left(\frac{x}{2\sqrt{Dt}}\right) \quad (2)$$

The inverse complementary error function of $2(c-c_2)/(c_1-c_2)$ versus x from the experimental data is plotted in Figure 3a. If the model of volume diffusion in a semi-infinite bulk is applied, the data should fall on a straight line with the diffusion coefficient determining the slope of the line. The data falls remarkably well on a straight line through the reaction zone and into the α_2 lamellae, but has a discontinuity around the depth 11-12 μm corresponding to the contrast difference in the SEM image, whose origin will rely on the future TEM study to clarify. Using the data points at depths less than that corresponding to the discontinuity to calculate the diffusion coefficient, a value of $D = 3.15 \times 10^{-15}$ m^2/s was obtained, which is comparable with the value obtained by the measurement of reaction zone thickness, 3.94×10^{-15} m^2/s.

The second model applied to analyze the data is the grain boundary diffusion model. For type B and C grain boundary diffusion kinetics, there is a numerical rule [17] that ln c is proportional to $x^{6/5}$. We plotted the same Ti concentration profile as ln c versus $x^{6/5}$ (see figure 3b). In this case, the data does not display an obvious linear aspect and the discontinuity is seen at a value of 18-19 μm$^{6/5}$, which suggests that the diffusion under these conditions in the PST system follows either type A or a mixture of type A and B kinetics, where the bulk diffusion effect cannot be ignored.

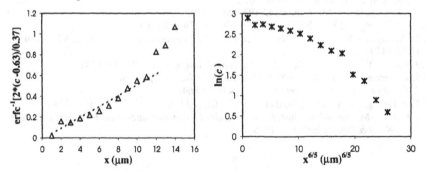

Figure 3. (a) *Composition profile in α_2 lamella fitted by complementary error function,*
(b) *Composition profile in α_2 lamella fitted by ln(c) vs. $x^{6/5}$.*

CONCLUSIONS

In diffusion couples of Ti and PST TiAl, a reaction layer forms with composition close to the stoichiometric α_2-Ti$_3$Al phase. Ti diffuses into the PST α_2 phase while the PST γ phase transforms into the α_2 phase. Diffusion coefficients obtained by directly measuring the thickness of the reaction layer are 3-4 orders higher than the self-diffusion coefficients from the tracer diffusion experiments in single phase polycrystalline material. The composition profiles in α_2 lamellae showed surprising features that rely on future TEM studies to solve.

ACKNOWLEDGEMENT

This research was supported by National Science Foundation grant no. DMR96-15228.

REFERENCES

1. F. H. Froes and C. Suryanarayana, in *Physical Metallurgy and Processing of Intermetallic Compounds*, N. S. Stoloff and V. K. Sikka, ed., Chapman & Hall, New York, 1996, Chapter 8.
2. S. C. Huang and J. C. Chesnutt, in *Intermetallic Compounds: Vol. 2, Practice*, J. H. Westbrook and R. L. Fleischer, ed., John Wiley & Sons, 1994, Chapter 4.
3. J. Horton, S. Hanada, I. Baker, R. D. Noebe and D. Schwartz, ed., *High Temperature Ordered Intermetallic Alloys VI* (MRS, Pittsburgh, 1995), **364**.
4. I. Baker, R. Darolia, J. D. Whittenberger and M. H. Yoo, ed., *High Temperature Ordered Intermetallic Alloys V* (MRS, Pittsburgh, 1993), **288**.
5. D. P. Pope, C. T. Liu and S. H. Whang, ed., *High Temperature Intermetallics, Mater. Sci. Eng. A* **192/193** (1995).
6. M. J. Blackburn, in *The Science, Technology and Application of Titanium*, Eds. R. I. Jaffee and N. E. Promisel, Pergamon, London (1970) 633.
7. H. Umeda, K. Kishida, H. Inui and M. Yamaguchi, *Mater. Sci. Eng. A* **239-240** (1997) 336.
8. E. L. Hall and S. C. Huang, *J. Mater. Res.* **4** (1989) 595.
9. Y. W. Kim, *J. Metall.* **46**(7) (1994) 30.
10. Y. W. Kim, *Mater. Sci. Eng. A* **192/193** (1995) 519.
11. L. M. Hsiung and T. G. Nieh, *Mater. Sci. Eng. A* **239-240** (1997) 438.
12. K. Maruyama, R. Yamamoto, H. Nakakuki and N. Fujitsuna, *Mater. Sci. Eng. A* **239-240** (1997) 419.
13. J. Breuer, T. Wilger, M. Friesel and Chr. Herzig, *Intermetallics* **7** (1999) 381.
14. Chr. Herzig, T. Przeorski and Y. Mishin, *Intermetallics* **7** (1999) 389.
15. J. Rüsing and Chr. Herzig, *Intermetallics* **4** (1996) 647.
16. Chr. Herzig, M. Friesel, D. Derdau and S. V. Divinski, *Intermetallics* **7** (1999) 1141.
17. I. Kaur, Y. Mishin and W. Gust, *Fundamentals of Grain and Interphase Boundary Diffusion*, John Wiley & Sons, Chichester, 1995.

Mat. Res. Soc. Symp. Proc. Vol. 646 © 2001 Materials Research Society

Control of Lamellar Orientation in γ-TiAl Based PST Crystal by Using Seed Crystals

Y. Yamamoto, H. Morishima[1], K. Koike[1], M. Takeyama and T. Matsuo
Department of Metallurgy and Ceramics Science, Tokyo Institute of Technology
2-12-1, Ookayama, Meguro-ku, Tokyo 152-8552, JAPAN
[1]Graduate Student, Tokyo Institute of Technology, Tokyo, JAPAN

ABSTRACT

Unidirectional solidification of Ti-48Al binary alloy using γ-TiAl single-phase seed crystals has been carried out by an optical floating zone method. The lamellar orientation of the grown PST crystal follows the orientation of the Ti-57Al seed crystal, while it fails to follow that in the case of the Ti-53Al seed. Microstructure analysis reveals that the seed crystal of Ti-57Al exhibits a flat liquid/solid interface in melting ($\gamma \rightarrow \gamma + L$) even after making contact with 48Al to grow, whereas the seed of Ti-53Al shows a cellular interface due to the peritectic reaction in melting ($\gamma \rightarrow \alpha + L$). At the 57/48 interface, an abrupt change of Al concentration was detected from the seed to the grown crystal, indicating an occurrence of composition travel to skip the peritectic reaction, which is responsible for the control of lamellar orientation of the grown PST crystals. The same attempt has been made by using the 57Al single crystal seed with a different orientation, and the lamellar orientation of the grown PST crystal was confirmed to follow the orientation of the seed.

INTRODUCTION

Gamma-TiAl based alloys have been developed for high temperature structural applications because of high specific strength at elevated temperatures [1]. Most of the currently developed alloys have an α_2/γ fully lamellar microstructure which is formed when high temperature α phase is slowly cooled to $\alpha_2 + \gamma$ two-phase region [2,3]. In case that the Ti-48at%Al alloy is unidirectionally solidified, the microstructure becomes a two-phase single crystal with fully lamellar microstructure (PST crystal) [4,5]. Matsuo et al. recently revealed that the creep resistance of Ti-48Al PST crystal is superior to that of the fully lamellar polycrystalline alloy when the lamellar plates are aligned parallel to tensile creep loading axis [6,7]. Thus, from engineering viewpoint, it is very important to establish the method to control the lamellar orientation of PST crystals. Yamaguchi et al. [8] have successfully made columnar grain structure with the lamellar orientation aligned parallel to the growth direction. However, as far as the authors are aware, no one has ever successfully controlled the lamellar orientation of the PST crystals.

Recently, we have attempted to control the lamellar orientation of Ti-48Al PST crystals in growing using seed crystals with various Al compositions [9,10]. In case of the 48Al PST as a seed, the grown crystals always become PST under certain conditions, but the lamellar orientation can never follow the orientation of the seed crystal, because the seed crystal becomes polycrystal due to γ → α phase transformation in heating to melt. However, the microstructure examination of the contact region between the seed and grown crystals revealed that the grown crystal follows the orientation of one of the newly formed α grain, indicating a possibility to control the lamellar orientation of the PST if the grain formation in heating is fully suppressed.

In this study, two gamma single phase alloys Ti-53Al and 57Al have been used as seeds to grow Ti-48Al PST by unidirectional solidification. Since we found that the lamellar orientation of the grown crystal follows that of the 57 Al seed, the way to control the lamellar orientation of Ti-48Al PST crystal is presented.

EXPERIMENTAL

Alloys used in this study were Ti-48at%Al for grown material, and Ti-53Al and 57Al for seed materials. These seed alloys show a γ-TiAl single phase with no solid/solid phase transformations, although the primary solidification phase in 53Al is α with a peritectic reaction (L+α→γ) and that in 57Al is γ, as shown in Fig. 1. The 48Al alloy was prepared by an induction skull melting, followed by centrifugal casting to bar with 14 mm in diameter and 150 mm in length. The seed alloys were prepared by arc melting, followed by drop casting to bar with 8 mm in diameter and 80 mm long. These bar ingots for seed were unidirectionally solidified (UDS) using an optical floating zone (OFZ) furnace under a flowing argon at a growth rate in the range of 5 to 30 mm/h. The crystal growth

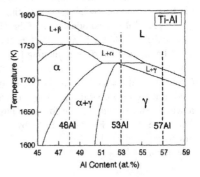

Figure 1. Ti-Al binary phase diagram, showing the alloy compositions of grown crystal(48Al) and seed materials (53Al and 57Al).

of 48Al was performed in OFZ method using the UDS seed crystals. Microstructures were examined by optical and scanning electron microscopes. The phase compositions were analyzed by electron probe microanalyzer (EPMA).

RESULTS & DISCUSSION

Crystal growth of 48Al with 53Al seed

Figure 2 shows a vertical section near the interface between 48Al grown and 53Al seed crystals (48/53). Many lamellar grains with diameter of less than 1 mm are observed near the interface region. Note that the seed of 53Al does not become a single crystal by UDS but becomes polycrystal with large grains of 3 mm in diameter, indicating that the observed grains at the interface region are formed by γ→α phase transformation due to the peritectic reaction

Figure 2. A vertical section near the contact region between the 48Al grown and 53Al seed crystals.

($\gamma \rightarrow \alpha + L$) in heating, as shown in Fig. 1. It should also be noted that, even so, the grown crystal becomes PST since one of the grains is selected to grow. These results show that the 53Al seed is not suitable for controlling the lamellar orientation in 48Al PST.

Crystal growth of 48Al with 57Al seed

The seed of 57Al after UDS was a bicrystal, and by using the seed the crystal growth of 48Al was performed. Figure 3 shows an optical microstructure of the vertical section showing the region near the grown and seed crystals. The contact interface of 57Al/48Al is fairly flat and no grains are formed at the interface. Interestingly, two PST crystals corresponding to each of the bicrystal seed are observed in the 48Al grown crystal. The right-hand side of the PST crystal in

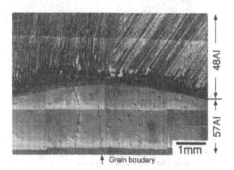

Figure 3. A vertical section near the contact region between the 48Al grown and 57Al seed crystals.

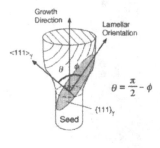

Figure 4. Schematic illustration showing the grown PST and the seed crystals: θ is the angle between normal to the lamellar plates and the growth direction.

Fig. 3 is observed parallel to the lamellae, so that the lamellar orientation ϕ with respect to the growth direction is identified to be 40°. Thereby, the angle θ between the normal to the lamellar plane and the growth direction becomes 50°, as shown in Fig. 4, since the lamellar plates are parallel to one of four {111} planes in γ phase. A back-reflection Laue analysis revealed that the right-hand side of the seed crystal has an orientation of [012] along the growth direction, and the angle between the directions of $[012]_\gamma$ and $<111>_\gamma$ becomes either 50° or 70° on the standard stereographic projection. These results strongly suggest that the grown crystal follows the orientation of the seed crystal.

Figure 5 shows a high magnification view of the interface region on the left-hand side of the crystal in Fig. 3, together with the composition profile of Al in γ phase across the interface along the black line. There exists a γ single phase region consisting of small grains just above the flat interface, but because of little difference in the contrast, these grains should be sub-grains with low angle boundaries. The composition of Al changes from 57 to 50 % within a limited region (300 µm) of the seed toward the interface, and remains almost unchanged from the interface to the PST crystal.

The growth following the seed orientation is probably attributed to the composition travel to skip the peritectic reaction to occur. As shown in Fig. 6, in case that the liquids of the 48Al bar and 57Al seed are mixed in contact, the average Al composition of the liquid should be in the range where the primary solidification phase is α ((1) in Fig. 6). Holding the liquid for a while makes the aluminum composition of the seed decreased ((2) in Fig. 6). And once the crystal growth begins, the primary α phase forms on top of the seed crystal ((3) in Fig. 6), with following the orientation of seed crystal. The formation of such primary α phase at the beginning of the crystal growth is the key for the grown crystal to follow the orientation of the seed crystal, even though the primary phase of the alloy to grow change to β during crystal growth, since we confirmed that the β phase transforms to α with following the orientation of the grown α phase [9,10].

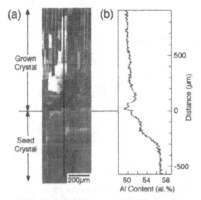

Figure 5. (a) A back-scattered electron image across the interface and (b) Al composition profile along the line of the image.

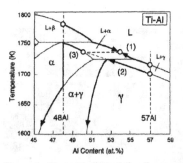

Figure 6. Composition travel for the crystal growth of 48Al alloy with 57Al seed crystal: (1) an average liquid composition in contact with the seed and 48Al to grow, (2) change in Al content of the seed in contact with 48Al, (3) change in the composition of primary α phase formed on top of the seed crystal.

The same attempt has been made using the 57Al single crystal seed with $[001]_\gamma$ orientation along the growth direction, in which the angle $\theta = 55°$ respect to any $<111>_\gamma$ direction. The resulting microstructure is shown in Fig. 7. The lamellar orientation ϕ in the grown PST crystal becomes 35°, as is expected from the relationship of $\theta = \pi/2 - \phi$.

Although there is still a question which of the four $<111>$ directions in the γ single crystal seed is followed, the present results clearly demonstrate a step forward in the method to control of lamellar orientation in γ-TiAl based PST crystal.

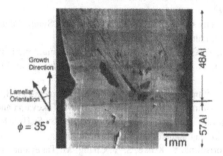

Figure 7. *An optical microstructure of the vertical section near the contact region between the 48Al grown crystal and the [001] oriented 57Al seed.*

SUMMARY

Unidirectional solidification of Ti-48at%Al alloy has been examined using γ-TiAl single-phase alloys as seed crystals, in order to control the lamellar orientation of the 48Al grown single crystal (PST crystal). The PST crystal can easily be grown regardless of the seed compositions; however, the 53Al seed does not work out because of the formation of newly oriented grain formation due to the peritectic reaction ($\gamma \rightarrow L+\alpha$) at the tip of the seed crystal when heating to melt. In contrast, the 57Al seed crystal works well to control the lamellar orientation of the 48Al PST crystal. The growth in the seed orientation is attributed to the composition travel to skip the peritectic reaction and result in primary solidification of α at the tip of the seed.

ACKNOWLEDGEMENTS

This research is supported by the research grant on "Research for the Future Program" from Japan Society for the Promotion of Science (96R12301), and in part by a grant-in-aid (11875149) for Scientific Research from Ministry of Education, Science and Culture, Japan.

REFERENCES

1. Y. W. Kim, JOM., **46**, 30 (1994).
2. Y. Yamabe and M. Kikuchi, Journal of the Japan Institute of Metals, **30**, 37 (1991).
3. M. Takeyama, T. Kumagai, M. Nakamura and M. Kikuchi in Proc. Intl. Symp. Structural Intermetallics, edited by R. Darolia, J. J. Lewandowski, C. T. Liu, P. L. Martin, D. B. Miracle and M. V. Nathal, (TMS, Pennsylvania, 1993) p.167.
4. T. Fujiwara, A. Nakamura, M. Hosomi, S. R. Nishitani, Y. Shirai and M. Yamaguchi, Philo. Mag. A, **61**, 591 (1990).
5. M. Takeyama, T. Hirano and T. Tsujimoto in Proc. Intl. Symp. Intermetallic Compounds (JIMIS-6), edited by O. Izumi, (JIM, Sendai, 1991) p.507.
6. T. Matsuo, T. Nozaki, T. Asai, S. Y. Chang and M. Takeyama, Intermetallics, **6**, 695 (1998).
7. N. Shiratori, S. Hirata, T. Asai, M. Takeyama and T. Matsuo, Key Engineering Materials, **171-174**, 639 (2000).
8. M. Yamaguchi, D. R. Johnson, H. N. Lee and H. Inui, Intermetallics, **8**, 511 (2000).
9. Y. Yamamoto, H. Morishima, K. Koike, S. Y. Chang, M. Takeyama and T. Matsuo in Proc. 5th Intl. Conf. Structural Functional Intermetallics, edited by S. H. Whang, C. T. Liu, D. P. Pope, H. Vehoff and M. Yamaguchi, to be published.
10. Y. Yamamoto, H. Morishima, K. Koike, S. Y. Chang, M. Takeyama and T. Matsuo, Report of the 123rd Committee on Heat-Resisting Materials and Alloys, Japan Society for the Promotion of Science, **41**, 213 (2000).

Nickel Aluminides

Mat. Res. Soc. Symp. Proc. Vol. 646 © 2001 Materials Research Society

Dislocation Processes and Deformation Behavior in <001>-Oriented Fe_x-Ni_{60-x}-Al_{40} Single Crystals

P. S. Brenner, R. Srinivasan, R. D. Noebe*, T. Lograsso** and M. J. Mills
Department of Materials Science and Engineering, The Ohio State University, Columbus, OH 43210
*NASA Glenn Research Center, Cleveland Ohio 44135
**111A Metals Development, Ames Laboratory, Ames, IA 50011

ABSTRACT

The mechanical properties and dislocation microstructure of single crystals with a range of compositions within the Fe_x-Ni_{60-x}-Al_{40} pseudobinary system have been investigated, with the purpose of bridging the behavior from FeAl to NiAl. Experiments are focused on the compression testing of <001> oriented single crystals with compositions where x = 10, 20, 30, 40, and 50 (in atomic percent). Observations of a<111> dislocation morphologies at room temperature and both a<111> and non-a<111> dislocation activity at elevated temperatures are reported and discussed. Measurements of the yield strength, elastic modulus and strain hardening rates are reported, and the variation of strength with composition is correlated with dislocation dissociation and overall dislocation morphology.

1. INTRODUCTION

Extensive research has been conducted with respect to the binary NiAl and FeAl intermetallic compounds having the B2 crystal structure [1-3]. Their attractive high temperature properties, relatively high yield strengths, and resistance to oxidation and corrosion have piqued interest in these systems. Unfortunately, poor low-temperature ductility, inadequate fracture toughness, and rapid loss of strength at high temperatures have been critical negatives for potential applications. Nevertheless, from a fundamental viewpoint, these alloys represent an ordered intermetallic system in which complex point defect and dislocation behavior play a strong role in governing the deformation mechanisms. NiAl alloys show higher room temperature yield strengths than FeAl, but little-to-no room temperature tensile ductility. Their deformation via a[001] dislocations provides only three independent slip systems, falling short of the five slips systems required for plastic flow (in polycrystals) by the Von Mises criterion. FeAl deforms with a lower room temperature yield strength, and exhibits significant (although somewhat variable) ductility. The larger ductility has been explained by the fact that active a<111> dislocations provide sufficient slip systems to satisfy the Von Mises criterion. There is also evidence that hydrogen embrittlement and weak grain boundaries are also important factors limiting the overall ductility [3].

Of particular interest in this study is the effect of large solute additions on the properties of ternary compositions within the (Ni,Fe)-40Al pseudobinary system, with which we are able to bridge the behavior from NiAl to FeAl binary B2 compounds. In early (Ni,Fe)-Al psuedobinary work by Nagpal and Baker [4], the hardness as a function of composition was measured for alloys containing 45%Al. A dramatic influence of quenching rate for Fe-rich compositions was also observed. Patrick, et al [5] performed a TEM study of dislocation structures in polycrystals over a range of compositions with 40%Al. They determined that a change from <111> to <100> slip occurs in the composition range between 30Fe-30Ni-40Al and 50Fe-10Ni-40Al. Most recently, Pike [6-7] has determined the site occupancy, point defect concentration, and hardness as a function of composition. Furthering these (Ni,Fe)-40Al pseudobinary studies with the correlation of mechanical properties and dislocation mechanisms should provide significant insight into the parameters that control deformation, and may provide solutions to the critical drawbacks inherent in the FeAl and NiAl systems. The present studies have focused on the behavior of a<111> dislocations in hard oriented [001] single crystals as a function of composition, both at room temperature and elevated temperatures. Motion of a<111> dislocations in [001] oriented crystals is induced because there is no resolved shear stress for the glide of the <001> slip vectors dominant in polycrystalline NiAl deformation. As the a<111> dislocation is the desired slip vector for ductility, a characterization of its nature versus composition and temperature is crucial to understand and optimize the mechanical properties of the pseudobinary system. A complementary set of polycrystalline samples are also being investigated in order to investigate their corresponding strengths and potential for ductility.

2. EXPERIMENTAL

Single crystals in the Fe_x-$Ni_{60-x}Al_{40}$ composition range were prepared at the Materials Preparation Facility of the AMES Laboratory at Iowa State University using the Bridgman method. Note that all compositions will be in atomic percent. Following homogenization at 1000°C for 72h, [001] oriented compression samples were electro-discharge machined in the form of 3 x 3 x 8 mm parallelepipeds. The side faces were {010} oriented and electropolished to facilitate optical slip trace analysis after mechanical testing. The compositions investigated were 10Fe-50Ni-40Al, 20Fe-40Ni-40Al, 30Fe-30Ni-40Al, 40Fe-20Ni-40Al, and 50Fe-10Ni-40Al. Single crystal mechanical data for Ni-40Al and Fe-40Al from Noebe [8] and Yoshimi [9] are reported for comparison purposes. The actual compositions and impurity levels for the single crystals are given in Table 1.

Table 1: Nominal and actual compositions and impurity levels for the single crystals.

Nominal Composition	at%						ppm by weight		
	Ni	Fe	Al	Si	Ti	Hf	C	O	N
10Ni-50Fe-40Al	10.9	50.2	38.9	160	<90	<20	<10	17	3
20Ni-40Fe-40Al	20.6	39.7	39.7	<70	<5	<10	24	72	4
30Ni-30Fe-40Al	31.5	30	38.5				23	10	<1
40Ni-20Fe-40Al	39.7	20.4	39.9	<50	<5	<10	13	34	1
50Ni-10Fe-40Al	49.6	10.6	39.9	70	<20	<5	44	15	3

Compression samples of the Fe-rich compositions (ranging from 10-30 % Fe) were sealed in quartz tubes evacuated (to 10^{-6} torr) and backfilled with high purity argon. These samples were subjected to vacancy reducing heat treatments at 450°C of varying durations (10-40 days), and hardness variations were monitored with microhardness measurements in order to insure that an equilibrium vacancy content had been obtained [10]. These vacancy annealing conditions have been previously established by Schneibel,et al. [11] as an effective treatment. Following suitable vacancy reducing heat treatment, the samples were tested in an ambient environment using an Instron model 1362 equipped with a MAR-M246 compression cage. Displacement was measured with LVDT's mounted on the plattens of the cage. The tests were all performed at a constant strain rate of 10^{-4}/s to a final plastic strain of 1-4%.

Thin foils were prepared for transmission electron microscopy (TEM) with foil normals either in the direction of the compression axis [001] or normal to an active slip plane. The foils were cut from the compression samples using a low speed saw equipped with a SiC abrasive wheel. Final electrojet polishing was performed at –40°C, at 25-30 mA and 4-8 V, on a Tenupol-3 twin-jet electropolishing unit using an electrolyte of 25% nitric acid in methanol for the Fe-rich samples and an electrolyte of 10% perchloric acid in methanol for the Ni-rich samples. Transmission Electron Microscopy (TEM) studies were performed on a Phillips CM200 LaB_6 electron microscope operating at 200kV.

3. RESULTS

3.1 Mechanical Behavior

The 0.2% yield strengths for each composition versus testing temperature is shown in Figure 1, along with the aforementioned data from Noebe [8] and Yoshimi [9]. This graph reveals the high yield strength of the 30Fe composition, the high to moderate yield strength of the 20Ni, 40Ni, and 50Ni compositions, and the moderate strength of the 10Ni composition. While the data as a function of temperature is rather sparse, it is clear that the variation of the yield strength with temperature is relatively modest at lower temperatures for all compositions. In addition, all five ternary alloys demonstrated a "knee" in yield strength at elevated temperatures, above which the temperature dependence is significantly larger. The appearance of such a "knee" has previously been correlated with a transition from <111> dislocations at lower temperatures to non-<111> activity at higher temperatures [12-15]. As discussed below, a similar transition appears to be occurring in the present case. The pronounced decrease in yield strength is evident in the approximate temperature window from about 600 to 850K; with the initial transition temperature varying significantly with composition. It is indeed interesting to note that all alloys show the slip transition (associated with the sharp decrease in yield strength) in a narrow temperature window, even though there are significant differences in the dislocation configurations in these alloys at lower temperatures, as will be discussed below. Finally, it is noted that in this temperature regime the ternary compositions are all significantly softer than the binary Ni-40Al compound.

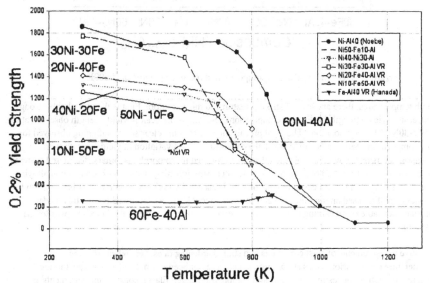

Figure 1: Yield strength and composition versus temperature for (Fe,Ni)-Al hard-oriented single crystals tested in compression. "VR" indicates samples subjected to a vacancy reduction heat treatment.

In order to more clearly see the variation of lower temperature strength with composition, Figure 2 presents the single crystal yield strength versus composition at both room temperature and 600K. At

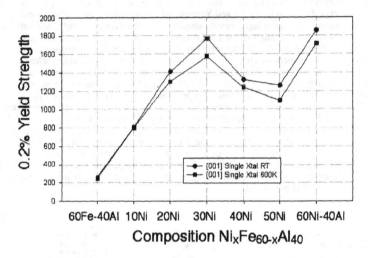

Figure 2: Yield strength versus composition at room temperature and 600K.

both temperatures, there are several striking features to this variation. First is the remarkably strong increase in yield strength for Fe-rich compositions. Second is the peak in yield strength observed at the 30Fe-30Ni-40Al composition. The third prominent feature is the clear solute softening observed for for compositions between Ni-40Al and 10Fe-50Ni-40Al. This variation in yield strength with composition is much more complicated than that observed in isomorphous, binary disordered systems such as Cu-Ni and Au-Ag [16]. In these cases, solute strengthening is always observed upon alloying the pure metals, and the increase in yield strength is classically understood in terms of both size and modulus misfit, based on the Fleisher solid solution model [17]. As discussed below, this model appears incapable of explaining the behavior observed at either extreme of the Fe-Ni-Al pseudobinary system.

Relative measurements of elastic modulus and strain hardening rates were taken from the room temperature compression stress strain curves, but have not been normalized with respect to the inherent moduli of the testing apparatus and compression cage. Elastic constant measurements are presently being conducted, and will be reported elsewhere. Nevertheless, all tests were performed in the same apparatus and thus should provide accurate relative comparisons. As shown in Figure 3a, the elastic modulus increases substantially for the 20Fe and 30Fe compositions relative to the Fe- and Ni-rich compositions. It is interesting to note that this maximum in the modulus values corresponds roughly with the maximum observed in the measured yield strength. The work hardening rates shown in Figure 3b were all measured at a plastic strain of about 2%. It should be noted that strain hardening

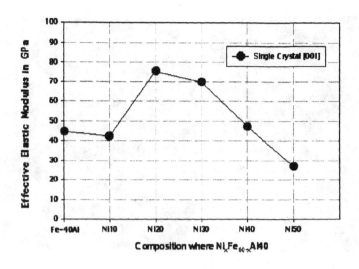

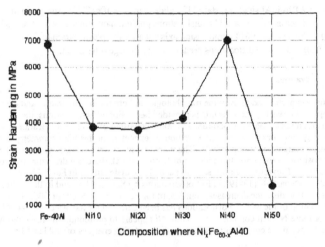

Figure 3: (a) Elastic modulus and (b) strain hardening rate (at ~ 2% plastic strain) measured from compression tests at room temperature.

was nearly linear for most compositions. With the exception of the binary Ni-40Al composition, the strain hardening rates are large for all compositions (ranging from ~ E/6 for Fe-40Al, ~E/10 for 20Fe-40Ni-40Al and ~ E/20 for 40Fe-20Ni-40Al).

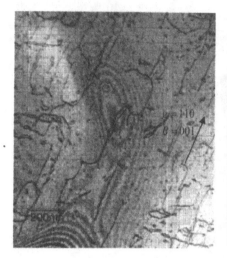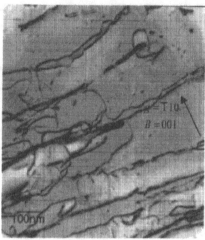

Figure 4: (a) 40Fe-20Ni-40Al single crystal {112} glide plane cut after 2% plastic strain in compression at room temperature. a<111> dislocations predominantly in screw orientation. (b) Ni50Fe10Al single crystal {001} compression axis cut after 2% plastic strain in compression at 600K. a <111> dislocations predominantly in edge orientation.

3.2 Dislocation Observations

Dislocation analysis was performed to examine morphologies at both room temperature and elevated temperatures. A brief summary of these results is provided here. For all compositions, the deformation microstructures at room temperature and 600K consisted of a<111> dislocations, with typically several slip systems operative in a given region, as expected for this high symmetry zone axis in the absence of strong latent hardening effects. There are two noteworthy variations in a<111> behavior with composition. First, upon the addition of Ni to Fe-40Al, there is a dramatic decrease in the dissociation distance. Previous work has shown that a<111> dislocations in Fe-40Al are dissociated into a/2<111> partials [18-19]. TEM observations by Savage [19], and in this work, have shown that the dissociation distance decreases from 3.5nm for Fe-40Al to 1.9nm for 50Fe-10Ni-40Al. No resolvable dissociation of the a<111> dislocations has been detected for the 40Fe-20Ni-40Al composition, nor for more Ni-rich compositions. Thus, Ni appears to profoundly increase the APB energy, as expected from simple consideration of the relative ordering energies of NiAl and FeAl.

A second noteworthy change with composition is in the preferred line direction of the a<111> dislocations. TEM images for compositions ranging from Fe-40Al to 30Fe-30Ni-40Al reveal a<111> dislocations predominantly near screw orientation (see Figure 4a). Somewhere between Ni30 and Ni50 there is a transition from observed <111> screws to <111> edge line lengths, as exemplified by the image in Figure 4b. This implies that the edge lengths are most mobile in Fe-rich samples, while the screw line direction is more mobile for Ni-rich compositions. This tendency is observed at both room temperature and 600K.

Several transitions are also observed as a function of temperature, the first involving the predominant slip plane. While the Fe-40Al and 50Fe-10Ni-40Al samples deform via slip on {110} planes at both room temperature and 600 K, all compositions with greater Ni content showed a predominance of {112} slip planes at room temperature. However, at 600 and 700K respectively, a transition to slip on {110} type planes is seen for the 40Fe-20Ni-40Al and 10Fe-50Ni-40Al compositions. This transition to {110} slip planes at higher temperatures is believed to be a trend for all Ni-rich alloys.

The second important transition with temperature is that from <111> slip at lower temperatures to a<100> climb at higher temperatures. This transition, which is common to all the ternary compositions, is exemplified by the dislocation structure shown in Figure 5 for the 10Fe-50Ni-40Al composition deformed at 750K, where several families of a<100> dislocations are observed. Note that this sample was deformed at a temperature only about 50K higher than the apparent temperature of the "knee" seen in the yield strength versus temperature plot of Figure 1, and yet no a<111> dislocations are present. The common transition to a<001> climb should be contrasted with previous work that has proven the operation of a<110> glide as the dominant deformation process for binary Ni-rich compositions [20]. Thus, the significantly larger yield strengths for Ni-40Al in the temperature range from 600K to 900K appears to be directly related to differences in the a<111> decomposition process, as well as to differences in the deformation mode above the "knee".

4. DISCUSSION

Figure 5: 10Fe-50Ni-40Al single crystal with foil normal parallel to [001] compression axis after 1.5% plastic strain at 750K. The dominant Burgers vector is a<001>.

The very large solute strengthening observed upon the addition of Ni to Fe-40Al, which persists up to 30% Ni additions is clearly associated with the influence of Ni on the critical resolved shear stress for a<111> dislocations. Note that these compositions have been carefully annealed in order to minimize the effect of quenched-in vacancies which is known to produce significant strengthening in FeAl [18,21-22]. The strengthening due to Ni does not appear to be consistent with a classical solute interaction for several reasons. Based on ALCHEMI measurements [23], Ni is expected to substitute for Fe on the Fe sublattice. However, there is only a modest size misfit between Ni and Fe based on their Goldschmidt radii [24]. In the Fleischer solute strengthening model [17], the effective misfit due to atomic size and modulus, ε_s is given by: $\varepsilon_s = |\varepsilon_G - \beta\varepsilon_b|$, where ε_G is the modulus misfit, ε_b is the size misfit and β is a term which accounts for the greater elastic interaction for dislocations with edge character. In this context it is noteworthy that in this composition range a<111> screw dislocations are the dominant line direction, indicating their limited mobility. In addition, the overall increase in the elastic modulus in this regime indicates that the shear modulus should also increase. Since the $\beta\varepsilon_b$ term is by convention taken as positive, adding an elastically "stiffer" atom will actually tend to *reduce* the value of ε_s.

The low mobility of screw dislocations, as evidenced by their long lengths in TEM observations, suggests that a classical Peierls mechanism is operative. It is tempting to further hypothesize that Ni may be altering the core structure of the a<111> dislocations, and thus changing the Peierls stress. A second observed trend however argues against this interpretation. For all compositions, the yield strength is only weakly dependent on temperature below the "knee". The observation of such yield strength "plateaus" which have large stress levels, is difficult to reconcile on the basis of a simple Peierls model for which it is expected that strength will continuously decrease with increasing temperature.

Takeuchi [25] has presented a modification of the Peierls mechanism in the smooth kink regime for dissociated dislocations. He has considered two cases: one in which kink-pair nucleation is correlated on the partial dislocations, and one in which kink pairs are nucleated independently. While in the former case, the yield strength falls continuously with temperature, in the latter case a critical, athermal stress is predicted. This critical stress is associated with the creation of a small additional fault area between a kink-pair on the leading partial dislocation in the case of uncorrelated kink-pair formation. For the present case of an a<111> dislocation dissociated into a/2<111> superpartials with an APB in between, this critical stress is given by:

$$\tau_c = (Gbd/4\pi w_e^2)\{1 + 2(d/w_e)\}$$

where G is the shear modulus, d is the period length of the Peierls potential and w_e is the equilibrium separation between partials. It can be seen that the critical stress varies roughly as the inverse of the square of dissociation distance. Recalling the observed decrease in dissociation distance upon the addition of Ni to Fe-40Al, this model is qualitatively consistent with the increase in the nearly plateau-like yield strength and the observed line orientation of the dislocations. The predicted increase in critical stress is however significantly less than that observed, particularly between Fe-40Al and 50Fe-10Ni-40Al compositions. The Takeuchi model is based on continuum elasticity, and does not include possible changes to the superpartial core structures. Thus, it would appear that the observed strengthening deserves further investigation from an atomistic approach. Nonetheless, the relatively simple model of Takeuchi [25] also appears to have some merit in explaining several aspects of the strengthening effect of Ni.

The solid solution softening effect observed from Ni-40Al to 10Fe-50Ni-40Al appears to be associated with the replacement of Ni antisite defects with Fe antisite defects on the Al sublattice in this composition range. Pike [6] has argued that considering the Goldschmidt radii of 1.24, 1.27, and 1.43 for Ni, Fe, and Al respectively, then replacing Fe atoms with Ni atoms actually reduces the size misfit on the Al sublattice. That this regime is controlled by more classical solute (or anti-site defect) interactions is indeed consistent with the dominance of edge dislocations in the Ni-rich compositions. This argument is also supported by a significant increase in lattice parameter with Fe additions. There are however several aspects of this behavior which remain difficult to reconcile on the basis of this explanation. First, the softening effect and observed lattice parameter increase is more dramatic than might be expected on the basis of the modest size misfit between Ni and Fe atomic radii. In addition, the measured yield strength for 10Fe-50Ni-40Al is slightly *lower* than that for Ni-50Al, based on previous measurements [26]. The recent work of Liu, et al [27] exploring magnetic effects on atomic size in these compounds appears to offer important, additional insight with regards to the magnitude of the softening due to Fe additions.

Only a preliminary explanation can be offered at this time for the composition dependence of the yield strength variation with temperature. With the exception of the 30Fe-30Ni-40Al composition, the transition from a<111> glide to a<100> climb appears to occur at about 700K. It should be noted that we have not yet determined the yield strength and microstructure for 30Fe-30Ni-40Al at 700K, however the knee clearly lies somewhere between 600K and 775K. Thus, a similar "knee" temperature may be exhibited by all of the ternary compositions, in spite of the radically different CRSS values for a<111> activity at lower temperatures.

These results indicate that the most significant improvement in strength at intermediate temperatures is offered by Ni antisite defects in the Ni-rich, binary compounds. The "knee" is delayed to significantly higher temperatures in comparison to all of the ternary compositions, which accounts for the higher strength in this regime. In Ni-44Al, it has been conclusively shown that the "knee" is associated with a decomposition of a<111> dislocations into a<110> and a<001> dislocations [20], and that a<110> dislocation glide processes dominate above the "knee" [28]. In contrast, in the ternary compounds, a<100> climb appears to dominate, even at temperatures just above the "knee". In this respect, the mechanism of the slip transition in the ternary compounds appears to be similar to that observed in binary, stoichiometric Ni-50Al. The position of the "knee" and the deformation processes at higher temperature should be strongly affected by vacancy content and mobility, since both a<100> climb and a<110> "glide" are diffusion mediated processes (the latter involving coordinated climb/glide of coupled a<100> dislocations, as described in detail elsewhere [28-30]). The importance of vacancy kinetics is emphasized by considering the recent work of Collins and co-workers [31]. Using the perturbed-angular correlation of gamma rays (PAC) method, they have determined that vacancies become mobile in Ni-50Al at about 623°C, which corresponds very well with its "knee" temperature. This favorable correlation suggests that a complete understanding of the effect of composition (including ternary additions) on the slip transition, the "knee" temperature and the deformation processes at higher temperature will require additional understanding of vacancy content and kinetics in these alloys.

5. CONCLUSIONS

The variation of yield strength with composition for <100> oriented, B2 single crystals in the Fe_x-Ni_{60-x}-Al_{40} pseudobinary system has been correlated with dislocation behavior at both room temperature and elevated temperatures. At room temperature, the yield strength varies in a complex way with composition, indicating complicated solute effects on the CRSS for a<111> dislocations, which dominate the deformation microstructures for all compositions at lower temperature. Dramatic solute strengthening is observed for Fe-rich compositions which can not be rationalized on the basis of classical solute strengthening theory alone. Additional factors may include an increase in the elastic modulus and the APB energy in this composition range. The possible relationship between the CRSS for a<111> superdislocation motion and APB energy has been discussed in the context of a model by Takeuchi [25] for uncorrelated kink-pair formation on the a/2<111> superpartials. This explanation is also consistent with the predominance of the screw line direction for compositions up to 30Fe-30Ni-40Al, which exhibits the highest yield strength of all the ternary compositions. The solute softening observed between 60Ni-40Al and 10Fe-50Ni-40Al is explained based on the preference of Fe to occupy the Al sublattice, thereby reducing the size misfit on this sublattice. Consistent with this view, the a<111> dislocations tend to align preferentially along edge line directions in this composition range, indicating that the elastic interactions with substitutional solutes or anti-site defects are important. However, the magnitude of the softening is difficult to justify based on the small difference in atomic radius between Ni and Fe. At elevated temperatures, all compositions exhibit a "knee" in the yield strength, above which rapid softening occurs. For all the ternary compositions studied to date, this softening is related to a transition to a<100> climb activity above the "knee".

ACKNOWLEDGEMENTS

This research has been supported by the US Department of Energy, Office of Basic Energy Sciences, Materials Sciences Division under Contract No. DE-FG0296ER45550 (for PRB, SR, MS and MJM), and Office of Basic Energy Sciences, and Contract No. W-7405-ENG-82 (for TL).

REFERENCES

1. D.B. Miracle, *Acta Metall Mater*, **41**, 649 (1993).
2. R.D. Noebe, R. R. Bowman and M. V. Nathal, *Int. Mater. Rev.* **38**, 193 (1993).
3. I. Baker and P. R. Munroe, *Int. Mater. Rev.* **42**, 181 (1997).
4. P. Nagpal and I. Baker, *Metall. Trans. A*, **21A**, 2281 (1990).
5. D. K. Patrick, K. M. Chang, D. B. Miracle and H. A. Lipsitt, *MRS Proceedings*, **213**, 267 (1991).
6. L. M. Pike, Jr., PhD. Thesis, University of Wisconsin-Madison, (1997).
7. L. M. Pike, Y. A. Chang and C. T. Liu, *Intermetallics*, **5**, 601 (1997).
8. R. D. Noebe, NASA TM 106534 (1992).
9. K. Yoshimi, S. Hanada, and M. H. Yoo, *Acta metall. mater.* **43**, 4141 (1995).
10. P. S. Brenner, M. S. Thesis, The Ohio State University (2000).
11. J.H. Schneibel, P.R. Munroe, and L.M. Pike, Materials Research Society, **460**, (1996).
12. R. T. Pascoe and C.W.A. Newey, *Metal Sci. J.* **5**, 50 (1971).
13. M. H. Loretto and R. J. Wasilewski, *Phil. Mag.* **23**, 1311 (1971).
14. J. T. Kim, PhD. Thesis, University of Michigan (1990).
15. E. P. George and I. Baker, *Phil. Mag. A* **77**, 737 (1998).
16. R. W. K. Honeycomb, *The Plastic Deformation of Metals* (Edward Arnold, London, 1984).
17. L. Fleischer, *Acta metall.*, **9**, 996 (1963).
18. M.A. Crimp and K.M. Vedula, Phil. Mag. A., **63**, 559 (1991).
19. M. F. Savage, R. Srinivasan, M. S. Daw, T. A. Lograsso and M. J. Mills, *Mater. Sci. and Eng. A* **258**, 20 (1998).
20. R. Srinivasan, J. Brown, M. S. Daw, R. D. Noebe and M. J. Mills, *Phil. Mag. A*, **80**, 2841 (2000).
21. P. Nagpal and I. Baker, *Metall. Trans. A*, **21A**, 2281 (1990).
22. P. Monroe, *Intermetallics*, **4**, 5 (1996).
23. I. M. Anderson, *Acta mater.*, **45** 3897 (1997).
24. F. Laves, *Theory of Alloy Phases* (ASM Publication, Cleveland, 1956).
25. S. Takeuchi, *Phil. Mag. A* **71**, 1255 (1995).
26. M. J. Mills, R. Srinivasan, R. D. Noebe, T. Lograsso and M. S. Daw, *Interstital and Substitutional Effects in Intermetallics*, eds. I. Baker, R. D. Noebe and E. P. George (TMS Publications, Warrendale, 1998), p. 99.
27. C. T. Liu, C. L. Fu and J. H. Schneibel, *MRS Proceedings*, this volume.
28. R. Srinivasan, M.S. Daw, R.D. Noebe and M. J. Mills, *Phil. Mag. A*, submitted for publication.
29. M. J. Mills and D. B. Miracle, *Acta Metall. Mater.*, **41**, 86 (1993).
30. M. J. Mills, R. Srinivasan, M. S. Daw, *Phil. Mag. A*, **77**, 801 (1998).
31. G. S. Collins, J. Fan and B. Bai, *Structural Intermetallics*, eds. M. V. Nathal, R. Darolia, C. T. Liu, P. L. Martin, D. B. Miracle, R. Wagner and M. Yamauchi (TMS Publications, Warrendale, 1997), p. 43.

Mat. Res. Soc. Symp. Proc. Vol. 646 © 2001 Materials Research Society

Atomistic Modeling Of Ternary And Quaternary Ordered Intermetallic Alloys

Guillermo Bozzolo[1], Joseph Khalil[2], Matthew Bartow[2], Ronald D. Noebe[2]
[1] Ohio Aerospace Institute, 22800 Cedar Point Road, Cleveland, OH 44142.
[2] National Aeronautics and Space Administration, Glenn Research Center, Cleveland, OH 44135.

ABSTRACT

The structure of ternary and quaternary NiAl-based ordered intermetallic alloys is studied using the BFS method for alloys. A simple calculational procedure, based on the determination of the energetics of local environments surrounding defect atoms is introduced and applied to the study of the defect structure and phase formation in NiAl-based systems. The procedure is illustrated with two different examples: 1) the phase structure of Ni-Al-Fe alloys, focusing on the concentration dependence of the site preference behavior of Fe in NiAl, and 2) the precipitation of a β' phase in NiAl(Ti,Hf) alloys, focusing on the role of Hf in lowering the solubility limit of Ti in NiAl, thus enhancing the precipitation of a Heusler $Ni_2Al(Ti,Hf)$ phase.

INTRODUCTION

The availability of computationally efficient and physically sound quantum approximate methods allows for the analysis of multicomponent systems at the atomic level. This in turn increases our knowledge-base and understanding of the structure of materials that is currently limited by the inherent difficulties in obtaining such detailed information from experimental studies. The Bozzolo-Ferrante-Smith (BFS) method for alloys [1] is free of many of the constraints, such as the number or type of elements considered, the crystal structure, or the presence of multiple phases, that have limited the use of other theoretical techniques to the application of problems concerning ternary or higher order intermetallic alloys. In this paper we apply the BFS method to the energetic analysis of 3- and 4- element systems with the goal of gaining a better understanding of the interaction between multiple alloying additions and their influence on the microstructure of NiAl-based alloys. This work is meant to complement recent experimental efforts [2,3] and to begin to provide detailed explanations for observed results.

THE BFS METHOD

The BFS method [1] is based on the concept that the energy of formation of a given atomic configuration is the superposition of the individual atomic contributions $\Delta H = \Sigma \varepsilon_i$. Each contribution ε_i is the sum of two terms: a strain energy ε_i^s, computed in the actual lattice as if every neighbor of the atom i was of the same atomic species i, and a chemical energy, ε_i^c, computed as if every neighbor of the atom i was in an equilibrium lattice site of a crystal of species i, but retaining their actual chemical identity. The computation of ε_i^s, using Equivalent Crystal Theory (ECT) [4], involves three pure element properties for atoms of species i: cohesive energy, lattice parameter and bulk modulus. We rely on calculations using the linear-muffin tin orbital method in the atomic sphere approximation (LMTO-ASA) [5,6] for the determination of these parameters. The use of ab-initio methods for the determination of input

Table I: Linear-muffin tin orbital (LMTO-ASA) [5,6] results for the lattice parameter, cohesive energy, and bulk modulus for the bcc phases of all the elements considered in this work, and the resulting Equivalent Crystal Theory (ECT) [4] parameters p, α, λ and l (see text). The BFS parameters are listed as a matrix, where the matrix element a_{ij} denotes the interaction BFS parameter Δ_{ij} between elements i and j.

	LMTO results (for all elements in bcc form)			ECT parameters			
	Lattice parameter (Å)	Cohesive energy (eV)	Bulk Modulus (GPa)	p	$\alpha(\text{Å}^{-1})$	$\lambda(\text{Å}^{-1})$	$l(\text{Å})$
Ni	2.752	5.869	249.2	6	3.067	0.7632	0.2716
Al	3.192	3.942	77.3	4	1.875	1.0383	0.3695
Fe	2.742	6.071	311.53	6	3.093	0.6955	0.2475
Ti	3.213	6.270	121.0	6	2.722	1.0476	0.3728
Hf	3.494	7.411	13.02	10	4.210	1.1302	0.4022

BFS parameters Δ_{ij} (Å^{-1})					
i\j	Ni	Al	Fe	Ti	Hf
Ni		-0.06078	-0.04107	-0.07315	-0.10972
Al	0.09160		0.29862	-0.07036	-0.10741
Fe	0.12180	-0.06717			
Ti	0.46037	0.20203			0.34333
Hf	0.42215	0.42260		-0.07523	

parameters, while necessary and unavoidable (i.e., there is no experimental input for the bcc phase of most of the elements used in this work), introduces a corresponding dependence on the type of method used. As the BFS method concentrates on the calculation of differences in energy, it is expected that such dependence will be minimized and that the predicted results will be, to a satisfactory extent, somewhat independent of the ab-initio method used for input. The chemical energy includes two BFS interaction parameters (i.e., Δ_{AB} and Δ_{BA}), that can be determined from any pair of alloy properties. In this work, Δ_{AB} and Δ_{BA} (A, B = Ni, Al, Ti, Hf, Fe) were computed from LMTO-ASA calculations by fitting to the calculated energy of formation and equilibrium lattice parameter of the corresponding B2 AB phases. Finally, the strain and chemical energy are linked with a coupling function g, which ensures the correct volume dependence of the BFS chemical energy contribution. All the single-element and BFS alloy parameters used in this work are listed in Table I. We refer the reader to Ref. 1 for a detailed discussion of the BFS method, its definitions, and operational equations.

In what follows, using the notation introduced in Ref. 1, X(A) indicates an X atom occupying a site in the A sublattice. X(A)A(B) represents an X(A) substitution followed by the displaced A atom occupying a B site (A(B)). The resulting X(A) and A(B) point defects can then occupy nearest-neighbor sites, denoted with a subscript '1' (i.e., $[X(A)A(B)]_1$), or locate themselves far away from each other so that the interaction between the defect atoms can be ignored, denoted with a subscript 'f'. For an atom X occupying an A site, X(A), we also introduce the concept of a 'local environment' (LE) which is used to denote a group of fifteen

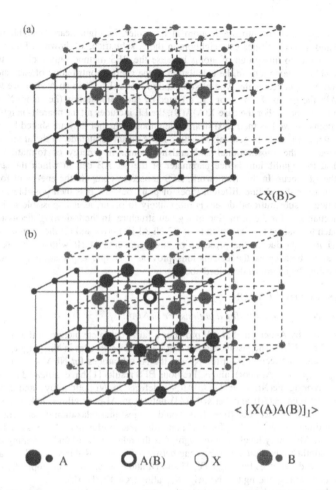

Figure 1: *(a) <X(B)> local environment: X atom (light grey) in an B site, surrounded by eight A nearest-neighbors (large black disks) and six B next-nearest-neighbors (large grey disks). (b) <[X(A)A(B))]₁>, with the substitutional X atom in an A site, and the displaced A atom (circle) in a B site. The two 'defect' atoms (the substitutional X and antistructure A) are nearest-neighbors and are surrounded by an environment of 20 atoms, accounting for all the nearest- and next-nearest-neighbors, some of which are shared between them. A or B atoms at distances greater than one lattice parameter from either one of the defect atoms are indicated with small black and grey disks, respectively.*

atoms (in a bcc lattice), with the X atom at its center, its eight nearest-neighbors in the B sublattice and its six next-nearest-neighbors in the A sublattice, as shown in Fig. 1.a. We can limit our analysis to this group of atoms because the BFS method only includes up to next-nearest-neighbor interactions in the calculation of the energy contribution of each atom. While the notation X(A) indicates the type of substitution made (i.e., X in an A site), we will denote with <X(A)> the group of atoms, or LE, affected by such substitution (i.e., atom X and its first and second neighbors). For the case X(A)A(B), the LE includes all the nearest- neighbors of the X and A atoms, as well as their next-nearest-neighbors, whether they are shared by X and A or not, as shown in Fig. 1.b. In general, an active LE is meant to include all the atoms affected by the presence of the substitutional and antisite atoms. Its energy of formation is always computed at the equilibrium lattice parameter of the B2 AB cell in which the substitutions defining the LE occur. In this framework, BFS is used to compare the energy of formation of the different LE's that define different types of point defects. It is then possible to determine why a particular substitutional defect is most likely to occur, as it is possible to identify the driving mechanisms for the formation of a given structure. In the following discussion, we use this approach to analyze (1) the structure of B2 Ni-Al-Fe alloys and (2) the change in solubility limit of Ti in NiAl due to the presence of small amounts of Hf with the formation of β' precipitates. In both cases the calculations provide a simple explanation for the observed structures using the approach described above.

RESULTS AND DISCUSSION

Site preference of ternary alloying additions: Fe in NiAl

Recent work by Anderson et al. [2] provides a fairly complete picture of the site preference behavior of Fe in NiAl. The results suggest that Fe occupies Ni sites in $Ni_{50-x}Al_{50}Fe_x$ (or Ni-poor) alloys, is divided between sites in the Ni and the Al sublattices for Ni:Al 1:1, and prefers Al sites in $Ni_{50}Al_{50-x}Fe_x$ (or Al-poor), alloys. Previous BFS results [1] have indicated a slight energy difference favoring Fe(Ni) over Fe(Al)Al(Ni) defects in Ni-poor alloys and even smaller differences between Fe(Al) and Fe(Ni)Ni(Al) defects in Al-poor alloys, indicating no strong preference of Fe for either sublattice. This would suggest that relaxation or temperature effects could be ultimately responsible for the particular site preference of Fe as a function of stoichiometry. An energy level diagram highlights the relative likelihood of finding one type of defect over another, based on the energy gap between the two possible states. For example, Fig. 2 displays these differences for single Fe, Ti and Hf atoms added to an isotropically relaxed 72-atom NiAl cell, corresponding to a $Ni_{50}Al_{48.61}X_{1.39}$ alloy (X = Fe, Ti, Hf).

From this traditional energy-minimization data alone, it is not transparent how each individual substitution (Fe(Al), Al(Ni), etc.) influences the final outcome. In order to understand their role, it is convenient to take a slightly different route: studying the energetics of the unrelaxed LE of each substitutional atom. In doing so, it is possible to understand the contribution of each individual substitution to the final energy balance, thus identifying the mechanisms that drive a specific behavior.

Analysis of the BFS energies of formation of the corresponding LE defining each type of defect indicate that the driving mechanism for the choice of site for Fe atoms resides mostly on the behavior of the neighboring Ni and Al atoms rather than in Fe itself. Table II lists the BFS energies of several LE's (<Fe(Al)>, <Fe(Ni)>, <Ni(Al)>, <Al(Ni)>) needed to explain the observed behavior. The LE's corresponding to Fe(Ni)Ni(Al) and Fe(Al)Al(Ni) extended defects

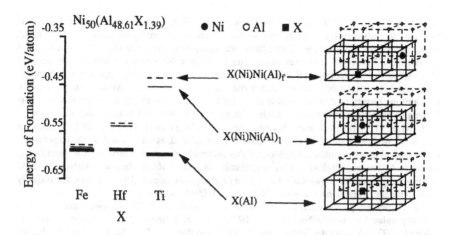

Figure 2: *Energy spectrum for various site occupancy schemes when alloying additions are made to $Ni_{50}(Al,X)_{50}$. The thick solid lines correspond to X(Al) defects, whereas thin solid lines and dashed lines correspond to X(Ni)Ni(Al) defects, as shown. The labels 'l' and 'f' denote the distance between the various substitutional and antisite defects created (see text). All energies are based on a computational cell of 72 atoms.*

Table II: BFS strain (ε^s), chemical ($g\varepsilon^c$) and total energy (e_τ) contributions to the energy of formation (in eV/atom) of several defects and their local environment, embedded in a 72-atom B2 cell. Δe_τ indicates the increase in energy of formation per bond with respect to the same group of atoms in a pure B2 NiAl alloy. For the double-centered LE, the center atoms are located at nearest-neighbor ($[]_1$), next-nearest-neighbor ($[]_2$), or greater than second-neighbor distance ($[]_f$). The subindex H ($[]_H$)indicates that the center atoms (Ti and Hf) locate themselves following a Heusler ($L2_1$) pattern in the Al-sublattice (i.e., sharing an Al atom as a next-nearest-neighbor).

	ε^s	$g\varepsilon^c$	e_τ	Δe_τ		ε^s	$g\varepsilon^c$	e_τ	Δe_τ
<Ni(Ni)>	0.3201	-0.9232	-0.6031	0.0000	<[Fe(Ni)Ni(Al)]$_f$>	0.3140	-0.8917	-0.5777	0.0254
<Ni(Al)>	0.3135	-0.9056	-0.5921	0.0110	<[Fe(Ni)Ni(Al)]$_1$>	0.3140	-0.8944	-0.5804	0.0227
<Al(Ni)>	0.3267	-0.8354	-0.5088	0.0943	<[Fe(Al)Al(Ni)]$_f$>	0.3206	-0.8112	-0.4906	0.1125
<Fe(Al)>	0.3140	-0.8989	-0.5849	0.0182	<[Fe(Al)Al(Ni)]$_1$>	0.3206	-0.8137	-0.4931	0.1099
<Fe(Ni)>	0.3206	-0.9093	-0.5887	0.0144	<[Ti(Al)+Hf(Al)]$_1$>	0.3669	-0.8542	-0.4872	0.1154
<Ti(Al)>	0.3262	-0.8691	-0.5429	0.0598	<[Ti(Al)+Hf(Al)]$_2$>	0.3735	-0.8997	-0.5162	0.0865

are also included. To make a fair comparison of the different single and extended LE's, it is necessary to embed each LE in a B2 cell so that all bonds affected by the presence of the defect are accounted for. To do so, we locate the atoms composing the LE at the center of a 72-atom equilibrium B2 NiAl cell and refer all energies to the pure AB version of that cell. Because all LE's, as well as the reference B2 cell, are evaluated at the same lattice parameter, the energy difference between the cell with the defect and the reference cell, Δe_T (last column in Table II), indicates the energy cost of performing the specific substitutions that characterize the defect, as it contains *only* the contributions from the atoms in the LE, and not the rest of the cell.

Clearly, <Fe(Ni)> is strongly favored (Δe_T = 0.0144 eV/atom versus 0.0182 eV/atom for <Fe(Al)>), due to the numerous Fe-Al bonds that are created, as represented by the larger negative chemical energy contribution. Most of this favorable chemical contribution is provided by the eight Al nearest-neighbors. This contribution is greatly diminished in <Al(Ni)>. The Al nearest-neighbors are responsible for the large change in chemical energy which ultimately leads to the unfavorable nature of <Al(Ni)> (Δe_T = 0.0943 eV/atom).

For Ni-rich alloys, the energy of formation of <Fe(Al)> is 0.0182 eV/atom, while the corresponding value for <[Fe(Ni)Ni(Al)]$_r$> is 0.0254 eV/atom, further reduced to a more favorable value of 0.0227 eV/atom if <Fe(Ni)> and <Ni(Al)> overlap (see Fig. 1.b). Thus Fe(Al) substitutions are more likely to occur than Fe(Ni) ones. It is interesting to see, however, that for <Fe(Ni)> alone, the energy of formation is just 0.0144 eV/atom, indicating that Fe itself would rather occupy a Ni site than an Al site. It is the creation of the antistructure Ni atom (Δe_T = 0.0110 eV/ atom) that raises the energy enough to favor Fe(Al) substitutions over Fe(Ni) ones. Still, the difference in energy between the two options is small (Δe_T = 0.0182 eV/atom for <Fe(Al)> vs. Δe_T = 0.0227 eV/atom for <[Fe(Ni)Ni(Al)]$_1$>), making the presence of both defects proportionally possible.

A similar analysis for Ni-poor alloys, involves the competition between Fe(Ni) and Fe(Al)Al(Ni) configurations. As noted above, the energy of formation of <Fe(Ni)> is 0.0144 eV/ atom, while the corresponding value for <Fe(Al)> is 0.0182 eV/atom. The preference for Ni sites is clear at this stage, but it is further enhanced, as seen previously, by the high energy cost of creating an antistructure Al atom. The large energy of formation of <Al(Ni)>, 0.0943 eV/atom, together with the energy of <Fe(Al)>, results in a total energy of 0.1125 eV/atom for <[Fe(Al)Al(Ni)]$_r$>, with little if any relief for overlapping Fe(Al) and Al(Ni) defects, <[Fe(Al)Al(Ni)]$_1$>, for which the energy of formation is 0.1099 eV/atom. Once again, the preference of Fe atoms for Ni sites in Ni-poor alloys is clear, only that in contrast with the Al-poor alloys, it is the highly unfavorable energy to form an Al(Ni) defect that allows Fe(Ni) to be the dominant defect.

This analysis can be summarized by stating that Fe preference for Ni sites is essentially common to both Ni- and Al-poor alloys. It is the energy cost of creating an antistructure Al or Ni atom that is the deciding factor on whether Fe takes the deficient element site or not. As a result, Fe(Ni) substitutions are greatly favored for Ni-poor alloys while there is little specific preference in Al- poor alloys, although a Fe(Al) defect is slightly more likely than the more complex substitution involving Fe(Ni)Ni(Al).

Interaction between alloying additions: Hf and Ti in NiAl

It has been shown that Ti additions at the expense of Al in NiAl alloys results in the formation of β' (Heusler) Ni$_2$AlTi precipitates above the solubility limit of Ti (~5 at.%) in NiAl

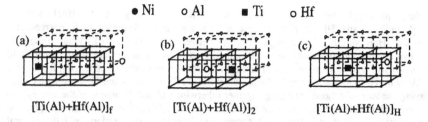

Figure 3: *Computational cells indicating the Ni (solid line) and Al (dashed line) sublattices in a B2 NiAl matrix with Ti and Hf atoms occupying (a) separated, (b) consecutive and (c) alternating sites in the Al sublattice. The subindices indicate far (f), second-neighbor distance (2), and third neighbor distance, occupying 'Ti' sites in a Heusler arrangement (H), respectively.*

[7]. Recent experimental work [3,8] shows that small additions of Hf reduce the solubility of Ti in NiAl resulting in the formation of Heusler precipitates at much lower Ti concentration (3%). It was also shown that Hf partitions to the β' phase. This result can be intuitively understood from the fact that Ni_2AlHf is itself a Heusler alloy and that the solubility of Hf in NiAl is exceedingly low [9]. While there is no experimental confirmation of the site preference of Hf in β', previous modeling results indicate that its behavior is similar to that of Ti, in preferring Al sites [1]. The fact that Hf shares the same individual site preference behavior in NiAl (i.e., marked preference for Al sites), would lead one to believe that Hf plays a role similar to Ti and that, due to its lower solubility limit, it nucleates precipitation of the β' phase. A clearer picture emerges when analyzing the BFS contributions within the scheme of interacting local environments.

In a previous BFS study [1], it was shown that both Ti and Hf show a clear preference for Al sites in Al-poor alloys (i.e., large energy gap, ~ 0.15 eV/atom, between X(Al) and X(Ni)Ni(Al) (X = Ti, Hf) substitutions), as seen in Fig. 1. Within the LE approach, it is therefore unnecessary to consider <Ti(Ni)> and <Hf(Ni)>, concentrating only on the possible distributions of Ti and Hf in the Al sublattice. Table II lists the BFS energies of the single defect and some double-defect (Ti,Hf) LE's. Consistent with the results shown in Ref. 1, the formation energies of the LE's confirm the preference of Ti and Hf for Al sites. The double-defect LE's that arise from the defects shown in Fig. 3 include (a) the case of separated (i.e., several lattice constants) Ti(Al) and Hf(Al) substitutions, $[Ti(Al)+Hf(Al)]_f$, (b) a configuration where Ti and Hf occupy adjacent corners in the Al-sublattice, $[Ti(Al)+Hf(Al)]_2$, and (c) a similar case where Ti and Hf opposing corners of a cube in the Al sublattice, $[Ti(Al)+Hf(Al)]_H$, the same ordering that defines a Heusler phase. The fact that $<[Ti(Al)+Hf(Al)]>$ has the highest formation energy of all three possible cases, indicates that Hf and Ti atoms would rather occupy neighboring sites than remain in solid solution in the NiAl matrix. The (b) and (c) cases represent two possible scenarios: (b) can be interpreted as Hf occupying an Al site in a β' Ni_2AlTi cell, whereas (c) can be seen as Hf substituting for a Ti atom in a perfect Ni_2AlTi cell. The lower energy of formation of $<[Ti(Al)+Hf(Al)]_2>$ and $<[Ti(Al)+Hf(Al)]_H>$ with respect to $<[Ti(Al)+Hf(Al)]_f>$ confirm the observed partitioning of Hf to the β' phase [3]. The fact that (c) has the lowest formation energy ($\Delta e_\tau = 0.0721$ eV/atom) clearly indicates that Hf will play the role of Ti in making the formation of β' energetically favorable. However, the small energy difference between (b) and (c) (~0.01 eV/atom) suggests that two independent processes favor the formation of β', and that they are almost as likely to occur. Hf will tend to substitute for Ti atoms in the Heusler precipitate (see (c)) but it is almost as likely for Hf to locate itself in 'Al' sites rather than 'Ti' sites within the Ti-Al sublattice, thus playing the role of a 'glue' bringing together otherwise isolated Ti atoms. If the energy gap between (b) and (c) were larger, there would be much less driving force to form a β' phase at low solute concentrations. The small energy gap between configurations and the general attraction of Hf and Ti atoms is therefore consistent with the observed lowering of the solubility limit of Ti in NiAl in the presence of Hf, as even small amounts of Hf would favor the process of nucleating dispersed Ti atoms in a Heusler environment.

CONCLUSIONS

The results in this study represent a brief survey of the general effect of different alloying additions on the structure of NiAl and the method by which such factors can be modeled. For NiAl+Fe, the experimentally observed behavior is reproduced: Fe(Ni) substitutions in Ni-poor alloys and a split between Ni and Al sites, with a preference for Al sites, otherwise. For NiAl+Ti+Hf, it was shown that the tendency of Hf to form a β' phase, together with its ability to attract isolated Ti atoms, explain the lowering of the solubility limit of Ti in NiAl, in agreement with experiment.

ACKNOWLEDGMENTS

Fruitful discussions with N. Bozzolo are gratefully acknowledged. We thank A. Wilson and J. Howe for experimental verification of NiAlTiHf behavior, and P. Abel and J. Garces for their support of the activity. This work was supported by the HOTPC program at NASA Glenn Research Center.

REFERENCES

1. G. Bozzolo and J. Ferrante, J. Computer-Aided Mater. Design 2 (1995) 113;
   G. Bozzolo, R. D. Noebe and F. Honecy, Intermetallics 8 (2000) 7.
2. I. M. Anderson, A. J. Duncan, J. Bentley, Intermetallics 7 (1999) 1017, and refs. therein.
3. A. W. Wilson, Ph.D. Thesis, University of Virginia, 1999.
4. J. R. Smith, T. Perry, A. Banerjea, J. Ferrante and G. Bozzolo, Phys. Rev. B 44 (1991) 6444;
   G. Bozzolo, J. Ferrante and A. M. Rodriguez, J. Comp.-Aided Mater. Design 1 (1993) 285.
5. G. Bozzolo, C. Amador, J. Ferrante and R. D. Noebe, Scripta Metall. 33 (1995) 1907.
6. O. K. Andersen, Phys. Rev. B 12 (1975) 3060.
7. A. W. Wilson, J. M. Howe, A. Garg and R. D. Noebe, Mat. Sci. Eng. A 289 (2000) 162.
8. A. Garg, R. D. Noebe, J. M. Howe, A. Wilson and V. Levit, in Proceedings of Microscopy and Microanalysis, 1996, Eds. G. W. Bailey, J. M. Corbett, R. V. W. Dimlich, J. R. Michael and N. J. Zaluzec, San Francisco Press, San Francisco, 1996, p. 998.
9. A. Garg and R. D. Noebe, Scripta Mater. 39 (1998) 437.

Mat. Res. Soc. Symp. Proc. Vol. 646 © 2001 Materials Research Society

Transition Metal Impurity-Dislocation Interactions in NiAl: Dislocation Friction and Dislocation Locking.

O.Yu. Kontsevoi[1], Yu.N. Gornostyrev[1, 2], and A.J. Freeman[1]
[1]Department of Physics and Astronomy, Northwestern University,
Evanston, IL 60208-3112, U.S.A.
[2]Institute of Metal Physics, Ekaterinburg, Russia.

ABSTRACT

The energetics of the interaction of the <100>{010} edge dislocation in NiAl with early $3d$ transition metal (TM) impurities was studied using the *ab initio* real-space tight-binding LMTO-recursion method with 20,000 atom clusters and up to 1,000 non-equivalent atoms in the dislocation core. The coordinates of the atoms in the core were determined within the Peierls-Nabarro (PN) model with restoring forces determined from full-potential LMTO total energy calculations. TM impurities were then placed in different substitutional positions near the dislocation core. For most positions studied, the interaction between impurities and the dislocation is found to be repulsive (dislocation friction). However, when the impurity is in the position close to the central atom of the dislocation core, the interaction becomes strongly attractive, thus causing dislocation locking. Since the size misfit between the Al atom and the substituting TM atom is very small, this locking cannot be explained by elastic (or size misfit) mechanisms; it has an electronic nature and is caused by the formation of the preferred bonding between the electronic states of the impurity atom and the localized electronic states appearing on the central atom of the dislocation core. The calculated results are then discussed in the scope of experimental data on solid solution hardening in NiAl.

INTRODUCTION

The improvement of the strength of materials due to doping by ternary additions has become a traditional alloy design approach. At impurity concentrations below the solubility limit and at low temperatures, solid solution hardening (SSH) [1], which is determined by the interactions of impurities with dislocations, is the main reason for an increase of the yield stress. Hence, the nature of elementary impurity-dislocation interactions is one of the key questions in the physics of the strength and plasticity of solid solutions.

Although impurity-dislocation interactions depend on many factors, according to the prevailing point of view [1, 2], the size misfit between the impurity and host atoms appears to make the main contribution in a majority of alloys. On the other hand, there are a number of cases, when the electronic structure of impurities plays a more important role than size misfit. In particular, it was found that sp-impurities and transition metal (d-) impurities with the same size misfit give very different contributions to SSH in Ni_3Al [3, 4] (the so-called "extra" solution hardening effect [3]). In NiAl [4, 5] alloyed with transition metal impurities, SSH differs significantly for the elements with similar atomic radii and correlates with electronic structure features rather than with size misfit [6].

In addition to parelastic interactions due to size misfit, other elementary mechanisms of impurity-dislocation interactions have been proposed [1, 2], such as: (i) the chemical interaction caused by the capture of the impurity atom by the stacking fault; (ii) the dielastic interaction due to the modulus misfit; and (iii) the "electrostatic" interaction. The last mechanism is connected with charge perturbations of the electronic subsystem caused by the impurity ion, and therefore should be most directly dependent on peculiarities of the electronic structure. This is actually the case for covalent [7] and ionic crystals. In metals, however, where itinerant electrons effectively screen the excess charge, estimates lead to only a minor role for the electrostatic mechanism [8].

Previously, the model tight-binding method was used for calculating the electronic structure and energy of a crystal containing the dislocation and impurity [9]. For the cases of a screw dislocation in Fe and a split edge dislocation in Ni, the "extra" electronic contribution to the impurity-dislocation interaction was found to be small. Later, the possibility of electron localization on some types of dislocations was demonstrated in the framework of the simple single band tight-binding model [10], and it was suggested that it could lead to radical changes in the energetics of impurity-dislocation interactions. Recently, the formation of localized electronic states on edge dislocations with <100> Burgers vector in NiAl was confirmed by first-principles real-space tight-binding linear muffin-tin orbital recursion (TB-LMTO-REC) calculations [11]. Here we present results of an *ab initio* investigation of the interactions of transition metal impurities with dislocations of this type.

METHOD OF CALCULATIONS

The electronic structure and energetics of the impurity-dislocation interactions was calculated using the first principles real-space TB-LMTO-REC method [12] within the local density approximation (LDA). An "embedded cluster" technique was applied as follows: the smaller cluster, which includes the atoms in the central part of the dislocation with the most lattice distortions and the impurity atom, was embedded in the large cluster, representing a piece of bulk material with a single dislocation and a single impurity. Within the embedded cluster, the radii of atomic spheres were calculated according to the charge neutrality condition and were adjusted automatically during the iterations toward self-consistency; outside of the inner cluster they were kept fixed. The size of the embedded cluster was chosen to be large enough to guarantee that perturbations of the potential due to the dislocation core are mostly confined within that cluster. This was done by comparing the potential parameters and local densities of states (LDOS) for atoms near the boundary of the embedded cluster with those of their nearest neighbors outside the embedded cluster. A more detailed description of the method of calculations is given in [11].

It is known from experiment that the primary deformation mode in NiAl is associated with <100> Burgers vector dislocations [5]. The results of our recent first-principles calculations [11] demonstrated that localized electronic states exist in the core of <100> edge dislocations in NiAl. Previous calculations for a simplified tight-binding model [10] suggested that localized states might give a significant contribution to the impurity-dislocation interaction. To verify this assumption for a realistic case, we performed *ab initio* calculations for 3d-impurities placed near the dislocation core.

The coordinates of the atoms were determined within the Peierls-Nabarro (PN) model with dislocation structure parameters obtained from first principles calculations of the generalized stacking fault (see [13] for details). The structure of the dislocation determined within this approach corresponds to the minimum energy configuration resulting from a balance of elastic and atomic restoring forces [14]. Despite the simplicity of the PN model, the use of an *ab initio* determination of the restoring forces allows one to obtain quite reasonable results for both the structure and mobility [13] of the dislocations, as can be judged from a comparison with the results of HRTEM observations and atomistic simulations [15].

The size of the cluster was taken to be about 20,000 atoms, which was proven to be large enough to minimize the effects of boundaries, and the embedded cluster (or self-consistency region) contained up to 1,000 non-equivalent atoms. Calculations were carried out self-consistently until the convergence of the total energy was within 1 mRy.

RESULTS AND DISCUSSION

Experimental evidence shows that alloying NiAl with early transition metals leads to (i) a strong SSH effect [5], and (ii) an anomalously high increase of the creep resistance [18]. We performed electronic structure and total energy calculations for early transition metal impurities Ti, V, Cr and Mn. Our estimates show that for these elements, the parelastic impurity-dislocation interaction is very small (less than 10^{-1} eV). Therefore the size-misfit mechanism cannot be responsible for the above mentioned alloying effects, and thus electronic mechanisms of impurity-dislocation interaction should play an essential role.

The impurities were placed in different substitution positions near the <100>{010} edge dislocation core in NiAl. It is known from both experiment and previous theoretical calculations ([6, 16, 17] and references therein) that early 3d-metal impurities occupy the Al sublattice in NiAl. A fragment of the model of the central part of the dislocation core is shown in Fig. 1. Impurities were placed in different positions marked 1-8, substituting for the corresponding Al atom. The ionic radii for early transition metals substitution impurities are close to the radius of Al in NiAl; therefore, the effect of relaxation around these impurities can be neglected.

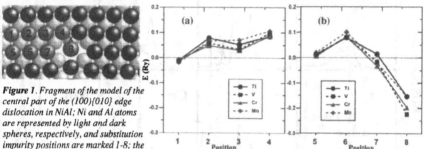

Figure 1. *Fragment of the model of the central part of the {100}{010} edge dislocation in NiAl; Ni and Al atoms are represented by light and dark spheres, respectively, and substitution impurity positions are marked 1-8; the central Ni atom of the dislocation core is marked "X"*

Figure 2. *Impurity-dislocation interaction energy for 3d- impurities in positions: (a) 1-4; (b) 5-8.*

The impurity-dislocation interaction energy was calculated as the difference between the total energy of the dislocation with the impurity near the center of the core and far from the center where the impurity-dislocation interaction becomes negligible and therefore its energy is assumed to be zero. In Fig. 2(a), the calculated impurity-dislocation energies are shown for impurities in positions 1-4 "above" the central atom of the dislocation core. Changes of the relative position of impurity and dislocation from position $1 \rightarrow 2 \rightarrow 3 \rightarrow 4$ simulate the motion of the dislocation toward the impurity: positions when impurities are close to the center of the dislocation are energetically less favorable. Therefore, in this case, during the dislocation glide "repulsion" between the dislocation and impurity takes place - an effect known as "dislocation friction" [1].

Now, if the impurity atom is placed in positions 5, 6, 7, 8 "below" the central atom of the dislocation core (Fig. 2(b)), the energetics of the impurity-dislocation interaction change: being at first repulsive when the impurity is in positions 5 and 6, it becomes strongly attractive, when the impurity is in position 8. This situation, when there is a direct attractive interaction between solute atoms and dislocations, can be characterized as "chemical locking".

To understand the mechanism of chemical locking, we consider the local densities of states (LDOS) for the TM impurities in position 8 in the <100>{010} edge dislocation core which are shown by solid lines in the lower panel of Fig. 3; for comparison, we also show the LDOS for TM impurities in bulk NiAl by dashed lines. In the upper panel of Fig. 3, solid lines show the LDOS for the central Ni atom of the dislocation core (this atom is marked "X" in Fig. 1); for comparison, dashed lines show Ni LDOS in bulk NiAl. It should be noted that an impurity atom in position 8 and a Ni atom in position "X" are nearest neighbors.

First, we consider the Ti impurity. Its LDOS is characterized by a sharp peak, or impurity level, at -0.2 Ry. As we reported previously [11], localized electronic states appear on the central Ni atom of the dislocation core, forming a sharp peak of the LDOS near -0.33 Ry. From the lower panel, one can see that the impurity level of Ti in position 8 in the dislocation core turns out to lie lower than in bulk NiAl, by -0.13 Ry: it is shifted to the position of the peak of the localized electronic states on its Ni nearest neighbor, forming a strong resonance with them. This is an indication of strong hybridization between the impurity electronic states and the Ni localized electronic states, showing that there is a strong chemical interaction between these two atoms.

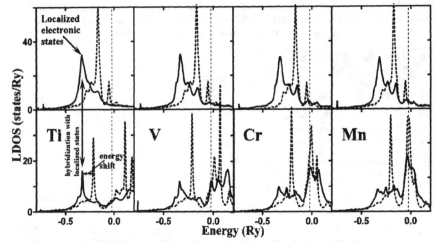

Figure 3. Dashed lines - the LDOS for impurity atoms in bulk NiAl (lower panel), and the Ni LDOS in bulk NiAl (upper panel); solid lines - the LDOS for the impurity atom in position 8 (see Fig.1) in the <100>{010} edge dislocation in NiAl (lower panel) and the LDOS for the central Ni atom of the dislocation core (upper panel).

Since the shape of the impurity peak in the Ti LDOS does not change significantly after the hybridization, the changes of the band (or one-electron) energy in the rigid-band approximation is approximately equal to the energy shift of the impurity peak. It is seen from Fig. 2(b) that the value of the shift of the impurity peak (-0.13 Ry) is close to the impurity-dislocation energy (-0.15 Ry); therefore, the one-electron contribution is the main contribution to the impurity-dislocation interaction energy. This is a clear indication that the electronic mechanism, namely the strong hybridization between the localized electronic states in the dislocation core and the impurity states, is the cause of the impurity-dislocation locking by the "chemical locking" mechanism.

For V, Cr and Mn impurities, the impurity-dislocation interaction energy is similar or even higher than for Ti. This can be a consequence of the larger number of valence electrons in these elements that results in even stronger hybridization. The broadening of the impurity peak and the formation of the impurity band also proves the stronger hybridization. The rigid-band approach to estimate the impurity-dislocation energy, however, becomes less accurate in this case.

Another important result of our calculations is the oscillatory behavior of the impurity-dislocation interaction energy; this is especially noticeable in Fig. 2(a) for impurities in positions 1-4, as well as the high interaction energy in positions 2 and 6, which are relatively far from the center of the dislocation core. The central atom of the dislocation core has significant charge excess due to electron localization on it, causing a large charge transfer and long-ranged Friedel oscillations of the charge density. In the computational scheme we employ, the charge transfer cannot be calculated directly, because of the use of the charge neutrality condition. However, the changes of the atomic radii do point out indirectly the presence of a significant charge transfer. Therefore the mechanism of impurity-dislocation interaction for atoms in positions 2 and 6 is similar to the "electrostatic locking" [8] and is caused by the long-range charge oscillations due to electron localization in the dislocation core. Thus, the <100> edge dislocation in bcc metals and intermetallics represent a unique example of unusually high "electrostatic locking" interactions, in contrast to other dislocations where the magnitude of this interaction is very small [8].

As regards the chemical locking with the impurity in position 8, the strong impurity-dislocation attraction may influence the macroscopic mechanical properties of NiAl in two ways. First, for dislocation glide during plastic deformation, this relative position will give a local energy minimum, and so more stress will be necessary to move the dislocation further. Since the impurity-dislocation interaction energy is large, this mechanism will contribute strongly to the solid solution hardening. Second, the segregation of impurities on the dislocation will occur for the dislocation at rest.

In contrast to the chemical locking mechanism, where the impurity has to be in a nearest-neighbor position to the center of the dislocation core, the electrostatic locking is relatively long-ranged (see Fig. 2). Because of this, the "dislocation friction" of the gliding dislocations resulting from the repulsive impurity-dislocation interaction should give an effective contribution to strengthening.

Besides its role in SSH, the addition of transition metal impurities can also improve the high-temperature properties of NiAl. As an example, Ti additions of the order of 2.5 to 3% result in a 200 to a 5,000-fold reduction in the creep rate as compared to that of binary NiAl [18]. Since this concentration is below the solubility boundary of Ti, which is 5% [19], the usually assumed mechanism of precipitate hardening is not acceptable, and so the nature of the large effect of Ti additions on the creep rate remained unclear. The results of our calculations suggest that due to the strongly attractive impurity-dislocation interaction, Ti segregation on edge dislocations should play an important role in this effect. This segregation was actually observed experimentally [18].

CONCLUSIONS

We have shown that dislocation locking, by both chemical locking and electrostatic locking mechanisms, takes place during the interaction of early transition metal (Ti, V, Cr, Mn) impurities with the <100>{010} edge dislocation in NiAl. This mechanism can be expected not only in NiAl, but also in other B2 intermetallics where the deformation is carried by <100> dislocations (CoTi, CoHf, and CoZr). We identified the nature of the chemical locking to be caused by the strong hybridization between impurity 3d-states and localized electronic states appearing in the dislocation core. The electrostatic locking is associated with long-range Friedel charge oscillations caused by the electron localization in the dislocation core. These findings are important for understanding the nature of solid solution hardening and the anomalous increase of creep resistance upon alloying in intermetallic alloys.

ACKNOWLEDGMENTS

This work was supported by the Air Force Office of Scientific Research under grant No.F49620-98-1-0321 and by the National Science Foundation under Cooperative Agreement No. ACI-9619019, through the University of Illinois; it utilized the Origin2000 at the NSF supported National Center for Supercomputing Applications, University of Illinois at Urbana-Champaign, and at the Naval Oceanographic Office (NAVO).

REFERENCES

1. P. Haasen, *Physical Metallurgy,* edited by R.W. Cahn and P. Haasen, (North Holland, Amsterdam, 1996), pp. 2009-2073.
2. H. Suzuki, *Dislocation in Solids,* edited by R.F.N. Nabarro (North Holland, Amsterdam, 1979), V.4, chapter 15.
3. Y. Mishima, S. Ochai, N. Hamano, Y. Yodogava, and T. Suzuki, *Trans. Jpn. Inst. Met.* **27**, 648 (1986).
4. V.O. Abramov and O.V. Abramov, *Dokl. Akad. Nauk SSSR* **318**, 883 (1991);
5. R.D. Noebe, R.R. Bowman, and M.V. Nathal, *Int. Mater. Rev.* **38**, 193 (1993).
6. N.I. Medvedeva, Yu.N. Gornostyrev, D.L. Novikov, O.N. Mryasov, and A.J. Freeman, *Acta Mater.* **46**, 3433 (1998).
7. S.G. Roberts, P. Pirouz, and P.B. Hirsch, *J. Mater. Sci.* **20**, 1739 (1985).
8. J. Friedel, *Dislocations* (Pergamon Press, Oxford, 1964).
9. T. Shinoda, K. Masuda-Jindo, Y. Mishima, and T. Suzuki, *Phys. Rev. B* **35**, 2155 (1987); *Phil. Mag. B* **62**, 289 (1990).
10. A.O. Anokhin, M.L. Galperin, Yu.N. Gornostyrev, M.I. Katsnelson, and A.V. Trefilov, *JETP Lett.* **59**, 369 (1994); *Phys. Metall. Metallogr.* **79**, 242 (1995); *Phil. Mag. B* **73**, 845 (1996).
11. O.Yu. Kontsevoi, O.N. Mryasov, Yu.N. Gornostyrev, A.J. Freeman, M.I. Katsnelson, and A.V. Trefilov, *Phil. Mag. Lett.* **78**, 427 (1998).
12. O.Yu. Kontsevoi, O.N. Mryasov, A.I. Liechtenstein, and V.A. Gubanov, *Soviet Phys. Solid St.* **34**, 154 (1992); O.Yu. Kontsevoi, O.N. Mryasov, and V.A. Gubanov, *ibid.* **34**, 1406 (1992).
13. N.I. Medvedeva, O.N. Mryasov, Yu.N. Gornostyrev, D.L. Novikov, and A.J. Freeman, *Phys. Rev. B* **54**, 13506 (1996).
14. J.P. Hirth, and J. Lothe, *Theory of Dislocations* (Wiley, New York, 1982).
15. M.J. Mills, M.S. Daw, S.M. Foiles, and D.B. Miracle, in *High Temperature Ordered Intermetallic Alloys V* edited by I. Baker, R. Darolia, J.D. Whittenberger, and M.H. Yoo, (Mater. Res. Soc. Proc. **288**, Pittsburgh PA, 1993) pp. 257-268.
16. O.Yu. Kontsevoi, O.N. Mryasov, Yu.N. Gornostyrev, and A.J. Freeman, in *Tight-Binding Approach to Computational Materials Science,* edited by P.E.A. Turchi, A. Gonis, and L.Colombo, (Mater. Res. Soc. Proc. **491**, Warrendale, PA, 1998) pp. 143-148.
17. G. Bozzolo, R.D. Noebe, and F. Honecy, *Intermetallics* **8**, 7 (2000).
18. P.H. Kitabjian, A. Garg, R.D. Noebe, and W.D. Nix, *Metall. Mater. Trans. A* **30**, 587 (1999).
19. G. Bozzolo, R.D. Noebe, J. Ferrante, A. Garg, and C. Amador, *NASA Technical Memorandum 113119* (1997).

Mat. Res. Soc. Symp. Proc. Vol. 646 © 2001 Materials Research Society

Microhardness Anisotropy and the Indentation Size Effect (ISE) on the (100) of Single Crystal NiAl

M.E. Stevenson, M.L. Weaver, R.C. Bradt
Department of Metallurgical and Materials Engineering, The University of Alabama,
Tuscaloosa, AL 35487-0202, USA

ABSTRACT

The Knoop microhardness anisotropy of single crystal NiAl, a B2 intermetallic compound, was investigated on the (001) plane at indentation test loads from 25 to 500 g. An energy balance analysis was applied to analyze the indentation size effect (ISE). The load independent, orientation independent Knoop microhardness was determined to be 220 kg/mm^2 for the (100) plane of NiAl.

INTRODUCTION

Indentation hardness testing is currently of extensive practical use in evaluating the mechanical properties of intermetallic compounds such as NiAl and other technically significant materials.[1] For these analyses a complete understanding of the indentation process is critical to properly interpret hardness data for applications to mechanical and wear properties. Two phenomena that assume important roles during indentation are single crystal anisotropy[2] and the indentation size effect, or ISE.[3] The microhardness anisotropy has been successfully addressed by comparison to the primary slip system activated in the particular crystal structure. With regards to the ISE, however, a unified theory has not been completely accepted, but rather several different mechanisms for its existence have been advanced. These mechanisms include: friction between the specimen and indenter facets,[4] elastic recovery after indentation,[5] the load required to initiate plastic flow, work hardening,[3] the energy required to nucleate dislocations,[5] statistical measurement errors[6] and strain gradient plasticity.[7] This paper addresses both the microhardness anisotropy and the indentation size effect in single crystal NiAl in terms of a fundamental energy balance, which leads to the prediction of a load independent Knoop microhardness that is also orientation independent on the (100) crystal plane.

EXPERIMENTAL PROCEDURES

One nominally stoichiometric NiAl single crystal slab was obtained from General Electric Aircraft Engines, Cincinnati, OH. The dimension of this slab was 25 mm x 32 mm x 100 mm. The slab was homogenized at 1589 K for 1 hour in argon followed by furnace cooling to room temperature. Crystal orientation was determined using the back reflection Laue technique. Single crystal specimens were prepared for microhardness measurements on the (001) crystal plane. Oriented samples were removed from the slab using a diamond saw. After removal, they were epoxy mounted and manually polished through 600 grit silicon carbide. Final polishing was conducted with a 0.06 µm colloidal silica suspension in a vibratory polisher.

Microhardness was measured using an automated testing machine* equipped with a Knoop indenter and fitted with a goniometer stage for precise definition of the angular orientations of the test specimen. The Knoop indenter geometry was chosen because it provides a long, shallow pyramidal indentation with essentially no elastic relaxation and a low propensity for cracking.[8] The Knoop indenter geometry also provides an excellent means to study the microhardness anisotropy of a material by orienting the long axis of the indenter parallel to a specific crystallographic direction, which is the directional specification in this study. Knoop microhardness is specified by the equation:

$$H_K = \frac{14.229 \cdot P}{d^2} \tag{1}$$

Where P is the applied indentation test load and d is the length of the long axis of the resulting pyramidal shaped indentation. The impression diagonal length was measured immediately after indentation and converted to hardness via Equation (1). To observe the complete symmetry of the (100) plane, microhardness measurements were taken every 22.5° from the [110] to the [$\bar{1}$10] on the (100). To assess the indentation size effect, Knoop microhardness measurements were determined at the following five indentation test loads: 25, 50, 100, 200, 300 and 500 grams. A loading dwell time of ten seconds was used for all measurements.

RESULTS

For the (001) plane of NiAl, the Knoop microhardnesses as a function of crystallographic orientation and indentation test load are shown in Figure 1. These results clearly illustrate the microhardness anisotropy on the (001) plane. The microhardness profile is a maximum for the <100> directions on this plane, with minima occurring for the <110> directions. These results are in agreement with the microhardness anisotropy in NiAl reported by Ebrahimi et al.[9] Their research reported the microhardness on the (100) plane for an indentation test load of 500 g and directional orientations similar to this study. Their results yielded microhardness values 276 kg/mm² for the <100> directions and 261 kg/mm² for the <110> directions.

The indentation test load dependence of the Knoop microhardness is shown in Figure 2. From these results it is clear that for the (100) plane of NiAl, there is a significant load dependence for the hardness. In the [100] direction on the (100), the lowest indentation test load of 25 g yields a hardness almost twice that of the 500 g indentation test load.

The Load Independent Knoop Microhardness
Many theories have been advanced to both explain and analyze the indentation size effect, but few are universally applicable.[10,11] One analysis is based on an energy balance between the indenter and the test specimen. The basis of that energy balance can be summarized for the work or energy of indentation by integrating the test load, P, as[12]:

$$Energy \ or \ Work = \int_0^h P dx = \alpha \cdot d^2 + \beta \cdot d^3, \tag{2}$$

---

* *Buehler* Micromet 2004

where α and β are surface and volume coefficients. Converting the indenter dimensions and the indentation depth, h, to the impression length, d, incorporating the geometry of the specific indenter reduces to:

$$\frac{P}{d} = a_1 + a_2 \cdot d. \qquad (3)$$

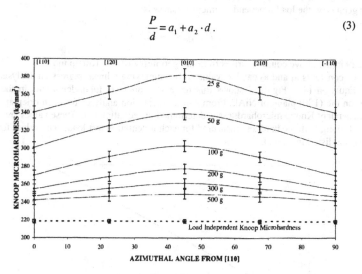

Figure 1. Knoop microhardness anisotropy on the (100) plane in single crystal NiAl. Error bars represent the 95% confidence intervals of a student "t" distribution.

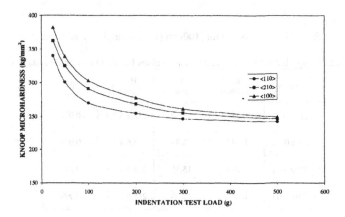

Figure 2. Load dependence of the Knoop microhardness on the (100) in single crystal NiAl.

Here a_1 and a_2 are constants for the particular indenter geometry of interest. As shown by Li and Bradt,[11] by applying the mathematical conditions for load independent hardness (dH/dP = 0), it is the constant a_2 that is directly related to the load independent microhardness. For the Knoop indenter geometry, the load independent microhardness is:

$$H_{K-LIH} = 14.229 \cdot a_2 \qquad (4)$$

The 14.229 in the above equation arises from the geometry of the Knoop indenter.

The constants a_1 and a_2 can be evaluated by employing a linear regression analysis of the form of Equation (4). Figure 3 shows this regression analysis for indentations in the [100] direction on the (100) plane of NiAl. From similar regression analyses for each orientation, the Load independent Knoop microhardness is determined via equation (4). These values are plotted in Figure 1 along with the hardness measured for each indentation test load. The values for a_1, a_2 and H_{K-LIH} are listed in Table I.

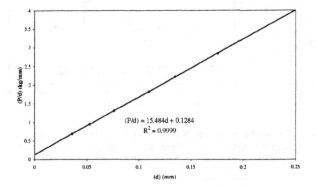

Figure 3. (P/d) vs. (d) for [100] on (100) in single crystal NiAl.

Table I. Energy balance analysis and H_{K-LIH} values for (001) in single crystal NiAl.

Orientation	a_1 (kg/mm)	a_2 (kg/mm²)	H_{K-LIH} (kg/mm²)	R^2
[110](001)	0.384	15.37	218.8 ± 5.1	0.998
[120](001)	0.342	15.36	218.6 ± 5.1	0.999
[010](001)	0.273	15.35	218.5 ± 4.8	0.999
[210](001)	0.342	15.36	218.6 ± 5.1	0.999
[110](001)	0.384	15.37	218.8 ± 5.1	0.998

As noted previously, many explanations have been advanced to explain the ISE. In almost all cases, however, the specific models do not hold for all types of materials. This suggests that the ISE is related to the indentation process rather than an effect related to the specimen. One aspect of the indentation process that is often neglected is the role of friction at the indenter/specimen interface. Several researchers have addressed this issue by lubricating during the indentation process.[1,13,14] In all cases, their results show an almost complete disappearance of the ISE. In terms of the analysis used in this experiment, the a_1 value is affected by frictional effects, with the a_2 value remaining constant.[4] These results indicate that the ISE is caused by surface frictional effects, and not attributed to the many other theories introduced previously.

From Figure 1 it is clear that the load independent Knoop microhardness, the H_{K-LIH}, does not vary as a function of crystallographic orientation. This is in contrast with the highly anisotropic nature of the hardness for the 25 g indentation test load. These two features can be explained by considering the nature of the plastic deformation and the size of the indentations at both low and high indentation test loads. At the lowest indentation test load, the sizes of the indentations are small and the deformation can be accommodated through slip on the primary slip system alone. This being the case, one would expect that the hardness anisotropy at the low test loads is dominated by slip on the primary slip system. As the indentation test load and the indent size increase, slip must be activated on additional slip systems in order for the deformation to be accommodated. This phenomenon has been documented to occur during the indentation process by a number of researchers.[15,16] The activation of slip on additional systems completely masks the directional anisotropy observed for low test loads and yield a load independent, orientation independent Knoop microhardness.

SUMMARY AND CONCLUSIONS

Knoop microhardness were measured on the (100) crystallographic plane of single crystal NiAl as a function of both crystal direction and indentation test load. Both a directional microhardness anisotropy and the presence of an indentation size effect (ISE) were observed. The microhardness anisotropy profile has maxima for the <100> type directions, with minima occurring for the <110> type directions. This anisotropy gradually decreases as the indentation test load is increased, showing little directional dependence at an indentation test load of 500 g.

The presence of an Indentation Size Effect, or ISE, in these microhardness measurements was addressed using an energy balance analysis applied to the indentation process. This analysis allows for calculation of the load independent Knoop microhardness, H_{K-LIH} for the (100) plane of single crystal NiAl. The (100) crystallographic plane of NiAl has a load independent, orientation independent microhardness of 220 kg/mm^2.

REFERENCES

1.   Bystrzycki, J. and R.A. Varin, "The Frictional Component in Microhardness Testing of Intermetallics". *Scripta Metallurgica et Materiala*, 1993. 29: p. 605-609.
2.   Daniels, F.W. and C.G. Dunn, "The effect of orientation on Knoop hardness of Single crystals of zinc and silicon ferrite". *Transactions of ASM*, 1949. 41: p. 419-422.

3.   Tate, D.R., "A comparison of microhardness indentation tests". *Transactions of ASM*, 1945. 35: p. 374-375.

4.   Ghosh, A., et al., "The frictional component of the ISE in microhardness testing". *Journal of Materials Research*, 1993. 8: p. 1028-1032.

5.   Hays, C. and E.G. Kendall, "An analysis of Knoop microhardness". *Metallurgy*, 1973. 6: p. 275-282.

6.   Yost, F.G., "On the Definition of Microhardness". *Metallurgical Transactions*, 1983. 14: p. 947-952.

7.   Nix, W.D. and H. Gao, "Indentation size effects in crystalline materials: A law for strain-gradient plasticity". *Journal of the Mechanics and Physics of Solids*, 1998. 46(3): p. 411-425.

8.   Li, H. and R.C. Bradt, "The effect of indentation induced cracking on the apparent microhardness". *Journal of Materials Science*, 1996. 31: p. 1065-1070.

9.   Ebrahimi, F., A. Gomez, and T.G. Hicks, "Nature of slip during knoop indentation on {100} surface of NiAl". *Scripta Metallurgica et Materials*, 1996. 34(2): p. 337-342.

10. Quinn, J.B. and G.D. Quinn, "Indentation Brittleness of Ceramics: A fresh approach". *Journal of Materials Science*, 1997. 32(16): p. 4331-4346.

11. Li, H. and R.C. Bradt, "Knoop Microhardness Anisotropy of Single Crystal LaB_6". *Materials Science and Engineering A*, 1991. 142: p. 51-61.

12. Stevenson, M.E. and R.C. Bradt, "Analysis of Microhardness Measurements". *Advanced Materials and Processes: Heat Treating Progress*, 2000: p. H41-H43.

13. Kaji, M., M.E. Stevenson, and R.C. Bradt, *To Be Published.*

14. Shi, H. and M. Atkinson, "A friction effect in low-load hardness testing of copper and aluminum". *Journal of Materials Science*, 1990. 25(2111-2114).

15. Brookes, C.A. and R.P. Burnand, "Hardness Anisotropy in Crystalline Solids", in *The Science of Hardness Testing and its Research Applications*, J. Westbrook, J. Shaw, and H. Conrad, Editors. 1973, ASM: Metals Park, OH.

16. Ma, Q. and D.R. Clarke, "Size dependent hardness of silver single crystals". *Journal of Materials Research*, 1995. 10: p. 853-862.

Mat. Res. Soc. Symp. Proc. Vol. 646 © 2001 Materials Research Society

Direct Molecular Dynamics Simulations of Diffusion Mechanisms in NiAl

D. Farkas and B. Soulé de Bas
Dept. of Materials Science and Engineering, Virginia Polytechnic Institute and State University, Blacksburg, VA 24061, USA

ABSTRACT

Molecular dynamics simulations of the diffusion process in ordered B2 NiAl at high temperature were performed using an embedded atom interatomic potential. Diffusion occurs through a variety of cyclic mechanisms that accomplish the motion of the vacancy through nearest neighbor jumps restoring order to the alloy at the end of the cycle. The traditionally postulated 6-jump cycle is only one of the various cycles observed and some of these are quite complex. A detailed sequential analysis of the observed 6-jump cycles was performed and the results are analyzed in terms of the activation energies for individual jumps calculated using molecular statics simulations.

INTRODUCTION

Contrary to the case of pure metals, where the self-diffusion mechanism is well established and consists of a nearest neighbor jumps (NN), the diffusion in B2 compounds is much more complex. Two main categories of mechanisms have been postulated to characterize the diffusion in B2 compounds: the mechanisms involving next nearest neighbor jumps (NNN), where the order is maintained at all times, and cyclic mechanisms involving nearest neighbor jumps, that destroy order temporarily.

The next nearest neighbor jump mechanism can be expected as energetically favorable based on the fact that there is no disorder created during the process. Thus, Donaldson and Rawlings [1], based on a study of the diffusion tracer, suggested a NNN mechanism for the Ni atoms in the NiGa B2 compound. Theoretical considerations and static computer simulation studies [2], performed for B2 NiAl also suggested the NNN mechanism may be energetically favorable. The nearest neighbor jump mechanisms can be argued to be less favorable because the atoms jump initially to a site in the wrong sublattice, creating partial disorder in the crystal in the form of antisites. Several mechanisms have been suggested where the partial order is recovered after a certain number of nearest neighbor jumps, constituting a diffusion cycle. The best known of these is the 6-jump cycle (Elcock and McCombie [3]) where the vacancy migrates along a definite path of 6 nearest neighbor jumps. Since this mechanism was first proposed in 1958, this has been widely accepted as a main diffusion mechanism in B2 ordered alloys. Wynblatt [4], in a study based on β-AgMg, found that the 6-jump cycles was energetically the most favorable. Investigations done with quasi-elastic Mössbauer spectroscopy [5] and nuclear resonant scattering [6] on FeAl showed that the diffusion of Fe in the B2 phase takes place via NN jumps. Similarly, studies done using nuclear neutron scattering [7] on NiGa showed that the Ni atoms diffuse via NN jumps. Other mechanisms have also been proposed, such as the anti-structure bridge [8], where the vacancy migrates through a "bridge" created by an existing antisite, or the antisite-assisted 6-jump cycles [9], where the presence of extra antisites lowers the activation energy barrier. These additional mechanisms are of interest mainly for the case of a non-stoichiometric and partially ordered alloys.

In the present study, our goal is to observe the mechanisms of diffusion in B2 ordered compounds by direct molecular dynamics. The B2 compound was simulated as an initially homogeneous alloy of stoichiometric composition, in which one vacancy was introduced. The potentials used [10] were developed to fit the properties of B2 NiAl. In the following, we first discuss the activation energies predicted by the potentials for the different expected mechanisms. We then present the statistical analysis of the direct observations of the molecular dynamics simulations. We finally report a detailed time evolution analysis of the 6-jump cycles observed.

COMPUTATIONAL PROCEDURE

Migration energies for the vacancy have been obtained by a simple energy minimization technique in a molecular statics framework (Mishin and Farkas [2]). This technique was applied using the potentials of reference [10] to NNN vacancy jumps within the same sublattice as well as to the NN vacancy jumps involved in traditional <110> 6-jump cycles. For these calculations we used a cubic block of 9 unit cells along each side.

All the statistical analysis reported below is based on simulations using standard molecular dynamics in a constant temperature algorithm. We used a block that is cubic in shape with repeating periodic boundary conditions in the three directions. Our simulation size was of 125 unit cells arranged in a 5ax5ax5a cube. The vacancy is initially introduced in a Ni site. The block has 249 atoms that are arranged in a perfectly ordered structure. The lattice parameter used was that of equilibrium at the temperature of the simulations, e.g. 1200 K. The potentials predict that the perfect stoichiometric equilibrium lattice parameter at this temperature is a_{1200}=2.9363 Å. The diffusion processes were studied over a range of 750000 steps using a time step of 2×10^{-3} ps. This corresponds to a study the diffusion phenomena during 1.5 ns. The details of the diffusion process were studied by monitoring the atom displacements initially every 50000 steps (0.1 ns). During a cycle type mechanism, the atomic displacements were monitored much more closely; every 50 to 100 steps or 0.1 to 0.2 ps. This allows the study of the details of the mechanism and the exact timing of the various jumps and jump attempts within a particular cycle.

STATISTICAL PROCEDURE

For a statistical analysis of the results we classified the diffusion event data according to the different mechanisms they represent. A diffusion event was taken to be one or more atomic jumps from initiation to the restoration of the perfect order of the lattice. The different types of events that we observed and report below are: 6-jump cycles, 10-jumps cycles, 14-jump cycles, and failed attempts with the vacancy returning to the original position. Thus a total of 93 events were studied starting with one Ni vacancy. The 6-jump cycles were the most common successful diffusion event. We considered the frequency of the different possible types of 6-jump cycles (e.g., the <110> cycle, the <100> straight cycle and the <100> bent cycle [11]). We also differentiated the uninterrupted 6-jump cycles from the interrupted ones, where a neutral atom (not involved in the resulting 6-jump cycles) jumps to the vacant site and goes back to its previous configuration with the rest of the sequence proceeding normally. The 10-jump cycle is a new mechanism observed in the present investigation, in which the vacancy follows a definite path of 10 jumps. As in the 6-jump cycle case, we differentiated the uninterrupted jump cycles from the interrupted ones. The 14-jump cycle is also a new mechanism and in this case, the vacancy follows a definite path of 14 jumps. Details of the new mechanisms will be reported elsewhere.

The category of "failed attempts" refers to events where one or more jumps to nearest neighbor sites occur and are then reversed. All atoms as well as the vacancy go back to their original positions. We report three different situations: the first being where one atom jumps and goes back. In the second type two atoms jump one after another and go back in the opposite order (the second atom goes back and then the first one), and finally the case where more than two atoms jump and go back in successive steps reversing the order of the initial attempt. We also had a few cases of undefined uncompleted cyclic mechanisms. These were not common, and may be an indication of the fact that thermodynamically, a small degree of disorder is expected since the order parameter at 1200 K is close to but not exactly one. Special attention was devoted to detecting possible next nearest neighbor jumps but, as discussed below, none were observed.

RESULTS

Using the molecular statics method [2], we obtained the activation energy for a next nearest neighbor jump. The calculated Ni-vacancy formation energy is 0.68 eV, and the Ni-vacancy migration energy for a next nearest neighbor jump is 2.07 eV. We also determined the energy-displacement curve for a <110> cycle performed by a Ni vacancy. The result is shown in figure 1. The highest migration energy, corresponding to the third jump is 1.34 eV. This value is lower than the one found for the next nearest neighbor jump. Based on this approach, the 6-jump cycle is expected to be more favorable than the next nearest neighbor mechanism for this particular potential. We stress that these numbers are strongly dependent on the potential used. The results of the MD simulations for a Ni vacancy at 1200 K are presented in table 1. A total of 67 events at this temperature are reported. We also ran simulations at 1150 K with very similar results. The 6-jump cycles are the primary mechanism involved in the diffusion. These account for about 40% of the total events observed. Some of the 6-jump cycles are interrupted by the inclusion of an extra jump at some point within the cycle. The process returns then to the normal path and ends in a configuration totally equivalent to the uninterrupted case.

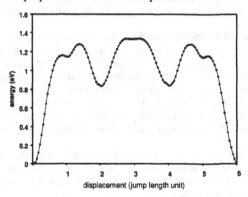

Figure 1. Energy-displacement curve for <110> jump cycle of a Ni vacancy

Table 1. The statistical occurrence of the various events observed for a Ni vacancy during 34.5 ns at 1200 K (17.25 million time steps)

Events	Specifications	Values	Total
6-jump cycles	Uninterrupted	32.8%	
	Interrupted	7.4%	
	[110] cycles	100%	
	[100] straight cycles	0%	
	[100] bent cycles	0.0%	40.2%
10-jump cycles	Uninterrupted	4.4%	
	Interrupted	1.5%	5.9%
14-jump cycles		1.5%	1.5%
Attempt returning to the original Position	Involving 1 atom	6%	
	Involving 2 atoms	26.9%	
	Involving more than 2 atoms	6%	38.9%
Other		13.5%	13.5%
		Total	100%

The interrupted 6-jump cycles correspond to 19% of all the 6-jump cycles. Differentiating the different types of 6-jump cycles, we see that they are mainly are <110> cycles. Thus at 1200 K we have 100% of <110> 6-jump cycles. At 1150 K, we observed one <100> bent cycle. These results agree with the reports of most studies that estimate the <110> cycle to be the most probable one. The 10-jump cycles correspond to about 6% of the events. These include the regular and the interrupted 10-jump cycles. The statistical occurrence of these events is lower than that corresponding to the 6-jump cycle. The 14-jump cycle is a rare event. The failed cycle attempts account for 38.9% at 1200 K. Most of these are cycle attempts that are reversed after the second jump. This can be understood on the basis of the potential energy curve in fig. 1. If the system has enough energy to accomplish both jump 1 and 2, it is in a local valley. It is then reasonable that the system has a significant probability of going back to its original position rather than completing jump 3 because the energy barrier for jump 3 is slightly higher than that needed for reversing jump 2.

In the final category "other", the final configuration of the run is complex and temporarily disordered. This probably corresponds to the fact that at this temperature the equilibrium state of the system is still very close to perfect order but the equilibrium order parameter is already somewhat lower than unity. These complex diffusion paths can thus be interpreted as resulting from the small deviation of perfect order, which already occurs at this temperature.

In the case of a Ni vacancy, according to the activation energy values (figure 1), the 6-jump cycle is expected to be more favorable than the next nearest neighbor mechanism. In agreement with this expectation, the latter mechanism was not observed in our simulations.

In figure 2 we show the x displacements of the three jumping atoms along the x direction as a function of the time, for two representative 6-jump cycles observed at 1200 K. For a detailed sequential analysis of the cycle it is sufficient to analyze displacements in only direction, since this indicates precisely when the jump occurs.

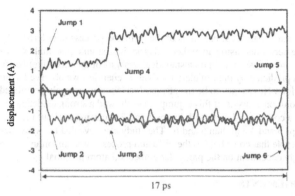

Figure 2a. First example of displacement versus time curve of a <110>jump cycle of a Ni-vacancy.

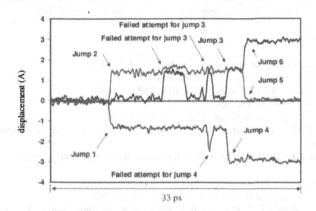

Figure 2b. Second example of a displacement versus time curve of a <110>jump cycle of a Ni-vacancy.

The total time that it takes to complete a cycle varies from 15 to 45 ps. This is much shorter than the average time elapsed between cycles, which is 840 ps. Our detailed analysis shows that within the sequence, some specific jumps are correlated. In the first sequence reported in fig. 2a, jumps 1-2, 3-4, and 5-6 clearly occur together and can be seen as coordinated. In the second sequence (fig.2b), we have again coordinated jumps. Jumps 1-2 appear clearly coordinated. Then there is a failed attempt for jump 3 by itself, followed by a failed attempt for jump 3-4 coordinated. Finally jumps 3-4 and 5-6 clearly occur together completing the sequence. On the basis of the activation energy curve reported in fig. 1, the fact of correlating the jump 1-2, 3-4, and 5-6 can be seen as a way of minimizing residence time at the high-activated states. The migration energy results clearly suggest that jumps 3 and 4 should occur together, since there is no valley in between these states. In the case of jumps 1-2 and 5-6 there is a valley between the two states but the jumps occur anyway in a coordinated manner.

CONCLUSIONS

Direct molecular dynamics simulations of the diffusion process in ordered B2 NiAl at high temperature were performed using an embedded atom interatomic potential. We were able to follow the diffusion process and obtain a statistical analysis of the most likely diffusion mechanisms. As predicted by the calculated activation energies, we observed a large majority of diffusion events for a Ni vacancy to be the standard <110> 6-jump cycle postulated in 1958. Detailed time evolution analysis of this 6-jump cycle showed a strong tendency in this type of cycle to occur in a coordinated way, with jumps 1 and 2 occurring at about the same time, followed by jumps 3 and 4 and then 5 and 6. The study also revealed two new mechanisms, the 10 and 14-jump cycle that contribute to the diffusion process. It is important to stress that these results are strongly dependent on the particular embedded atom potential used.

ACKNOWLEDGEMENTS

This work was supported by the National Science Foundation, under grant DMR 97-53243. We also acknowledge many helpful discussions with Dr. Yuri Mishin.

REFERENCES

1. A. T. Donaldson and R. D. Rawlings, Acta Metall. **24**, 285 (1976)
2. Y. Mishin and D. Farkas, Phil. Mag. A. **75**, 1 (1997)
3. E. W. Elcock and C.W. McCombie, Phys. Rev. **109**, 6 (1958)
4. P. Wynblatt, Acta Metall. **15**, 1453 (1967)
5. G. Vogl and B. Sepiol, Acta Metall. Mater. **42**, 3175 (1994)
6. B. Sepiol, C.Czihak, A. Meyer, G. Vogl, J. Metge, and R. Rüffer, Hyperfine Interact. **113**, 449 (1998)
7. M. Kaisermayr, J. Combet, H. Ipser, H. Schicketanz, B. Sepiol, G. Vogl, Phys. Rev. B **61**, 18 (2000)
8. C. R. Kao and Y. A. Chang, Intermetallics **1**, 237 (1993)
9. M. Athène, P. Bellon and G. Martin, Phil. Mag. A **76**, 3 (1997)
10. D. Farkas, B. Mutasa, C. Vaihlé, and K. Ternes, Modelling Simul. Mater. Sci. Eng. **3**,201 (1995)
11. M. Arita, M. Koiwa, and S. Ishioka, Acta Metall. **37**, 1363 (1989)

Mat. Res. Soc. Symp. Proc. Vol. 646 © 2001 Materials Research Society

Long-Term Creep and Oxidation Behavior of a Laves Phase-Strengthened NiAl-Ta-Cr Alloy for Gas Turbine Applications

Martin Palm and Gerhard Sauthoff
Max-Planck-Institut für Eisenforschung GmbH
Max-Planck-Str. 1
D-40237 Düsseldorf
Germany

ABSTRACT

A Laves phase strengthened NiAl-Ta-Cr alloy (IP 75) has been developed for structural applications in gas turbines as well as in heat exchangers. It has a lower density and a higher thermal conductivity as well as a higher melting point than conventional superalloys. Different methods for alloy production have been established including investment casting (IC) and a powder metallurgical method (PM). Creep properties have been determined in compression and tension. Tension tests up to 10000 hours were performed at 900 °C and 1000 °C for both PM and IC materials. The oxidation behavior of the PM and IC material was studied for up to 1000 hours by thermogravimetry in air at constant temperature between 600 °C and 1300 °C. The results are compared with results obtained from long-term isothermal and cyclic oxidation experiments in air at 1200 °C for 17900 hours. Microstructures and scales were examined by light optical and scanning microscopy, X-ray powder diffraction and electron probe microanalysis.

INTRODUCTION

One of the most demanding applications for materials is in gas turbines, where materials have to withstand temperatures above 1000 °C in an oxidizing environment under heavy stresses for long times. For such applications, Ni-base superalloys are used which are restricted to temperatures below 1200°C.

For higher temperatures NiAl-base alloys are considered which show not only higher melting temperatures and higher thermal conductivities, but also lower densities than Ni-base superalloys. Current developments include eutectic alloys, oxide-dispersion strengthened alloys and alloys which are strengthened by precipitates of second phases. The two latter concepts should yield a high creep resistance at and above 1000°C, which is a prerequisite for high-temperature applications.

Following the concept of strengthening by precipitates a new NiAl-base alloy of nominal composition 45 Ni, 45 Al, 7.5 Cr and 2.5 Ta (always at.%) – IP 75 for short – has been developed [1-4]. Compared to IN 738 LC, a superalloy currently used for gas turbine applications, IP 75 has a much higher melting point of 1638°C (IN 738 LC: < 1315°C), a lower density (6.3 vs. 8.1 g/cm^3) and a higher thermal conductivity (57 vs. 27 W/mK at 1000°C), while coefficients of thermal expansion, specific heat capacities and Young's moduli are of the same order. Though properties for IP 75 have been well established, long-term creep and oxidation data have not yet been available. The present work is directed at clarifying the long-term creep and oxidation behavior at high temperatures in oxidizing environments, which is part of a materials development program in progress.

EXPERIMENTAL

Rectangular creep samples of 5 mm x 5 mm x 10 mm for compression tests and "dog-bone" tensile specimens of 4 mm diameter and 35 mm length were prepared by electro-discharge machining (EDM). All creep tests were carried out in air with the loading direction along the length of the samples. Plates of 10 mm x 10 mm x 2 mm for thermogravimetric tests were also prepared by EDM. Oxidation tests were carried out using a Setaram SETSYS 16 thermobalance. Tests were performed in synthetic air (20.5% O_2, 79.5% N_2) at a flow rate of $1 \bullet 10^{-4}$ m^3/s. Scanning electron microscopy (SEM) investigations were performed with a CamScan S4 instrument equipped with an Oxford type 5108 detector for energy dispersive analysis. X-ray diffraction patterns in the 2Θ range 25° to 125° were recorded with Co-Kα_1 radiation.

CREEP

Secondary creep rates of material, which was produced by industrial-scale investment casting [5] (grain size: 100 μm), were studied in compression with step-wise increasing load in the temperature range 900°C to 1300°C (Figure 1). Stress exponents varied between 9 at 900°C to 5 at 1300°C. In Figure 2 the creep resistance of IP 75 (in compression; grain size: 75 μm) is compared with that of IN 100 (in tension). IN 100 shows a higher creep resistance only up to 1000°C whereas above this temperature the IN 100 creep resistance drops steeply and the creep resistance of IP 75 extends to higher temperatures. In Figure 3 the time to rupture (in tension) is shown for the directionally solidified Ni-base superalloy IN 738 DS (DS: directionally solidified) and for IP 75 (grain sizes: 75 μm for IC 75, 10 μm for PM 75) at 900°C and 1000°C. At both temperatures the creep strengths of the Ni-base superalloy are superior.

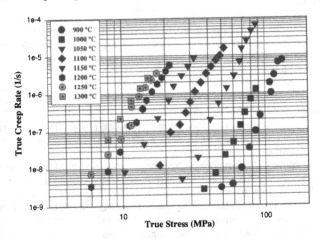

Figure 1. *Stress dependence of secondary creep rate of IC 75 (industrial-scale investment casting) [5] in compression with step-wise load increases at various temperatures.*

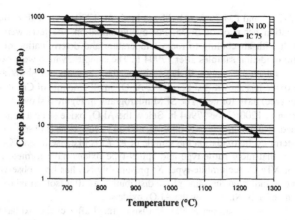

Figure 2. *Creep resistance of IC 75 (in compression) and IN 100 (in tension) as a function of temperature (at $10^{-7} s^{-1}$ secondary creep rate).*

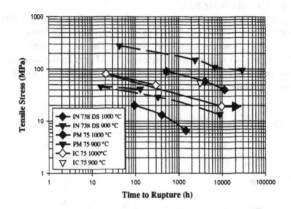

Figure 3. *Stress dependence of time to rupture for IP 75 and IN 738 at 900°C and 1000°C (in tension).*

OXIDATION

The oxidation behavior of IP 75 has been studied by thermogravimetry in the temperature range 600°C to 1300°C. In Figure 4 the results of the thermogravimetric tests are shown for the temperature range 1000°C to 1200°C. At all temperatures a parabolic time law has been observed. SEM investigations of cross sections of the thermogravimetric samples as well as X-ray diffraction experiments revealed that adherent Al_2O_3 scales were formed on all samples. No preferential oxidation of the Laves phase has been observed. Parabolic rate constants for the curves in Figure 4 vary from 4.6×10^{-14} $g^2 cm^{-4} s^{-1}$ to 3.5×10^{-13} $g^2 cm^{-4} s^{-1}$ between 1000°C and 1200°C. These values are in good agreement with data by [6] for the formation of α-Al_2O_3 scales on NiAl-Cr alloys.

Samples of IC 75 and PM 75 were oxidized in air for 17904 h to analyze the long-time static and cyclic oxidation behavior of IP 75. For cyclic oxidation, the tests were interrupted after periods of 336 h with air cooling to room temperature and determination of weight. Figure 5 shows long-term oxidation samples after 17904 h. The samples, which were exposed at constant temperature, are intact, i.e. they still have sharp edges and do not show any cracks. X-ray analysis of the oxide has shown that besides α-Al_2O_3 small amounts of $CrTaO_4$ (rutile-type) and $NiAl_2O_4$ (spinel-type) have formed. The SEM micrograph in Figure 6 shows a cross section of IC 75 after oxidation at 1200°C for 17904 h. Below the Al_2O_3 oxide scale a zone of about 500 μm width has formed, which consists only of NiAl, and no Laves phase is observed in this zone.

After long-term cyclic oxidation the sample of IC 75 is likewise intact. X-ray analysis of the oxide has shown that more rutile-type and spinel-type oxides have formed, and in addition a small amount of XTa_2O_6 (columbite-type; X: presumably Ni) has been observed. SEM analysis of the cross section indicated the same phase distribution as after isothermal oxidation (Figure 6) again with a zone of only NiAl below the Al_2O_3 oxide scale.

In contrast to this, the sample of PM 75 disintegrated after cyclic oxidation (Figure 5) and SEM analysis of a cross section revealed that it has completely transformed into oxides. This contrasting oxidation behavior of PM 75 was checked in more detail. Figure 7 shows the weight gain of IC 75 and PM 75 during cyclic oxidation. The samples of PM 75 indeed showed a steep weight gain increase after about 10000 h.

SEM investigation of cross sections of the sample after 11540 h showed that the sample contains three different zones (Figure 8a). The oxide (Al_2O_3) does not form a uniform scale, but cone-like tongues protrude into the sample (Figure 8b). The metallic matrix of the sample consists only of NiAl (Figure 8c) with small particles of Ni_3Al next to the oxide protrusions (Figure 8b). In the sample center a considerable amount of AlN has formed, which is surrounded by Ni_3Al and NiAl zones (Figure 8d).

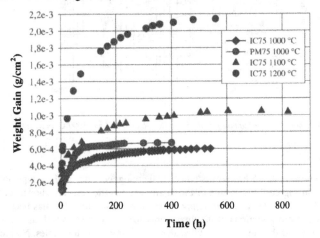

Figure 4. *Weight gain as a function of time for IP 75 alloys during isothermal oxidation in synthetic air at various temperatures.*

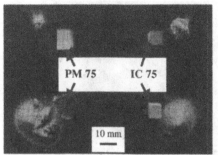

Figure 5. *Samples of IP 75 after isothermal (above) and cyclic (below) oxidation for 17904 h at 1200°C with small piles of oxides which cracked off during (repeated) cooling.*

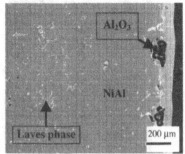

Figure 6. *SEM micrograph of a sample cross section after isothermal oxidation for 17904 h at 1200°C (BSE contrast).*

It is concluded that oxidation of IP 75 occurs by the formation of an alumina scale under both isothermal and cyclic oxidation conditions. This results in Al depletion of the matrix and the overall composition of the alloy becomes more Ni-rich near the surface. This, however, leads to an increased solid-solubility for Ta in NiAl [7] leading to the dissolution of the Ta-rich Laves phase below the alumina scale. Since the grains and the Laves phase precipitates are comparatively coarse in the cast alloy (> 10 μm), dissolution of the Laves phase is slow and limited to the surface zone of the alloy. In the fine-grained PM 75 alloy the Laves phase dissolves more quickly. After the dissolution of the Laves phase and further Al depletion, Ni_3Al forms giving rise to accelerated oxidation of the alloy. This rapid oxidation process is apparently accompanied by an inward diffusion of nitrogen with AlN formation in the central part of the sample (Figure 8d). The formation of AlN causes Al depletion of the matrix whereby Ni_3Al (light regions in Fig. 8d) is formed. As a result of the combined attack, rapid disintegration of the sample occurs.

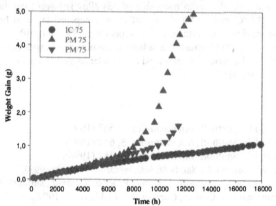

Figure 7. *Weight gain as a function of time for IC 75 and PM 75 (two samples) during cyclic oxidation at 1200°C in air.*

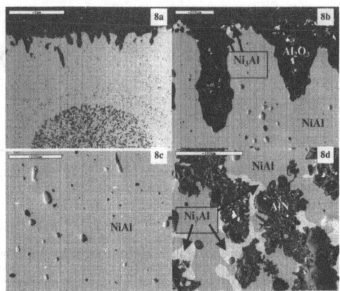

Figure 8. SEM micrographs (BSE) of PM 75 after cyclic oxidation at 1200°C for 11540h in air: sample cross section (8a), surface zone with oxide protrusions (8b), NiAl matrix (8c), central region with formation of AlN (8d).

CONCLUSIONS

As part of a materials development in progress for application in stationary gas turbines, the long-term creep and oxidation behavior of the Laves phase-strengthened NiAl-base alloy IP 75 has been studied in detail. The creep resistance of this alloy is lower than that of conventional superalloys up to 1000°C, but it still shows sufficient creep resistance at higher temperatures above the possible service temperatures of the superalloys. The alloy IP 75 also offers a high oxidation resistance due to the formation of adherent alumina scales with parabolic growth. This advantageous oxidation behavior was observed even under long-term cyclic oxidation conditions at 1200°C for the cast material.

REFERENCES

1.   B. Zeumer and G. Sauthoff, *Intermetallics*, **5**, 563 (1997).
2.   B. Zeumer and G. Sauthoff, *Intermetallics*, **5**, 641 (1997).
3.   B. Zeumer and G. Sauthoff, *Intermetallics*, **6**, 451 (1998).
4.   B. Zeumer, W. Sanders, G. Sauthoff, *Intermetallics*, **7**, 889 (1999).
5.   M. Palm, J. Preuhs, G. Sauthoff, *J. Mater. Process. Technol.*, submitted.
6.   M.W. Brumm and H.J. Grabke, *Corrosion Sci.*, **33**, 1677 (1992).
7.   M. Palm, W. Sanders, G. Sauthoff, *Z. Metallkde.*, **87**, 390 (1996).

Financial support by the BMBF under grant 03M3028 is gratefully acknowledged.

Mat. Res. Soc. Symp. Proc. Vol. 646 © 2001 Materials Research Society

1300 K Compressive Properties of Directionally Solidified Ni-33Al-33Cr-1Mo

J. Daniel Whittenberger[1], S.V. Raj[1], and Ivan. E. Locci[2]
[1]NASA-Glenn Research Center, Cleveland, OH 44135, John.D.Whittenberger@grc.nasa.gov
[2] Case Western Reserve University at the NASA-Glenn Research Center, Cleveland, OH 44135

ABSTRACT

The Ni-33Al-33Cr-1Mo eutectic has been directionally solidified by a modified Bridgeman technique at growth rates ranging from 7.6 to 508 mm/h to produce grain/cellular microstructures containing alternating plates of NiAl and Cr alloyed with Mo. The grains had sharp boundaries for slower growth rates (≤ 12.7 mm/h), while faster growth rates (≥ 25.4 mm/h) lead to cells bounded by intercellular regions. Compressive testing at 1300 K indicated that alloys DS'ed at rates between 25.4 to 254 mm/h possessed the best strengths which exceed that for the as-cast alloy.

INTRODUCTION

Directional solidification (DS) of NiAl-X systems has shown promise for the simultaneous improvement of elevated temperature strength and room temperature toughness [1-4]. In general it was believed that these benefits could only be possible when the structure was perfectly aligned and fault free. Unfortunately such ideal microstructures tend to demand very slow grow rates which would be impracticable for commercialization. To determine if faster growth rates could produce materials with acceptable elevated temperature strength properties, the eutectic system Ni-33Al-33Cr-1Mo (at %) has been directionally solidified at rates ranging from 7.6 to 508 mm/h. This system was chosen since it forms a lamellar eutectic microstructure [1-3] comprised of NiAl and Cr alloyed with Mo {Cr(Mo)} which has demonstrated a room temperature toughness of 17.3 MPa·√m [3]. This paper presents the alloy chemistry, microstructure and 1300 K compressive behavior for both as-cast and directionally solidified materials. Compression was utilized to determine mechanical properties because of the ease of machining and testing specimens and the measurement of identical 1300 K compressive and tensile flow stress - strain rate properties in several DS'ed NiAl-X systems [5,6].

EXPERIMENTAL PROCEDURES

A detailed discussion of the techniques employed to produce directionally solidified rods of 33Ni-33Al-33Cr-1Mo is presented in refs. 4 and 6. In short 19 mm diameter as-cast alloy bars were prepared by induction melting in Al_2O_3 crucibles and casting into a copper chill mold. Such bars were then placed in high purity alumina open-ended tubes for directional solidification in a modified Bridgeman apparatus under flowing high purity argon. Preferential solidification was accomplished by pulling the tube through a hole in a fixed position water-cooled copper baffle which yielded thermal gradients at the liquid/solid interface of about 8-10 K/mm. A total of seven DS'ed rods were produced at growth rates of 7.6, 12.7, 25.4, 50.8, 127, 254 and 508 mm/h. Samples for chemistry and metallography examination were taken from each as- cast bar as well as the aligned region of each DS'ed rod. Chemical analysis to determine both major and minor solute metallic elements was performed by an inductively coupled plasma (ICP) technique. The concentrations of nitrogen and oxygen were determined by an inert gas fusion method, while the carbon level was measured by the combustion extraction method. Transverse and longitudinal sections of selected as-cast bars and all seven DS'ed rods were metallographically prepared and examined by light optical techniques.

Parallelepiped compression samples 8 x 4 x 4 mm in size with the long axis parallel to the casting or DS growth direction were electrodischarge machined from several as-cast bars and the aligned region of each DS'ed rod. Both constant load and constant velocity compressive testing was undertaken in air at 1300 K. Lever arm test machines were utilized for constant load creep experiments, where deformation was determined as a function of time by measuring the relative positions of the ceramic push bars applying the load to the specimen. Constant velocity tests were conducted in a universal machine at crosshead rates ranging from 1.7 x 10^{-2} to 1.7 x 10^{-6} mm/s. All the acquired test data were normalized to the final specimen length, and true stresses, strains and strain rates were determined with the assumption of constant volume.

RESULTS AND DISCUSSION

Alloy Composition -- The average, maximum and minimum values and the standard deviation for Al, Cr and Mo are reported for the as-cast bars and directionally solidified rods in Table 1.

Table. 1 Composition of as-cast and directionally solidified Ni-33Al-31Cr-1Mo alloys

	Aluminum, at. %				Chromium, at. %				Molybdenum, at. %			
	Avg.	Max.	Min.	Std. Dev.	Avg.	Max.	Min	Std. Dev.	Avg.	Max.	Min.	Std. Dev.
As-Cast	33.00	33.26	32.43	0.28	33.39	33.56	33.29	0.10	1.03	1.05	1.02	0.01
DS'ed	32.85	33.71	31.21	0.68	33.47	34.90	32.88	0.54	1.02	1.04	0.99	0.016

The alloys also contained (at. %) about 0.009Cu, 0.010Fe, 0.005Si, 0.07C, 0.002N, 0.03O and 0.001S as impurities. The results in Table 1 indicate the control of the alloy chemistry for all seven as-cast bars and DS'ed rods was quite good, and the average values are close to the intended Ni-33Al-31Cr-1Mo composition. While there is little change between the Mo levels in the as-cast and DS'ed alloys, surprisingly directionally solidification appeared to increase the difference between the maximum and minimum values for both Al and Cr, and thus Ni, over those measured in the as-cast bars. Such changes are, in turn, reflected by the larger standard deviations for the DS'ed rods.

Alloy Microstructure -- Photomicrographs illustrating the transverse structure of as-cast and DS'ed Ni-33Al-33Cr-1Mo are presented in Fig. 1. Relatively rapid solidification during casting into a Cu-chill mold produced NiAl dendrites scattered among grains containing both NiAl and Cr(Mo) (Fig.1(a)). This can be compared to the regions (grains) of parallel Cr(Mo) and NiAl plates delineated by sharp boundaries (Fig. 1(b)) after DS'ing at 12.7 mm/h; cells enclosed by relatively thick intercellular regions where each cell contains lamella in a radial pattern (Fig. 1(c)) after DS'ing at 127 mm/h; and cells containing (Cr,Mo) fibers which are surrounded by a thick, coarse structured intercellular border after DS'ing at 508 mm/h (Fig. 1(d)). The general between 50.8 to

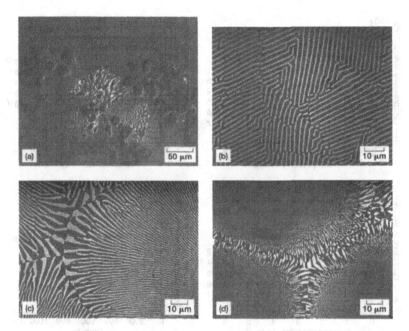

Figure 1. *Transverse microstructure of as-cast and directionally solidified NiAl-(Cr,Mo) eutectics. (a) As-cast Ni-33.1Al-33.3Cr-1.03Mo, (b) Ni-32.16Al-34.06Cr-1.03Mo grown at 12.7 mm/h, (c) Ni-32.59Al-33.40Cr-1.04Mo grown at 127 mm/h and (d) Ni-32.86-33.30-1.03Mo grown at 508 mm/h. NiAl is the dark phase and (Cr,Mo) is the bright phase.*

appearance of the transverse microstructure in Fig. 1(c)) is typical of that found after DS'ing between 50.8 to 254 mm/h. The structure after the slowest growth rate (7.6 mm/h) is similar to that shown in Fig. 1(b) except that occasional cells are partially composed of Cr(Mo) fibers.

In spite of some difference in chemistry among all the as-cast bars and DS'ed rods (Table 1), no third phases were found nor was there any evidence of NiAl or Cr(Mo) dendrites in the DS'ed regions. Overall the microstructures of the present Ni-33Al-33Cr-1Mo rods as a function of DS rate are similar to those found in Ni-33Al-31Cr-3Mo [4,7] as a function of growth rate. The structure in Fig. 1(b) is also in agreement with that for NiAl-32.4Cr-1Mo grown at 12.7 mm/h by Cline, et al. [8] utilizing a Bridgeman technique. On the other hand Yang et al. [3] were able to maintain the sharp boundary cells comprised of the parallel plate type of microstructure (Fig. 1(b)) at growth rates of both 50 and 100 mm/h in Ni-33Al-33Cr-1Mo through an Edge-defined Film-fed Growth method.

<u>1300 K Compressive Properties</u> -- Examples of the compressive stress - strain curves obtained from constant velocity testing and the creep curves measured under constant load conditions are given in Fig. 2. The stress - strain curves (Fig. 2(a-c)) indicated that all materials underwent work hardening over the first one percent stain followed by continued flow at a more or less

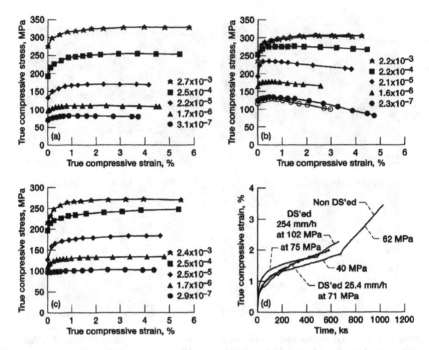

Figure 2. *True compressive stress - strain curves as a function of strain rate from 1300 K constant velocity testing of Ni-33Al-33Cr-1Mo: (a) As-cast, (b) DS'ed at 25.4 mm/h and (c) DS'ed at 508 mm/h; and true compressive creep curves from constant engineering stress creep testing of three forms of Ni-33Al-33Cr-1Mo: As-cast at 40 MPa, DS'ed at 25.4 mm/h at 70 MPa and DS'ed at 254 mm/h at 75 and 102 MPa.*

constant stress. As indicated by the two pairs of filled and open symbols in Fig. 2(b), testing under nearly identical conditions produced essentially the same stress-strain curves. Overall the constant velocity data (Figs. 2(a-c)) illustrate that 1300 K strength of the alloys decreases as the imposed deformation rate decreases, and at slower strain rates the as-cast material (Fig. 2(a)) is weaker than the DS'ed alloys (Fig. 2(b,c)). The creep curves for all the alloys (Fig. 2(d)) displayed normal behavior with work hardening during primary creep followed by steady state flow. This figure also illustrates the advantage of directional solidification, as like amounts of creep strain were accumulated over ~600 ks in spite of the much higher stresses on the two DS'ed alloys than on the as-cast material.

The 1300 K plastic flow stress - strain rate properties of as- cast and DS'ed Ni-33Al-33Cr-1Mo are presented in Fig.3, where flow strength was taken as the stress at 1 % from the constant velocity test results (Fig. 2(a-c)) and the average stress over the steady state regime from the creep curves (Fig. 2(d)). The results from two as-cast Ni-33Al-33Cr-1Mo bars are presented in Fig. 3(a), and they illustrate that both bars possessed alike strengths. While all the flow stress - strain rate data (Fig. 3(b)) for the seven directionally solidified rods falls within a

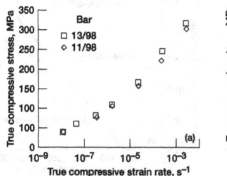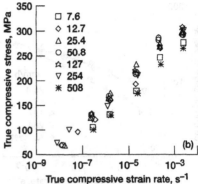

Figure 3. *True 1300 K compressive flow stress - strain rate behavior for Ni-33Al-33Cr-1Mo. (a) two as-cast bars and (b) seven DS'ed rods as a function of growth rate.*

well defined band, visual examination of the results suggests that the properties of slowest (7.6 mm/h) and fastest (508 mm/h) DS'ed rods were inferior to Ni-33Al-33Cr-1Mo DS'ed at intermediate growth rates. This contention was statistically tested utilizing an exponential stress law in combination with a dummy variable, where it was verified that 1300 K deformation characteristics of the rods DS'ed between 25.4 and 254 mm/h were equivalent. This group of 4 rods was statistically superior in strength compared to that of the 7.6 and 508 mm/h materials and marginally stronger than the rod DS'ed at 12.7 mm/h.

The 1300 K behavior of the best DS'ed rods of Ni-33Al-33Cr-1Mo are compared to the as-cast alloy in Fig. 4(a). This figure also illustrates the linear regression fits of the flow stress (σ in MPa) - strain rate ($\dot{\varepsilon}$ in s^{-1}) data for both sets of material. Because of the log-log format of Fig. 4(a), the exponential fit for the DS'ed alloys, $\dot{\varepsilon} = 3.13 \times 10^{-10}\exp(0.051\sigma)$, shows curvature; whereas the power law description of the as cast alloy, $\dot{\varepsilon} = 1.61 \times 10^{-18}\sigma^{6.00}$, is a straight line. Taken together the data show that directional solidification yields a very positive (~2x) strength advantage over simple casting at strain rates < 10^{-5} s^{-1}, but there is no advantage at faster deformation rates.

As the 1300 K properties of DS'ed Ni-33Al-33Cr-1Mo are not dependent on growth rates ranging from 25.4 to 254 mm/h, changes in microstructual parameters, such as a refinement in interlamellar spacing, cell diameters and intercellular regions, are either unimportant or act in a manner to counter balance each other. In comparison to the previous study by Yang, et al. [3], structure could be important since testing of their planar eutectic Ni-33Al-33Cr-1Mo at $\dot{\varepsilon}$ ~ 5 x 10^{-4} s^{-1} resulted in ultimate tensile strengths (UTS) of 420 MPa at 1255 K and 348 MPa at 1366 K. Linear extrapolation of these results to 1300 K suggests a UTS of about 390 MPa which is approximate 100 MPa greater than the best flow stress measured in the current study (Fig. 3(b)). While a potential for improvement might exist, the curves in Fig. 4(b) reveal that the 1300 K slow strain rate compressive properties of the current DS'ed Ni-33Al-33Cr-1Mo are slightly better than those of DS'ed NiAl-34Cr and equivalent to those of DS'ed Ni-33Al-28Cr-6Mo [2].

In summary, we were able to reproducibly melt and cast Ni-33Al-33Cr-1Mo and directionally solidify this NiAl + Cr(Mo) two phase eutectic by the Bridgeman technique at growth rates ranging from 7.6 to 508 mm/h. Compressive testing at 1300 K indicated that alloys

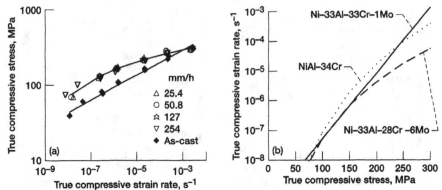

Figure 4. *Comparison of the 1300 K strength properties of the strongest DS'ed Ni-33Al-33Cr-1Mo rods to those for the (a) As-cast alloy and (b) DS'ed NiAl-Cr(Mo) alloys [2].*

DS'ed at rates between 25.4 to 254 mm/h possessed the best strengths; hence it might be possible to produce materials with acceptable elevated temperature strength at a reasonably fast DS rate. Future work will focus on the 1200 and 1400 K behavior to determine if the same dependency on growth rate is maintained; additionally the room temperature toughness of as-cast and DS'ed alloys will be measured to determine if an optimum combination of properties can be achieved.

SUMMARY OF RESULTS

Seven bars of Ni-33Al-33Cr-1Mo have been directionally solidified at rates ranging from 7.6 to 508 mm/h which generally produced alternating plates of NiAl and (Cr,Mo) in grains with sharp boundaries ($\leq$ 12.7 mm/h) or cells surrounded by intercellular regions ($\geq$ 25.4 mm/h). Compressive testing at 1300 K indicated that rods DS'ed at rates ranging from 25.4 to 254 mm/h had the best strength which, in turn, was substantially better than the properties of the as-cast alloy.

REFERENCES

1. J. L. Walter and H. E. Cline, *Metall. Trans.*, 1 (1970) 1221-1229.
2. D.R. Johnson, X.F. Chen, B.F. Oliver, R.D. Noebe and J.D. Whittenberger. *Intermetallics* 3 (1995) 99-113.
3. J. M. Yang, S. M. Jeng, K. Bain and R. A. Amato, *Acta Mater.*, 45 (1997) 295-305.
4. J.D. Whittenberger, S.V. Raj, I.E. Locci and J.A. Salem. *Intermetallics* 7 (1999) 1159-1168.
5. J.D. Whittenberger, R.D. Noebe, D.R. Johnson and B.F. Oliver. *Intermetallics* 5 (1997) 173-184.
6. S. V. Raj, I. E. Locci and J. D. Whittenberger, *Creep Behavior of Advanced Materials for the 21st Century* (edited by R. S. Mishra, A. K. Mukherjee and K. Linga Murty), TMS, Warrendale PA, 1999, pp. 295-310.
7. S.V. Raj and I.E. Locci, "Microstructural Characterization of a Directionally Solidified Ni-33Al-31Cr-3Mo Eutectic Alloy as a fuction of Withdrawal Rate" submitted for publication.
8. H. E. Cline, J. L. Walter, E. Lifshin and R. R. Russell, *Metall. Trans.*, 2 (1971) 189-194.

Mat. Res. Soc. Symp. Proc. Vol. 646 © 2001 Materials Research Society

Fracture Strength of Grain Boundaries in Ni_3Al

Jian-Qing Su, M. Demura and T. Hirano
National Research Institute for Metals, 1-2-1 Sengen, Tsukuba, Ibaraki 305-0047, Japan

ABSTRACT

The fracture strength of grain boundaries in Ni_3Al was quantitatively measured by performing tensile tests on miniature bicrystal specimens with various grain boundary types. A fairly good relationship between fracture mode, fracture strength and Σ value was established. The $\Sigma1$, $\Sigma3$ and $\Sigma9$ boundaries were found to be strongly crack-resistant but the adjacent bulk fractured. In contrast, $\Sigma5$, $\Sigma7$, $\Sigma11$, $\Sigma13$ and random boundaries were found to be less crack-resistant and fractured at lower stresses than the fracture strength of the bulk. In the latter case, the fracture strength of the $\Sigma11$ and $\Sigma13$ boundaries was higher than that of the $\Sigma5$ and $\Sigma7$ boundaries. Most random boundaries were as weak as the $\Sigma5$ and $\Sigma7$ boundaries, but some exhibited high fracture strength comparable to that of the $\Sigma11$ and $\Sigma13$ boundaries.

INTRODUCTION

Previous work suggests that fracture behavior of grain boundaries in Ni_3Al is likely to depend on grain boundary type. For the first time, Hanada et al. found no cracks along $\Sigma1$ (low angle) and $\Sigma3$ boundaries in bent specimens of recrystallized stoichiometric Ni_3Al, indicating that these two boundaries are crack-resistant compared to the other boundaries [1]. Later, Lin and Pope confirmed this tendency in bent specimens of melt-spun Ni-24.8at%Al-0.2at%Ta ribbons [2]. These studies qualitatively indicate a dependence of grain boundary strength on Σ value. Therefore, it is necessary to establish a quantitative relation between the fracture properties of grain boundary and Σ value.

However, it is not easy to measure the fracture strength, primarily because of the difficulty in preparing bicrystal tensile specimens with various grain boundary types. So far there is no systematic measurement of the fracture strength as a function of Σ value. In our previous study, we developed a technique for fabricating large-grained sheets of binary stoichiometric Ni_3Al by skin pass cold-rolling and subsequent recrystallization [3]. This technique enabled us to prepare bicrystal tensile specimens, even though they were small in size. The objective of this study was to establish a relationship between fracture properties and Σ value using these miniature bicrystal specimens.

EXPERIMENTAL

Large-grained sheets of binary stoichiometric Ni_3Al (Ni-24.5 at% Al) whose grain boundaries traversed perpendicularly through the sheet were fabricated by recrystallization [3]. Miniature bicrystal tensile specimens having a gauge section of $0.8 \times 1 \times 2$ mm were cut from the sheets by ultrasonic cutting machining. The grain boundary was arranged normal to the tensile axis. Orientations of the bicrystal specimens were measured by Laue X-ray back reflection. The Σ values of the grain boundaries were calculated according to Brandon's criterion

[4]. If the deviation angle $\Delta\theta_d$ from an exact CSL relation was smaller than the critical value $\Delta\theta_c = 15/\Sigma^{1/2}$, i.e. $\Delta\theta_d /\Delta\theta_c < 1$, the boundary was referred to as a CSL boundary with that Σ value. In this study, the boundaries with Σ values exceeding 25 were designated as random boundaries (RBs). A total of 44 bicrystal specimens were prepared for tensile testing: 13 specimens possessed CSL boundaries and the others RBs.

Tensile tests were performed horizontally in shoulder grips at a strain rate of 4.2×10^{-3} /s on a micro tensile machine with a maximum load of 50 kg. Load was applied perpendicularly to the grain boundary, the normal opening mode for intergranular crack. The testing environment was in air at room temperature. The fracture surface was examined by scanning electron microscopy (SEM).

In our previous Auger study, no impurities were detected on the fractured boundary [3], and thus the effect of impurities on tensile properties can be ignored.

RESULTS

Table I summarizes the grain boundary parameters of all the CSL and four representative RB specimens with the measured tensile properties, e. g. Σ value, grain boundary planes and the deviation from an exact CSL relation ($\Delta\theta_d /\Delta\theta_c$). All the boundaries were of the mixed type having both tilt and twist characteristics. The multiple $\Sigma3$, $\Sigma5$, $\Sigma13$ and RBs have different grain

Table I. Mechanical properties of the tensile samples.

Specimen No.	Σ	Grain boundary Planes	$\Delta\theta_d /\Delta\theta_c$	Fracture stress (MPa)	Elongation (%)	Fracture mode*
1	1	(1 -3.2 0)/(1 9.5 3)	-	549.1	52.0	T F
2	3	(3.2 -7.7 -1)/(3 6.6 -1)	0.09	524.3	73.7	T F
3	3	(4.8 7 -2)/(2 -10 1)	0.24	651.2	66.5	T F
4	3	(3 7 -4.6)/(-7 3 5.3)	0.54	> 545.1	> 48.7	-
5	3	(2.8 2 -2)/(-2 3 3)	0.12	> 601.7	> 11.9	-
6	5	(5 8 2)/(-8 1.2 6.2)	0.75	131.9	9.4	I F
7	5	(4 5 1.6)/(-1 8 1.1)	0.93	71.1	3.6	I F
8	7	(5 5 -4)/(4.8 1 -3)	0.64	132.2	5.8	I F
9	9	(1 3 0)/(-2 3.7 3.1)	0.01	> 624.2	> 16.3	-
10	11	(-4 6.4 5)/(2 8.8 0)	0.98	367.2	16.8	I F
11	13	(0 5 -1)/(0.5 8.6 0)	0.51	370.5	39.1	I F
12	13	(-1 1 1)/(-7 4.4 3)	0.91	360.9	11.5	I F
13	13	(4 3 -4)/(5 2.3 -2)	0.94	149.4	10.1	I F
14	RB	(-4 5.5 3)/(-3.8 4.8 3)	-	282.5	14.9	I F
15	RB	(7 -16 3)/(0 1 0)	-	203.0	13.6	I F
23	RB	(5.7 5.4 -5)/(-5.3 9 3)	-	107.5	6.1	I F
41	RB	(-2 9.2 0.4)/(4.3 9 1)	-	55.0	1.9	I F

*T F: transgranular fracture; I F: intergranular fracture.

boundary planes, and without exception the planes are non-symmetrical planes. In varying degrees, the CSL boundaries deviate from the exact CSL relations.

Figure 1 shows the typical stress-strain curves of some bicrystal specimens. For two Σ3 (Nos. 4 and 5) and Σ9 (No. 9) specimens, the test was terminated before fracture because the limit of the load cell (50kg) was reached. All the specimens deformed plastically after yielding with linear work hardening rate up to fracture. In some samples showing large plastic deformation (e. g. Σ3 samples), a significant crystallographic rotation was found. We also noted that the work hardening rates for a given boundary type are different, indicating orientation dependence of work hardening rate. A more refined investigation on the deformation behavior of bicrystal is at present under way. The tensile properties are summarized in Table I. The fracture strength of all the 44 samples was plotted as a function of the Σ value in Fig. 2. It is found that fracture properties are related to the Σ value. The boundaries are clearly classified into two groups according to fracture mode: Σ1, Σ3 and Σ9 boundaries which do not fracture and the other boundaries which do fracture.

The first group specimens containing Σ1, Σ3 and Σ9 boundaries exhibited high fracture strengths of over 520 MPa levels comparable to that seen in the bulk single crystals. These specimens were ductile and showed a high degree of fracture elongation of 52~74%. The fracture mode was typical of transgranular fracture. For example, Fig. 3a shows the fracture surface of Specimen No. 3 containing a Σ3 boundary. It shows a highly faceted feature, which is characteristic of transgranular fracture in Ni₃Al. Specimens Nos. 4, 5 (containing Σ3 boundary) and No. 9 (containing Σ9 boundary) did not fracture even when the stress increased to the fracture strength of the bulk, and therefore they are expected to fracture transgranularly. From the results of multiple Σ3 boundaries, it is, therefore, considered that whatever the grain boundary plane and the deviation from exact CSL relation are, Σ3 boundary would not fracture, and this

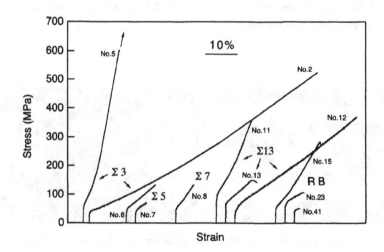

Figure 1. Typical stress-strain curves of some bicrystal specimens.

may hold for $\Sigma 1$ and $\Sigma 9$ boundaries. The strong crack-resistance of the $\Sigma 1$ and $\Sigma 3$ boundaries is consistent with the very low cracking tendency of these boundaries observed in polycrystalline Ni$_3$Al [1, 2]. Interestingly, the high fracture strength of the $\Sigma 9$ specimen was unanticipated.

The second group specimens containing $\Sigma 5$, $\Sigma 7$, $\Sigma 11$, $\Sigma 13$ boundaries and RBs fractured intergranularly. As shown in Fig. 3b, the fracture surface is basically flat. Multiple $\Sigma 13$ boundaries and RBs fractured intergranularly, which indicates that the second group boundaries

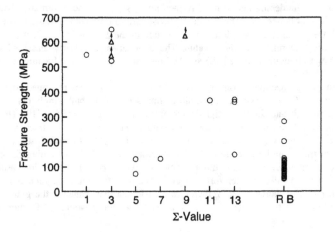

Figure 2. *Plot of the fracture strength as a function of Σ value. Arrows mean that the tensile tests were terminated before fracture.*

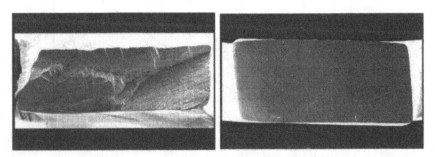

Figure 3. *Fracture surface of (a) specimen No. 3 containing a $\Sigma 3$ boundary and (b) specimen No. 23 containing a random boundary.*

fracture irrespective of the grain boundary plane or the deviation from the exact CSL relation. The fracture strength, which was lower than that of the first group, was not identical within this group but dependent on the Σ value, with some scatter noted. The $\Sigma11$ and $\Sigma13$ specimens fractured at the high stress level of $360 \sim 370$MPa in this group with one exception in the $\Sigma13$ specimens (No. 13). The fracture elongation of $10 \sim 40\%$ was rather large. In comparison, the $\Sigma5$ and $\Sigma7$ specimens fractured at the low stress level of $70 \sim 135$ MPa, which was slightly above the yield stress. Most of the RB specimens fractured at low stress levels similar to those seen in $\Sigma5$ and $\Sigma7$ specimens, while some showed higher fracture strength. For example, the highest fracture stress, 283 MPa (No. 14), was nearly five times larger than that of the weakest RB specimens. The fracture elongation of the RB specimens ranged from zero to 15%.

DISCUSSION

Crack propagation is believed to be the results of low cohesion and difficulty in producing local shear [5], and we think this is equally true for Ni_3Al. It is, therefore, considered that the fracture strength of grain boundaries is closely related to their cohesive strength. Thus the relation between fracture strength and Σ value can be understood in a certain extend in terms of grain boundary energy γ_{gb}, which is inversely related to grain boundary cohesive strength. It is expected that $\Sigma1$, $\Sigma3$ and $\Sigma9$ boundaries would have very low γ_{gb}, the γ_{gb} of $\Sigma5$, $\Sigma7$ boundaries and most of RBs would be much larger, and that of $\Sigma11$ and $\Sigma13$ intermediate between the two. Unfortunately, there have been no measurements of the γ_{gb} in Ni_3Al so far. Only Chen et al. [6] have systematically calculated the γ_{gb} of the symmetric tilt [001] boundaries as a function of misorientation for stoichiometric and off-stoichiometric boundaries using interatomic potential. They showed that the γ_{gb} is at a minimum for $\Sigma1(100)$ and increases in the order of $\Sigma13(320)$, $\Sigma5(210)$, $\Sigma5(310)$, $\Sigma17(530)$, $\Sigma65(740)$, $\Sigma17(410)$, $\Sigma29(520)$ and $\Sigma29(730)$. This order is qualitatively consistent with the measured fracture strength for $\Sigma1$, $\Sigma13$, $\Sigma5$ and RB in this study, i.e. the fracture strength decreases with increasing γ_{gb}, which proves the relation between the Σ value, γ_{gb}, and the fracture strength. Since there are no other CSL boundaries around the [001] axis, it is not possible to discuss the γ_{gb} of the $\Sigma3$, $\Sigma7$, $\Sigma9$ and $\Sigma11$ boundaries. The only reference to these CSL boundaries is the results in Al whose fcc structure is similar to the $L1_2$ structure in Ni_3Al. Hasson and Goux [7] experimentally and theoretically determined the γ_{gb} for the symmetric tilt [001] and [011] boundaries in Al. The results show a relatively low γ_{gb} at the $\Sigma3(111)$ and $\Sigma11(311)$ boundaries with deep cusps and a high γ_{gb} at the other boundaries, which is roughly consistent with the above prediction for γ_{gb} in Ni_3Al. Thus, the Σ value dependence of the fracture strength can be qualitatively understood in terms of the γ_{gb} value.

It is well known that the Σ value does not precisely specify the geometry of the grain boundaries. The macroscopic geometry of grain boundaries is described at least by five independent parameters: three for the relative orientations of the two adjacent grains, i.e. the Σ value in the CSL model [8], and two for the inclination of the grain boundary plane. In addition, actual boundaries deviate from the exact CSL relation. We noticed that the intergranular fracture strength for the same Σ value specimens showed a certain scatter (see Fig. 2). One of the reasons for this scatter can be ascribed to the freedom of the grain boundary plane existing for a given Σ value, as well as the deviation from the exact CSL relation. It is likely that a higher fracture strength in the scatter is linked to a lower γ_{gb} and vice versa. However, it is noted that the fracture mode is not controlled by grain boundary plane and deviation from exact CSL relation, it is exclusively determined by Σ value.

The extrinsic factor - environmental embrittlement - has been demonstrated to be a major cause of low ductility and brittle intergranular fracture of polycrystalline Ni_3Al [9]. It is considered that the present results are associated with the effect of the moisture-induced hydrogen embrittlement. Still it is noted that the fact that $\Sigma 1$, $\Sigma 3$ and $\Sigma 9$ boundaries did not fracture in air indicates that these three boundaries are insensitive to environmental embrittlement.

CONCLUSIONS

By measuring the fracture strength of Ni_3Al bicrystals, the following results were obtained.
(1) The fracture mode closely depends on the Σ value of the grain boundaries. $\Sigma 1$, $\Sigma 3$ and $\Sigma 9$ boundaries do not fracture, whereas $\Sigma 5$, $\Sigma 7$, $\Sigma 11$, $\Sigma 13$ and random boundaries fracture at lower stress than the fracture strength of the bulk.
(2) Among the boundaries which fracture, the fracture strength of the $\Sigma 5$ and $\Sigma 7$ boundaries is very low and slightly above the yield stress, while that of the $\Sigma 11$ and $\Sigma 13$ boundaries is larger than that of $\Sigma 5$ and $\Sigma 7$ boundaries.
(3) Most random boundaries are as weak as the $\Sigma 5$ and $\Sigma 7$ boundaries. However, some show high fracture strength comparable to the $\Sigma 11$ and $\Sigma 13$ boundaries.

ACKNOWLEDGEMENTS

We thank Dr. K. Kishida for his helpful discussions. This research has been carried out under the Japanese Science and Technology Agency (STA) Fellowship Program. J. Q. Su expresses his gratitude for the award of an STA Post-doctoral Fellowship.

REFERENCES

1. S. Hanada, T. Ogura, S. Watanabe, O. Izumi and T. Masumoto, *Acta Metall.*, **34**, 13(1986).
2. H. Lin and D. P. Pope, *Acta Metall.*, **41**, 553(1993).
3. T. Hirano, M. Demura, E. P. George and O. Umezawa, *Scripta Mater.*, **40**, 63(1999).
4. D. G Brandon, *Acta Metall.*, **14**, 1479(1966).
5. M. L. Jokl, V. Vitek, C. J. McMahon Jr and P. Burgers, *Acta metall.*, **37**, 87(1989).
6. S. P. Chen, A. F. Voter and D. J. Srolovitz, *Mat. Res. Soc. Symp. Proc.*, ed. N. S. Stoloff, C. C. Koch, C. T. Liu and O. Izumi, MRS, Pittsburgh, PA, **81**, 45(1987).
7. G Hasson and C. Goux , *Scripta Metall.*, **5**, 889(1971).
8. H. Mykura, in *Grain Boundary Structure and Kinetics,* ed. R. W. Bulluffi, AMS, Metals Park, OH, p.445, 1980.
9. C. T. Liu, *Scripta Metall.*, , **27**, 25(1992).

Mat. Res. Soc. Symp. Proc. Vol. 646 © 2001 Materials Research Society

Nonstoichiometry and Defect Mechanism in Intermetallics with L1₂-Structure

Herbert Ipser, Olga P. Semenova, Regina Krachler, Agnes Schweitzer, Wenxia Yuan[1], Ming Peng[1], and Zhiyu Qiao[1]
Inst. f. Anorganische Chemie, Universität Wien, Währingerstraße 42, A-1090 Wien, Austria
[1] Dept. of Physical Chemistry, Univ. of Science and Technology Beijing, P.R. China 100083

ABSTRACT

A statistical-thermodynamic model was derived which allows to describe thermodynamic activities in intermetallic compounds with L1₂-structure as a function of composition and temperature. The energies of formation of the four types of point defects (anti-structure atoms and vacancies on both sublattices) were used as adjustable parameters. The model was applied to the three compounds Ni_3Al, Ni_3Ga, and Pt_3Ga, and it permitted to estimate for the first time the defect formation energies for Ni_3Ga and to provide initial estimates for Pt_3Ga.

INTRODUCTION

Intermetallic compounds with the cubic L1₂-structure have attracted considerable scientific interest in recent years. One prominent example is the nonstoichiometric compound Ni_3Al which has gained technological importance in the development of so-called superalloys due to some of its unique properties as, for example, high-temperature strength and excellent corrosion resistance [1-3]. Obviously, many of the outstanding properties of Ni_3Al can be related to the type and amount of defects present in thermodynamic equilibrium and to the variation of different defect concentrations with temperature and composition.

One possible way to obtain information about types of point defects which are present in the crystal lattice as well as their concentrations is the combination of statistical thermodynamics and accurate experimental thermodynamic data, as shown previously by two of the authors for B2- and B8-phases [4-6]. Thus it has been the aim of the present research to derive a statistical model for L1₂-phases which would be able to describe the thermodynamic activities as a function of temperature and composition and to apply the corresponding model equations to experimental activity data for the intermetallic compounds Ni_3Al, Ni_3Ga, and Pt_3Ga which crystallize in this structure type.

THEORETICAL MODEL

The cubic L1₂ crystal structure can be divided into two sublattices, the α-sublattice (face centered positions) and the β-sublattice (corner positions); in the ideally ordered crystal, all α-sites are occupied by A-atoms and all β-sites by B-atoms, thus yielding the A_3B stoichiometry. Four types of point defects are allowed in the lattice, both as thermal defects at temperatures $T > 0$ K and as constitutional defects which are responsible for the deviation from stoichiometry: anti-structure atoms and vacancies on both sublattices. The possibility of interstitial defects is neglected.

For the derivation of the statistical model the A_3B crystal is taken as an *open* system which is allowed to exchange both energy and matter with its surrounding. Consequently the grand partition function Ξ and the corresponding grand potential Ω have to be used. In this derivation we follow the ideas of Wagner and Schottky [22] assuming that the addition or removal of one defect of any type causes the same change in internal energy independently of the numbers of defects that are already present. All details have already been outlined elsewhere [7]; it should suffice to point out that the energies of formation of the four possible point defects are taken to be constant (see above), and their values at the stoichiometric composition are used as adjustable parameters in the calculation of the thermodynamic activities. If such defect formation energies are available from the literature, the activity can be calculated as function of temperature and composition in a straightforward way. If they are unknown, experimental activity data can be used to estimate their values by a simple curve fitting procedure. With the obtained values it is possible to compute the concentrations of all four types of point defects, again as functions of temperature and composition.

EXPERIMENTAL DETAILS

Whereas for Ni_3Al both experimental Al activities [8,9] and values of the defect formation energies [10-15] were available from the literature, the corresponding information was rather scarce for Ni_3Ga and Pt_3Ga: only a limited number of Ga activity values had been reported for the two compounds [16,17]. Therefore it was decided to determine Ga activities for these alloys by means of an emf method using stabilized zirconia as a solid electrolyte. The obtained data should be evaluated by means of the statistical-thermodynamic model to estimate the four defect formation energies.

The following cell arrangement was employed:

$$Pt \,|\, M_3Ga, Ga_2O_3 \,|\, ZrO_2 \text{ (stab.)} \,|\, Fe, Fe_xO \,|\, Pt \qquad (1)$$

(where M = Ni, Pt) and the resulting emf values were combined with those of the cell

$$Pt \,|\, Ga, Ga_2O_3 \,|\, ZrO_2 \text{ (stab.)} \,|\, Fe, Fe_xO \,|\, Pt \qquad (2)$$

which had been reported by Katayama et al. [16]. From this the partial Gibbs energies of Ga and the corresponding Ga activities a_{Ga} were obtained by the following relationship

$$\Delta \overline{G}_{Ga} = RT \ln a_{Ga} = -3EF \qquad (3)$$

where E is the combined emf of cells (1) and (2) and F is the Faraday constant; R and T have the usual meaning. From the temperature dependence of E the partial enthalpies and entropies of Ga can be obtained. All experimental details can be found in Ref. [18].

RESULTS AND DISCUSSION

The Compound Ni_3Al

For the intermetallic compound Ni_3Al, the different sets of defect formation energies reported in the literature for $x_{Al} = 0.25$ were used to calculate the thermodynamic activity of Al at 1400 and 1600 K as a function of composition, and the corresponding curves were compared with experimental values by Steiner and Komarek [8] and Hilpert et al. [9]. Figure 1 shows the results for 1400 K: it can be seen that the best agreement between experimental values [8] and calculated activity curve is obtained with the defect formation energies by Debiaggi et al. [10]. A similarly good agreement was obtained at 1600 K with experimental values by Hilpert et al. [9]. Therefore it was concluded that the defect formation energy values by Debiaggi et al. [10] should be the most reliable ones:

$$E_f(Ni^{Al}) = 0.31 \text{ eV}, E_f(Al^{Ni}) = 0.66 \text{ eV}, E_f(V^{Ni}) = 1.48 \text{ eV}, E_f(V^{Al}) = 2.14 \text{ eV}$$

where $E_f(Ni^{Al})$ and $E_f(Al^{Ni})$ are the energies of formation of an Ni atom on an Al site and vice versa, $E_f(V^{Ni})$ and $E_f(V^{Al})$ are the energies of formation of an Ni or Al vacancy, respectively. It can be seen that the energies to form vacancies are considerably higher than those to form anti-structure defects. This means that the main type of defects will be anti-structure atoms whereas the concentrations of vacancies will be smaller by several orders of magnitude (see [7]); any deviation from stoichiometry will, in principle, be caused by anti-structure atoms only.

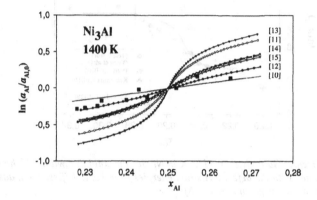

Figure 1. *Comparison of the calculated Al activities (referred to the activity value at the stoichiometric composition, $a_{Al,0}$) in Ni$_3$Al at 1400 K using the defect formation energies from Refs. [10-15] with experimental data (■) by Steiner and Komarek [8].*

The Compound Ni$_3$Ga

The results of our emf-measurements for the Ni$_3$Ga-phase are shown in Figure 2; they are compared with literature data by Katayama et al. [16], Pratt and Bird [19], and Kushida et al. [20]. Relying on our own activity values, a statistical-theoretical curve was fitted through the data points with the defect formation energies listed in Table 1. Using these values, the concentrations of the different types of point defects were calculated as shown in Figure 3. Again it can be seen that the concentrations of anti-structure atoms are much higher than the vacancy concentrations (which are of the order of 10^{-11} to 10^{-7}) and that the derivation from stoichiometry is due to anti-structure defects.

Recently defect formation energies for Ni_3Ga were obtained by *ab-initio* calculations by Wolf et al. [21], their values are listed in the second row of Table 1. It can be seen that the agreement with our values is excellent with the exception of $E_f(V^{Ga})$.

Table I. Defect formation energies in Ni_3Ga; in eV

$E_f(Ni^{Ga})$	$E_f(Ga^{Ni})$	$E_f(V^{Ni})$	$E_f(V^{Ga})$	
0.60	0.60	1.50	2.00	present results
0.57	0.57	1.50	2.71	*ab-initio*, Ref. [21]

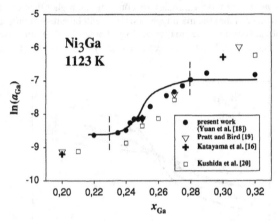

Figure 2. *Experimental Ga activity at 1123 K in Ni_3Ga: the solid line within the homogeneity range was calculated with the defect formation energies from Table 1 (first row); dashed lines denote phase boundaries as given in Ref. [18]*

The Compound Pt_3Ga

Activity measurements were started recently in the Pt_3Ga-phase which also crystallizes in the $L1_2$-structure type. Figure 4 shows the corresponding results at 1100 K together with the data by Katayama et al. [17]. Although the number of data points is still limited a curve was fitted through them to obtain a preliminary estimate of the defect formation energies. The corresponding values are:

$$E_f(Pt^{Ga}) = 1.0 \text{ eV}, E_f(Ga^{Pt}) = 1.2 \text{ eV}, E_f(V^{Pt}) = 1.5 \text{eV}, E_f(V^{Ga}) = 2.0 \text{ eV}$$

It can be seen that the energies necessary to form anti-structure defects are considerably higher than for Ni_3Al and Ni_3Ga but still smaller than the energies of formation for vacancies. As a consequence, Pt_3Ga exhibits less disorder than the other two compounds at comparable temperatures.

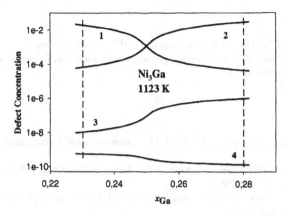

Figure 3. *Defect concentrations (referred to the total number of lattice sites) in Ni_3Ga at 1123 K as a function of composition: 1, anti-structure Ni-atoms (Ni^{Ga}); 2, anti-structure Ga-atoms (Ga^{Ni}); 3, Ni vacancies (V^{Ni}); 4, Ga vacancies (V^{Ga}); dashed lines denote phase boundaries as given in Ref. [18]*

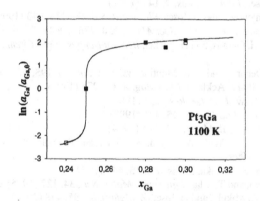

Figure 4. *Experimental Ga activity at 1100 K in Pt_3Ga (■) together with data by Katayama et al. [17] (□); the solid line was calculated with the defect formation energies given in the text*

CONCLUSION

Statistical-thermodynamic modeling has been used in combination with experimental activity data to obtain information on type and concentrations of the point defects present in the three L1$_2$-compounds Ni$_3$Al, Ni$_3$Ga, and Pt$_3$Ga. At the same time it has been possible to estimate the corresponding defect formation energies for Ni$_3$Ga and (tentatively) also for Pt$_3$Ga.

ACKNOWLEDGMENT

Financial support of the Austrian Science Foundation (FWF, project number P12962-CHE) and of the National Natural Science Foundation of China (NSFC, project number 29871005) is gratefully acknowledged.

REFERENCES

1. M.J. Donachie, in: *Superalloys Source Book*, ed. M.J. Donachie (ASM International, 1984), p.3.
2. in: *Superalloys 1988*, ed. D.N. Duhl, G. Maurer, S. Antolovich, L. Lund, and S. Reichmann (The Metallurgical Society, Inc., 1988).
3. in: *Intermetallic Compounds - Principles and Practice*, Vol. 1 and 2, ed. J.M. Westbrook and R.S. Fleischer (John Wiley, 1995).
4. R. Krachler, H. Ipser, and K. L. Komarek, *J. Phys. Chem. Solids,* **50**, 1127 (1989); **51**, 1239 (1990).
5. H. Ipser and R. Krachler, in: *Design Fundamentals of High Temperature Composites, Intermetallics, and Metals-Ceramic Systems*, ed. R.Y. Lin, Y.A. Chang, R.G. Reddy, and C.T. Liu (The Minerals, Metals, and Materials Society, 1996), p. 187.
6. R. Krachler and H. Ipser, *Intermetallics*, **7**, 141 (1999).
7. R. Krachler, O. P. Semenova, and H. Ipser, *Phys. Stat. Sol. (b)*, **216**, 943 (1999).
8. Steiner and K.L. Komarek, *Trans. Met. Soc. AIME*, **230**, 786 (1964).
9. K. Hilpert, M. Miller, H. Gerads, and H. Nickel, *Ber. Bunsenges. Phys. Chem.*, **94**, 40 (1990).
10. S.B. Debiaggi, P.M. Decorte, and A.M. Monti, *Phys. Stat. Sol. (b)*, **195**, 37 (1996).
11. F. Gao, D.J. Bacon, and G.J. Ackland, *Phil. Mag. A*, **67**, 275 (1993).
12. S.M. Foiles and M.S. Daw, *J. Mater. Res.*, **2**, 5 (1987).
13. C.L. Fu and G.S. Painter, *Acta Mater.*, **45**, 481 (1997).
14. J. Sun and D. Lin, *Acta Metall. Mater.*, **42**, 195 (1994).
15. H. Schweiger, E. Moroni, W. Wolf, W. Püschl, W. Pfeiler, and R. Podloucky, *Nat. Res. Soc. Proc.*, **552**, KK5.151 (1999).
16. I. Katayama, S. Igi, and Z. Kozuka, *Trans JIM*, **15**, 447 (1974).
17. I. Katayama, T. Makino, and T. Iida, *High Temp. Mater. Sci.*, **34**, 127 (1995).
18. W. Yuan, O. Diwald, A. Mikula, and H. Ipser, *Z. Metallkde.*, **91**, 448 (2000).
19. J.N. Pratt and J. M. Bird, *J. Phase Equil.*, **14**, 465 (1993).
20. A. Kushida, T. Ikeda, H. Numakura, and M. Koiwa, *J. Japan Inst. Metals*, **64**, 202 (2000).
21. W. Wolf, H. Schweiger, and R. Podloucky, ongoing research, University of Vienna, Austria (2000).
22. C. Wagner and W. Schottky, *Z. Physik. Chem.*, **B11**, 163 (1931).

Titanium Aluminides III

Mat. Res. Soc. Symp. Proc. Vol. 646 © 2001 Materials Research Society

Static and Dynamic Strain Ageing in Two-Phase Gamma Titanium Aluminides

U. Christoph, F. Appel
GKSS Research Centre, Institute for Materials Research, D-21502 Geesthacht, Germany

ABSTRACT

The deformation behaviour of two-phase titanium aluminides was investigated in the intermediate temperature interval 450-750 K where the Portevin-LeChatelier effect occurs. The effect has been studied by static strain ageing experiments. A wide range of alloy compositions was investigated to identify the relevant defect species. Accordingly, dislocation pinning occurs with fast kinetics and is characterized by a relatively small activation energy of 0.7 eV, which is not consistent with a conventional diffusion process. Furthermore, the strain ageing phenomena are most pronounced in Ti-rich alloys. This gives rise to the speculation that antisite defects are involved in the pinning process. The implications of the ageing processes on the deformation behaviour of two-phase titanium aluminide alloys will be discussed.

INTRODUCTION

Gamma based titanium aluminide alloys are promising candidates for high temperature structural applications. Current investigations focus on Ti-rich alloys close to the stoichiometric composition of γ(TiAl) with modest amounts of several ternary elements [1,2] which are mainly composed of the intermetallic phases α_2(Ti$_3$Al) and γ(TiAl). The deformation behaviour of the two-phase alloys is characterized in the temperature interval 450-750 K by discontinuous yielding and a negative strain-rate sensitivity. These attributes have been associated in several studies with the Portevin-LeChatelier effect [3-8]. The effect has been closer investigated by static strain ageing experiments. Accordingly, the diffusion mechanism which underlies ageing is characterized by an activation energy of about $Q_a = 0.7$ eV [5-8]. On the other hand, the self diffusion energy for Ti in γ(TiAl) has been estimated as $Q_{sd} = 3.01$ eV [9] while recent measurements have led to $Q_{sd} = 2.6$ eV [10]. This energy is significantly higher than Q_a and indicates that the defect transfer onto the mobile dislocations cannot be explained by a conventional vacancy exchange mechanism. Unfortunately no values for the migration energies of fast diffusing elements in γ(TiAl) have been definitely established yet. Thus it is difficult to speculate about the transfer of impurity atoms onto the dislocations. Therefore further investigations have been necessary to identify the defect species responsible for ageing. Information on this matter can be obtained through the variation of the defect densities by varying the alloy compositions. Therefore a wide range of alloy compositions has been investigated.

EXPERIMENTAL DETAILS

The compositions (in at. %) and thermomechanical treatments of the alloys are described in detail elsewhere [11]. The alloys are inserted in the TiAl phase diagram (Fig. 1) in accordance with the volume fractions of the phases α_2(Ti$_3$Al) and γ(TiAl) and the temperatures of thermomechanical treatment thereby indicating that single-phase α_2, two-phase $\alpha_2+\gamma$, and single-phase γ alloys were involved in the investigations. Some of the alloys contained ternary and quaternary elements such as Nb, Ga, Cr, Mn, Si, B, and C which are important with regard to technical applications. The concentrations of the impurity elements N and O were typically 200 and 1000 ppm, respectively.

The influence of composition on strain ageing was investigated by static strain ageing experiments as indicated by the stress strain curve in Fig. 2 a. The samples were deformed in

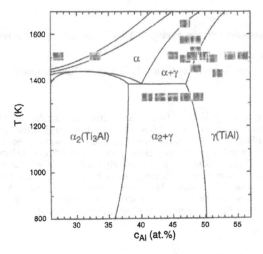

Fig. 1. Phase composition of the alloys. Each alloy is represented in the TiAl phase diagram in accordance with the volume fractions of the intermetallic phases $\alpha_2(Ti_3Al)$ and $\gamma(TiAl)$ and the temperature of the thermomechanical treatment. Phase diagram after [12].

compression. After different strains the machine was stopped so that stress relaxations occured. After different time intervals t_a the samples were reloaded and the resulting stress increments $\Delta\sigma_a$ were measured. As indicated by the double-logarithmic plot in Fig. 2 b, the stress increments saturate, the saturation times t_a being temperature dependent. For the experiments the deformation conditions T = 573 K and t_a = 7200 s were chosen to assure that the diffusion processes are terminated. The experiments were performed at a strain of ε = 1.25 % to avoid any influence of the strain dependency of the stress increments. The saturated stress increment $\Delta\sigma_s$ is considered a measure of dislocation locking during ageing.

RESULTS

Off-stoichiometry and antisite defects

No influence of the elements Nb, Cr, Mn, Si, B, C, N, and O on $\Delta\sigma_s$ has been found [11] thereby indicating that these elements are not involved in the ageing process. However, $\Delta\sigma_s$ significantly depends on the Al-concentration of the alloys as indicated in Fig. 3. The maximum of the effect occurs for Ti-rich alloys which contain the phases α_2 and γ. From this the question arises which of these phases is more important with regard to ageing. In two-phase ($\alpha_2+\gamma$) alloys deformation is mainly confined to $\gamma(TiAl)$. This has been attributed to the strengthening of the α_2-phase by absorbing interstitial impurity elements from the γ-phase [13]. Consequently, ageing should be mainly determined by the γ-phase.

The γ-phase in two-phase alloys is formed during cooling in two stages [14]. At temperatures corresponding to the ($\alpha+\gamma$)-phase field (Fig. 1) the ordered γ-phase with tetragonal $L1_0$-structure precipitates from the disordered α-phase. Below the eutectoid point at about 1400 K the remaining α-phase decomposes into the ordered α_2- and γ-phases. Accordingly the γ-phase is formed non-isothermally. Furthermore the cooling rates of the alloys studied have been typically 10 K/min. These cooling rates are certainly too fast to establish thermal equilibrium in the materials. In accordance with the shape of the borders between the $\gamma/(\alpha+\gamma)$-

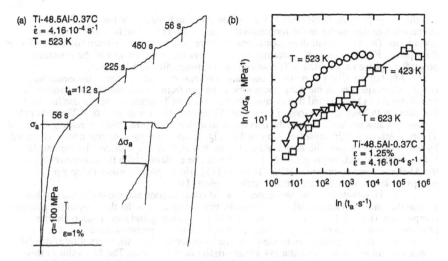

Fig. 2. (a) Sequence of a strain ageing experiment performed under a relaxing stress on a Ti-48.5Al-0.37C alloy. The stress increments $\Delta\sigma_a$ were measured as the difference in stress before ageing and the upper yield point occuring on reloading. Different time intervals t_a between unloading and subsequent reloading are indicated. Evidently, the stress increments are strain-dependent. (b) The double-logarithmic plot indicates that the stress increments $\Delta\sigma_a$ saturate, the saturation times t_a being temperature dependent.

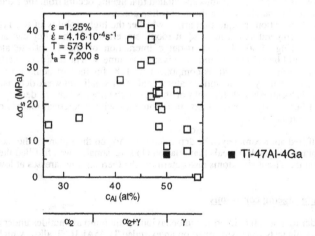

Fig. 3. Dependence of the saturated stress increment $\Delta\sigma_s$ on Al-concentration or the sum of the Al- and Ga-concentration in the case of the Ti-47Al-4Ga alloy, respectively, which is emphasized by a black point. For explanation see text.

and the $\gamma/(\alpha_2+\gamma)$-phase fields (Fig. 1) the γ-phase thus can have a wide range of Ti-concentrations depending on thermomechanical treatment. For illustration a Ti-47Al alloy is considered. The Ti-concentration approximately corresponds to the maximum solubility of Ti in the γ-phase (Fig. 1). After a heat treatment at a temperature in the vicinity of the eutectoid point, the γ-phase may contain a Ti-concentration between 50 and 53 percent.

Most of the alloys were thermomechanically treated in the vicinity of the eutectoid point (Fig.1). Consequently, as the alloy composition changes from stoichiometric composition to Ti-rich alloys it can be expected that on average, the excess of Ti-atoms in the γ-phase increases. For alloys with Ti-concentrations of more than 53 percent the average excess of Ti-atoms in the γ-phase remains unchanged since the maximum solubility is reached. With increasing excess of Ti-atoms $\Delta\sigma_s$ steeply increases, as indicated by Fig. 3, thereby suggesting that ageing is caused by the surplus Ti-atoms. The surplus Al-atoms are less effective, as indicated by the low $\Delta\sigma_s$-values for the Al-rich single-phase γ-alloys. This can be explained with the lower mobility of the Al-atoms when compared with the Ti-atoms [15]. The surplus Ti-atoms in the γ-phase occupy Al-lattice sites thereby forming antisite defects [16].

The Ti-antisite atom density depends on Al-concentration and thermal treatment but also on the content of ternary alloying elements. For example, consider the alloy of the nominal composition Ti-47Al-4Ga. The alloy has, according to scanning and transmission electron microscope investigations, a single-phase γ-microstructure in correspondence with the occupation of Al-sublattice sites by Ga [17]. The Ti-concentration varies very little and is 50 atomic percent in average so that a low antisite defect density results. The $\Delta\sigma_s$-value of this alloy is very low (Fig. 3) thereby supporting the idea that ageing is caused by Ti-antisite atoms.

Dislocation pinning mechanisms by Ti-antisite atoms

It can be expected that the diffusion energy of the Ti-antisite atoms approximately corresponds to the self diffusion energy of Ti-atoms. Because of the comparatively low activation energy of the ageing process therefore long-range diffusion of the antisite atoms to the dislocations is not likely. It is thus speculated that ageing occurs from the exchange of Ti-antisite atoms and vacancies within the dislocation cores.

In this connection the question arises whether the high mobility of the Ti-antisite atoms at intermediate temperatures is reduced at room temperature so that they can act as localized glide obstacles (Fig. 4 a). This short-range interaction of the Ti-antisite atoms with the dislocations should be manifested in the activation volume V. The reciprocal activation volume is proportional to the thermal stress component σ^* [18] for its part being proportional to the square root of the density of the antisite atoms and other small and weak defects, which can be overcome by dislocations with the help of thermal activation. Since the density of the antisite atoms should be linearly related to $\Delta\sigma_s$ in accordance with theoretical models of static strain ageing [19], the relation

$$\Delta\sigma_s \sim 1/V^2 \qquad (1)$$

should be fulfilled approximately. The dependency of $\Delta\sigma_s$ on the square of the reciprocal activation volume (Fig. 4 b) reveals that relation (1) is reasonably well fulfilled thereby indicating that the Ti-antisite atoms also determine the thermal flow stresses at low temperatures.

Ductility under ageing conditions

In order to assess the deformation behaviour of titanium aluminides under ageing conditions, tensile tests were performed on an extruded Ti-45Al-10Nb alloy, which exhibits a strong ageing effect. Fig. 5 shows a stress strain curve of an experiment which was performed at 573 K. At 1 % strain the machine was stopped for 2 h so that static strain ageing occured. The sample did not fail immediately after reloading, but deformation continued up to 1.8 %. The high strength of these alloys, however, has mainly been attributed to their fine microstructure

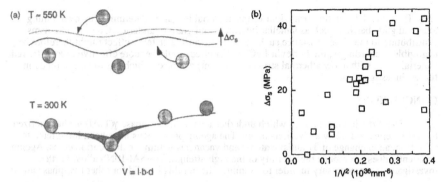

Fig. 4. Dislocation pinning mechanisms by Ti-antisite atoms at low and intermediate temperatures. (a) The Ti-antisite atoms within the dislocation cores are very mobile at intermediate temperatures so that they can form defect atmospheres. On the other hand, at low temperatures the Ti-antisite atoms are immobile and therefore operative as localized glide obstacles. (b) Dependence of the saturated ageing stress increment $\Delta\sigma_s$ on $1/V^2$.

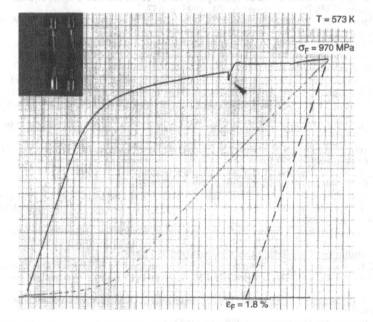

Fig. 5. Ductility of an extruded Ti-45Al-10Nb alloy under ageing conditions. The tensile test sample (the sample is indicated in the insert before and after testing) was deformed up to a plastic strain of 1 %. Then the machine was stopped for 2 h so that stress relaxation occured (arrow). Subsequently the sample was reloaded and deformed to failure. The deformation temperature, fracture stress, and fracture strain are indicated.

[20]. Thus, the deformation behaviour of this material is mainly determined by overcoming athermal glide barriers such as lamellar interfaces and grain boundaries. Other mechanisms contributing to the glide resistance such as the Portevin-LeChatelier effect might therefore be negligible. Accordingly, the Portevin-LeChatelier effect could be more effective in low strength materials (i. e. with a low athermal stress component) a subject which has to be examined in further investigations.

CONCLUSIONS

The diffusion mechanism which underlies ageing in two-phase γ(TiAl) is characterized by fast kinetics and a low activation energy. The ageing phenomena have been interpreted to arise from an exchange of Ti-antisite atoms and vacancies within the dislocation cores. Ageing apparently does not reduce the ductility of the high-strength Ti-45Al-10Nb alloy. Further investigations are necessary in order to examine whether this is valid for other two-phase alloys.

REFERENCES

1. D. M. Dimiduk in *Gamma Titanium Aluminides* , ed. by Y.-W. Kim, R. Wagner and M.Yamaguchi, (TMS Proc., Warrendale, PA, 1995) p. 3.
2. Y.-W. Kim in *Gamma Titanium Aluminides*, ed. by Y.-W. Kim, R. Wagner and M. Yamaguchi, (TMS Proc., Warrendale, PA, 1995) p. 637.
3. F. Appel and R. Wagner in *Gamma Titanium Aluminides*, ed. by Y.-W. Kim, R. Wagner and M. Yamaguchi, (TMS Proc., Warrendale, PA, 1995) p. 231.
4. A. Bartels, C. Koeppe, T. Zhang and H. Mecking in *Gamma Titanium Aluminides*, ed. by Y.-W. Kim, R. Wagner and M. Yamaguchi, (TMS Proc., Warrendale, PA, 1995) p. 655.
5. M. A. Morris, T. Lipe and D. G. Morris, Scripta Materialia 34, 1337 (1996).
6. D. G. Morris, M. M. Dadras and M. A. Morris-Munoz, Intermetallics 7, 589 (1999).
7. U. Christoph, F. Appel and R. Wagner in *High Temperature Ordered Intermetallic Alloys VII*, ed. by C. C. Koch, N. S. Stoloff, C. T. Liu and A. Wanner, (MRS Proc., Pittsburg, PA, 1997) p. 207.
8. U. Christoph, F. Appel and R. Wagner, Materials Science and Engineering A **239-240**, 39 (1997).
9. S. Kroll, H. Mehrer, N. Stolwijk, Ch. Herzig, R. Rosenkranz and G. Frommeyer, Zeitschrift für Metallkunde **83**, 8 (1992).
10. Ch. Herzig, T. Przeorski and Y. Mishin, Intermetallics 7, 389 (1999).
11. U. Christoph and F. Appel, submitted to Acta Materialia.
12. U. R. Kattner, J.-C. Lin and Y. A. Chang, Metallurgical Transactions A 23, 2081 (1992).
13. V. K. Vasudevan, M. A. Stucke, S. A. Court and H. L. Fraser, Philosophical Magazine Letters 59, 299 (1989).
14. C. McCullough, J. J. Valencia, C. G. Levi and R. Mehrabian, Acta Metallurgica 37, 1321 (1989).
15. Y. Mishin and Ch. Herzig, Acta Materialia 48, 589 (2000).
16. M. H. Yoo and C. L. Fu in *Structural Intermetallics*, ed. by R. Darolia, J. J. Lewandowski, C. T. Liu, P. L. Martin, D. B. Miracle and M. V. Nathal (TMS Proc., Warrendale, PA, 1993) p. 283.
17. C. J. Rossouw, C. T. Forwood, M. A. Gibson and P. R. Miller, Philosophical Magazine A 74, 77 (1996).
18. G. Schöck, Physica Status Solidi 8, 499 (1965).
19. J. P. Hirth J. and Lothe in *Theory of Dislocations* (second edition, Krieger Publishing Company, Malabar, 1992).
20. J. D. H. Paul, F. Appel and R. Wagner, Acta Materialia 46, 1075 (1998).

Mat. Res. Soc. Symp. Proc. Vol. 646 © 2001 Materials Research Society

Microstructure and Property Change of Ti-49Al Induced by Hydrogen Charging

E. Abe, K.W. Gao*, M. Nakamura
National Research Institute for Metals, 1-2-1 Sengen, Tsukuba 305-0047, JAPAN
*University of Science & Technology Beijing, Beijing, 100083 P.R.China.

ABSTRACT

We have investigated an effect of hydrogen gas-charging on the microstructure and the mechanical property of a Ti-49at.%Al alloy. After hydrogen-charging performed under an atmospheric pressure of hydrogen gas at 1023K for 3 hours, the alloy with γ-single phase has become completely brittle, while this hydrogen-induced embrittlement is suppressed for that with $(\gamma+\alpha_2)$ two-phase microstructure composed of lath-precipitates in the γ matrix. A significant microstructural change was found to occur for the two-phase alloy (approximately 340ppm hydrogen in the alloy); a thin amorphous layers with a few nm thickness appear at the pre-existing γ/α_2 interfaces in the lath-precipitates after hydrogen-charging. In-situ TEM observation confirmed that the amorphous region transforms to a nano-crystalline state after heating to 1000K at which the hydrogen could be removed (degassed), indicating that the amorphous phase is not a binary Ti-Al phase but a ternary Ti-Al-H one. This, in turn, suggests that the γ/α_2 interface in the lath packets act as the most preferential sites for hydrogen storage. Therefore, the *scavenging* is expected to occur effectively for the microstructure composed of $\gamma-\alpha_2$ fine lamellae in which a large number of γ/α_2 interfaces exist. It is worthwhile mentioning that the fine-scale of the lamellae makes it possible to have a large number of interfaces for a given volume of the α_2 phase.

INTRODUCTION

Intermetallic alloys based on γ-TiAl are now regarded as promising candidates as high-temperature light-weight structural materials. An environmental embrittlement for γ-base alloys is known to be significant; the alloys are embrittled in air (or hydrogen), in spite of the fact that they are ductile in vacuum [1,2]. After a number of studies, it was found that this environmental embrittlement can be suppressed for the γ-base alloys with a dual-phase microstructure, namely, $\gamma-\alpha_2$-Ti$_3$Al two-phase microstructure which is realized for the Ti-rich side below the equi-atomic composition of TiAl. For a Ti-49at.%Al alloy, both the γ-single phase and $(\gamma+\alpha_2)$ two-phase microstructures are obtainable by a simple heat treatment [3], and thus the alloy is a good candidate to study a role of the α_2 phase on the property change.

We report here that a hydrogen-embrittlement can be suppressed for the two-phase Ti-49Al alloy with a fine $\gamma-\alpha_2$ lamellae-lath containing microstructure [4], while the alloy with γ-single phase revealed a significant hydrogen-embrittlement. A significant microstructural change was observed for the two-phase alloy; a thin amorphous layer appears at the pre-existing γ/α_2 interfaces after hydrogen-charging [5]. The formation process of the amorphous phase is discussed in terms of a hydrogen behavior during charging. A likely mechanism of the embrittlement-suppression for the two-phase alloy is proposed based on a *scavenging* effect [6-8] of the γ/α_2 interfaces that could allow hydrogen to be dissolved preferentially.

EXPERIMENTAL PROCEDURE

An alloy ingot with a nominal composition of Ti-49at.%Al was prepared by arc-melting under an argon cover gas. The ingot was hot-isostatically-pressed (HIP'ed) at 1473K under an argon pressure of 120MPa, then homogenized at 1473K for 24h, followed by isothermal forging at this temperature to a thickness reduction of approximately 80%. Tensile specimens with a gauge section of 15 × 4 × 1.5mm were cut from the forged alloy and then polished mechanically. To obtain the two-phase microstructure composed of the γ-α_2 lamellae-lath precipitates, the specimens were wrapped with Ta foils, heat treated at 1573K for 2h, followed by Ar gas-cooling. Hydrogen

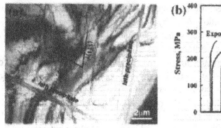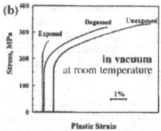

Fig. 1. (a) TEM micrograph of a γ-α_2 two-phase Ti-49at.%Al alloy, showing formation of packets of lath precipitates in the γ grain interior. This is taken from the hydrogen-charged sample. (b) Typical tensile stress-strain curves of the two-phase Ti-49Al alloy with and without hydrogen.

charging was performed under an atmospheric pressure of hydrogen gas at 1023K for 3 hours, followed by furnace-cooling in the flowing hydrogen gas. Hydrogen content measured by a weight difference before and after a thermal desorption spectroscopy (TDS) analysis was approximately 340wppm. Thin foils for transmission electron microscopy (TEM) were prepared by cutting the tensile specimen to about 300mm slices, grinding to the thickness of about 100mm, and standard twin-jet electropolishing using a 10% perchloric acid-ethanol solution at 253K. HRTEM observations were performed on a 400kV electron microscope (Jeol JEM-4000EX).

EXPERIMENTAL RESULTS & DISCUSSION

Figure 1(a) shows a representative microstructure of the two-phase Ti-49Al alloy, in which lath precipitates are formed parallel to the {111} planes of the γ matrix. These laths are composed of the γ-α_2 lamellae with an orientation relationship {111}γ // (0001)α_2 and <110>γ // <1120>α_2. The formation

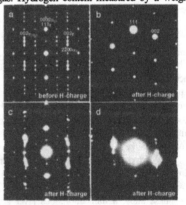

Fig. 2. Electron diffraction patterns obtained from the sample (a) before and (b)-(d) after hydrogen charging. These were taken from the lath precipitates (a), (c), (d) and the γ matrix (b).

process of the lath-lamellae follows that described in ref.[2]. First, during isothermal annealing at

1573K, the α phase precipitates as laths on every {111}γ planes. Subsequently, these α platelets transform into the γ-α₂ two-phase lamellae through the α → α₂+γ transformation during continuous cooling (Ar gas-cooling). As shown in fig.1(b), the alloy with this microstructure revealed a certain amount of elongation even after hydrogen-charging, although it decreases significantly as compared with that before hydrogen-charging. We note that the Ti-49Al alloy with the γ-single phase becomes completely brittle (no elongation) after hydrogen-charging [4]. Therefore it can be said that the two-phase Ti-49Al alloy is a hydrogen-embrittlement-resistant material. By removing the hydrogen in solution, the ductility recovers significantly to a good level, also shown in fig.1(b).

Before and after hydrogen-charging, there was no detectable change in the γ matrix by TEM observation; no hydrides are formed as confirmed by the corresponding electron diffraction (ED) pattern of fig.2(b). However, a significant microstructural change was found to occur for the γ-α₂ laths in the sample after hydrogen charging (hereafter, denoted as H-TiAl alloy). As shown in fig.2(c), the SAED pattern reveals a halo-ring typical of an amorphous state, which does not exist in the pre-charged sample of fig. 2(a). It is also noticeable in fig. 2(c) that the diffraction spots of the g and a₂ phases are broadened, indicating a considerable amount of strain in the phases. The halo pattern was confirmed to appear commonly for the lath-precipitates in the H-TiAl alloy. It seems that the intensity of the halo-ring becomes weaker after tilting the specimen, as shown in fig. 2(d), suggesting an anisotropic nature of the amorphous region.

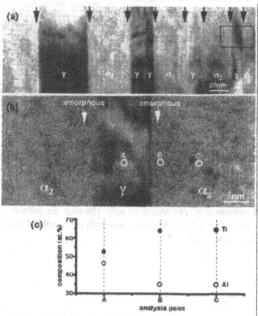

Fig. 3. (a) High-resolution TEM image of the lath packet, showing γ-α₂ fine lamellae nature of the packet. (b) Enlargement of the rectangle region in (a). (c) Microchemical analysis by EDS on the regions marked A, B, C in (b), using an electron probe with diameter less than 1nm. Note that Ti/Al ratios of the amorphous and the α₂ phases are estimated to be approximately the same (Ti-36.0±2.0at.% Al).

Figure 3(a) shows a HRTEM image taken along the orientation in the ED pattern of fig.2(c). From this image, it is seen that the lath-precipitate is composed of the α_2 and γ plates with a thickness of a few tens of nanometers (γ/γ interfaces are characterized in terms of twin and variant relations). We note that the lattice fringe contrasts become unclear around the γ/α_2 interfaces, as indicated by arrows, whereas these are visible in both the γ and α_2 plates. This is seen clearly in fig. 3(b) which is an enlarged view of the rectangle region in fig.3(a). From this image, one notices that the region with fewer fringes around the interfaces corresponds to the amorphous region with a thickness of a few nanometers. Note that this lamellae morphology of the amorphous region explains well the intensity change of the halo-ring in figs.2(c) and (d). As shown in fig.3(c), microchemical analysis by EDS revealed that the Ti/Al ratio of the amorphous phase is approximately the same as that of the α_2 phase, suggesting that the amorphization has occurred selectively at the pre-existing α_2 phase.

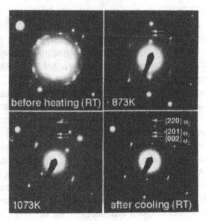

Fig. 4. Electron diffraction patterns during an in-situ heating TEM experiment taken at (a) room temperature (b) 873K (c) 1073K and (d) again room temperature after cooling in TEM.

The present TEM investigation has clearly shown that an amorphous region appears at the γ/α_2 interfaces after hydrogen-charging at atmospheric pressure. It is generally known for the Ti-Al system that the free energy of the amorphous state is much higher than that of crystalline state over a wide composition range [9]. In fact, the amorphous Ti-Al phase can be obtained only by sputter deposition (extremely high cooling rate) [10]. Thus, it naturally follows that the present amorphous phase exists not as a binary Ti-Al phase but as a ternary Ti-Al-H one. That is, hydrogen in solution stabilizes the amorphous phase. In order to check this possibility, we employed an in-situ TEM to investigate a microstructural change during heating; the TDS analysis showed that almost all of hydrogen in solution can be removed (degassed) by heating up to about 1000K. Figure 4(a)-(d) show the change in a ED pattern of the γ-α_2 lath-lamellae during in-situ heating in TEM. It is clearly seen that the single halo-ring becomes double- at 873K (fig.4(b)) and triple-rings at 1073K (fig.4(c)), suggesting that the amorphous phase transforms to a poly-crystalline state. The rings are indexed well by the α_2 phase, as shown in fig.4(d). Figure 5 is a HRTEM image taken with the sample of fig.4(d), confirming a nano-crystalline state at the γ/α_2 interfaces.

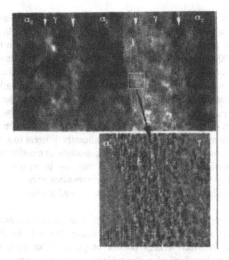

Fig. 5. High-resolution TEM image of the lath packet after the in-situ heating in TEM, confirming that the amorphous layers at the γ/α_2 interfaces has been transformed to nano-crystalline state.

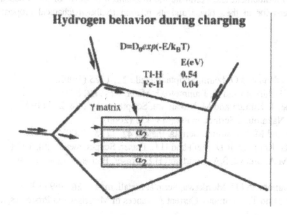

Fig. 6. Schematic drawing of the hydrogen behavior during charging to the two-phase Ti-49Al alloy

These observations suggest strongly that the hydrogen-charging causes amorphization of the pre-existing α_2 phase near the γ/α_2 interface regions, and the hydrogen in solution is seem to be enriched at these amorphous regions. On this basis, it is considered that hydrogen first invades into the bulk sample probably via grain boundaries and then preferentially into the lath packets via the γ-α_2 interfaces (rapid-diffusion route). After that, hydrogen selectively dissolves into the α_2 lamellae in the γ-α_2 lath packets (fig.6). During this process, the α_2 phase near the interface region could transform to the amorphous phase. At this stage, there is no clear answer on why the amorphization occurs around the interfaces, instead of formation of hydrides [11]. However, it is especially noted that the local concentration of hydrogen in the α_2 lamellae could be extremely high; assuming that the α_2 volume fraction in the present sample is approximately 10%, the hydrogen concentration in the α_2 lamellae is estimated be in a order of ~1000wppm (this is derived from the fact that approximately 230wppm hydrogen is detected by TDS for Ti-49at.%Al alloy with γ-single phase [4]). Furthermore, the local concentration could be much higher at the amorphous regions if the hydrogen is remarkably enriched at these regions. This high-concentration situation and the rapid-diffusion (probably faster than in the α_2 phase) of hydrogen around the γ/α_2 interface would cause the local amorphization observed. It is worth pointing out that a bulk-diffusion of hydrogen in the Ti-Al alloys is not so fast as compared that in the Fe-base alloys, because their activation energies would be significantly different (e.g. 0.54eV for Ti-H and 0.04eV for Fe-H), which controls a hydrogen-induced α_2-amorphous transformation. The above interpretation on effect of the interface - rapid-diffusion route - may also be supported by the fact that, in case of a two-phase Ti-Al alloy with α_2 particles at the grain boundaries, only the fcc-type hydride is alternatively formed when a considerable amount of hydrogen is charged at relatively high-pressure gas atmosphere (>10MPa) [11].

Now we know that the near γ/α_2 interface in the lath packets act as the most preferential sites for hydrogen storage. Therefore, the *scavenging* is expected to occur effectively for the microstructure composed of γ-α_2 fine lamellae in which a large number of γ/α_2 interfaces exist. It is worthwhile mentioning that the fine-scale of the lamellae makes it possible to have a large number of interfaces for a given volume of the α_2 phase. These considerations in turn suggest that a hydrogen in solution in the γ phase should cause a embrittlement and dominate the mechanical property of the γ-based materials. Details of hydrogen behavior in the γ phase and its relation to the mechanical properties will be discussed elsewhere.

REFERENCES

1. T. Kawabata, M. Tadano and O. Izumi, Scripta. metall., **22**, 1725 (1988).
2. M. Nakamura, K. Hashimoto and T. Tsujimoto, J. Mater. Res., **68**, 8 (1993).
3. T. Kumagai, E. Abe, T. Kimura and M. Nakamura, Scripta. mater., **34**, 235 (1996).
4. K.W. Gao and M. Nakamura, Scripta. mater., **43**, 135 (2000).
5. E. Abe, K.W. Gao and M. Nakamura, Scripta. mater., **42**, 1113 (2000).
6. M.J. Kaufman, D.G. Konitzer, R.D. Shull and H.L. Fraser, Scripta. metall., **20**, 103 (1986)
7. V.K. Vasudevan, M.A. Stucke, S.A. Court and H.L. Fraser, Philos. Mag. Lett., **59**, 299 (1989).
8. R. Uemori, T. Hanamura and H. Morikawa, Scripta. metall. mater., **26**, 969 (1992).
9. H. Onodera, T. Abe and T. Tsujimoto, Current Advances in Materials and Processes, **6**, 627 (1993).
10. E.Abe, M. Onuma and M. Nakamura, Acta. mater., **47**, 3607 (1999).
11. K. Li, T.M. Pollock, A.W. Thompson and M. De Graef, Scripta. metall. mater., **32**, 1009 (1995).

Mat. Res. Soc. Symp. Proc. Vol. 646 © 2001 Materials Research Society

Effects of a pre-anneal on the mechanical properties of γ-TiAl

Fabienne Grégori[1], Philippe Penhoud and Patrick Veyssière
LEM, CNRS-ONERA, BP 72, 92322 Châtillon cedex, France.
[1]LPMTM, Institut Galilée, 99 av. Clément, 93430 Villetaneuse, France.

ABSTRACT

The temperature, T_P, and the stress, σ_P, at which the yield stress of γ-TiAl alloys peaks are influenced by factors such as alloy composition and load orientation. Available data indicates that in the absence of adequate heat treatments, T_P and σ_P are shifted towards values significantly higher than those of samples pre-annealed in order to precipitate interstitial atoms. Conditions under which precipitation influences the anomalous regime are revealed by dedicated tests. In addition to pointing out the clear effect of interstitials, these tests suggest that further ordering may contribute to the strength of Al-rich γ-TiAl.

INTRODUCTION

The many sets of deformation experiments conducted on single phase γ-TiAl alloys all reveal a more or less pronounced positive yield stress dependence on temperature. Tests carried out on single crystals show in addition that the anomaly is not restricted to the operation of a given slip system but that all three potential slip systems ($1/2<110]\{111\}$, $<011]\{111\}$ and $1/2<112]\{111\}$), behave anomalously and this occurs within a comparable range of temperature [1-7]. It should be kept in mind that determinations of operative slip systems are however not all fully ascertained.

Based on transmission electron microscope analyses of the conditions of dislocation locking there is an active debate as to the intrinsic *vs.* extrinsic origin of the yield stress anomaly. In brief, some authors believe that the anomaly originates entirely from core properties of the various families of dislocations while others argue that, in single phase γ-TiAl alloys, the behaviour of at least some dislocation families is influenced by sub-nanometric, extrinsic particles. Presumably related to interstitial atoms, the nature of these particles is essentially unknown. A clear experimental indication that the stability of γ-TiAl is capable of affecting the plastic behaviour of this alloy was provided by early creep tests [8] that subsequently motivated the design of thermal treatments in order to avoid such problems during deformation at a constant strain rate [9-10].

In the present paper, we review published data on the temperature dependence of the yield stress depending upon whether or not alloys were subjected to stabilising heat treatments. The review is complemented by dedicated tests designed to demonstrate the finite contribution of extrinsic factors to the yield stress anomaly.

REVIEW OF AVAILABLE DATA

In order to limit uncertainties due to the variety of operative slip systems, the present survey is to a large extent focused on experiments conducted on single crystals. It should be kept in mind that, because of difficulties in growing single phase γ-TiAl single crystals near the two-phase region, most of the experiments referred to in the following were conducted on Al-rich single crystals (i.e. 54-56 at.% Al).

Beyond approximately 54 at.% Al, Al atoms in excess substitute for Ti forming a superstructure of Ti_3Al_5 [11] with the nose of the TTT curve located somewhere between 400 and 600°C for 54-55 at.% Al [9]. Additional information on properties of the Ti_3Al_5 phase can be found in [12-16].

On the other hand, interstitial atoms of C, N and O precipitate forming at least two phases known as the H and P phases. The H phase with composition Ti_2AlX (X = C, N and/or O) is hexagonal-based [17-18], while the P phase forms on a perovskite cell with composition $Ti_3AlX_{0.5}$ [19-20]. These phases precipitate at temperatures higher than for the Ti_3Al_5 superstructure and this may depend on alloy composition. In 54 at.% Al γ-TiAl, the H phase segregates around 900-1100°C. Very little is known on the domain of existence of the P phase in Al-rich γ-TiAl which has been however detected in as-grown 54.3 and 55 at.% Al deformed directly at 800-900°C [21]. In nearly stoichiometric TiAl, the P phase is found in alloys solution annealed at 1050°C and then further baked at 800-900°C.

TEMPERATURE DEPENDENCE OF THE YIELD STRESS OF Al-RICH γ-TiAl

In order to reduce possible extrinsic effects of a second phase on the strength of near-54 at.% Al γ-TiAl polycrystals, Hug *et al.* [10] designed the following specific procedure. The alloy is first baked at 1300°C for 48 h and then quenched forming a homogeneous L1$_0$ state. It is subsequently annealed at 1000°C for 144 h to segregate interstitial atoms in the form of large, distant precipitates of H phase. The latter condition of distance is essential for it leaves alloy strength unaffected. As to a possible contribution of the Ti_3Al_5 phase, Hug *et al.* chose not to explore temperatures located between 20°C and 600°C.

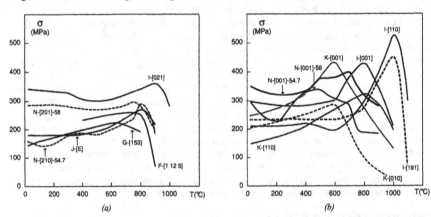

Figure 1. Temperature dependence of the flow stress of Al-rich γ-TiAl single crystals. (a) A comparison between several orientations aimed at activating ordinary dislocations. (b) Rather symmetrical orientations including several independent tests along a near-[001] orientation that inhibits slip of ordinary dislocations. Note that, in each pair of tests conducted in the [110] and in the near-[010] orientations, the peak is shifted over several hundreds °C from one measurement to another. Letters in labels refer to first authors (F: Feng, G: Grégori, I: Inui, J: Jiao, K: Kawabata, N: Nakano).

The variety of mechanical responses of single crystals of γ-TiAl is exemplified in Figure 1 regardless of thermal history. Clearly, the yield stress peak is scattered both in temperature and strength and this holds true even for samples deformed under similar load orientations. The effect is particularly striking in the near-[001] orientation Figure 1(*b*).

To further study the effects of thermal history, data are sorted out in Figure 2 according to whether deformation tests on single crystals had been preceded by a heat treatment between 900 and 1050°C in view of precipitating interstitial atoms. Further information on these experiments is provided in [22]. Figure 2 reveals the good correlation between the flow stress peak position and thermal history, with a few singularities though.

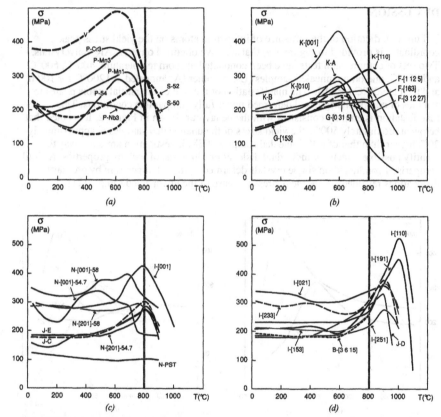

Figure 2. Dependence of the flow stress upon temperature in selected samples of γ-TiAl. (a) & (b) Samples all pre-annealed between 900°C and 1050°C. (a) Selected data on polycrystalline alloys (P: Phan, the notation P-Xᵧ indicates the presence of y at.% of a ternary element X = Cr, Mn or Nb, S: Sriram , V: Viguier). (b) Tests on single crystal samples. (c) & (d) Samples not annealed between 900°C and 1050°C prior to deformation. (c) Samples that exhibit a peak below or at 800°C. (d) All peaks are located beyond 800°C (B: Bird).

As to pre-annealed samples, 6 peaks are located at the borderline, however none beyond that temperature. The two peaks S-50 and S-52 in Figure 2(a) correspond to stoichiometric and near-stoichiometric polycrystals. In Figure 2(b) the three F-plots were obtained under close load orientations that actually differ from the load orientation for K-[110]. Regarding inadequately annealed samples, the borderline-like behaviour for N-[201]-54.7 and N-[201]-58 is unexplained (Figure 2(c)). Three out of the additional five exceptions, N-[001]-54.7, N-[001]-58 and I-[001], may result from the activation of twinning at the temperature of the peak. The tests J-C and J-E were conducted after a pre-anneal at 1100°C that may not have achieved full alloy solution.

DISCUSSION

Further indication of the influence of interstitial atoms on the yield stress peak coordinates is provided in Figure 3(a) that we have obtained on γ-Ti$_{45}$Al$_{55}$ polycrystals. Two sets of compression tests have been conducted at room temperature, 400°C, 600°C and 800°C, one on unannealed samples (U), the other (A) annealed at 1200°C for 100 h and then at 900°C for 11 h. The manifestation of the role of interstitial atoms relates to samples of common origin and deformed under fully comparable conditions. This manifestation is entirely consistent with the behaviour shown in Figure 2. It is noted that below approximately 500°C the yield stress of the unannealed samples is approximately 20% higher than that of fully annealed samples. TEM investigation are underway to identify possible effects on individual dislocations in terms of locking properties. A similar comparison conducted on single crystals deformed at room temperature by ordinary dislocations revealed a 33% yield stress increase of the unannealed sample [21].

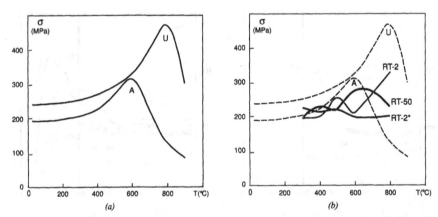

Figure 3. Influence of a pre-anneal on the yield stress of γ-Ti$_{45}$Al$_{55}$ polycrystals. (a) Deformation of annealed (A) and unannealed (U) samples is conducted at RT, 400°C, 600°C and 800°C. (b) Deformation at room temperature of samples annealed 100h at 1200°C then baked 2 h (RT-2) and 50 h (RT-50) at the temperatures indicated in abcissa; in the series lablled RT-2* the first anneal was conducted at 1000°C. The dashed curves, which are those in (a), are indicated for comparison.

Another factor that may effect alloy strength is the duration of the annealing time just before deformation. This was investigated in three series of tests at room temperature:

- series RT-2 samples were homogenized at 1200°C, oil quenched and then annealed at 300, 400, 500, 600, 700 and 800°C for 2 hours (one sample per test),
- RT-50, same as RT-2 save for the anneals that lasted 50 hours,
- RT-2*, same as in RT-2 but for the homogenization temperature lowered to 1000°C.

Up to 500°C, the RT yield stresses compare rather well with yield stresses reported for A. The steep yield stress increase of RT-2 beyond 600°C is reminiscent of the behaviour of the U series indicating that a fraction of interstitials is not segregated after homogenization for 2 h at 1200°C. Beyond 700°C, the decrease suggests that a larger fraction of interstitials is segregated, though not all (compare the peaks of A and RT-50). These results are entirely consistent with the influence of a pre-anneal in the creep response of γ-TiAl [8].

The RT-2* series shows the modest influence of the second annealing temperature save for a hump around 500°C which appears also to take place between 400°C and 500°C in RT-2 and RT-50. The variations of yield stress within these humps are slightly larger than experimental uncertainties, suggesting a true, extrinsic effect presumably of Ti_3Al_5. For the composition studied here, this corresponds mainly to short-range order and is clearly manifested in samples deformed in single slip by a striation contrast [16] which is, however, hard to detect in samples deformed by multiple slip.

It should be mentioned that limited TEM investigations conducted on several of the above-mentioned samples have confirmed the absence of visible precipitates whereas annealed samples contain frequent particles of H-phase. There is also significant indication that the pinning of screw ordinary dislocations is more pronounced after deformation at room temperature in the unannealed sample.

CONCLUSION

The present contribution provides evidence that the yield stress peak can be dramatically affected by transformations that may or may not take place in the alloy depending on thermal treatments prior to deformation. Typically, the peak is raised to larger stresses and shifted by 200-300°C towards higher temperatures when samples are deformed in the as-grown state or else after being pre-annealed beyond approximately 1100°C. What occurs then is that during deformation within the 600-1100°C range, interstitial atoms of C, N and/or precipitate to form a H-phase or a P-phase . In adequately annealed samples, interstitial atoms are fully precipitated in the form of large and distant particles which do not strengthen the alloy. In a temperature range relevant to the flow stress anomaly, the yield stress of unannealed samples is 20% to 33% larger than that of fully annealed samples.

Hence, it may not be entirely safe to rationalize the yield stress of Al-rich γ-TiAl exclusively via the thermally-activated dislocation locking in a single-phase $L1_0$ lattice.

REFERENCES

Due to lack of space, a significant fraction of relevant contributions is not accounted for in the following list, in particular those related to the issue of extrinsic vs. intrinsic dislocation locking. The authors should like to express their apologies to the missing contributors. A more extensive bibliographic coverage is nevertheless provided in [22].

1. T. Kawabata, T. Kanai and O. Izumi, *Acta Metall.*, **33**, 1355 (1985).
2. T. Kawabata and O. Izumi, *High Temperature Aluminides and Intermetallics*, Eds. S. H. Whang, C. T. Liu, D. P. Pope and J. O. Stiegler (The Minerals, Metals & Materials Society, Warrendale, PA, 1990) pp. 403.
3. H. Inui, M. Matsumoro, D.-W. Wu and M. Yamaguchi, *Phil. Mag. A*, **75**, 395 (1997).
4. S. Jiao, N. Bird, P. B. Hirsch and G. Taylor, *Phil. Mag. A*, **78**, 777 (1998).
5. Q. Feng and S. H. Whang, *High Temperature Ordered Intermetallics VIII*, Eds. E. P. George, M. J. Mills and M. Yamaguchi (Materials Research Society, Warrendale, PA, 1999), vol. 552, pp. KK1.10.1.
6. S. Jiao, N. Bird, P. B. Hirsch and G. Taylor, *Phil. Mag. A*, **79**, 609 (1999).
7. S. Jiao, N. Bird, P. B. Hirsch and G. Taylor, *Phil. Mag. A*, **80**, in press (2000).
8. A. Loiseau and A. Lasalmonie, *Mater. Sci. Engng*, **67**, 163 (1984).
9. G. Hug, Ph.D. Thesis, Paris-Sud (Orsay) (1988).
10. G. Hug, A. Loiseau and P. Veyssière, *Phil. Mag. A*, **57**, 499 (1988).
11. R. Miida, S. Hashimoto and D. Watanabe, *Japan. J. Appl. Phys.*, **21**, L59 (1982).
12. T. Nakano, K. Matsumoto, T. Seno, K. Oma and Y. Umakoshi, *Phil. Mag. A*, **74**, 251 (1996).
13. T. Nakano, T. Seno, K. Hayashi and Y. Umakoshi, *THERMEC'97*, Eds. T. Chandra and T. Sakai (The Minerals, Metals & Materials Society, Warrendale, PA, 1997) pp. 1497.
14. H. Inui, K. Chicugo, K. Kishida and M. Yamaguchi, *Electron Microscopy 1998*, Eds. H. A. Calderon Benavides and M. J. Yacaman (Institute of Physics Publishing, Bristol, 1998), vol. II, pp. 175.
15. T. Nakano, K. Hagihara, T. Seno, N. Sumida, M. Yamamoto and Y. Umakoshi, *Phil. Mag. A*, **78**, 385 (1998).
16. F. Grégori and P. Veyssière, *Phil. Mag. A*, **79**, 403 (1999).
17. M. J. Kaufman, D. G. Konitzer, R. D. Shull and H. L. Fraser, *Scripta Metall.*, **20**, 103 (1986).
18. G. Hug and E. Fries, *Gamma Titanium Aluminides*, Eds. Y.-W. Kim, D. M. Dimiduk and M. H. Loretto (Minerals, Metals & Materials Society, Warrendale, 1999) pp. 125.
19. W. H. Tian, T. Sano and M. Nemoto, *Phil. Mag. A*, **68**, 965 (1993).
20. W. H. Tian and M. Nemoto, *Gamma Titanium Almuminides*, Eds. Y.-W. Kim, R. Wagner and M. Yamaguchi (The Minerals, Metals & Materials Society, Warrendale, PA, 1995) pp. 689.
21. F. Grégori, Ph. D. Thesis, University of Paris VI (1999).
22. F. Grégori, P. Penhoud and P. Veyssière, *Phil. Mag. A*, **81**, in press (2001).

Mat. Res. Soc. Symp. Proc. Vol. 646 © 2001 Materials Research Society

Mechanical Properties of γ-TiAl Based Alloys at Elevated Temperatures

M. Weller[1], A. Chatterjee[1], G. Haneczok[2], F. Appel[3] and H. Clemens[3]

[1]Max-Planck-Institut für Metallforschung, Seestraße 92,
D-70174 Stuttgart, Germany

[2]Institute of Physics and Chemistry of Metals,
Silesian University, Katowice, Poland

[3]Institut for Materials Research, GKSS Research Centre, Max-Planck-Strasse, D-21502
Geesthacht, Germany

ABSTRACT

Mechanical loss (internal friction) and creep experiments were carried out on specimens of a Ti-46.5at.%Al-4at.%(Cr,Nb,Ta,B) alloy with differently spaced fully lamellar microstructures. The creep tests were performed in a temperature range of 970 K to 1070 K at 175 MPa. For the mechanical loss measurements a low frequency subresonance torsion apparatus was applied, operating in the frequency range of 0.01 Hz to 10 Hz. The mechanical spectra show two phenomena: (i) A loss peak of Debye-type at 900 K (0.01 Hz) which is controlled by an activation enthalpy of 3.0 eV. The loss peak is related to thermally activated (reversible) motion of dislocation segments which are pinned at the lamellae interface and within gamma lamellae. (ii) A viscoelastic high temperature background above 1000 K with an activation enthalpy of 3.8 eV. This value agrees well with the activation enthalpy of 3.6 eV from creep experiments. Both high temperature background as well as creep are assigned to diffusion controlled climb of dislocations.

INTRODUCTION

γ-TiAl based alloys of technical importance for high temperature applications [1,2] are based on two-phase alloys consisting of γ-TiAl (ordered tetragonal face-centered $L1_0$ structure) and about 15 vol.% α_2-Ti₃Al (ordered hexagonal DO_{19} structure). By appropriate heat treatments fully lamellar microstructures are obtained in which the α_2-phase is arranged within colonies of parallel (γ+α_2) laths with lamellar spacings typically below 1 μm [3,4,5]. At elevated temperatures such fully lamellar specimens exhibit superior creep resistance in combination with high yield strength and an elastic modulus which is retained at elevated temperatures.

Recent experiments on various materials, including intermetallic compounds [6,7] and ceramics [8] showed that mechanical loss (internal friction) experiments give another access for the study of the mechanical behavior at elevated temperatures. In the present paper we present combined mechanical loss and creep experiments.

γ-TiAl material with differently spaced fully lamellar microstructures was studied by combined internal friction and creep experiments to get more insights of its high-temperature mechanical properties.

MATERIAL

The starting sheet material had a nominal composition of Ti-46.5at.%Al-4at.%(Cr,Nb,Ta,B) and was fabricated employing a hot-rolling process [3]. The as-rolled microstructure consists predominately of equiaxed γ-grains with an average grain diameter of 15-20 µm [4]. In order to adjust fully lamellae microstructures with a narrow lamellar interface spacing and a colony size smaller than 150 µm, heat-treatments were conducted within the single α-phase field. Due to different cooling rates the mean interface spacing, including γ/α$_2$ as well as γ/γ interfaces, could be varied between 1.2 µm to 140 nm [5,9].

Fig. 1a shows a light-optical image of a fully lamellar microstructure with a colony size of approximately 130 µm. Fig. 1b exhibits alternate γ- and α$_2$-lamellae within a colony. Obviously, γ/α$_2$ interfaces can act as dislocation sources [9] which emit dislocation loops marked by an arrow. The mean interface spacing (λ_{mean}) was measured by transmission electron microscopy (TEM) applying the line intersection method [5].

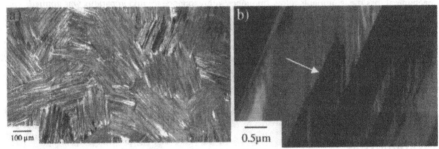

Fig. 1: a) Fully lamellar microstructure of Ti-46.5at.%Al-4at.%(Cr,Nb,Ta,B) sheet material (light optical microscopy).The average colony size is approximately 130 µm. The colonies consist of alternate γ- and α$_2$-laths. b) TEM bright field image taken from a lamellar colony. Note that interfaces can act as dislocation sources. The arrow indicates a dislocation loop.

EXPERIMENTAL

For the internal friction measurements, specimens with dimensions of 50x5x1 mm^3 were prepared from heat treated Ti-46.5at.%Al-4at.%(Cr,Nb,Ta,B) sheet. The mechanical loss measurements were carried out with a subresonance torsion apparatus using forced vibrations in the frequency range of 10^{-3} Hz to 50 Hz [10]. The mechanical loss angle φ was determined from the phase shift between applied stress and strain. The maximum shear strain was about 2x10^{-5}

corresponding to a maximum shear stress of 1 MPa. The internal friction measurements were carried out in 1 mbar He atmosphere in the temperature range from 300 K to 1280 K.

For the creep experiments specimens with an overall length of 50 mm and a gauge area of 30 mm x 3 mm (thickness 1mm) were cut out from the fine grained sheet by spark erosion before the heat-treatment was applied. Creep specimens with differently adjusted fully lamellar microstructures were tested in air in a temperature range of 970 K to 1070 K in tension at a constant load of 525 N, which corresponded to an initial stress of 175 MPa.

RESULTS

Mechanical Loss Experiments

Figure 2 shows mechanical loss measurements obtained on a fully lamellar specimen with a mean interface spacing of 0.29 μm. Depicted is the internal friction $Q^{-1} = \tan \phi$ versus temperature T for various frequencies (f = 0.04, 0.1 and 0.4 Hz). The loss spectra $Q^{-1}(T)$ show two phenomena: (i) a loss peak in the temperature range of 900 K to 1050 K and (ii) a high-temperature background above 1000 K. The inset shows the loss peak at 900 K in a more expanded scala. Figure 3 shows the 950 K peak for various lamellar spacings of 0.14 μm, 0.29 μm and 1.2 μm (f=0.1 Hz). The peak increases with decreasing lamellae spacing and is shifted to lower temperatures.

Figure 2 shows that with increasing measuring frequency both the 950 K peak and the high-temperature background are shifted to higher temperatures. This indicates that both phenomena are due to thermally activated relaxation processes. The relaxation time τ is then determined by an Arrhenius equation :

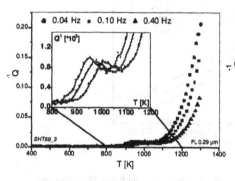

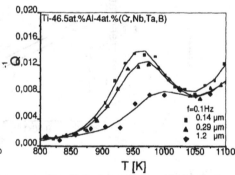

Fig. 2: Internal friction Q^{-1} vs. T for different frequencies. Fully lamellar specimen: $\lambda_{mean} = 0.29\ \mu m$

Fig.3: Internal friction Q^{-1} vs. T for various lamellae spacings (λ_{mean}: 0.14 μm, 0.29 μm and 1.2 μm).

$$\tau^{-1} = \tau_\infty^{-1} \exp\left(\frac{-H}{kT}\right), \qquad\qquad (1)$$

where H is the activation enthalpy, k is the Boltzmann's constant and τ_x^{-1} represents an atomic attempt frequency. The loss peak in figures 2 and 3 is of Debye-type. The activation enthalpy can be determined from the temperature shift of the peak with frequency. Evaluation of the data for the three specimens with different interface spacings has shown that the activation enthalpy for the peak is independent of the microstructure with H = 3.0 ± 0.1 eV. The attempt frequency is in the range of $\tau_x^{-1} = 1?10^{15}\,s^{-1}$. The high-temperature background in Fig. 2 increases at constant T proportional to 1/f which is characteristic for viscoelastic relaxation. Analysis of the data by applying a Maxwell rheological model, for which $Q^{-1} = 1/(2\pi f\tau)$, gives for the differently spaced fully lamellar specimens H = 3.8 ± 0.2 eV (for details of the evaluations see [7,8]).

Creep tests
The results obtained from creep test on Ti-46.5at.%Al-4at.%(Cr,Nb,Ta,B) specimens with fully lamellar microstructure indicate that both the primary creep strain and minimum creep rate decrease with decreasing mean interface spacing. The apparent activation enthalpy was estimated to approximately $H_c = 3.6$ eV from $\dot{\varepsilon}$ versus 1/T plots in a temperature range of 973 K to 1073 K and an applied stress of 175 MPa, as shown in Fig. 4. A stress exponent n of about 5 was determined within a stress range of 100 to 220 MPa at 1070 K, independent of the mean lamellae spacing.

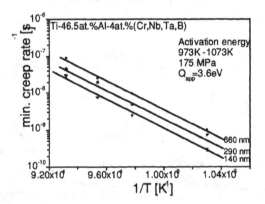

Fig. 4: Arrhenius plot for the determination of the apparent activation enthalpy H_c from creep experiments conducted on fully lamellar Ti-46.5at.%Al-4at.%(Cr,Nb,Ta,B) material with three different interface spacings.

DISCUSSION

The high-temperature properties of the investigated Ti-46.5at%Al-4at%(Cr,Nb,Ta,B) Al sheet material exhibit two phenomena:

(i) The damping spectra show a peak around 900 K which exhibits anelastic behavior. The activation enthalpy of 3.0 eV is significantly lower than that of the (viscoelastic) high-temperature background.The peak has other origin than a high-temperature peak observed by Hirscher et. al.[11] in single phase γ-TiAl polycrystals with high Al contents of 54.1 at.%. The peak is assigned to stress induced reversible, local motion of dislocation segments anchored at their ends either at lamellar interfaces (see Fig. 1b) or at obstacles present within the γ-lamellae [9]. The peak properties can be calculated according to the dislocation theory. For a regular dislocation network with uniform loop lengths L the relaxation strength can be approximated by $\Delta = 2 Q_m^{-1} = 1/20 \rho L^2$ (see e.g. [12], p. 360). By assuming that the length of the bowing dislocation segments is about equal or smaller than that of the mean lamellar interface spacing, i.e. $L \approx \lambda_{mean}$, a lower limit for the density of the dislocations contributing to relaxation can be estimated from the peak heights for various lamellae distances. The following values were obtained: $\lambda_{mean} = 0.14\ \mu m$, $\rho = 3 \cdot 10^9\ cm^{-2}$; $\lambda_{mean} = 0.29\ \mu m$, $\rho = 6 \cdot 10^8\ cm^{-2}$, $\lambda_{mean} = 1.2\ \mu m$, $\rho = 2 \cdot 10^7\ cm^{-2}$.The values and their variations are with lamellae spacings are in good agreement with TEM observations [13].

The activation enthalpy of 3.0 eV indicates that the local motion of the dislocations occurs with lower activation enthalpy than diffusion controlled climb of dislocations.

(ii) The activation enthalpy of the high-temperature background (H=3.8 eV) correlates well with that from creep experiments (H_c = 3.6 eV). Similar results were obtained for γ-TiAl alloys of same composition, but different microstructure (fine-grained near γ) [7] (H = 3.9 eV, H_c = 3.6 eV). The activation enthalpies determined by mechanical spectroscopy and creep tests are in the range of those obtained from self diffusion experiments. The activation enthalpy for Ti self diffusion in γ-TiAl (Q_{Ti}^{*} = 2.6 eV) was determined recently by Herzig et al. [14] using a radiotracer method. Separately, an activation enthalpy for Al self diffusion of Q_{Al}^{*} = 3.71 eV was evaluated by these authors by combining their self diffusion coefficients with interdiffusion coefficients determined from Sprengel et. al. [15].

From this we conclude that the same type of diffusion controlled deformation mechanism is observed by both methods, i.e. mechanical spectroscopy and creep testing. Diffusion assisted creep is expected to be the rate-controlling mechanism at high temperatures and low stresses. Whether volume or grain boundary diffusional creep is predominating can be judged from creep parameters (T, σ and grain size). If grain boundary diffusion is the rate-controlling mechanism, a lower activation enthalpy is anticipated than for volume diffusion. Since this is not observed in the present case, grain boundary diffusion controlled deformation can be ruled out. On the other hand, the observed stress exponent in the range of 5 indicates a creep process which is controlled by dislocation climb (for diffusional creep a stress exponent in the range of 1-2 is expected). This leads to the conclusion that both the high temperature damping background as well as creep are dominated by dislocation climb occurring by (volume) self diffusion.

CONCLUSIONS

Mechanical spectroscopy and creep testing conducted on an engineering two-phase Ti-46.5at%Al-4at%(Cr,Nb,Ta,B) alloy with differently spaced fully lamellar microstructure give comparable activation enthalpies (3.8 eV and 3.6 eV, respectively) which are in the range of those determined for self diffusion. These results indicate that both properties (high-temperature background and creep) are controlled by volume diffusion assisted climb of dislocations.

Additionally, the mechanical spectra exhibit a loss peak around 950 K. This is assigned to glide of dislocation segments which are pinned at γ/α_2 interface and γ/γ interfaces as well as within γ lamellae. The bowing out of the dislocation segments occurs with a lower activation enthalpy (H = 3.0 eV) than that obtained for diffusion-controlled climb of dislocations (3.6 - 3.8 eV). Analysis of the peak heights allows to give an estimate on the densities of locally mobile dislocations, which are in the range of 10^7 to $10^9 cm^{-2}$ and dependent on microstructure.

ACKNOWLEDGEMENTS

We thank Prof. Dr. E. Arzt for helpful discussions and Plansee Aktiengesellschaft (Reutte, Austria) for providing sample material.

REFERENCES

[1]   Y.-W. Kim, *J. Met*, **46**,30 (1994).

[2]   H. Clemens and H. Kestler, *Advanced Engineering Materials*, **9**, 551 (2000).

[3]   H. Clemens, W. Glatz, N. Eberhard, H.P. Martinz, and W. Knabl, *Mat. Res. Soc. Symp. Proc.* 460 (1997).

[4]   H. Clemens and F. Jeglitsch, *Pract. Metallography*, **37**, 194(2000).

[5]   A. Chatterjee, U. Bolay, U. Sattler, and H. Clemens in: *Intermetallics and Superalloys*, Vol 10 (Eds. D.G. Morris, S. Naka, P. Caron), Wiley VCH-Weinheim, 233 (2000).

[6]   M. Weller, M. Hirscher, E. Schweizer, and H. Kronmüller, *J. de Physique*, IV **6**, C8,231 (1996).

[7]   M. Weller, A. Chatterjee, G. Hanczok, and H. Clemens, *J. of Alloys and Compounds* **310**, 134 (2000).

[8]   A. Lakki, R. Herzog, M. Weller, H. Schubert, C. Reetz, O. Görke, M. Kilo, and G. Borchardt, *J. European Ceramic Society* **20**, 285 (2000).

[9]   F. Appel, B.A. Beaven and R. Wagner, *Acta Metall. Mater.*, **41** ,1721 (1993).

[10]   M. Weller: *J. de Physique IV* **5**, C7, 199 (1995).

[11]   M. Hirscher, D.Schaible, H. Kronmüller, *Intermetallics* **7**, 347 (1999).

[12]   A.S. Nowick and B.S. Berry: *Anelastic Relaxation in Crystalline Solids* (Academic Press N.Y., 1972).

[13]   A. Chatterjee, unpublished results, (2000).

[14]   Ch. Herzig, T. Przeorski, and Y. Mishin, *Intermetallics* **7**, 389 (1999).

[15]   W. Sprengel, N. Oikawa, and H. Nakajima, *Intermetallics* **4**, 185 (1996).

Mat. Res. Soc. Symp. Proc. Vol. 646 © 2001 Materials Research Society

Phase Stability and High Temperature Tensile Properties of W doped gamma-TiAl

Keizo Hashimoto[1], Hirohiko Hirata[1] and Youji Mizuhara[2]

[1]Dept. of Materials Science & Eng. Teikyo University,
Utsunomiya, 320-8551, JAPAN
[2]Advanced Technology Research Labs. Nippon Steel Corp.,
Hikari, 743-8510, JAPAN

ABSTRACT

Tungsten (W) doped γ-TiAl is one of promising alloys among many other proposed TiAl base alloys, for the purpose of structural applications at elevated temperatures. Ingots of W doped γ-TiAl were produced by plasma arc melting, followed by homogenizing heat treatment and isothermal forging to control their microstructures. The phase stability of W doped γ-TiAl has been studied quantitatively, using the specimens quenched from 1273 K. Equilibrium compositions of consisting phases were analyzed by means of EDS analysis in a TEM. An isothermal cross section of the Ti-Al-W ternary phase diagram at 1273K has been proposed based on the experimental observations. Small amounts of W addition (< 1at%) to Ti-48at%Al cause a phase shift from $\alpha_2 + \gamma$ to $\alpha_2 + \beta + \gamma$, which suggests that W is the strongest β stabilizer among transition metals, such as Cr and Mo. Mechanical property measurements of W doped γ-TiAl show that the high temperature tensile strength has been improved by the W addition. Relationships between the microstructures and the mechanical properties of W doped γ-TiAl have been discussed.

1. INTRODUCTION

Gamma titanium aluminides (γ-TiAl) have been studied extensively, since they have been considered as a candidate material for advanced jet engines, automobile exhaust valves, turbo-chargers and so on. Alloy design of γ-TiAl has progressed in developing advanced alloy compositions and microstructures. Many researchers have reported that the mechanical properties of γ-TiAl are improved by micro alloying [1-4]. The first generation of γ-TiAl developed by S.C.Huang [5,6] is Ti-48at%Al-2.0at%Cr-2.0at%Nb (48-2-2), which is a castable material having moderate ductility at room temperature. Both Cr and Nb belong to the transition metal group in the periodic table and are also β phase stabilizing elements in Ti. Other reported third alloying elements to γ–TiAl such as V, Mn, Ta, Mo, and W are also β phase stabilizing elements in Ti. On the other hand, it is known that Al stabilizes the α phase of Ti. Consequently, phase stability of α, β and γ phases is one of the most crucial parameter for developing the optimized phases and microstructures of γ-TiAl base alloys. In order to assess the third element additions in γ-TiAl, it is necessary to consider the Ti-Al-X ternary phase diagrams, phase stability and microstructure of γ-TiAl which change depending on the compositions and the process temperatures [7,8].

Recently, various types of W doped γ-TiAl have been developed [9-11]; however, the phase stability of the Ti-Al-W system has not been fully understood. The present report, therefore, aims at presenting experimental data on Ti-Al-W ternary systems. Based on the experimental observations, a Ti-Al-W ternary phase diagram has been proposed and the relationships between microstructures and mechanical properties have been investigated.

2. EXPERIMENTAL PROCEDURE

Tungsten (W) doped γ-TiAl ingots were prepared by the plasma arc melting (PAM) technique. Since W has an extremely high melting temperature, a Ti-40wt.%W ingot was prepared by PAM and used as a master alloy to produce the various compositions of the Ti-Al-W specimens. These specimens were annealed at 1323 K for 3.5×10^5s in a vacuum and then their compositions were analyzed. The chemical compositions of W doped γ-TiAl are listed in Table 1. The compositions of bulk specimens were determined by X-ray fluorescence spectrometry using standards whose compositions were determined by inductively coupled plasma (ICP) spectrometry and chemical analysis. The oxygen contents of these specimens were kept less than 350wtppm. In order to break up the casting microstructure, isothermal forging (ITF) were carried out. Cylindrical specimens were deformed at 1573K with an initial strain rate 5×10^{-4} s^{-1}, until 70% reduction.

Table 1   Compositions of W doped γ-TiAl

Specimen No.	Ti(at%)	Al(at%)	W(at%)	Ti/Al ratio
118	52.4	44.7	2.9	1.17
84	52.3	46.7	1.0	1.12
108	54.9	44.5	0.65	1.27
106	52.0	47.4	0.64	1.09
105	51.7	47.9	0.32	1.07
107	54.9	44.8	0.32	1.23

Blocks of specimens ($7 \times 7 \times 40$mm^3) were encapsulated in a quartz tube, heat-treated at 1273K for 1.08×10^6s, and subsequently quenched in ice water. Microstructures and existing phases of quenched specimens were examined by optical microscope (OM) and X-ray diffractometry (XRD). For EDS analysis, thin foil specimens were prepared by jet polishing at 253K using electrolyte (6 vol% perchloric acid, 34vol% n-butyl alcohol and 60vol% methanol). TEM observations were performed using a JEM-2000FX microscope equipped with EDS analysis system at 200keV.

Tensile test specimens (gauge section; $3 \times 13.5 \times 2$ mm^3) were machined from the isothermally forged sample (ITF) and the heat treated sample (HT) with an electro-discharge machine, then polished with emery papers. Tensile tests were carried out in a vacuum with a strain rate 5×10^{-4} s^{-1} at 1373K.

3. RESULTS AND DISCUSSION

3.1 Ti-Al-W ternary phase diagram

Quenched specimens were examined by OM and XRD. According to XRD analysis, α, β and γ peaks were identified from all of the six specimens although β phase peak intensity is very small in specimens #105 and #107. Microstructure observations also suggested $\alpha+\beta+\gamma$ three phases. In order to determine the existing phases and compositions simultaneously, TEM observations were carried out. Four adjacent grains that are designated A, B, C and D in Ti-48.0at%Al-0.3at%W were studied. Based on electron diffraction analysis, the A grain is identified as α phase, and the B, C and D grains are identified to be γ phase. A, B, C and D grain's composition have been determined by EDS. Table 2 shows the grain No, thickness, Al-Kα value, analyzed compositions and phase type. The same kinds of observations have been conducted for all of the 1273K quenched specimens. Table 3 shows the analyzed compositions of four grains (A,B,C and D) in Ti-46.7at%Al-1.0at%W, the D grain (β phase) is adjacent to the A grain (γ phase). Based on EDS analysis, γ–α and γ–β tie lines can be drawn.

Table 2 Results of EDS quantitative analysis from grain A, B, C and D in Ti-48.0at%Al-0.3at%W

Grain No.	Film Thickness/nm	Al-Kα	Compositions by SQMTF method			Phase Type
			Ti(at%)	Al(at%)	W(at%)	
C	254	1.2836	48.2	51.3	0.51	γ
A	473	1.2836	57.2	42.0	0.87	α
B	362	1.2836	47.8	51.6	0.64	γ
D	254	1.2836	47.3	52.1	0.58	γ

Table 3   Results of EDS quantitative analysis from grain A, B, C and D in Ti-46.7at%Al-1.0at%W

Grain No.	Film Thickness/nm	Al-Kα	Compositions by SQMTF method			Phase Type
			Ti(at%)	Al(at%)	W(at%)	
D	853	1.2836.	45.2	41.0	13.71	β
A	485	1.2836	45.2	53.2	1.40	γ
B	270	1.2836	44.7	54.3	1.02	γ
C	991	1.2836	42.2	56.9	0.94	γ
E	677	1.2836	46.3	52.4	1.33	γ

Figure 1 describes the experimentally determined isothermal cross section of Ti-Al-W ternary phase diagram at 1273K. Since the determined tie lines (solid line) are very close to the phase boundary of the α+β+γ coexisting region (triangle), phase boundaries (dotted line) have been proposed. The most striking features of the proposed Ti-Al-W phase diagram are as follows. i) Small amounts of W addition (less than one at%) to TiAl causes a phase shift from α+γ to α+β+γ three phase co-existing region. ii) The α to γ phase boundary is parallel to the Ti-Al binary side. When Ti-Al-Cr, Ti-Al-Mo and Ti-Al-W phase diagrams are compared [12], the α+β+γ three-phase co-existing region shifts toward the Ti-Al binary side in order of atomic number. Ti-Al-W case is the closest. W is thus shown to be the strongest β stabilizing element in Ti among other transition elements.

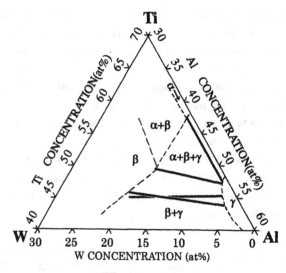

Fig.1   Determined compositional tie-line and proposed Ti-Al-W ternary phase diagram at 1273K

3.2 High temperature tensile strength of W doped TiAl

Phase stability is greatly influenced by the microstructure of W doped γ-TiAl. As a result, the mechanical properties of this alloy are markedly improved by W additions. Figure 2 shows the microstructures of Ti-46.2at%Al-3.0at%W after the two different fabrication processes (a) cast + heat-treated (HT), (b) cast + heat-treated + isothermally forged (ITF). HT specimen shows relatively large grain size (120μm) and fully lamellar microstructure with small amount of β phase (white line). On the other hand, ITF specimen shows a fine grain size (10μm) of γ+α₂ duplex microstructure plus equiaxed β grains (white grains). These morphologies are quite similar with those of Cr and Mo doped γ-TiAl.

The effect of W additions on tensile strength at 1373K is presented in Fig. 3. The tensile strength of HT specimens, which have fully lamellar microstructure, is higher than that of ITF specimens, which have fine $\alpha_2+\beta+\gamma$ duplex grains. According to Fig.3, all sets of data indicate that tensile strength increases with increasing W concentration from 0.3 to 0.6 at%, however, from 1.0 to 3.0at%W additions, the tensile strength is almost constant. The Ti/Al ratio also affects the high temperature tensile strength. Ti/Al ratio of 1.25 (45at% Al) is stronger than a ratio of 1.09 (47.4at%Al). High temperature tensile properties of Ti-Al-W specimens, which were produced by vacuum induction melting (VM) using a CaO crucible, were examined. The tensile strength of Ti-47at%Al-1.0at%W specimens is 332MPa at 1373K as the results of precipitation hardening by very fine Ti-W-C particles. TEM observations revealed that their size was 100nm rectangular shape and density was 40 particles /μm^2 .

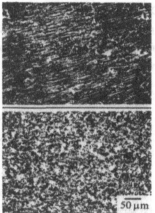

Fig.2 Microstructures of Ti-46.2at%Al-3.0at%W
    specimens
(a) cast and heat-treated at 1323K for 3.5×10^5s (HT)
(b) cast and heat-treated then isothermally forged
    at1573K (ITF)

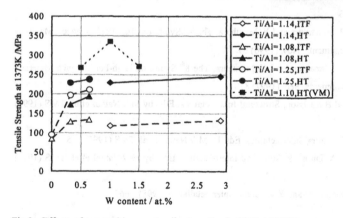

Fig.3   Effects of compositions on tensile strength of γ-TiAl at 1373K

4. CONCLUSIONS

The phase stability and mechanical properties of various compositions of W doped γ-TiAl have been examined:

1. W is the strongest β phase stabilizer among VIa group elements. 10 volume percent of β phase precipitation can be caused by 1at%W addition.

2. Mechanical properties of W doped γ-TiAl are strongly influenced by phase stability. High temperature strength and deformability are significantly improved by W additions.

3. High temperature strength of W doped γ-TiAl mainly depends on their microstructure, Ti/Al ratio and fine particle precipitation. The $\alpha_2+\gamma$ lameller structure shows a higher tensile strength than the fine duplex microstructure.

ACKNOWLEDGEMENTS

A part of this work was performed under the management of R&D Institute of Metals and Composites for Future Industries supported by the New Energy and Industrial Technology Development Organization.

REFERENCES

1. H.A. Lipsitt, D. Shechtman and R.E. Schafrik, Met.Trans., 6A, (1975) 1991

2. M.J. Blackburn and M.P. Smith, US Patent 4,294,615 (1981)

3. T. Kawabata, T. Tamura and O. Izumi, MRS Symp. Proc. 133, (1989) 329

4. Y-W. Kim, JOM 41, No.7, (1989) 24

5. S.C. Huang, US Patent 4,879,092 (1989)

6. C.M. Austin and T.J. Kelly, Structural Intermetallics. Eds. by R. Darolia et al. (Seven Springs) TMS (1993) 143

7. K. Hashimoto, T. Hanamura and Y. Mizuhara, The 7th Symp. on High-Performance Materials for Severe Environments. Nov. (1996) 69

8. K. Hashimoto, T. Hanamura and Y. Mizuhara, The 8th Symp. on High-Performance Materials for Severe Environments. Sep. (1997) 81

9. D.Lundstrom and B.Karlsson, Structural Intermetallics. Eds. by M.V.Nathal et al. TMS (1997) 461

10. V.Lupinc et al, Structural Intermetallics. Eds. by M.V.Nathal et al. TMS (1997) 515

11. Y-W Kim and D.M.Dimiduk, Structural Intermetallics. Eds. by M.V.Nathal et al. TMS (1997) 531

12. K.Hashimoto, M.Kimura and Y.Mizuhara, Intermetallics 6 (1998) 667

Mat. Res. Soc. Symp. Proc. Vol. 646 © 2001 Materials Research Society

Measurement of the Tension/Compression Asymmetry Exhibited by Single Crystalline γ-Ti 55.5at%Al

Marc Zupan and K.J. Hemker
Johns Hopkins Univ., Dept. of Mechanical Engineering, Baltimore, MD 21218, U.S.A

ABSTRACT

Microsample test specimens of single crystalline γ-Ti 55.5at%Al oriented near the [001], and [-110] crystal axes have been deformed in tension and compression at 973K. From these experiments, measurements of the 0.2% offset flow stress have been made as a function of temperature, crystal orientation and sense of the applied load. A measurable violation of Schmid's law was observed and a significant tension/compression asymmetry has been observed at this temperature for the above listed orientations. Single-cycle loading experiments designed to measure the tension/compression asymmetry of the yield strength on the same sample have been conducted along the ~[-110] and [001] orientations. The flow strength measurements from the specimens, which underwent the fully reversed cyclic-loading experiment, fall near those of the monotonically loaded microsamples deformed at the same temperature, which suggests that the same deformation mode was active in both tension and compression.

INTRODUCTION

Titanium aluminides have received considerable attention in recent years and are considered to be strong candidates for use as structural materials for intermediate-temperature applications in gas turbine engines [1,2]. Commercial alloys of TiAl will most likely have a lamellar two-phase structure, but the present study has focused on single phase γ with the anticipation that a fundamental understanding of its mechanical properties will promote the development of the more advanced two-phase alloys. To date, the mechanical testing of single crystalline γ-TiAl has been limited by the fact that sizeable single crystals of γ-TiAl are very hard to obtain. Compression testing has been used to measure the compressive flow strength of γ-TiAl at a variety of crystallographic orientations and temperatures [3-15]. These studies have reported anomalous yielding as a function of temperature, and the activity of superdislocations in the anomalous yielding regime. By comparison, the limited crystal size and lack of room temperature ductility displayed by aluminum rich γ-TiAl has precluded macro scale tensile testing.

The microsample testing machine developed by Sharpe [16,17] has greatly facilitated the room temperature evaluation of Young's modulus, and 0.2% flow strength on specimens that have overall dimensions of 1 mm x 3 mm and typical gage sections 400 μm thick, 300 μm wide and 250 μm long. The development of high temperature microsample testing now allows for microsample mechanical testing at temperatures up to and in excess of 1373K [18]. The work presented in this paper describes the efforts undertaken to conduct microsample testing at elevated temperatures in order to make accurate measurements of the compressive and tensile flow strength ($\sigma_{0.2\%}$) of single crystalline γ-TiAl as a function of temperature and crystallographic orientation in the anomalous yielding regime. Furthermore, fully reversed single-cycle compression/tension and tension/compression tests were carried out to evaluate the possibility of a tension/compression yielding asymmetry in γ-TiAl.

TEST CONDITIONS AND PROCEDURES

Single crystals 6-10 mm in diameter of γ-Ti 55.5%Al have been grown for this study using the optical float zone furnace at the University of Pennsylvania Laboratory for Research on the Structure of Matter. The as grown crystals were heat treated at 1300°C for 24 hours, furnace cooled to 1000°C, held for an additional 100 hours and furnace cooled to room temperature to remove point defects and inhomogeneities [8]. The crystals were oriented using selected area electron diffraction and cut into microsample specimens near the [001], and [-110] crystallographic orientations using a sinking EDM. In each case, the loading axis falls ~4° off of [001], and [-110] crystallographic orientations, with the exact orientation avoided because of the multiple slip that could occur due to the crystal symmetry. The top and bottom faces of the microsamples were mechanically polished to a mirror finish, and a focused ion beam at the University of Virginia, Department of Material Science and Engineering was used to deposit Pt lines 250µm apart on both surfaces. These Pt lines serve as the reflective gage markers for the laser based non-contact interferometric strain/displacement gage that was used to measure strain directly in the microsample gage section [16,19].

Microsamples of single crystal γ-Ti 55.5at%Al have been tested in both tension and compression at a strain rate of 10^{-4} s^{-1} in air. The specimens were deformed to 0.2% nominal plastic strain at 973K and the 0.2% offset flow stress was measured. The test temperature was chosen to be in the anomalous yielding regime so that the active deformation mechanisms could be linked to the anomaly. Tension/compression samples were also used to perform a fully-reversed single cycle test, where the specimen first yielded in compression, then the load was reversed and the sample yielded a second time in tension. For all of these tests, the Young's modulus was measured from the linear elastic loading region, and compared with values calculated from the stiffness matrix to provide a measure of the validity of the microsample stress and strain measurements, and these results are reported elsewhere [20].

EXPERIMENTAL RESULTS AND DISCUSSION
Compression and Tension Flow Stress Measurements

Compression tests on single crystalline γ-Ti 55.5at%Al have been conducted for the [001], and [-110] orientations, and the stress vs. strain response from these experiments are displayed in Fig 1(a) and Fig 2(a), respectively. The monotonically loaded microsample flow strength measurements from this study are reported in Table 1, and show good agreement with the published values of Inui et al. [13] at similar orientations and temperature. The study by

Table 1: Microsample compression and tension flow stress measurements and the maximum resolved shear stresses for superdislocation activity for single crystalline γ-Ti 55.5at%Al oriented near the [001], and [-110] crystallographic directions at 973K.

Loading Axis	Flow Stress (MPa)	Literature Values, Inui et al. [13] (MPa)	Active b_{super} Slip System	CRSS (MPa)
~[001] Comp	-391	-439, -374	(111)<-101]	-168
~[001] Ten	+168,+186,+174	----		+72,+79,+74
~[-110] Comp	-231	-219	(-111)<0-11]	-99
~[-110] Ten	+397,+371	----		+170,+159

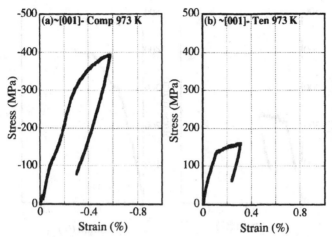

Figure 1: Compression and tension tests results for the near [001] orientation of single crystalline γ-Ti 55.5at%Al at 973K. The 0.2% flow strength for this alloys have been extracted from these curves and are displayed in Table 1.

Inui *et al.* [13] reported the prevalence of superdislocation activity up to the peak temperature in the anomalous yielding regime. In light of these observations, it has been concluded that superdislocations are also responsible for the plastic deformation obtained from the microsample experiments, and the fact that the microsample compression data is comparable to the other published data attests to the validity of this study's experiments.

Figure 1(b) displays the tensile results of single crystalline γ-Ti 55.5at%Al at 973K along the ~[001] crystallographic orientation. The 0.2% flow stress was measured from this curve and is reported in Table 1. Figure 1 reveals that a very pronounced tension compression asymmetry exists for the [001] orientation at this temperature, and inspection of the two curves shows that the 0.2% flow strength in compression is larger than that in tension. The ~[-110] crystal orientation was tested in tension at 973K, Fig 2(b), and the 0.2% flow stress was measured and is reported in Table 1. Similar to the results seen for the near [001] crystal orientation, a marked tension/compression asymmetry is displayed for single crystalline γ-Ti 55.5at%Al for this orientation and temperature. However, unlike the ~[001] orientation, for the ~[-110] crystal orientation the tensile flow strength is higher than the compressive flow strength and the asymmetry has reversed between the two orientations

The 0.2% flow stress for the tests shown in Figs. 1 and 2 have been multiplied by the largest Schmid factor for superdislocation activity for the two loading axes investigated. These maximum resolved shear stresses and active superdislocation slip systems are listed in Table 1. For the orientations tested in this study, the resolved shear stress does not collapse to a single critically resolved shear stress (CRSS) for either tensile or compressive loading. This finding suggests a violation of Schmid's law for γ-Ti 55.5at%Al at 973K.

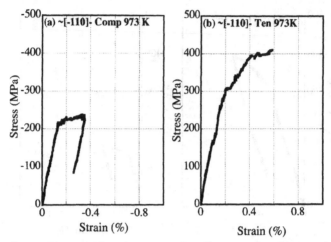

Figure 2: Microsample compressive and tensile stress vs. strain curves for near [-110] oriented single crystal γ-Ti 55.5at%Al. From the curves the 0.2% flow strength was measured to be -231 MPa in compression and 397 MPa in tension.

Single-Cycle Fully Reversed Tests

Experiments designed to measure the tension/compression asymmetry of the yield strength on the same sample have been conducted along the ~[-110] and [001] orientations to confirm the tension/compression asymmetry measured on the monotonically loaded specimens. The single-cycle loading experiments were conducted in two parts, with a specimen first loaded in compression to approximately 0.2% plastic strain, unloaded, and the same specimen then reloaded until it yielded a second time in tension.

Figure 3 shows the stress-strain curve for microsample specimen tested in a cyclic loading experiment at 973K along the ~[-110] loading axis. The microsample specimen was compressed and yielded at -234 MPa and then was reloaded in tension where yielding occurred at 388 MPa. The difference in the two yield values is indicative of the tension/compression asymmetry exhibited in single crystalline γ-Ti 55.5at%Al. The flow stress measurements from the single-cycle fully reversed experiment along the near [-110] loading axis are comparable to those reported in Table 1 for the monotonically loaded microsamples. A second cyclic loading experiment, not shown here, was conducted at 973K along the ~[001] loading axis. The microsample specimen was compressed and yielded at -392 MPa, the sense of applied load was then reversed and the specimen subsequently yielded in tension at 219 MPa.

The flow stress measurements from the single-cycle loading experiments show that apart from the usual work hardening from the previous cycle and experimental error, the tensile and compressive flow strength measurements are the same between the two tests. The fact that the

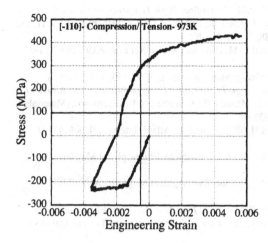

flow stress measurements from the specimens that underwent the fully reversed cyclic-loading experiments fall near those of the monotonically loaded samples deformed at the same temperature is clear evidence of the tension/compression asymmetry. Since anomalous yielding is exhibited for both tension and compression the results of the single-cycle experiments suggest that the same deformation mode was active in both tension and compression, and that twinning, which would be active for only one sense of applied load, is not controlling deformation.

Figure 3: Stress-strain curve for the cyclic loading of a single crystal along the near [-110] orientation. In the first loading excursion the microsample yielded at 234 MPa in compression. Subsequently at the same temperature the specimen was reloaded in tension and yielded at and 388 MPa.

SUMMARY AND CONCLUSION

The use of microsamples has provided a viable means to measure the high temperature deformation response of single crystalline γ-TiAl in both tension and compression. Values of the 0.2% flow stress have been measured as a function of temperature and sense of applied load on monotonically loaded microsamples. Single crystal microsample compressive flow stress measurements correlate well with the literature values along similar orientations and temperatures, see Table 1. A measurable tension/compression asymmetry is exhibited by single crystal γ-TiAl in the anomalous yielding regime for the near [001], and [-110] orientations. Single-cycle reversed tests have been used to verify the directions of the tension/compression asymmetry which resulted in flow stress measurement which are comparable to monotonically loaded specimen at the same temperatures.

REFERENCES

1.   Y.-W. Kim and D.M. Dimiduk, JOM **8**, 40-7 (1991).
2.   D.M. Dimiduk, D.B. Miracle and C.H. Ward, Materials Science and Technology **8**, 367-75 (1992).
3.   T. Kawabata, T. Kanai and O. Izumi, Acta Metall. **33**, 1355-66 (1985).
4.   T. Kawabata, T. Abumiya, T. Kanai and O. Izumi, Acta metall. mater. **38**, 1181-9 (1990).
5.   T. Kawabata, T. Kanai and O. Izumi, Philosophical Magazine A **63**, 1291-8 (1991).
6.   T. Kawabata, T. Kanai and O. Izumi, Philosophical Magazine A **70**, 43-51 (1994).
7.   Z.X. Li and S.H. Whang, Mater. Sci. Eng. **152**, 182-8 (1992).

8. M.A. Stucke, D.M. Dimiduk and P.M. Hazzledine, *High-Temperature Ordered Intermetallic Alloys V*, edited by. I. Baker, R. Darolia, J.D. Whittenberger and M.H. Yoo (Materials Research Society, **288,** 1993) 471-6.

9. M.A. Stucke, V.K. Vasudevan and D.M. Dimiduk, Mater. Sci Eng. **A192/193,** 111-9 (1995).

10. Z.-M. Wang, Z.X. Li and S.H. Whang, Mater. Sci. Eng. **A192/193,** 211-6 (1995).

11. N. Bird, G. Taylor and Y.Q. Sun, *High-Temperature Ordered Intermetallic Alloys VI,* edited by. J.A. Horton, I. Baker, S. Hanada, R.D. Noebe and D.S. Schwartz (Materials Research Society, **364,** 1995) 635-40.

12. Z.-M. Wang, C. Wei, Q. Feng, S.H. Whang and L.F. Allard, Intermetallics **6,** 131-9 (1998).

13. H. Inui, M. Matsumuro, D.-H. Wu and M. Yamaguchi, Philosophical Magazine A **75,** 395-423 (1997).

14. S. Jiao, N. Bird, P.B. Hirsch and G. Taylor, Philosophical Magazine A **78,** 777-802 (1998).

15. F. Gregori, *Plasticite de l'alliage Gamma-TiAl Role Des Dislocations Ordinaries et Superdislocations Dans L'anomalie de Limite D'Elasticite,* Doctoral thesis, Universite Paris 6, Paris. 186 (1999).

16. W.N. Sharpe Jr., *An Interferometric Strain/Displacement Measurement System,* Mechanics and Materials Branch NASA Langley Research Center, Rep. 101638 (1989).

17. W.N. Sharpe Jr. and R.O. Fowler, *Small Specimen Test Techniques Applied to Nuclear Reactor Vessel Thermal Annealing and Plant Life Extension,* edited by. W.R. Corwin, F.M. Haggag and W.L. Server (American Society for Testing and Materials, Philadelphia, Pa., 1993) 386-401.

18. M. Zupan, M.J. Hayden, C.J. Boehlert and K.J. Hemker, Experimental Mechanics Submitted, 14 pages (2000).

19. W.N. Sharpe Jr., Experimental Mechanics **8,** 164-70 (1968).

20. M. Zupan, *Microsample Characterization of the Tensile and Compressive Mechanical Properties of Single Crystalline Gamma-TiAl,* Doctoral thesis, The Johns Hopkins University, Baltimore, MD, USA(2000).

Mat. Res. Soc. Symp. Proc. Vol. 646 © 2001 Materials Research Society

On the Stress Anomaly of γ-TiAl

Marc C. Fivel[1], Francois Louchet[2], Bernard Viguier[3] and Marc Verdier[2]

[1] GPM2 / CNRS / INPG , B.P. 46, 38402 - St Martin d'Hères, FRANCE
[2] LTPCM / INPG / UJF, B.P. 75, 38402 - St Martin d'Hères, FRANCE
[3] LIMAT, ENSCT, INP de Toulouse, 118 rte de Narbonne, 31077 Toulouse cedex 4, FRANCE

ABSTRACT

A 3D mesoscopic simulation of dislocation behaviour is adapted to the case of γ-TiAl. It shows that, in the temperature range of the stress anomaly, ordinary dislocation motion essentially proceeds through a series of pinning and unzipping processes on screw dislocations. Pinning points are cross-slip generated jogs, whose density increases with temperature. Unzipping of cusps restores the screw character of the dislocation. The balance between pinning and unzipping becomes increasingly difficult as temperature raises, and results in dislocation exhaustion. All these features agree with the so-called "Local Pinning Unzipping" mechanism.

INTRODUCTION

Generally ascribed to ordinary dislocations [1-6], the precise origin of the anomaly in γ-TiAl has already been thoroughly debated [2-6]. Though some models are based on a variation with temperature of the density or efficiency of extrinsic obstacles (solutes, precipitates [5,6]), there are strong microscopic evidences that intrinsic obstacles related to lattice friction and (or) generated by the dislocation motion mechanism itself [1-4] play a prominent role in the anomalous mechanical behaviour. At the microscopic scale, ordinary dislocations exhibit a series of pinning points analysed as jogs amazingly aligned along the screw direction, whose density increases with temperature in the anomaly domain [2-4], and on which cusps are readily formed. The determination of the subsequent unpinning mechanism is of fundamental importance in the interpretation of the stress anomaly. Screw dislocations can indeed keep moving through either sideways unzipping of cusps (Local Pinning Unzipping or LPU), as initially proposed in the Louchet-Viguier (LV) model [2,3], but also by jog dragging (JD) or dipole dragging (DD) as mentioned by Sriram et al. [4]. Rows of prismatic loops also aligned parallel to the screw direction are often observed [7], that might be remains of the unzipping process, but the presence of dipoles is occasionally reported. A recent computation of the anomalous characters of these different mechanisms [8] showed that DD may have an anomalous character, but for stresses significantly larger than those observed experimentally; LPU also exhibits an anomalous character, for stresses in fair agreement with experiment. However, JD has a normal decreasing stress / temperature dependence, that reaches reasonable stress values at high temperatures. It can be concluded from this last point that JD should be ruled out as far as the stress anomaly is concerned, but that it might account for the high temperature alternative mechanism operating above the stress peak.

The present work aims at exploring the specific dislocation mechanisms expected in the $L1_0$ structure through an adaptation of the 3d mesoscopic dislocation simulation initially developed for FCC metals by Kubin et al.[9], at checking whether one or several of the above mentioned jog mechanisms are observed, and which one may be responsible for the stress anomaly.

PRINCIPLES OF THE SIMULATION

The general principles of the simulation can be found in [10] and can be summarised as follows. Dislocations are represented as interconnected small straight segments in an elastic continuum. Both time and space are discretised. Local stresses resulting from applied loading and internal stresses are computed on each of those segments. The driving force is obtained from the resulting Peach-Köhler force corrected by solid friction and line tensions. Dislocation motion is determined from the combination of this driving force with a viscous drag and (or) an Arrhenius velocity law. The internal stress field is computed at each step.

In the $L1_0$ version set up here, the space discretisation length is taken of the order of the Burgers vector **b**. Starting with a straight screw dislocation, kink pair (kp) discretisation lengths are determined by dividing in two parts any segment larger than a critical length close to the kink pair critical nucleation length $L_c = V^*/b^2$, where V^* is the corresponding activation volume. Stresses on primary and cross slip planes are computed on each segment, from which corresponding kp thermally activated nucleation probabilities are derived. They are compared to a random number between 0 and 1, which decides if kp nucleation is allowed, and into which plane. All neighbour kink pairs nucleated in the same plane are merged. Screw parts are allowed to move a distance $v_s \delta t$, where δt is the time step, and where the screw velocity is given by:

$$v_s = v_D \frac{b^2}{L_c} \frac{L}{L_c} \exp(-\Delta G_0 / kT) \sinh(\tau^* V^* / kT) \qquad (1)$$

where v_D is the Debye frequency, b the Burgers vector magnitude, L the dislocation length, ΔG_0 the activation energy for double kink nucleation, τ^* the effective resolved shear stress, k the Boltzmann constant and T the absolute temperature.

The L/L_c factor accounts for the enhancement of the dislocation velocity due to the competition between kink pairs nucleated in parallel, which otherwise should not be directly taken into account by the simulation.

Edge parts are also allowed to move, but on a distance $v_e \delta t$, where the edge velocity is controlled by viscous drag:

$$v_e = \tau^* b / B \qquad (2)$$

τ^* being the effective stress and B the phonon drag coefficient.

Three dimensional line tensions are then computed and dislocation shapes corrected accordingly before the next computation step. The numerical values used in the simulation are as follows:

Shear modulus μ = 65.2 GPa, Poisson's ratio ν = 0.236, Burgers vector modulus b = 0.28 nm, Phonon drag coefficient B = 1.5 10^{-5} Pa.s, Peierls stress τ_p = 35 10^{-5} μ = 22.8 MPa, activation volume V*=60 b^3, activation energy ΔG_0=0.45eV, Debye frequency ν_D=10^{13} s^{-1}, temperatures: 570, 900 and 1100 K, time step δt = 3 10^{-11} s.

Temperatures explored in this work are only indicative, and might be rescaled through an adjustment of activation energy ΔG_0. Owing to the stress anomaly, no experimental measurement of the activation energy is indeed available. The value used here for ΔG_0 is taken is the range explored in our previous analytical calculations [8], corrected by the work $\tau^* V^*$ of the resolved shear stress.

RESULTS

Two stress orientations have been tested so far: i) single slip conditions, and ii) non symmetrical double slip conditions, each of them at different temperatures. The general trends are that in all cases pinning points are spontaneously generated on screws, and that their density increases with temperature. Depending on loading conditions, more specific features are observed, as detailed hereafter.

<u>Single slip conditions</u> (pure shear on the primary plane):

At low temperatures (570K) a few pinning points are generated, leading to cusp formation. Some dipoles may be generated by screw propagation on both sides of such pinning points (figure 1a). Both cusps and dipoles are easily released by unzipping of edge segments, leaving prismatic loops. This is the only case where edge dipole formation could be observed.

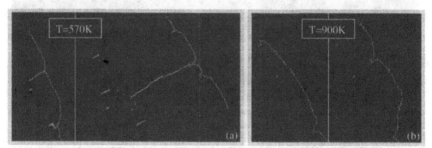

Figure 1. Single slip conditions: (a) dipole dragging is observed only at the lowest temperature (570K); (b)at high temperatures (900K), screw dislocations roughen and cusp density increases.

At higher temperatures (900K), dipoles are no more observed. Unzipping still occurs, and restores straight dislocations along the screw direction. Formation of prismatic loops aligned along the screw direction in the wake of the zipping jog is very often observed (figures 1b and

2a), giving features quite comparable to TEM observations (figure 2b). By contrast with the low temperature case, dislocations in figure 1b are now significantly cusped.

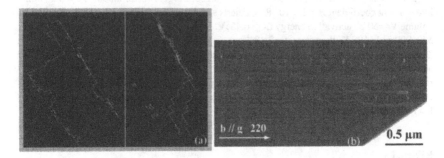

Figure 2. Single slip conditions, T=570K: (a) the simulation started with two opposite screw dislocations. Cusp unzipping is observed, and results in the formation of a row of prismatic loops. (b) an example of prismatic loops observed by TEM in the anomaly temperature range (T=773K) [11].

Non symmetrical double slip (10^0 off symmetrical double slip conditions):
At low temperatures (570K), unzipping still takes place (see figure 3a), and prismatic loops are still observed to be formed, in spite of much stronger pinning which slows down screw dislocation motion. Nonetheless, screw dislocations resulting from unzipping have significantly rougher shapes than in single slip conditions.

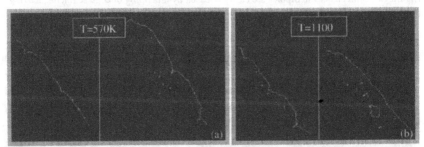

Figure 3. "Double slip" conditions:(a) at the lowest temperature (570K), the dislocation is slowed down by jogs. (b) at elevated temperatures (1100K) it may eventually stop after some sluggish motion (exhaustion). A few unlocking events are observed (last picture).

At high temperatures (1100K), the pinning rate is so large that dislocations are readily exhausted (figure 3b left). A few unlocking events of exhausted dislocations are nevertheless (though scarcely) observed (figure 3b right). The cusped character of dislocations already observed at the same temperature in single slip conditions is even more enhanced here.

DISCUSSION

Owing to the fact that the simulation does not contain any provision for diffusion, the jog dragging mechanism is not expected to be evidenced by this tool. Nevertheless, the analytical calculation mentioned above [8] shows that the required stresses are by far too large in the anomaly temperature range; this mechanism, that was shown to have a normal stress temperature dependence, is thus likely to be the alternative mechanism operating at high temperatures, i.e. above the stress peak. The present simulation is thence only focussed at the moment on the comparison of the relative efficiencies of the dipole dragging and LPU mechanisms.

Though a few dipole dragging events were observed, they appear to occur under very specific conditions (single slip and low temperatures). The reason is that these conditions decrease the cross slip frequency into the cross slip plane, and the two screw segments that drag the edge dipole can hardly cross slip back or again to form prismatic loops or 3d configurations. Dipole dragging being restricted to these specific conditions, it can hardly be invoked to account for the stress anomaly.

By contrast, the simulated microstructures are in very good agreement with TEM observations of TiAl after deformation [7] in the anomaly temperature range: the simulation confirms that cusped dislocations in screw orientation, and alignments of small prismatic loops observed in TEM are the signature of the Local Pinning Unzipping mechanism proposed several years ago in the LV model [2,3]. The dynamic observations are also fully consistent with the LPU mechanism.

At high temperatures, the screw dislocation exhaustion predicted by the LV model is also observed. Such a thermally activated exhaustion was a key point in this model, in that, through a balance with dislocation multiplication, it automatically results in both the stress anomaly and a strain rate sensitivity close to zero. It therefore appears that this last feature, which is systematically observed in association with stress anomalies in many structures, can be accounted for without necessarily invoking dynamic strain ageing. However, further segregation of solute atoms into dislocation cores previously immobilised through such an exhaustion process may be contemplated, in agreement with recent observations of extrinsic pinning [12,13].

The simulation also reveals a new feature that was not included in the LV model: dislocations are not only exhausted, but their average velocity is also reduced at constant stress as temperature increases, which is expected to further increase the amplitude of the anomaly.

The observation of some scarce unlocking events at the highest temperatures may have some consequences on flow stability in the vicinity of the stress peak, as previously shown in Ni_3Al [14].

CONCLUSIONS AND PERSPECTIVES

The present paper shows the very first results of a 3-d dislocation simulation for $L1_0$ crystals. Using simple rules for kink pair nucleation and propagation on screw dislocations, the simulation reproduces the main features of the LPU mechanism (pinning on cross-slip generated jogs, cusp unzipping, formation of aligned prismatic loops, ...), which thus appears to be the relevant mechanism at intermediate temperatures, i.e. as long as jog dragging is not allowed to operate. In particular, extrinsic obstacles do not seem to be necessary to account for both the observed microstructures nor the anomaly. Moreover, instead of being only responsible for a slowing down of individual dislocation velocities as in the other proposed models, pinning on jogs is shown to lead to an increasing exhaustion as temperature raises. This agrees with the LV model in which the balance between exhaustion and multiplication is also responsible for the low SRS value associated with the anomaly.

Further work is in progress, in particular through an increase of the crystal size in order to allow several unzipping steps for the same dislocation. The simulation will also be run on starting configurations made of several interacting dislocations in order to check a possible collective behaviour as experimentally observed in [1]. Finally, a possibility for climb should be introduced in order to investigate the competition between LPU and jog dragging as temperature increases.

REFERENCES

1. S. Farenc, and A. Couret,, in *High Temperature Ordered Intermetallic Alloys*, MRS Symposium Proc., Edited by Baker I., Dariola R., Whittenberger J.D. and Yoo M.N, **288**, 465 (1993).

2. F. Louchet, and B. Viguier, , Scripta Met. **31**, 369 (1994).

3. F. Louchet, and B. Viguier, Phil. Mag. A **71**, 1313 (1995).

4. S. Sriram, D.M. Dimiduk, P.M. Hazzledine, and V.K. Vasudevan, Phil. Mag. A **76**, 965 (1997).

5. D. Haussler, M, Bartsch, M. Aindow, I.P. Jones, and U. Messerschmidt, Phil. Mag. A **79**, 1045 (1995).

6. F. Gregori, P. Penhoud and P. Veyssiere, Phil. Mag, in press.

7. B. Viguier, K.J. Hemker, J. Bonneville, F. Louchet, and J.L. Martin, Phil. Mag. A **71**, 1295 (1995).

8. F. Louchet and B. Viguier, Phil. Mag. A **80**, 765 (2000).

9. L.P. Kubin, G.R. Canova, M. Condat, B. Devincre, V. Pontikis and Y. Bréchet, in Solid State Phenom. **23 & 24**, 455 (1992).

10. M. Verdier, M. Fivel and I. Groma, in Modelling Simul. Mater. Sci. Eng. **6**, 755-770, (1998).

11. B. Viguier, Ph D. thesis n°1375, EPFL, May 2nd 1995.

12. A. Couret, Phil. Mag. A **79**, 1977, (1999).

13. S. Zghal, A. Menand and A. Couret, Acta Mater. **46**, 5899, (1998).

14. F. Louchet and M. Lebyodkin, in *Proc. 4th International Conference on High Temperature Intermetallics, San Diego, CA (USA), April 27th-May 1st 1997*, Mat.Sci. Eng. A239-240, 804 (1997) (Edited by D.P. Pope, C.T Liu, S.H. Wang and M. Yamaguchi, Elsevier, 1997).

Mat. Res. Soc. Symp. Proc. Vol. 646 © 2001 Materials Research Society

Mechanisms contributing to the pinning of ordinary dislocations in γ-TiAl

Fabienne Grégori[1] and Patrick Veyssière

LEM, CNRS-ONERA, BP 72, 92322 Châtillon cedex, France.
[1]LPMTM, Institut Galilée, 99 av. J.B. Clément, 93430 Villetaneuse, France.

ABSTRACT

Properties of ordinary dislocations and related debris are investigated in single crystals of γ-TiAl deformed in single slip at 20°C and 400°C. Typical of this temperature range are prismatic loops isolated or belonging to arrays, together with multipoles. The interaction of these with mobile ordinary dislocations generates various configurations including trailing-like hairpins. Certain features are consistent with an extrinsic origin for point pinning.

INTRODUCTION

Over the whole range of positive temperature dependence of the flow stress of γ-TiAl, ordinary dislocations ($b = 1/2<110]$) exhibit a strong preference for the screw orientation together with a marked tendency to take pinning points and trailing-like dipolar configuration [1]. Attempts have been made at generating models of the flow stress anomaly and the two quantitative theories available rely on the idea that the pinning is caused by jogs, themselves created by cross-slip. In one case, it is the density of obstacles that increases with temperature [2], while the other available model postulates that the anomaly is determined by obstacle strength [3]. The models are consistent with selected TEM observations. A recent paper [4] compares the two models and concludes that the pinning point density thesis reflects well the physical situation encountered in γ-TiAl.

The microstructural properties of γ-TiAl deformed by ordinary dislocations are under-documented, limited by the facts that single crystal growth is difficult and facilities scarce. On the other hand, using polycrystals makes it difficult to distinguish the jogs that do not result from the intersection with forest dislocations. Another difficulty with polycrystals arises from the lack of knowledge of the stress tensor, thus on the conditions under which cross-slip operates. Finally, it is rare than thin foils sliced from polycrystalline samples offer the opportunity to compare dislocations portions of reasonable length along every dislocation character. As illustrated by the following results, this last constraint may bias the identification of the conditions of slip impediment in γ-TiAl.

The present paper focuses on dislocation anchoring at obstacles.

EXPERIMENTAL PROCEDURE

The experimental strategy was intended to explore how ordinary dislocations expand in their slip plane and what sort of debris this may generate. For this purpose, single crystals of Al-rich γ-TiAl, pre-annealed for 100 h at 1200°C then 11 h at 900°C, were deformed along a near-[153] load orientation as this belongs to the narrow region of the standard stereographic projection within which ordinary dislocations are known to be activated [5]. This particular orientation was chosen in order to deform γ-TiAl in single slip, hence to avoid ambiguities brought about by intersection with other slip systems. Thin foils were sliced parallel to the operative (-111) slip plane. It was found that no planes other than the intended plane were actually activated save for a limited activity of the primary slip direction ($\mathbf{b} = 1/2[110]$) in the (-11-1) cross-slip plane.

Curiously, the several dislocations families that are available in (-111), i.e. [101], [0-11] and 1/2[1-12], all coexisted in the (-111) planes. However, none of these generates sessile jogs by intersection with [110](-111). The operation of [101](-111) is not fully surprising since this slip system was in fact submitted to a resolved shear stress slightly lower than its own critical resolved shear stress. It was marginally activated, presumably assisted by stress concentrations. Unexpected were the two latter slip directions for which the resolved shear stresses were dramatically less than those of equivalent systems inclined to (-111). This seemingly paradoxical situation originates from the potential of the various actors of deformation for composing their Burgers vectors. Firstly, the two mostly stressed dislocations react to form 1/2[1-12] dislocations (reaction (1)). Then, the latter are again transformed into [101] dislocations by impacting ordinaries (reaction (2)). Reactions (1) and (2) are in fact a two-stepped variant for the $L1_0$ structure of the reaction between coplanar dislocations in fcc-related structures (reaction (3), $m = 1/2$ for fcc crystals and 1 in the $L1_2$ structure).

$$[0\text{-}11] + 1/2[110] \rightarrow 1/2[1\text{-}12] \tag{1}$$
$$1/2[1\text{-}12] + 1/2[110] \rightarrow [101] \tag{2}$$
$$m\,[0\text{-}11] + m\,[110] \rightarrow m\,[101] \tag{3}$$

It should be noted that the population of 1/2[1-12] dislocations can be also increased by direct decomposition of each of the two families of <011] dislocations [6].

OBSERVATIONS

Microstructures were studied in post mortem in samples deformed at room temperature, 400°C, 600°C and 800°C. The present contribution reports on the former temperatures for they are particularly well adapted to the study of dislocation locking in γ-TiAl.

In practice, the temperature at which samples were deformed could be unambiguously identified by direct inspection of the micrographs. Rather than in the density of pinning points, differences are found in the frequency at which locked screw segments straighten (considerably less after deformation at room temperature than at 400°C). In common are debris consisting in elongated dipoles, prismatic loops, multipoles and hairpin features connected to meandering ordinary dislocations. Salient properties are as follows.

1. The characteristics of *dislocation pinning* are more complex than discussed in [1-2] in that all segments are pinned and that the density of pinning points lies within the same order of magnitude irrespective of dislocation character. Another essential property is that some locked screw portions are totally unpinned as attested by their straightness under weak-beam conditions (Figure 1(*a*), F). By contrast, pinning points can also be aligned in the screw directions over several μm (Figure 1(*d*), u).

2. *Hairpins* are not as simple as previously envisioned. They may assume configurations that are consistent with a trailing process [3]. Several counterexamples are however

currently observed. In particular, the same dislocation line may well be linked to a hairpin on both sides, that is, by debris located simultaneously ahead and behind the line. Moreover, dislocations are sometimes sitting à *cheval* on a hairpin-like feature while clearly interconnected to this feature (not shown here).

3. The *multipoles* encountered in γ-TiAl are significantly distinct from those are found in fcc crystals with a low stacking fault energy or else in alloys containing short-range order [7-8]. They are comprised of a few branches of both signs, usually 6 to 10. They are in general rather short, of the order of 1-2 μm. Their extremities are often aligned in the screw direction. Although multipoles are much less frequent than prismatic loops their presence is systematic and their characteristics reproducible, indicative of a simple formation mechanism related to the specificity of dislocation organization in γ-TiAl.

4. *Prismatic loops* elongated in the edge direction are by far the most abundant of all debris generated during deformation. The loops are usually very narrow so that (i) they can be easily mistaken for dislocations intersecting the free surfaces unless appropriate foil orientations and imaging conditions are employed and (ii) images of loops can be so markedly weakened that they may well remain unnoticed.

5. At variance from what can be inferred at first sight, *prismatic loops* are not all distributed at random. In fact, a significant fraction of these is actually organized in arrays that can be of several types.

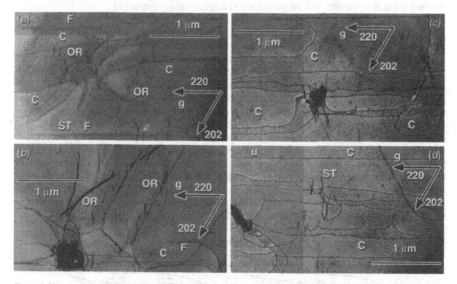

Figure 1. Negative prints of WB micrographs of 1/2[110] ordinary dislocation features after deformation at 20°C (a)-(b) and 400°C (c)-(d). ST : sawtooth, C : cusps, F : flat, OR : orientation relationship (see text). (a) All dislocation characters are represented but there is some marked elongation in the screw direction. (b) Same characteristics as in (a), which a lesser frequency of the screw direction. Some arrays of loops (OR) are clearly visible. They are comprised of several strings. (c)-(d) Ordinary dislocations are generally aligned in the screw orientation. The segment labelled "u" in (d) is rectilinear including a large density of small cusps "C" while certain neighbouring segments are much more loosely spaced. Mixed segments are cusped in places.

One type — extremely rare at the deformation temperatures investigated — consists of loops roughly aligned in the screw direction, reminding of the alignment of pinning points along screw portions [1]. A second type is again an alignment in the screw direction however in such a way that the origin of a given loop is aligned with the end of one of its nearest neighbours (Figure 1(b), OR). Strings incorporate up to 17 loops all inter-related by this particular orientation relationship (OR). An upper limit of loops per string is hard to asses since, however close the foil section to the slip plane, elongated features must eventually emerge at free surfaces.

Loop aggregates involving several strings are frequent [9]. They are unevenly distributed in foils (Figure 1(a), OR). Dense shoals of loops threaded by meandering ordinary dislocations are formed in places while nearby areas are almost loop free.

DISCUSSION

The several properties listed in the above point 1 are consistent with an extrinsic origin for the pinning of ordinary dislocations [10]. They suggest that screw dislocations assume a stable sessile core configuration and that they take pinning points upon ageing in the locked stage. A more precise analysis that will incorporate observations made after deformation at 600°C together with a full analysis of the temperature dependence of the pinning point distribution will be published in a forthcoming paper [9].

The OR (point 5) indicates that loop arrays originate from several cross-slip annihilations of a given dipole. The annihilations take place dynamically one after another as the two dislocations of opposite signs impinge forming a dipole. Surrounding obstacles govern the velocities at which the two branches approach each other. Differences in velocities determine the local orientation of the dipole. When the character is near screw, the dipole is truncated by cross-slip forming two hairpin segments, each closed by a jog in the cross-slip plane. The hairpins reorient spontaneously towards the edge character in order to reduce the total stored elastic energy. During the reorientation the two closing jogs glide in the same cross-slip plane imposing the OR (Figure 2). The reorientation results in dipolar branches with an increased edge component reducing the probability for a second cross-slip annihilation to take place. It appears therefore unlikely that a pre-existing dipole be segmented more than once which returns us to the formation of loop arrays in the course of dipolar impingement. The formation of loop arrays is at the origin of a great deal of the debris observed in γ-TiAl deformed at room temperature and 400°C. What occurs upon the impact of an array by a mobile ordinary dislocation depends essentially on two parameters

- the geometry of the array, itself determined by the relative velocities of the two impinging dislocations to form a dipole;
- the angle of incidence of impacting dislocations with respect to the loop axis.

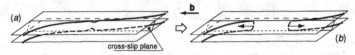

Figure 2. The mechanism of dipole truncation by cross-slip. The two hairpins relax towards the edge character in keeping the closed extremities in the screw orientation, giving rise to the observed orientation relationship (OR).

A prismatic loop, which generally is near edge in character, has but little interactions with a screw dislocation unless the slip plane of the latter intersects the loop. By contrast, the interaction with an edge dislocation is much stronger and depends on loop length.

1. In case of a *near-edge* dislocation, two distinct situations may take place :
    – The loops are rather close and the impacting dislocation is captured by the array. It assumes a meandering shape that reflects its interaction with the array. This is in turn propitious to a second series of annihilations with an incident dislocation of opposite sign forming a second loop string intermixed with the first. As the process repeats itself, the density of loops increases locally so does the probability of capture of incoming dislocations. This provides a means to form veins. Beyond a certain loop density in the veins, the captured dislocations can become so immersed in the shoal of loops that their likeliness of being annihilated decreases. This is how dislocation patterning should take place in γ-TiAl and possibly in other materials.
    – The loops forming a given array are distant enough for the impacting dislocation to by-pass the obstacles by bowing out in between two neighbouring loops (one by-passing is actually sufficient in order to overcome the entire loop array). The debris then expected are multipoles arranged in strings and related by an OR [9].

2. In case of an impacting *near-screw* dislocation a junction may form with the loop when the dislocation arrives at contact with the jogs. It is in this particular case that sawtoothed configurations are formed (ST, Figure 3). Since junctions can be made with either jogs of a given loop, certain hairpins will sit behind and others ahead of the impacting dislocation (point 2). A dislocation may straddle a rather narrow loop when its slip plane coincides with the habit plane of the attractive branch of the loop.
We have observed all these situations in γ-TiAl after deformation at RT and 400°C.

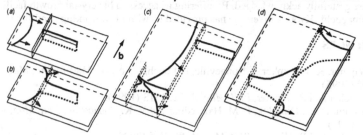

Figure 3. Interaction between a mobile screw segment and loops. (a) Before intersection. (b) The junction and related relaxation with the first jog of the first loop. (c) The dislocation has escaped from the first jog. It impacts the second jog of the first loop and the first jog of the second loop of the array. (d) The resulting sawtoothed configuration.

At the origin of loop arrays is the formation of dipoles and this implies in turn the operation of double cross-slip; however, under certain conditions. If dislocations were generated by sources distributed at random in the crystal or else if the free flight distance in the cross-slip plane were large, then the probability of dipole formation would certainly not allow the high densities observed in γ-TiAl. More likely is the generation of dipoles by a dislocation generated by a given source and that undergoes double cross-slip manoeuvres. At some stage, the segments that loop around the jogs in the cross-slip plane glide backward (which is equivalent to taking an opposite sign). They meet portions of the same dislocation that continued to slip in the plane from where double cross-slip departed forming dipoles. The process may repeat itself yielding multipoles when the free-flight distance in the cross-slip plane remains of the order of the cross-section for dipolar capture.

CONCLUSION

The origin of the locking of ordinary dislocations along the screw direction is unclear. It takes place in the absence of pinning points suggesting directional bonding, which is however inconsistent with an increasing frequency of screw portions with test temperature.

As suggested by the presence of rectilinear screw portions, cusped mixed and edge segments and alignment of tens of pinning points in the screw direction over several μm, pinning point formation involves at least in part an extrinsic contribution. The presence of invisible particles along the locked dislocation line may assist cross-slip in places [11-12].

The several microstructural elements described above, i.e. loop arrays, multipoles and dislocations decorated by hairpin configurations, are part of the same wholesale evolution that takes its origin in the truncation of a dipole into a loop string. Sawtooth configurations result from the intersection between a mobile dislocation and a prismatic loop.

Loop arrays provide sites for the heterogeneous nucleation of dislocation patterning. The process should apply to a number of other materials. It is noted that the existence of loop arrays has been reported in Cu-Ni deformed monotonously at low temperatures [13].

ACKNOWLEDGEMENTS

The invaluable contributions of Philippe Penhoud (LEM) to Fabienne Grégori's PhD Thesis are gratefully acknowledged. By offering us access to his crystal growth facilities, Prof. Yamaguchi (Kyoto University) has provided us with ideal working conditions.

REFERENCES

1. B. Viguier, K. J. Hemker, J. Bonneville, F. Louchet and J.-L. Martin, *Phil. Mag. A*, **71**, 1295 (1995).
2. F. Louchet and B. Viguier, *Phil. Mag. A*, **71**, 1313 (1995).
3. S. Sriram, D. M. Dimiduk, P. M. Hazzledine and V. K. Vasudevan, *Phil. Mag. A*, **76**, 965 (1997).
4. F. Louchet and B. Viguier, *Phil. Mag. A*, **80**, 765 (2000).
5. H. Inui, M. Matsumoro, D.-W. Wu and M. Yamaguchi, *Phil. Mag. A*, **75**, 395 (1997).
6. F. Grégori and P. Veyssière, *Phil. Mag. A*, **80**, 2913 (2000).
7. P. M. Hazzledine, *J. Phys. Paris*, **27**, C3 (1966).
8. P. M. Hazzledine, *Can. J. Phys.*, **45**, 765 (1967).
9. F. Grégori and P. Veyssière, *Phil. Mag. A*, **81**, submitted (2001).
10. A. Couret, *Phil. Mag. A*, **79**, 1977 (1999).
11. S. Zghal, A. Menand and A. Couret, *Acta Mat.*, **46**, 5899 (1998).
12. D. Häussler, M. Bartsch, M. Aindow, I. P. Jones and U. Messerschmidt, *Phil. Mag. A*, **79**, 1045 (1999).
13. U. Essmann and H. Mughrabi, *Phil. Mag. A*, **40**, 731 (1979).

AUTHOR INDEX

SUBJECT INDEX

Printed in the United States
By Bookmasters